Texts and Monographs in Physics

Springer
Berlin
Heidelberg
New York
Barcelona
Hong Kong
London
Milan
Paris
Singapore
Tokyo

Texts and Monographs in Physics

From Microphysics to Macrophysics I + II Methods and Applications of Statistical Physics By R. Balian

Multi-Hamiltonian Theory of Dynamical Systems By M. Błaszak

Quantum Mechanics: Foundations and Applications 3rd enlarged edition By A. Böhm

Operator Algebras and Quantum Statistical Mechanics I + II 2nd edition By O. Bratteli and D. W. Robinson

Statistical Methods in Quantum Optics 1. Master Equations and Fokker–Planck Equations By H. J. Carmichael

Geometry of the Standard Model of Elementary Particles By. A. Derdzinski

Scattering Theory of Classical and Quantum *N*-Particle Systems By. J. Derezinski and C. Gérard

Effective Lagrangians for the Standard Model By A. Dobado, A. Gómez-Nicola, A. L. Maroto and J. R. Peláez

Quantum The Quantum Theory of Particles, Fields, and Cosmology By E. Elbaz

Quantum Relativity A Synthesis of the Ideas of Einstein and Heisenberg By D. R. Finkelstein

Quantum Mechanics I + II By A. Galindo and P. Pascual

The Elements of Mechanics By G. Gallavotti

Local Quantum Physics Fields, Particles, Algebras 2nd revised and enlarged edition By R. Haag

Supersymmetric Methods in Quantum and Statistical Physics By G. Junker

***CP* Violation Without Strangeness** Electric Dipole Moments of Particles, Atoms, and Molecules By I. B. Khriplovich and S. K. Lamoreaux

Quantum Groups and Their Representations By A. Klimyk and K. Schmüdgen

Statistical Mechanics of Lattice Systems Volume 1: Closed-Form and Exact Solutions 2nd, revised and enlarged edition By D. A. Lavis and G. M. Bell

Statistical Mechanics of Lattice Systems Volume 2: Exact, Series and Renormalization Group Methods By D. A. Lavis and G. M. Bell

Fields, Symmetries, and Quarks 2nd, revised and enlarged edition By U. Mosel

Quantum Entropy and Its Use By M. Ohya and D. Petz

Generalized Coherent States and Their Applications By A. Perelomov

Path Integral Approach to Quantum Physics An Introduction 2nd printing By G. Roepstorff

Renormalization An Introduction By M. Salmhofer

Finite Quantum Electrodynamics The Causal Approach 2nd edition By G. Scharf

From Electrostatics to Optics A Concise Electrodynamics Course By G. Scharf

The Mechanics and Thermodynamics of Continuous Media By M. Šilhavý

Large Scale Dynamics of Interacting Particles By H. Spohn

The Theory of Quark and Gluon Interactions 2nd completely revised and enlarged edition By F. J. Ynduráin

Relativistic Quantum Mechanics and Introduction to Field Theory By F. J. Ynduráin

David A. Lavis George M. Bell

Statistical Mechanics of Lattice Systems

Volume 2: Exact, Series and Renormalization Group Methods

With 47 Figures and 7 Tables

Springer

Dr. David A. Lavis
Department of Mathematics
King's College, University of London
Strand
London WC2R 2LS, United Kingdom

Professor George M. Bell †
Department of Mathematics
King's College, University of London
Strand
London WC2R 2LS, United Kingdom

ISSN 0172-5998
ISBN 3-540-64436-9 Springer-Verlag Berlin Heidelberg New York

Library of Congress Cataloging-in-Publication Data applied for.

Die Deutsche Bibliothek - CIP-Einheitsaufnahme
Lavis, David A.: Statistical mechanics of lattice systems / D.A. Lavis ; G.M. Bell. - Berlin ; Heidelberg ; New York ; Barcelona ; Hong Kong ; London ; Milan ; Paris ; Singapore ; Tokyo : Springer
(Texts and monographs in physics) 2. Exact, series and renormalization group methods : with 7 tables. - 1999 ISBN 3-540-64436-9

Typesetting: Camera ready by the first author using a TeX macro package
Cover design: *design & production*, Heidelberg
SPIN: 10552520 55/3144/di - 5 4 3 2 1 0 – Printed on acid-free paper

Preface

Most of the interesting and difficult problems in statistical mechanics arise when the constituent particles of the system interact with each other with pair or multiparticle energies. The types of behaviour which occur in systems because of these interactions are referred to as *cooperative phenomena* giving rise in many cases to phase transitions. This book and its companion volume (Lavis and Bell 1999, referred to in the text simply as Volume 1) are principally concerned with phase transitions in lattice systems. Due mainly to the insights gained from scaling theory and renormalization group methods, this subject has developed very rapidly over the last thirty years.[1] In our choice of topics we have tried to present a good range of fundamental theory and of applications, some of which reflect our own interests.

A broad division of material can be made between exact results and approximation methods. We have found it appropriate to include some of our discussion of exact results in this volume and some in Volume 1. Apart from this much of the discussion in Volume 1 is concerned with mean-field theory. Although this is known not to give reliable results close to a critical region, it often provides a good qualitative picture for phase diagrams as a whole. For complicated systems some kind of mean-field method is often the only tractable method available. In this volume our main concern is with scaling theory, algebraic methods and the renormalization group. Although these two volumes individually contain a less than comprehensive range of topics, we have attempted to ensure that each stands alone as treatment of its material. References to Volume 1 in this volume are normally given alongside other alternative sources.

A first edition of Volume 1 was published by Ellis Horwood (Bell and Lavis 1989). In April 1993, while work was still under way on this volume, my friend and co-author George Bell suddenly died. Although the chapters which were George's responsibility were largely complete, they were still in hand-written form and some of my part of the book had yet to be written. In the normal course of events we would have discussed and amended all the material extensively. That this could not happen must be to the detriment of the book. It also led to some delay which made the production of a revised

[1] An account of the historical development of the subject from its beginnings in the nineteenth century is given by Domb (1985).

version of Volume 1 desirable. I am very grateful to Professor Elliott Lieb who encouraged and facilitated this edition of both volumes by Springer. I should also like to thank Anna Lavis, for drawing the diagrams from which the postscript versions were prepared. Geoffrey Bellringer of King's College library provided me with valuable help in compiling the bibliography.

The exchange of ideas and information with friends and co-workers has contributed to this book. Parts of this volume were first written for an M.Sc. course at King's College and many students, by their questions and comments, have helped to reduces errors and obscurities. I should like to thank Colin Bushnell, Michael Freeman, Byron Southern and Nicholas Williams for reading and commenting on parts of the text. I am also indebted to the last of these for computing two of the figures. Conversations with Geoffrey Joyce and Ivan Wilde have been a great help in understanding difficult points. Above all I acknowledge my debt to George Bell, with whom I had the good fortune to work over many years. Finally I wish to express my gratitude to Professor Wolf Beiglböck and the staff at Springer for their patience and advice.

London, December 1998. *David Lavis*

Table of Contents

1. Thermodynamics and Statistical Mechanics

1.1 Introduction

There is a detailed discussion of the thermodynamics and statistical mechanics needed for the theory of lattice models in Chaps. 1 and 2 of Volume 1. Here we give a brief review and introduce the *coupling-density representation* of thermodynamics, which is useful for scaling theory and the renormalization group. For other presentations of thermodynamics the reader is referred to Pippard (1967), ter Haar and Wergeland (1966) or Landsberg (1978) and of statistical mechanics to Huang (1963), Ma (1985), Chandler (1987) or Thompson (1988).

Thermodynamics is concerned with the laws governing the macroscopic properties of systems subject to thermal change. The only explicit recognition of the atomic nature of matter which occurs in the theory is the inclusion, for systems with a variable number of particles, of the chemical potential, or energy, which a particle brings on entry to the system. Otherwise a thermodynamic system is regarded as a 'black box' with a set of variables related by thermodynamic laws. When the system is at equilibrium the variables will all assume time-independent values.

The principle aim of statistical mechanics is to derive the macroscopic properties of a system from the laws governing the microscopic behaviour of its constituent particles. In particular, statistical mechanics gives rise to the equations of thermodynamics once the thermodynamic variables have been given an interpretation within the theory. Since this is done in terms of the expectation values associated with a probability distribution, fluctuations of dependent variables are predicted for finite systems. Exact correspondence to a thermodynamic system at equilibrium occurs only in the thermodynamic limit (see Sects. 1.4 and 4.2) of an infinite system, when the relative fluctuations of the dependent variables disappear.

In the rest of this chapter we summarize the basic formulae of thermodynamics and statistical mechanics in a form convenient for our subsequent use and consistent with the presentation in Volume 1.

1.2 Thermodynamic Formulae and Variables

Thermodynamic variables fall into two classes, intensive and extensive. Intensive variables again are of two types, fields such as the temperature $\widetilde{T}$, the pressure $\widetilde{P}$, the magnetic field $\widetilde{\mathcal{H}}$ and the chemical potential μ and densities which are defined below. For fixed values of the intensive variables, the extensive variables such as entropy $\widetilde{S}$, volume $\widetilde{V}$, magnetic moment $\widetilde{\mathcal{M}}$ and number of molecules M are proportional to the size of the system. The reason for attaching the tilde symbol to these variables (apart from μ and M) is that, as in Volume 1, we shall find it convenient to define all fields so that they have the dimensions of energy and extensive variables so that they are dimensionless. (It should be noted that μ and M are already of these types.) For temperature $\widetilde{T}$, pressure $\widetilde{P}$ and magnetic field $\widetilde{\mathcal{H}}$ we define the fields

$$T = k_{\mathrm{B}}\widetilde{T}\,, \qquad P = \widetilde{P}v_0\,, \qquad \mathcal{H} = \widetilde{\mathcal{H}}m_0\,, \tag{1.1}$$

where k_{B} is Boltzmann's constant, v_0 is a standard volume and m_0 is the dipole moment of a single molecule. The extensive variables conjugate to fields are dimensionless. Corresponding to those of (1.1) we define, in terms of the entropy $\widetilde{S}$, volume $\widetilde{V}$ and magnetic moment $\widetilde{\mathcal{M}}$, the extensive variables

$$S = \widetilde{S}/k_{\mathrm{B}}\,, \qquad V = \widetilde{V}/v_0\,, \qquad \mathcal{M} = \widetilde{\mathcal{M}}/m_0\,. \tag{1.2}$$

The internal energy U of a system is an extensive variable arising from the kinetic energies of the molecules and the energies of interaction between them. For a system with $n_{\mathrm{f}}+1$ fields T, ξ_i, $i=1,\ldots,n_{\mathrm{f}}$, with conjugate extensive variables S, Q_i, $i=1,\ldots,n_{\mathrm{f}}$

$$\mathrm{d}U = T\mathrm{d}S + \sum_{i=1}^{n_{\mathrm{f}}} \xi_i \mathrm{d}Q_i\,. \tag{1.3}$$

It may then be shown (see, for example, Volume 1, Sect. 1.7) that

$$U = TS + \sum_{i=1}^{n_{\mathrm{f}}} \xi_i Q_i \tag{1.4}$$

and

$$S\mathrm{d}T + \sum_{i=1}^{n_{\mathrm{f}}} Q_i \mathrm{d}\xi_i = 0\,. \tag{1.5}$$

A subset, $Q_1,\ldots,Q_\eta$, of the extensive variables can now be replaced as independent variables by their conjugate fields $\xi_1,\ldots,\xi_\eta$ to give, using a Legendre transformation, the enthalpy

$$H_\eta = U - \sum_{i=1}^{\eta} \xi_i Q_i = TS + \sum_{i=\eta+1}^{n_{\mathrm{f}}} \xi_i Q_i\,, \tag{1.6}$$

with

$$\mathrm{d}H_\eta = T\mathrm{d}S - \sum_{i=1}^{\eta} Q_i \mathrm{d}\xi_i + \sum_{i=\eta+1}^{n_\mathrm{f}} \xi_i \mathrm{d}Q_i \,. \tag{1.7}$$

The replacement of $Q_1, \ldots, Q_\eta$ by their conjugate fields $\xi_i, \ldots, \xi_\eta$ means that the system is open to energy interchange by way of these fields. The appropriate thermodynamic potential (or free energy) is now F_η, where

$$F_\eta = H_\eta - TS = U - TS - \sum_{i=1}^{\eta} \xi_i Q_i = \sum_{i=\eta+1}^{n_\mathrm{f}} \xi_i Q_i \,, \tag{1.8}$$

$$\mathrm{d}F_\eta = -S\mathrm{d}T - \sum_{i=1}^{\eta} Q_i \mathrm{d}\xi_i + \sum_{i=\eta+1}^{n_\mathrm{f}} \xi_i \mathrm{d}Q_i \,. \tag{1.9}$$

From (1.8) and (1.9),

$$\begin{aligned} Q_i &= -\frac{\partial F_\eta}{\partial \xi_i}\,, \qquad i = 1, \ldots, \eta\,, \qquad S = -\frac{\partial F_\eta}{\partial T}\,, \\ \xi_i &= \frac{\partial F_\eta}{\partial Q_i}\,, \qquad i = \eta + 1, \ldots, n_\mathrm{f}\,, \qquad H_\eta = \frac{\partial (F_\eta / T)}{\partial (1/T)}\,. \end{aligned} \tag{1.10}$$

1.3 Statistical Mechanical Formulae

The formalism described in the previous section can be regarded as a macroscopic perspective on a system with a microscopic structure consisting of a collection of atoms or molecules of one or more types. The numbers of particles of each type, or component, will be among the set of extensive variables $Q_1, \ldots, Q_{n_\mathrm{f}}$. This book, in common with Volume 1, will be concerned largely with the statistical mechanics of classical (as distinct from quantum) particles. The exception to this is the work of Chap. 7, which includes series methods for the quantum Heisenberg model. In the present section we consider only classical statistical mechanics; the modifications needed for the quantum case are given in Appendix A.9.

A configuration of the particles of the system is called a *microstate* and is denoted by $\boldsymbol{\sigma}$. For each microstate there is a value $E(\boldsymbol{\sigma})$ of the statistical mechanical internal energy and values $\widehat{Q}_i(\boldsymbol{\sigma})$, $i = 1, \ldots, \eta$, of the dependent statistical mechanical extensive variables. The Hamiltonian $\widehat{H}_\eta(\boldsymbol{\sigma})$, defined by

$$\widehat{H}_\eta(\boldsymbol{\sigma}) = E(\boldsymbol{\sigma}) - \sum_{i=1}^{\eta} \xi_i \widehat{Q}_i(\boldsymbol{\sigma})\,, \tag{1.11}$$

is a crucial quantity in the statistical mechanics of the system with independent variables $T, \xi_1, \ldots, \xi_\eta, Q_{\eta+1}, \ldots, Q_{n_\mathrm{f}}$. It is the microscopic equivalent of the thermodynamic enthalpy defined by (1.6). The probability function $p_\eta(\boldsymbol{\sigma})$, for the distribution of $\boldsymbol{\sigma}$ is given by

$$p_\eta(\boldsymbol{\sigma}) = \frac{\exp[-\widehat{H}_\eta(\boldsymbol{\sigma})/T]}{Z_\eta}\,, \tag{1.12}$$

where Z_η is the partition function, given by,

$$Z_\eta(T,\xi_1,\ldots,\xi_\eta,Q_{\eta+1},\ldots,Q_{n_\mathrm{f}}) = \sum_{\{\sigma\}} \exp[-\widehat{H}_\eta(\boldsymbol{\sigma})/T]\,. \tag{1.13}$$

We now make the identifications

$$H_\eta = \langle \widehat{H}_\eta \rangle = \sum_{\{\sigma\}} p_\eta(\boldsymbol{\sigma})\widehat{H}_\eta(\boldsymbol{\sigma}) = -\frac{\partial \ln Z_\eta}{\partial(1/T)}\,, \tag{1.14}$$

$$Q_i = \langle \widehat{Q}_i \rangle = \sum_{\{\sigma\}} p_\eta(\boldsymbol{\sigma})\widehat{Q}_i(\boldsymbol{\sigma}) = T\frac{\partial \ln Z_\eta}{\partial \xi_i}\,, \quad i = 1,\ldots,\eta, \tag{1.15}$$

$$\xi_i = \left\langle \frac{\partial \widehat{H}_\eta}{\partial Q_i} \right\rangle = \sum_{\{\sigma\}} p_\eta(\boldsymbol{\sigma}) \frac{\partial \widehat{H}_\eta(\boldsymbol{\sigma})}{\partial Q_i} = -T\frac{\partial \ln Z_\eta}{\partial Q_i}\,,$$

$$i = \eta+1,\ldots,n_\mathrm{f}\,. \tag{1.16}$$

The 'averaging' relations (1.14)–(1.16) become identical to (1.10) if we make the identification

$$F_\eta = -T \ln Z_\eta(T,\xi_1,\ldots\xi_\eta,Q_{\eta+1},\ldots,Q_{n_\mathrm{f}})\,. \tag{1.17}$$

In the special case $\eta = 0$, $\widehat{H}_0(\boldsymbol{\sigma})$ is the energy $E(\boldsymbol{\sigma})$, H_0 is the internal energy U and F_0 is the Helmholtz free energy, usually denoted by A. The second derivatives of F_η with respect to the independent variables $T,\xi_1,\ldots,\xi_\eta$ are the *response functions*. It is not difficult to show that the response function with respect to the variables ξ_i,ξ_j is related to the covariance of the random variables $\widehat{Q}_i$ and $\widehat{Q}_j$ by

$$\frac{\partial^2 F_\eta}{\partial \xi_i \partial \xi_j} = -\frac{\partial Q_j}{\partial \xi_i} = -\frac{\partial Q_i}{\partial \xi_j} = -\frac{1}{T}\mathrm{Cov}[\widehat{Q}_i,\widehat{Q}_j]\,. \tag{1.18}$$

1.4 The Field-Density and Coupling-Density Representations

The formulation of thermodynamics given by (1.3)–(1.9) is referred to in Volume 1 as the *field-extensive variable representation*, where the next step in the development is to derive the *field-density representation* in which extensive variables are replaced by a set of intensive variables called *densities*. The importance of these is that they remain finite in the thermodynamic limit of an infinite system. The conditions for the existence of the thermodynamic limit for a one-component lattice fluid are given in Sect. 4.2.[1]

1.4.1 Thermodynamic Formalism

From (1.4) and (1.8) we see that, in order for F_η not to be identically zero, we must have $\eta < n_\mathrm{f}$ and thus Q_{n_f} is an independent extensive variable in terms of which the densities can be defined by

$$s = \lim_{Q_{n_\mathrm{f}} \to \infty} \frac{S}{Q_{n_\mathrm{f}}}, \qquad \rho_i = \lim_{Q_{n_\mathrm{f}} \to \infty} \frac{Q_i}{Q_{n_\mathrm{f}}}, \qquad i = 1, \dots, n_\mathrm{f} - 1. \tag{1.19}$$

For the independent densities ρ_i, $i = \eta + 1, \dots, n_\mathrm{f} - 1$, equation (1.19) is simply a matter of ensuring that in the limit $Q_{n_\mathrm{f}} \to \infty$, $Q_i \to \infty$ in such a way that the ratio Q_i/Q_{n_f} remains constant. The definition of the dependent densities depends on the existence of the free energy density f_η, defined by the thermodynamic limit

$$f_\eta = \lim_{Q_{n_\mathrm{f}} \to \infty} \frac{F_\eta}{Q_{n_\mathrm{f}}}. \tag{1.20}$$

Assuming the existence of this limit the dependent densities and fields are, from equations (1.10) and (1.20), then given by

$$\begin{aligned} s &= -\frac{\partial f_\eta}{\partial T}, \qquad \rho_i = -\frac{\partial f_\eta}{\partial \xi_i}, \qquad i = 1, \dots, \eta, \\ \xi_i &= \frac{\partial f_\eta}{\partial \rho_i}, \qquad i = \eta + 1, \dots, n_\mathrm{f} - 1, \end{aligned} \tag{1.21}$$

giving

$$\mathrm{d}f_\eta = -s\mathrm{d}T - \sum_{i=1}^{\eta} \rho_i \mathrm{d}\xi_i + \sum_{i=\eta+1}^{n_\mathrm{f}-1} \xi_i \mathrm{d}\rho_i \tag{1.22}$$

and, from (1.8),

$$f_\eta = \xi_{n_\mathrm{f}} + \sum_{i=\eta+1}^{n_\mathrm{f}-1} \xi_i \rho_i . \tag{1.23}$$

From (1.4) or (1.8) the internal energy density u is given by

$$u = sT + \xi_{n_\mathrm{f}} + \sum_{i=1}^{n_\mathrm{f}-1} \rho_i \xi_i \tag{1.24}$$

and, from (1.22)–(1.24),

---

[1] For a more general presentation the interested reader is referred to Ruelle (1969) or Griffiths (1972).

$$du = T ds + \sum_{i=1}^{n_f - 1} \xi_i d\rho_i \,, \tag{1.25}$$

$$0 = s dT + d\xi_{n_f} + \sum_{i=1}^{n_f - 1} \rho_i d\xi_i \,. \tag{1.26}$$

The enthalpy density h_η is given, from (1.6), by

$$h_\eta = u - \sum_{i=1}^{\eta} \rho_i \xi_i \,, \tag{1.27}$$

and thus

$$dh_\eta = T ds - \sum_{i=1}^{\eta} \rho_i d\xi_i + \sum_{i=\eta+1}^{n_f - 1} \xi_i d\rho_i \,. \tag{1.28}$$

To develop the coupling-density representation we observe that in statistical mechanics the internal energy $E(\boldsymbol{\sigma})$ is in general of the form

$$E(\boldsymbol{\sigma}) = -\sum_{\ell=1}^{n_e} E_\ell(\boldsymbol{\sigma}) \varepsilon_\ell \,, \tag{1.29}$$

for a set of energy parameters ε_ℓ.[2] These represent the energies of interaction between groups of particles, so that in fact the coupling-density representation goes a little beyond pure thermodynamics. It follows, from (1.11), that the Hamiltonian $\widehat{H}_\eta(\boldsymbol{\sigma})$ is a linear function of the sets ε_ℓ, $\ell = 1, \ldots, n_e$ and ξ_i, $i = 1, \ldots, \eta$.

From (1.14), the enthalpy H_η is also a linear function of these parameters and the internal energy U and internal energy density u can be expressed in the form

$$U = \langle E(\boldsymbol{\sigma}) \rangle = -\sum_{\ell=1}^{n_e} \langle E_\ell(\boldsymbol{\sigma}) \rangle \varepsilon_\ell = -\sum_{\ell=1}^{n_e} U_\ell \varepsilon_\ell \,, \tag{1.30}$$

$$u = -\sum_{\ell=1}^{n_e} \mathfrak{u}_\ell \varepsilon_\ell \,, \tag{1.31}$$

where

$$\mathfrak{u}_\ell = \lim_{Q_{n_f} \to \infty} \frac{U_\ell}{Q_{n_f}} \,, \qquad \ell = 1, \ldots, n_e \,. \tag{1.32}$$

The expression (1.31) replaces u in (1.24) and we replace du in (1.25) by

$$du = -\sum_{\ell=1}^{n_e} \varepsilon_\ell d\mathfrak{u}_\ell \,. \tag{1.33}$$

[2] The choice of sign for the energy parameters is convenient when considering internal and external couplings and in relation to the Ising model.

The *internal* and *external couplings* are defined by

$$K_\ell = \varepsilon_\ell / T\,, \qquad \ell = 1, \ldots, n_e\,, \tag{1.34}$$

$$L_i = \xi_i / T\,, \qquad i = 1, \ldots, n_\mathrm{f}\,, \tag{1.35}$$

respectively and it can easily be shown (Example 1.1) that

$$s = -\sum_{\ell=1}^{n_\mathrm{e}} \mathrm{u}_\ell K_\ell - L_{n_\mathrm{f}} - \sum_{i=1}^{n_\mathrm{f}-1} \rho_i L_i\,, \tag{1.36}$$

$$\mathrm{d}s = -\sum_{\ell=1}^{n_\mathrm{e}} K_\ell \mathrm{d}\mathrm{u}_\ell - \sum_{i=1}^{n_\mathrm{f}-1} L_i \mathrm{d}\rho_i\,, \tag{1.37}$$

$$0 = \mathrm{d}L_{n_\mathrm{f}} + \sum_{\ell=1}^{n_\mathrm{e}} \mathrm{u}_\ell \mathrm{d}K_\ell + \sum_{i=1}^{n_\mathrm{f}-1} \rho_i \mathrm{d}L_i\,. \tag{1.38}$$

The dimensionless free energy density is given by

$$\phi_\eta = f_\eta / T\,, \tag{1.39}$$

from which it follows (Example 1.2) that

$$\phi_\eta = L_{n_\mathrm{f}} + \sum_{i=\eta+1}^{n_\mathrm{f}-1} L_i \rho_i = -s - \sum_{\ell=1}^{n_\mathrm{e}} \mathrm{u}_\ell K_\ell - \sum_{i=1}^{\eta} L_i \rho_i\,, \tag{1.40}$$

$$\mathrm{d}\phi_\eta = -\sum_{\ell=1}^{n_\mathrm{e}} \mathrm{u}_\ell \mathrm{d}K_\ell - \sum_{i=1}^{\eta} \rho_i \mathrm{d}L_i + \sum_{i=\eta+1}^{n_\mathrm{f}-1} L_i \mathrm{d}\rho_i\,. \tag{1.41}$$

The main advantage for statistical mechanics of using a density representation is that all the variables remain finite in the thermodynamic limit. The relative merit of working with fields or couplings is not so clear-cut. Fields have the advantage of being physically more understandable. Important quantities measured by experimentalists are the response functions such as susceptibility, heat capacity and compressibility. These are given in the field-extensive variable representation by the second-order partial derivatives of the free energy F_η. They can in consequence be expressed easily in the field-density representation in terms of the second-order partial derivatives of the free energy density f_η. Their forms in the coupling-density representation are more complicated. On the other hand it will be seen, from (1.13), (1.17), and (1.39), that

$$\phi_\eta = -\frac{1}{Q_{n_\mathrm{f}}} \ln Z_\eta(L_1, \ldots, L_\eta, K_1, \ldots, K_{n_\mathrm{e}}, Q_{\eta+1}, \ldots, Q_{n_\mathrm{f}})\,. \tag{1.42}$$

In the limit $Q_{n_\mathrm{f}} \to \infty$ this yields

$$\phi_\eta = \phi_\eta(L_1, \ldots, L_\eta, K_1, \ldots, K_{n_\mathrm{e}}, \rho_{\eta+1}, \ldots, \rho_{n_\mathrm{f}-1})\,. \tag{1.43}$$

The dimensionless free energy density is a function of the independent couplings and densities, without explicit reference to the temperature. This makes it a convenient tool for the formulation of scaling theory and renormalization group methods.

1.4.2 Lattice Systems

The coupling-density representation gives a particularly elegant formulation of thermodynamics in the special case $\eta = n_{\mathrm{f}} - 1$ when there are no independent densities. The function $\phi_{n_{\mathrm{f}}-1}$ is dependent only on couplings and, from (1.40), is itself equal to the coupling $L_{n_{\mathrm{f}}}$. We shall in the rest of the chapter concentrate on this case, omitting subscripts indicating the number of independent external couplings. From equations (1.11), (1.29), (1.34) and (1.35)

$$\widehat{H}(\sigma)/T = \widehat{H}(L_i, K_\ell; \sigma) = -\sum_{\ell=1}^{n_{\mathrm{e}}} E_\ell(\sigma) K_\ell - \sum_{i=1}^{n_{\mathrm{f}}-1} \widehat{Q}_i(\sigma) L_i \,. \tag{1.44}$$

As will be seen in the discussion of scaling in Chap. 2, the distinction between the internal couplings K_ℓ and the external couplings L_i is physical rather than mathematical. The latter arise from interactions between the system and external fields like the magnetic field $\mathcal{H}$ or the chemical potential μ, whereas the former arise from the energies of interaction between the microsystems. The external couplings are thus, in general, open to change by the experimenter, both by individual variations of the fields ξ_i and together by variations of T. The internal couplings can be changed only by means of a change of T. This physical difference is exemplified by equation (1.33), where we have assumed, in taking the differential of (1.31), that the energy parameters are kept constant.

To make this distinction more precise consider a d-dimensional lattice with N, the number of lattice sites being the one independent extensive variable $Q_{n_{\mathrm{f}}}$. There is microsystem at each site of the lattice and the state of the microsystem at site $\boldsymbol{r}$ is specified by the variable $\sigma(\boldsymbol{r})$. The symbol $\boldsymbol{\sigma}$ represents the collection or vector specifying a set of values of all the microsystems on the lattice. The extensive variables $\widehat{Q}_i$ can now be expressed in the form

$$\widehat{Q}_i(\sigma) = \sum_{\{\boldsymbol{r}\}} \mathfrak{q}_i(\boldsymbol{r}), \qquad i = 1, \ldots, n_{\mathrm{f}} - 1 \,, \tag{1.45}$$

where $\mathfrak{q}_i(\boldsymbol{r})$ is a function of $\sigma(\boldsymbol{r})$.

Unless the states of the microsystems are very complicated there will be only a rather small number of independent extensive variables and different functions $\mathfrak{q}_i$. In the case of the spin-s *Ising model* (Volume 1, Chap. 2) there are $2s+1$ possible states, labelled with the variable $s_z(\boldsymbol{r}) = -s, -s+1, \ldots, s-1, s$, at the site $\boldsymbol{r}$. By analogy with the spin quantum number, s is allow to take positive integer or half-integer values and $\sigma(\boldsymbol{r}) = s_z(\boldsymbol{r})$ if s is integer, $\sigma(\boldsymbol{r}) = 2s_z(\boldsymbol{r})$ if s is half-integer. For the spin-$\frac{1}{2}$ Ising model[3] $\sigma(\boldsymbol{r}) = \pm 1$ and the only possibility is $\mathfrak{q}_1(\boldsymbol{r}) = \sigma(\boldsymbol{r})$. The two independent fields for this system are the temperature T and a magnetic field $\xi_1 = \mathcal{H}$, which tends to align the spins with value ± 1 according as $\mathcal{H} \gtrless 0$. The extensive variable conjugate to $\mathcal{H}$ is the statistical mechanical magnetization

$$\widehat{\mathcal{M}} = \sum_{\{\boldsymbol{r}\}} \sigma(\boldsymbol{r}), \tag{1.46}$$

with the corresponding magnetization density $\hat{m} = \widehat{\mathcal{M}}/N$. When $\sigma(\boldsymbol{r}) = -1, 0, 1$ as in the case of the dilute, or spin-1 Ising model (see Sect. 3.4), there is the additional possibility $\mathfrak{q}_2(\boldsymbol{r}) = \sigma^2(\boldsymbol{r})$.

The extensive internal energy variables E_ℓ in the Hamiltonian (1.44) involve interactions between groups of more than one site. They can be expressed in the form

$$E_\ell(\boldsymbol{\sigma}) = \sum_{\{\boldsymbol{r}\}} \mathfrak{e}_\ell(\boldsymbol{r}), \qquad \ell = 1, \ldots, n_{\mathrm{e}}, \tag{1.47}$$

where $\mathfrak{e}_\ell(\boldsymbol{r})$ is a function of $\sigma(\boldsymbol{r})$ and the state variables of some set of sites around $\boldsymbol{r}$. In most cases of interest the functions $\mathfrak{e}_\ell(\boldsymbol{r})$ are taken to be *local* in the sense that they involve only sites quite close to the site $\boldsymbol{r}$; nearest neighbours, second neighbours or the sites on the elemental plaquettes of the lattice are typical. Again the simplest example is the spin-$\frac{1}{2}$ Ising model, for which

$$\mathfrak{e}_1(\boldsymbol{r}) = -\frac{1}{\mathrm{z}} \sum_{\{\boldsymbol{r}'\}} \sigma(\boldsymbol{r})\sigma(\boldsymbol{r}'), \tag{1.48}$$

where the summation is over all nearest-neighbour sites $\boldsymbol{r}'$ of $\boldsymbol{r}$ and z is the coordination number of the lattice. The one internal energy parameter $\varepsilon_1 = J$, is chosen so that $J > 0$ encourages the alignment of neighbouring spins. Summing $\mathfrak{e}_1(\boldsymbol{r})$, given by (1.48), over all the sites of the lattice and adding the field term (1.46) the Hamiltonian (1.44) becomes, for the spin-$\frac{1}{2}$ Ising model,

$$\widehat{H}(L, K; \sigma(\boldsymbol{r})) = -K \sum_{\{\boldsymbol{r},\boldsymbol{r}'\}}^{(\mathrm{n.n.})} \sigma(\boldsymbol{r})\sigma(\boldsymbol{r}') - L \sum_{\{\boldsymbol{r}\}} \sigma(\boldsymbol{r}), \tag{1.49}$$

where $L = \mathcal{H}/T$ and $K = J/T$.

The values of the two sets of quantities $\mathfrak{q}_i(\boldsymbol{r})$, $i = 1, \ldots, n_{\mathrm{f}} - 1$ and $\mathfrak{e}_\ell(\boldsymbol{r})$, $\ell = 1, \ldots, n_{\mathrm{e}}$ are determined by the microstate $\boldsymbol{\sigma}$ of the system. We shall therefore refer to them respectively as *external* and *internal state operators.* It is clear from the above discussion that the sets $\{Q_i, L_i, \widehat{Q}_i(\boldsymbol{\sigma}), \mathfrak{q}_i(\boldsymbol{r})\}$ and

[3] Usually referred to simply as 'the Ising model' since this was the case which was solved for $d = 1$ by Ising (1925).

$\{U_i, K_i, E_i(\boldsymbol{\sigma}), \mathfrak{e}_i(\boldsymbol{r})\}$ play the same mathematical role in the development of the theory. We shall, therefore, use the generic set $\{C_j, \zeta_j, \widehat{C}_j(\boldsymbol{\sigma}), \mathfrak{c}_j(\boldsymbol{r})\}$ to stand for members of either the external or internal sets with

$$\widehat{C}_j(\boldsymbol{\sigma}) = \sum_{\{\boldsymbol{r}\}} \mathfrak{c}_j(\boldsymbol{r}) \,. \tag{1.50}$$

1.5 Correlation Functions and Symmetry Properties

The state operators $\mathfrak{q}_j(\boldsymbol{r})$, $i = 1, \ldots, n_\mathrm{f} - 1$, $\mathfrak{e}_\ell(\boldsymbol{r})$, $\ell = 1, \ldots, n_\mathrm{e}$, with $\boldsymbol{r}$ ranging over all N sites of the lattice and $n = n_\mathrm{f} + n_\mathrm{e} - 1$ form a set of nN random variables distributed according to the probability function (1.12), where the Hamiltonian is now given by (1.44), (1.45) and (1.47). Let $\mathfrak{c}_1(\boldsymbol{r}_1), \ldots, \mathfrak{c}_\tau(\boldsymbol{r}_\tau)$ be some choice of a subset of these variables at τ sites of the lattice.

1.5.1 Correlation Functions

A number of different τ-point correlation functions can be defined:

(i) The *total correlation function* $\langle \mathfrak{c}_1(\boldsymbol{r}_1) \cdots \mathfrak{c}_\tau(\boldsymbol{r}_\tau)\rangle$.

(ii) The *fluctuation correlation function* $\langle \delta\mathfrak{c}_1(\boldsymbol{r}_1) \cdots \delta\mathfrak{c}_\tau(\boldsymbol{r}_\tau)\rangle$ where

$$\delta\mathfrak{c}_j(\boldsymbol{r}) = \mathfrak{c}_j(\boldsymbol{r}) - \langle \mathfrak{c}_j(\boldsymbol{r})\rangle \,, \qquad j = 1, 2, \ldots, \tau \,, \tag{1.51}$$

is a measure of the fluctuation of $\mathfrak{c}_j$ at site $\boldsymbol{r}$ about its mean.

(iii) The *net* or *connected correlation function*, which is defined recursively by

$$\begin{aligned} \Gamma_\tau(\mathfrak{c}_1(\boldsymbol{r}_1) \cdots \mathfrak{c}_\tau(\boldsymbol{r}_\tau)) = \langle \mathfrak{c}_1(\boldsymbol{r}_1) \cdots \mathfrak{c}_\tau(\boldsymbol{r}_\tau)\rangle \\ - \sum \Gamma_{p_1}(\mathfrak{c}_{j_1}(\boldsymbol{r}_{j_1}), \ldots, \mathfrak{c}_{p_1}(\boldsymbol{r}_{p_1})) \\ \cdots \Gamma_{p_k}(\mathfrak{c}_{j_k}(\boldsymbol{r}_{j_k}), \ldots, \mathfrak{c}_{p_k}(\boldsymbol{r}_{p_k})), \end{aligned} \tag{1.52}$$

$$\Gamma_1(\mathfrak{c}_1(\boldsymbol{r}_1)) = \langle \mathfrak{c}_1(\boldsymbol{r}_1)\rangle, \tag{1.53}$$

where the sum in (1.52) is over products of the connected correlation functions for all partitions of the τ lattice sites.

For $\tau = 1$ the total and connected correlation functions are the same and the fluctuation correlation function is zero. For $\tau = 2$,

$$\begin{aligned} \Gamma_2(\mathfrak{c}_1(\boldsymbol{r}_1), \mathfrak{c}_2(\boldsymbol{r}_2)) &= \langle \mathfrak{c}_1(\boldsymbol{r}_1)\mathfrak{c}_2(\boldsymbol{r}_2)\rangle - \langle \mathfrak{c}_1(\boldsymbol{r}_1)\rangle\langle \mathfrak{c}_2(\boldsymbol{r}_2)\rangle, \\ &= \langle [\mathfrak{c}_1(\boldsymbol{r}_1) - \langle \mathfrak{c}_1(\boldsymbol{r}_1)\rangle][\mathfrak{c}_2(\boldsymbol{r}_2) - \langle \mathfrak{c}_2(\boldsymbol{r}_2)\rangle]\rangle \end{aligned} \tag{1.54}$$

and, for $\tau = 3$,

$$\begin{aligned}\Gamma_3(\mathfrak{c}_1(\boldsymbol{r}_1),\mathfrak{c}_2(\boldsymbol{r}_2),\mathfrak{c}_3(\boldsymbol{r}_3)) &= \langle \mathfrak{c}_1(\boldsymbol{r}_1)\mathfrak{c}_2(\boldsymbol{r}_2)\mathfrak{c}_3(\boldsymbol{r}_3)\rangle \\ &\quad - \langle \mathfrak{c}_1(\boldsymbol{r}_1)\rangle\Gamma_2(\mathfrak{c}_2(\boldsymbol{r}_2),\mathfrak{c}_3(\boldsymbol{r}_3)) \\ &\quad - \langle \mathfrak{c}_2(\boldsymbol{r}_2)\rangle\Gamma_2(\mathfrak{c}_3(\boldsymbol{r}_3),\mathfrak{c}_1(\boldsymbol{r}_1)) \\ &\quad - \langle \mathfrak{c}_3(\boldsymbol{r}_3)\rangle\Gamma_2(\mathfrak{c}_1(\boldsymbol{r}_1),\mathfrak{c}_2(\boldsymbol{r}_2)) \\ &\quad - \langle \mathfrak{c}_1(\boldsymbol{r}_1)\rangle\langle \mathfrak{c}_2(\boldsymbol{r}_2)\rangle\langle \mathfrak{c}_3(\boldsymbol{r}_3)\rangle \\ &= \langle[\mathfrak{c}_1(\boldsymbol{r}_1)-\langle \mathfrak{c}_1(\boldsymbol{r}_1)\rangle][\mathfrak{c}_2(\boldsymbol{r}_2)-\langle \mathfrak{c}_2(\boldsymbol{r}_2)\rangle] \\ &\quad \times[\mathfrak{c}_3(\boldsymbol{r}_3)-\langle \mathfrak{c}_3(\boldsymbol{r}_3)\rangle]\rangle\,. \end{aligned} \tag{1.55}$$

The two- and three-point fluctuation and connected correlation functions are the same. That this identity does not persist for large values of τ can be seen if we expand the general τ-point fluctuation correlation function

$$\begin{aligned}\langle \delta\mathfrak{c}_1(\boldsymbol{r}_1)\cdots\delta\mathfrak{c}_\tau(\boldsymbol{r}_\tau)\rangle = &\sum_{\{j_1,\dots,j_s\}} (-1)^{\tau-s}\langle \mathfrak{c}_{j_1}(\boldsymbol{r}_{j_1})\cdots \mathfrak{c}_{j_s}(\boldsymbol{r}_{j_s})\rangle \\ &\times\langle \mathfrak{c}_{j_{s+1}}(\boldsymbol{r}_{j_{s+1}})\rangle\cdots\langle \mathfrak{c}_{j_\tau}(\boldsymbol{r}_{j_\tau})\rangle\,,\end{aligned} \tag{1.56}$$

where the sum is over all subsets $\{j_1,\dots,j_s\}$ of the indices $\{1,2,\dots,\tau\}$. Each term in the expansion is of the form of an s-point total correlation function multiplied by $\tau - s$ one-point functions. As can be seen from (1.52), the expansion for $\Gamma_\tau(\mathfrak{c}_1(\boldsymbol{r}_1)\cdots\mathfrak{c}_\tau(\boldsymbol{r}_\tau))$ in terms of total correlation functions will contain terms which are products of p- and q-point total correlation functions where *both* p and q are greater than unity when $\tau > 3$.

From (1.43), (1.45) and (1.46)

$$\widehat{H}(L_i,K_\ell;\boldsymbol{\sigma}) = -\sum_{\{\boldsymbol{r}\}}\left[\sum_{\ell=1}^{n_\mathrm{e}}\mathfrak{e}_\ell(\boldsymbol{r})K_\ell + \sum_{i=1}^{n_\mathrm{f}-1}\mathfrak{q}_i(\boldsymbol{r})L_i\right]. \tag{1.57}$$

By distinguishing the couplings at each lattice site we obtain the generalized Hamiltonian

$$\widehat{H}^{(\mathrm{G})}(L_i(\boldsymbol{r}),K_\ell(\boldsymbol{r});\boldsymbol{\sigma}) = -\sum_{\{\boldsymbol{r}\}}\left[\sum_{\ell=1}^{n_\mathrm{e}}\mathfrak{e}_\ell(\boldsymbol{r})K_\ell(\boldsymbol{r}) + \sum_{i=1}^{n_\mathrm{f}-1}\mathfrak{q}_i(\boldsymbol{r})L_i(\boldsymbol{r})\right]. \tag{1.58}$$

Then reverting to the use of generic variables and with the corresponding generalization $Z^{(\mathrm{G})}$ of the partition function (1.13),

$$\langle \mathfrak{c}_1(\boldsymbol{r}_1)\cdots\mathfrak{c}_\tau(\boldsymbol{r}_\tau)\rangle = \frac{1}{Z^{(\mathrm{G})}}\frac{\partial^\tau Z^{(\mathrm{G})}}{\partial\zeta_1(\boldsymbol{r}_1)\cdots\partial\zeta_\tau(\boldsymbol{r}_\tau)}\,. \tag{1.59}$$

It can also be shown by induction (Binney et al. 1993) that

$$\Gamma_\tau(\mathfrak{c}_1(\boldsymbol{r}_1),\dots,\mathfrak{c}_\tau(\boldsymbol{r}_\tau)) = -N\frac{\partial^\tau\phi^{(\mathrm{G})}}{\partial\zeta_1(\boldsymbol{r}_1)\cdots\partial\zeta_\tau(\boldsymbol{r}_\tau)}\,, \tag{1.60}$$

for connected correlation functions. Returning now to the situation where the couplings ζ_j are site independent

$$\frac{\partial^\tau \phi}{\partial\zeta_1\cdots\partial\zeta_\tau} = \sum_{\{\boldsymbol{r}_1,\ldots,\boldsymbol{r}_\tau\}} \frac{\partial^\tau \phi^{(\mathrm{G})}}{\partial\zeta_1(\boldsymbol{r}_1)\cdots\partial\zeta_\tau(\boldsymbol{r}_\tau)}$$
$$= -\frac{1}{N}\sum_{\{\boldsymbol{r}_1,\ldots,\boldsymbol{r}_\tau\}} \Gamma_\tau(\mathfrak{c}_1(\boldsymbol{r}_1),\ldots,\mathfrak{c}_\tau(\boldsymbol{r}_\tau)), \tag{1.61}$$

where, in the second summation, the reduction to site-independent couplings is applied after differentiation. In the case of the connected two-point or *pair* correlation function this result can be derived directly without the use of (1.60). From (1.18), (1.50), and (1.54)

$$\frac{\partial^2\phi}{\partial\zeta_1\zeta_2} = -\frac{1}{N}\mathrm{Cov}[\widehat{C}_1,\widehat{C}_2] = -\frac{1}{N}\sum_{\{\boldsymbol{r}_1,\boldsymbol{r}_2\}} \Gamma_2(\mathfrak{c}_1(\boldsymbol{r}_1),\mathfrak{c}_2(\boldsymbol{r}_2)). \tag{1.62}$$

The quantity on the left of equation (1.62) is, like the second derivative of the free energy with respect to the fields in Sect. 1.3, a type of response function and this equation is an example of a *fluctuation-response function relation.* Such relations play an important role in statistical mechanics, relating microscopic and macroscopic properties of a system. When the two variables in the covariance in equation (1.62) are the same then the covariance becomes the variance (see Volume 1, Sect. 2.5).

1.5.2 Symmetry Properties

Suppose that the lattice is d-dimensional hypercubic with lattice spacing a, unit lattice vectors $\hat{\mathbf{\imath}}^{(i)}$, $i = 1,\ldots,d$ and $N^{(i)}$ sites in the direction $\hat{\mathbf{\imath}}^{(i)}$. Then

$$N = \prod_{i=1}^{d} N^{(i)}. \tag{1.63}$$

With periodic boundary conditions applied to the lattice, the dependence of all properties of the system are invariant under the transformation

$$\boldsymbol{r} \to \boldsymbol{r} + a\big(m^{(1)}N^{(1)},\ldots,m^{(d)}N^{(d)}\big), \tag{1.64}$$

where $m^{(1)},\ldots,m^{(d)}$ have any integer values. The τ-point connected correlation function then has the Fourier transform

$$\Gamma_\tau^*(\mathfrak{c}_1,\ldots,\mathfrak{c}_\tau;\boldsymbol{k}_1,\ldots,\boldsymbol{k}_\tau) = \sum_{\{\boldsymbol{r}_1,\ldots,\boldsymbol{r}_\tau\}} \Gamma_2(\mathfrak{c}_1(\boldsymbol{r}_1),\ldots,\mathfrak{c}_\tau(\boldsymbol{r}_\tau))$$
$$\times \exp\left[-\mathrm{i}\left(\boldsymbol{k}_1\cdot\boldsymbol{r}_1+\cdots+\boldsymbol{k}_\tau\cdot\boldsymbol{r}_\tau\right)\right], \tag{1.65}$$

with

$$\Gamma_\tau(\mathfrak{c}_1(\boldsymbol{r}_1),\ldots,\mathfrak{c}_\tau(\boldsymbol{r}_\tau)) = \frac{1}{N}\sum_{\{\boldsymbol{k}_1,\ldots,\boldsymbol{k}_\tau\}} \Gamma_\tau^*(\mathfrak{c}_1,\ldots,\mathfrak{c}_\tau;\boldsymbol{k}^{(1)},\ldots,\boldsymbol{k}^{(\tau)})$$
$$\times \exp\left[\mathrm{i}\left(\boldsymbol{k}_1\cdot\boldsymbol{r}_1+\cdots+\boldsymbol{k}_\tau\cdot\boldsymbol{r}_\tau\right)\right], \tag{1.66}$$

where the wave vectors are of the form

$$\boldsymbol{k}_j = \frac{2\pi}{a}\left(\frac{k_j^{(1)}}{N^{(1)}}, \ldots, \frac{k_j^{(d)}}{N^{(d)}}\right), \qquad k_j^{(i)} = 0, \ldots, N^{(i)} - 1 \tag{1.67}$$

(see Appendix A.1).[4] It follows from (1.61) and (1.65) that

$$\frac{\partial^\tau \phi}{\partial \zeta_1 \cdots \partial \zeta_\tau} = -\frac{1}{N}\Gamma_\tau^*(\mathfrak{c}_1, \ldots, \mathfrak{c}_\tau; \mathbf{0}, \ldots, \mathbf{0}). \tag{1.68}$$

In may cases of interest the system has the property of invariance under the *translations* of the form

$$\boldsymbol{r} \to \boldsymbol{r} + a\big(m^{(1)}, \ldots, m^{(d)}\big), \tag{1.69}$$

for any integers $m^{(1)}, \ldots, m^{(d)}$. The difference between (1.64) and (1.69), is that the former can be applied to each of the vectors $\boldsymbol{r}_1, \ldots, \boldsymbol{r}_\tau$ of $\Gamma_\tau(\mathfrak{c}_1(\boldsymbol{r}_1), \ldots, \mathfrak{c}_\tau(\boldsymbol{r}_\tau))$ independently, whereas (1.69) must be applied to them all simultaneously. One translation of the form (1.69) takes $\boldsymbol{r}_1$ to the origin to give

$$\Gamma_\tau(\mathfrak{c}_1(\boldsymbol{r}_1), \ldots, \mathfrak{c}_\tau(\boldsymbol{r}_\tau)) = \Gamma_\tau(\mathfrak{c}_1(\mathbf{0}), \mathfrak{c}_2(\bar{\boldsymbol{r}}_1), \ldots, \mathfrak{c}_\tau(\bar{\boldsymbol{r}}_{\tau-1})), \tag{1.70}$$

where $\bar{\boldsymbol{r}}_j = \boldsymbol{r}_{j+1} - \boldsymbol{r}_1$, $j = 1, \ldots, \tau - 1$ are the relative lattice vectors. The one-point connected correlation function, given from (1.53) by $\langle \mathfrak{c}_1(\boldsymbol{r}_1) \rangle$ is independent of the site location $\boldsymbol{r}_1$. We also see, from (1.52) and (1.56), that the total and fluctuation correlation functions have a similar dependence on relative vectors. With translational invariance one of the summations in the final form of (1.61) can be performed to give

$$\frac{\partial^\tau \phi}{\partial \zeta_1 \cdots \partial \zeta_\tau} = -\sum_{\{\bar{\boldsymbol{r}}_1, \ldots, \bar{\boldsymbol{r}}_{\tau-1}\}} \Gamma_\tau(\mathfrak{c}_1(\mathbf{0}), \mathfrak{c}_2(\bar{\boldsymbol{r}}_1), \ldots, \mathfrak{c}_\tau(\bar{\boldsymbol{r}}_{\tau-1})), \tag{1.71}$$

and, if $\bar{\Gamma}_\tau^*(\mathfrak{c}_1, \ldots, \mathfrak{c}_\tau; \bar{\boldsymbol{k}}_1, \ldots, \bar{\boldsymbol{k}}_{\tau-1})$ is the Fourier transform defined with respect to the relative vectors,

$$\frac{\partial^\tau \phi}{\partial \zeta_1 \cdots \partial \zeta_\tau} = -\bar{\Gamma}_\tau^*(\mathfrak{c}_1, \ldots, \mathfrak{c}_\tau; \mathbf{0}, \ldots, \mathbf{0}). \tag{1.72}$$

Finally we note that if, in additional to translational invariance, the system has $N^{(1)} = N^{(2)} = \cdots = N^{(d)}$ and all its properties invariant under *rotation* between lattice directions the two-point correlation function is given by

$$\begin{aligned}\Gamma_2(\mathfrak{c}_1(\boldsymbol{r}_1), \mathfrak{c}_2(\boldsymbol{r}_2)) &= \Gamma_2(\mathfrak{c}_1(\mathbf{0}), \mathfrak{c}_2(\boldsymbol{r}_2 - \boldsymbol{r}_1)) \\ &= \Gamma_2(\mathfrak{c}_1, \mathfrak{c}_2; |\boldsymbol{r}_2 - \boldsymbol{r}_1|).\end{aligned} \tag{1.73}$$

[4] In the case of lattices other than the hypercubic the wave vectors are defined in terms of reciprocal lattice vectors.

Examples

1.1 Prove equation (1.36) by substituting u, given by (1.31) into (1.24), dividing by T and using (1.34) and (1.35). Prove equation (1.37) by substituting from (1.33) into (1.25), dividing by T and using (1.34) and (1.35). By taking the differential of (1.36) and subtracting (1.37) establish (1.38).

1.2 Prove equation (1.40) by dividing (1.23) by T, and using (1.24), (1.31), (1.34) and (1.35). Take the differential of (1.40) and substitute from (1.37) to establish (1.41).

2. Phase Transitions and Scaling Theory

2.1 Introduction

Most physical systems can exist in a number of different *phases*, distinguished by their different types of molecular or atomic order. This order may be in the spatial configurations of one or more kinds of microsystems or it may be in the orientations or conformations of the microsystems themselves. (See Ziman (1979) for a discussion of the wide variety of possible types of order.) In the case of the vapour (gas), liquid and solid phases of, for example, water or hydrogen, the order is spatial with no order in the vapour, a short-range clustering type of order in the liquid and long-range lattice order in the solid. For water this is not the complete picture. There are at least nine different ice phases distinguished by their lattice structures and proton configurations (Volume 1, Appendix A.3, Eisenberg and Kauzmann 1969). The most well-known example of orientational order occurs in magnetic systems where ferromagnetism corresponds to the alignment of the magnetic dipoles of the microsystems. Although a simple magnetic system may possess just one type of ferromagnetic phase, more complex ferrimagnetic systems can have a large number of different magnetic phases. In the case, for example, of cerium antimonide fourteen different phases have been identified by neutron diffraction experiments (Fischer et al. 1978) and specific heat analysis (Rossat-Mignod et al. 1980). These phases exhibit different forms of anisotropic spatial ordering. The occurrence of different types of anisotropic ordering is also a feature which distinguishes between the nematic, smectic and columnar phases in liquid crystals (de Gennes and Prost 1993). The boundaries, or *transition regions*, between phases are characterized by discontinuous or singular behaviour in one or more of the thermodynamic variables associated with the system. A classification of phase transitions according to whether particular derivatives of the thermodynamic potential have singularities or discontinuities across the transition region was proposed by Ehrenfest (1933) (Volume 1, Sect. 3.1, Pippard 1967). However, it is now known than many transitions do not fit into this classification. The attempt to place all phase transitions into classes according these kinds of criteria has been abandoned and the more modest classification of Fisher (1967) has been largely adopted. In both Ehrenfest's and Fisher's terminologies regions of *first-order phase transitions* are those across which densities are discontinuous. These can also be identified as regions of phase coexistence. In place of the more detailed proposals of Ehrenfest, Fisher suggested that all transition regions across which the densities are continuous should be called *continuous transitions.* These may, of course, involve discontinuities or singularities in second or higher derivatives of the thermodynamic potential. In keeping with the nomenclature of Volume 1 we shall augment Fisher's scheme by referring to continuous transitions for which one or more of the thermodynamic response functions is singular as *second-order* and other continuous transitions as *higher-order.* The onset of ferromagnetism in iron is accompanied by a singularity in the susceptibility and is thus a second-order transition. The freezing of water is accompanied by a discontinuous reduction in density implying a first-order transition. In some of the discussion in this chapter we shall need to make a distinction between different types of transition regions. The term *coexistence region* will be used for a region of first-order transitions and *critical region* for any region in which the transition is continuous.

Ehrenfest's classification of transitions, referred to above, is one attempt to produce a taxonomy of phase behaviour. With a rather different emphasis, the earlier formulation of the law of corresponding states (see Sect. 2.3) is another. Scaling theory, the main subject of this chapter, can be viewed in a similar light. It develops from an assumed form for the free energy density, or equation of state, close to the transition region and the underlying idea is that the probability distribution of microstates is invariant under a uniform change of length scale. As will be described in this chapter, the asymptotic behaviour of thermodynamic functions as a transition region is approached can be characterized by a set of *critical exponents* (see also Volume 1, Sect. 3.4). Using purely thermodynamic arguments, inequalities between these exponents can be derived (see, for example, Buckingham 1972). One of salient features of scaling theory is that, subject to the validity of its assumptions, these inequalities become equalities, the so-called *scaling laws.* This means that a smaller set of variables, the *scaling exponents*, one for each independent field, are sufficient to characterize the leading asymptotic behaviour at the transition. Although scaling theory is unable to provide numerical values for scaling exponents, the generalization of the theory given by conformal invariance is able to do so in two dimensions. A conformal transformation can be understood as a generalization of a scaling transformation in which the rescaling factor varies continuously over the physical space of the system.

The idea of scaling was proposed independently by a number of authors (Widom 1965, Domb and Hunter 1965, Kadanoff 1966, Patashinskii and Pokrovshii 1966). The presentation used in this chapter is based on the subsequent developments by Hankey and Stanley (1972), Hankey et al. (1972) and Nightingale and 'T Hooft (1974). The application of conformal invariance to critical phenomena was first made by Polyakov (1970) and has been much developed in recent years (Cardy 1987, Christe and Henkel 1993).

The main purpose of this chapter is to develop the ideas of scaling theory. Sect. 2.2 presents a brief description of some of the kinds of combinations of transition regions which may occur in phase spaces of up to three dimensions and this is followed in Sect. 2.3 by a discussion of the ideas which contributed to the development of scaling theory. Sects. 2.4 to 2.7 contain the general theory with particular applications to a critical point and a tricritical point in Sects. 2.8–2.11 and 2.12–2.13 respectively. The corrections to scaling theory arising from dangerous irrelevant variables are considered in Sect. 2.14 and the relation between the scaling and universality hypotheses in Sect. 2.15. In Sect. 2.16 the modifications to bulk scaling which occur in systems which are finite in one or more dimensions is discussed and in Sect. 2.17 the generalization of scaling obtained from conformal invariance is briefly described.

2.2 The Geometry of Phase Transitions

We consider a system in the coupling-density representation in which, as in Sect. 1.4.2, there are no independent densities ($\eta = n_{\mathrm{f}} - 1$). From (1.40) and (1.41),

$$\phi = L_{n_{\mathrm{f}}} = f/T\,, \tag{2.1}$$

$$\phi = -s - \sum_{\ell=1}^{n_{\mathrm{e}}} \mathrm{u}_\ell K_\ell - \sum_{i=1}^{n_{\mathrm{f}}-1} L_i \rho_i\,, \tag{2.2}$$

$$\mathrm{d}\phi = - \sum_{\ell=1}^{n_{\mathrm{e}}} \mathrm{u}_\ell \mathrm{d}K_\ell - \sum_{i=1}^{n_{\mathrm{f}}-1} \rho_i \mathrm{d}L_i\,, \tag{2.3}$$

where, for the sake of brevity, the subscript $n_{\mathrm{f}} - 1$ has been omitted from f and ϕ. The system has $n = n_{\mathrm{f}} + n_{\mathrm{e}} - 1$ independent couplings. The free energy density, or potential, ϕ can be plotted as an n-dimensional hypersurface in the $(n+1)$-dimensional space of the variables $\phi, K_1, \ldots, K_{n_{\mathrm{e}}}, L_1, \ldots, L_{n_{\mathrm{f}}-1}$. This space and hypersurface are unaffected by the choice of $Q_{n_{\mathrm{f}}}$ in Sect. 1.3, since $\phi = L_{n_{\mathrm{f}}}$ and a different choice of an independent extensive variable would correspond simply to a relabelling of the axes. From (2.3) it can be seen that components of the gradient of the surface at any point correspond to the densities and the elements of the curvature matrix are thermodynamic response functions like the one appearing on the left-hand side of equation (1.62).

The potential $\phi(K_1, \ldots, K_{n_{\mathrm{e}}}, L_1, \ldots, L_{n_{\mathrm{f}}-1})$ is, for most values of its arguments, a smooth function.[1] The exceptional points are associated with some kind of critical or phase coexistence behaviour. Phase boundaries correspond to $(n-1)$-dimensional regions on the hypersurface of ϕ in which some kind

---

[1] A *smooth function* is defined to be one which has finite continuous partial derivatives of all orders in all of its arguments.

of non-smooth behaviour occurs. These are the transition regions of highest dimension. In general a transition region may have any dimension between $n-1$ and zero (a point). The projection of all these transition regions into the n-dimensional space of couplings yields the phase diagram of the system. The simplest form of a phase boundary is when some or all of the densities are discontinuous across the boundary and we have a first-order transition or region of two-phase coexistence. In a similar way a region of p-phase coexistence is in general of dimension $n+1-p$. Since this region must have a non-negative dimension we might conclude that largest number of phases which can coexist for n independent couplings is $n+1$. This restriction is the *Gibbs phase rule.* There are exceptions to this rule, which arise when the phase diagram has some special symmetry. One of these is the triangular ferrimagnetic Ising model (see, for example, Volume 1, Sect. 4.5) for which the phase diagram has reflectional symmetry about the zero-field axis. A compensation point occurs on this axis where four first-order transitions meet. However, mean-field (Bell 1974) and renormalization group (Lavis and Quinn 1983, 1987) calculations indicate that, at the compensation point, the discontinuities in density are zero, with the phases forming two pairs of the same ordering structure. A second example, described below, is the case of a metamagnet with a staggered field. Here the phase diagram is three-dimensional with reflectional symmetry about the plane on which the staggered field is zero. Mean-field calculations indicate (Kincaid and Cohen 1975) that, for a certain parameter range, this plane contains a line of four-phase coexistence; one more phase than would be allowed by the Gibbs phase rule. We now consider some examples of the kind of critical and coexistence phenomena which can occur for $n=2$ and $n=3$.

2.2.1 A Two-Dimensional Phase Space

For $n=2$, phases can be separated by lines of transitions. A *critical point* occurs at a second-order transition terminating a line of first-order transitions (Fig. 2.1(a)). Examples of this are the Curie point in a magnet and the termination of the vapour–liquid transition in a fluid system. A *triple point* occurs at the confluence of three lines of first-order transitions. An example of this is the point where the first-order transitions between ice, liquid water and steam meet. Other types of behaviour, which are grouped under the general heading of *multicritical phenomena* can also occur. When a line of first-order transitions is continued by a line of second-order transitions the two lines are said to meet at a *tricritical point* (Fig. 2.1(b)). When a line of second-order transitions terminates on a line of first-order transitions at an intermediate point this meeting point is called a *critical end-point* (Fig. 2.1(a)). Tricritical points and critical end-points occur for different parameter ranges when mean-field theory is applied to a metamagnetic model (see Volume 1, Sects 4.4). In that model spins are arranged in parallel planes

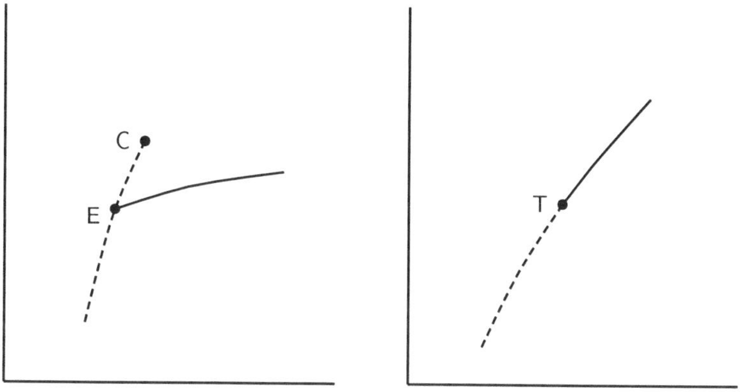

Fig. 2.1. **(a)** A critical point C terminating a line of first-order transitions and a critical end-point E where a line of second-order transitions terminates on a line of first-order transitions. **(b)** A tricritical point T where a line of first-order transitions meets a line of second-order transitions. First-order transitions are indicated by broken lines and second-order transitions by continuous lines.

with a ferromagnetic interaction within planes and an antiferromagnetic interaction between neighbouring planes. $FeCl_2$ is an experimental example of this kind of system. In this case the second-order transition line terminates at the Néel point on the zero-field axis and meets a first-order transition line at a point in the plane, with this latter terminating on the zero-temperature axis. The meeting of the two transition curves can be a tricritical point or a critical end-point according to the relative strengths of the ferromagnetic and antiferromagnetic interactions. The same type of behaviour occurs in the temperature-chemical potential plane of a dilute Ising, or Blume–Emery–Griffiths, model (see Volume 1, Sects. 6.8) though overall the phase pattern is somewhat different from that obtained for the metamagnet. The name 'tricritical point' in the context described above seems slightly odd since we are apparently considering a situation with only two phases present. However, as we explain below and in Sect. 2.12 (see also Lawrie and Sarbach 1984 and Volume 1, Chap. 6), a full understanding of tricritical points needs a phase space of more dimensions. A further type of multicritical behaviour which can occur in a two-dimensional phase space is when the critical lines of two phase transitions meet at a point, which is also the end of a coexistence line of first-order phase transitions separating the two ordered phases. The meeting point is called a *bicritical point*. Bicritical points have been observed most clearly in weakly anisotropic antiferromagnets such as $GdAlO_3$ and MnF_2 in carefully oriented magnetic fields (Rohrer 1975). A theoretical realization of a bicritical point occurs in the temperature-electric field plane of a six-vertex ferroelectric (Volume 1, Fig. 10.10 or Lieb and Wu 1972, Fig. 30.)

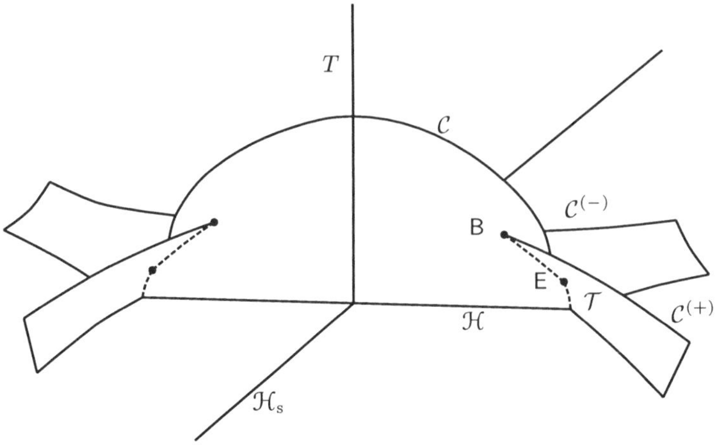

Fig. 2.2. Phase diagram for a metamagnetic model in the space of temperature T, magnetic field $\mathcal{H}$ and staggered field $\mathcal{H}_s$. The critical curve $\mathcal{C}$ meets the triple line $\mathcal{T}$ at the critical end-point E. The critical curves $\mathcal{C}^{(+)}$ and $\mathcal{C}^{(-)}$ meet on the plane $\mathcal{H}_s = 0$ at the bicritical end-point B. (Reprinted from Kincaid and Cohen (1975), p. 114, by permission of the publisher Elsevier Science.)

2.2.2 A Three-Dimensional Phase Space

When the phase space acquires a third dimension, so that $n = 3$, the boundaries between pairs of phases become surfaces. A first-order surface may be bounded by a line of critical points or *critical line.* Three first-order surfaces, separating three distinct phases can meet at a *triple line.* With these basic components phase diagrams can exhibit a rich diversity of structure. Many of these appear when a *staggered field,* alternating in sign between layers, is added to the metamagnetic model referred to above. Figure 2.2 reproduces a phase diagram for this model obtained by Kincaid and Cohen (1975) using mean-field methods. This form of phase diagram occurs for a particular range of the ratio between the ferromagnetic interaction within layers and the antiferromagnetic interaction between layers (see Sect. 2.12). Three phase boundaries meet along a triple line $\mathcal{T}$, which lies between the zero-temperature plane and the critical end-point E, where it meets the critical curve $\mathcal{C}$ and a line of *four-phase coexistence* BE. The point B is a *bicritical end-point,* where the critical curves $\mathcal{C}^{(+)}$ and $\mathcal{C}^{(-)}$ meet on the plane $\mathcal{H}_s = 0$. By varying the parameter ratio so that B and E coalesce a tricritical point can be achieved and the phase diagram takes the form shown in Fig. 2.8 below. At the precise parameter value for which this occurs the point of coincidence becomes a special four-phase critical point (see Kincaid and Cohen 1975). The form of the phase diagram Fig. 2.8 makes clear the reason why Griffiths (1970) chose the name tricritical for a point of coincidence of a first-order and critical line in a plane.

It will be clear from these brief remarks that even within the constraints of a phase space of at most three dimensions, not all the possible types of phase behaviour have been covered. (For a review of theoretical and experimental work on multicritical phenomena see Pynn and Skjeltorp (1984) and for tricritical phenomena, with further examples, the review by Lawrie and Sarbach (1984).)

2.3 Universality, Fluctuations and Scaling

As indicated above, one aim of the theory of critical phenomena is to formulate a classification of phase transitions. Criteria are sought in terms of which it can be said that certain phase transitions are 'of the same type'. A early attempt of this kind arose from the observation by van der Waals that his equation[2] could be expressed in terms of variables scaled with respect to their critical values. A *law of corresponding states* was formulated suggesting that a wide class of physical systems would, in terms of appropriately scaled variables, obey this same equation of state. The experimental fact that the critical ratio $P_c V_c / M_A T_c$, where M_A is Avogadro's number, is not the same for all gases caused some concern and was the first sign that the law of corresponding states has only limited applicability. However, as has been pointed out by Levelt Sengers (1974):

> Although this law has limited validity experimentally, and is not obeyed by most other analytic equations of state, its role in the development of the physical, chemical and engineering sciences dealing with fluids can hardly be exaggerated.

It led to many attempts to construct more general analytical equations of state and ultimately to the theory of Landau (1937) described in Chap. 3. The first steps towards the development of Landau theory were taken by van der Waals (1893). In his article on the theory of capillarity he investigated the asymptotic critical behaviour of his equation by expanding about the critical point in terms of the volume and temperature deviations from their critical values. This yielded the first calculation of the critical exponent, now known as β, for which he obtained the value $\frac{1}{2}$. This exponent measures the curvature of the coexistence curve in the vicinity of the critical point and, around the turn of the century, experimental results began to appear indicating disagreement with the theoretical value $\beta = \frac{1}{2}$. The initial reaction to these results was to suppose that the experimental measurements were not taken sufficiently close to criticality. In 1900 Verschaffelt was, however, able to show by analyzing the results of Young (1892) for isopentane,

---

[2] This was first presented in his doctoral thesis (van der Waals 1873). For accounts of the way that it stimulated the development of the theory of phase transitions see Rigby (1970), Rowlinson (1973), de Boer (1974) and Levelt Sengers (1974).

that $\beta = 0.3434$, with no evidence of the value tending towards $\frac{1}{2}$ as the critical point was approached. This then was the beginning of the realization that critical exponents (both β and the other critical exponents associated with the asymptotic properties of thermodynamic functions, see Volume 1, Sect. 3.4 and Sect. 2.8, below) were crucial for the classification of phase transitions. The development of generalized forms of the equation of state incorporating fractional exponents took another 65 years. By this time there was ample evidence that experimental systems yield critical exponents with values different from those predicted by classical (Landau) theory. Theoretical impetus was provided by the Onsager (1944) solution of the two-dimensional Ising model which demonstrated non-analytic behaviour at the critical point. With the benefit of hindsight we shall not trace the details of the early development of generalized equations of state leading to scaling theory. Instead we shall concentrate on trying to distinguish those properties of a system which affect its critical behaviour.

2.3.1 Universality

At points near a critical region in phase space a system will be in different stable phases. This entails the existence of fluctuations, which can be regarded as the short-time and local appearance of fragments of one of the phases in another. An example of this is the critical opalescence present in a fluid near the critical point of the liquid-vapour transition. Thus fluctuations are the appearance of locally ordered regions. These persist for a certain mean length of time $\mathfrak{t}$ called the *correlation time* and extend over a mean distance $\mathfrak{r}$ called the *correlation length.* Roughly speaking, dynamic and static critical behaviours are governed by $\mathfrak{t}$ and $\mathfrak{r}$ respectively. Since we are concerned exclusively with static critical behaviour it is the correlation length, measuring the mean extension of correlations, which will be of interest to us. Given that, in a neighbourhood of a critical region, correlations extend over distances large with respect to the detailed nature (short-range interactions, lattice structure etc.) of the system, it is intuitively easy to understand that this microstructure will not affect the characteristics of the system. This led to the formulation of a *universality hypothesis* which can be expressed in the following way (see, for example, Kadanoff 1976a):

All critical problems can be divided into *universality classes* differentiated by:

(a) the (physical) dimension d of the system,

(b) the symmetry group of the *order parameter*,

(c) perhaps other criteria.

Within each class, the critical properties, and in particular the critical exponents, are supposed to be identical or, at worst, to be continuous functions of very few parameters. Although, in theoretical models, systems of any dimension can be easily realized it might be supposed that experiment is relevant only to the case $d = 3$. However, polymers (linear macromolecules) and solids made up of chains of weakly coupled molecules provide examples of one-dimensional, or quasi one-dimensional systems and two-dimensional experimental systems can be realized in films, adsorbed phases, weakly coupled planes and surface phenomena. The order parameter is not always easy to define. In a fluid system undergoing a liquid-gas transition the order parameter is a density difference and of dimension one. In a magnetic transition it is the magnetization which may be of dimension one, two or three according to the constraints imposed on the system. The change of the nature of a phase transition according to the symmetry group of the order parameter is illustrated in the context of Landau theory in Chap. 3. The inclusion of item (c) by Kadanoff (1976a) could be viewed as a means of providing for exceptional cases which will almost inevitably arise in any kind of classification of this type. Two of these were discussed in some detail in Volume 1. In the bond-dilute Ising model (see Volume 1, Sect. 9.2) the order parameter is a scalar quantity, so one would, on the basis of (a) and (b) alone, expect the exponents to be those of the pure Ising model, for the corresponding value of d. It can, however, be shown (Volume 1, Sect. 9.2) that instead of the Ising exponents α and γ for the heat capacity and susceptibility respectively the actual exponents α_r and γ_r are given in terms of their Ising values by $\alpha_r = -\alpha/(1-\alpha)$ and $\gamma_r = \gamma/(1-\alpha)$. This change of exponent, whenever $\alpha \neq 0$, ($\alpha = 0$ for $d = 2$) is called *exponent renormalization* (Fisher 1968). It arises whenever the fraction of secondary sites occupied by solute atoms is treated as an independent variable. In the six-vertex model in an electric field (see Volume 1, Sect. 10.12) the order parameter is the scalar polarization. It may, however, be shown that, at the transition to the polarized state, the heat capacity diverges with an exponent $\alpha = \frac{1}{2}$ compared to the zero value of the corresponding Ising model. These two different types of transition are also exhibited by dimer models. They are discussed in Sect. 8.5 and by Nagle et al. (1989) who call them K-type (for Kasteleyn) and O-type (for Onsager). Finally the eight-vertex model should be mentioned. This yields continuously varying critical exponents (Sect. 5.9). The way in which such exponents can arise within the context of scaling theory is investigated in Sect. 2.6 and the particular case of the eight-vertex model is considered in Sect. 5.12. It should be emphasized that the concept of a universality class is associated with a phase transition and not with a physical or theoretical system as a whole. Systems with sufficient complexity may exhibit different phase transitions, in different parts of their phase spaces, lying in different universality classes.

A brief account of scaling theory is presented in Volume 1, Sect. 3.5. This will be complemented by a more general treatment using the approach of Hankey and Stanley (1972) in Sect. 2.4, below. This method has the virtues of mathematical elegance and generality. The basic premises do, however, seem to be rather arbitrary without the provision of some intuitive insight. This we shall now try to provide, using the example of the ferromagnetic Ising model, following the ideas of Kadanoff (1966).

2.3.2 Kadanoff's Scaling Method for the Ising Model

We consider the spin-$\frac{1}{2}$ Ising model on a d dimensional hypercubic lattice of N sites and lattice spacing a with Hamiltonian given by (1.49). The density conjugate to L is the magnetization density $m = \mathcal{M}/N$, where $\mathcal{M} = \langle\widehat{\mathcal{M}}\rangle$ and $\widehat{\mathcal{M}}$ is given by (1.46). The ferromagnetic Ising model is believed to have, for all $d > 1$, a first-order phase transition along the zero-field axis ($L = 0$) in the interval $K_c < K \leq \infty$, where the critical point is $K = K_c$, $L = L_c = 0$. The order parameter for this critical region is m and the variance of $\hat{m}$ is given, from (1.53) and (1.62), by

$$\mathrm{Var}[\hat{m}] = \frac{1}{N} \sum_{\{\boldsymbol{r}_1,\boldsymbol{r}_2\}} \Gamma_2(\sigma(\boldsymbol{r}_1),\sigma(\boldsymbol{r}_2)), \tag{2.4}$$

where

$$\Gamma_2(\sigma(\boldsymbol{r}_1),\sigma(\boldsymbol{r}_2)) = \langle\sigma(\boldsymbol{r}_1)\sigma(\boldsymbol{r}_2)\rangle - \langle\sigma(\boldsymbol{r}_1)\rangle\langle\sigma(\boldsymbol{r}_2)\rangle . \tag{2.5}$$

is the *spin-spin (connected pair) correlation function.* When the system has translational and rotational symmetry the spin-spin correlation function satisfies (1.73) (with $\mathfrak{c}_1 = \mathfrak{c}_2 = \sigma$) and $\Gamma_2(\sigma;r)$ is a measure of the amount of short-range order (spin alignment) in the system between spins separated by a distance r. It is, therefore, proportional to the probability of a cluster of aligned spins forming with spatial extent r. The correlation length $\mathfrak{r}(\sigma)$, defined by,

$$\mathfrak{r}^2(\sigma) = c(\mathfrak{d}) \frac{\sum_{\{\boldsymbol{r}_1,\boldsymbol{r}_2\}} r^2 \Gamma_2(\sigma(\boldsymbol{r}_1),\sigma(\boldsymbol{r}_2))}{\sum_{\{\boldsymbol{r}_1,\boldsymbol{r}_2\}} \Gamma_2(\sigma(\boldsymbol{r}_1),\sigma(\boldsymbol{r}_2))}, \tag{2.6}$$

is a measure of the largest clusters which will form. The prefactor $c(\mathfrak{d})$ is a function of the number of dimensions $\mathfrak{d} \leq d$ in which the system is infinite (see Sect. 2.16). Its exact form is unimportant for most purposes and it is sometimes omitted (see, for example, Ziman 1979, p. 26). However, in Sect. 3.3.2, it is shown that, within the Gaussian approximation, the choice $c(\mathfrak{d}) = 1/(2\mathfrak{d})$ is necessary for consistency.

Consider now a path in the phase diagram with small positive field (L small and positive) beginning at low temperature (K large). At this point thermal effects will be small and most spins will be aligned in the positive direction. As K is decreased thermal fluctuations will mean that random spin flips will show an increasing tendency to occur and the interaction energy between spins will encourage the formation of clusters based around these

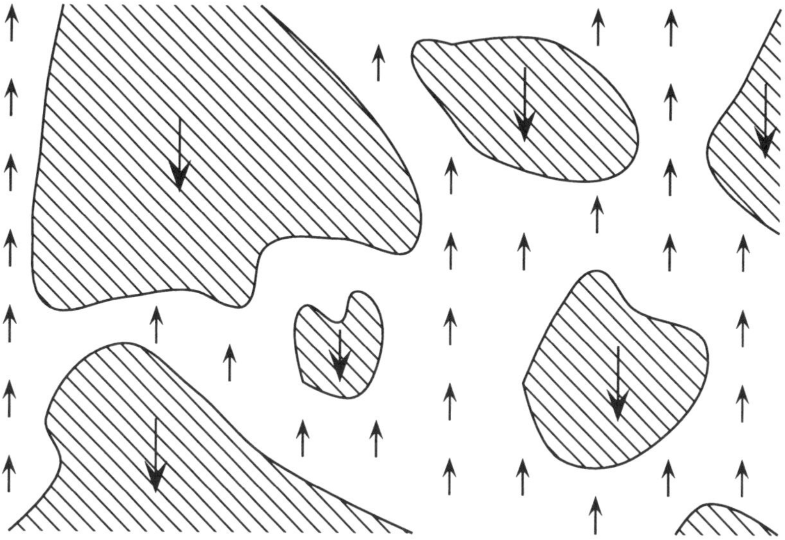

Fig. 2.3. Clusters of down spins forming in a sea of up spins. Clusters form of all sizes up to lengths of the order of $\mathfrak{r}(\sigma)$.

spins. As long as the correlation length is quite small these clusters will appear and disappear at random. However, as K continues to decrease towards the critical point, the spin-spin correlation function for a particular fixed r will increase. Correlations will grow stronger over larger distances and the correlation length and largest cluster size of down spins will become very large (see Fig. 2.3), finally diverging at the critical point. Near the critical point the correlation length $\mathfrak{r}(\sigma)$ is much greater than a, so it is possible to find a number λ much greater than unity, but such that λa is still much less than $\mathfrak{r}(\sigma)$. Imagine that the system is divided into cells λa on each side and containing λ^d spins (see Fig. 2.4). Near the critical point there will be a large degree of local ordering in the system. The variable $\tilde{\sigma}(\tilde{\boldsymbol{r}})$, defined for all $\tilde{\boldsymbol{r}}$ at the centres of the cells, by summing over the spins in the cell and dividing by λ^d, will have a value very close to $+1$ or -1. The idea of Kadanoff (1966) is that the variable $\tilde{\sigma}(\tilde{\boldsymbol{r}})$ plays the role of a single spin in a new model defined on the sites $\tilde{\boldsymbol{r}}$. The correlation length $\mathfrak{r}(\sigma)$ is the same in the new picture as in the original model if it is measured in units of a, but, if it is now measured in units of $\tilde{a} = \lambda a$, it has the value $\mathfrak{r}(\sigma)/\lambda$. The thermodynamic state of the original model is defined by the variables $\triangle K = K - K_{\mathrm{c}}$ and L. The rescaled model, in which the correlation length is smaller by a factor λ^{-1}, will appear to be further from the critical point. It will, therefore, have effective values of $\triangle K$ and L larger than those in the original model. These rescaled values we denote by $\lambda^{y_K} \triangle K$ and $\lambda^{y_L} L$, for some unknown positive scaling exponents y_K and y_L. The important *assumption* made at this stage is that, for points sufficiently close to the critical point in phase

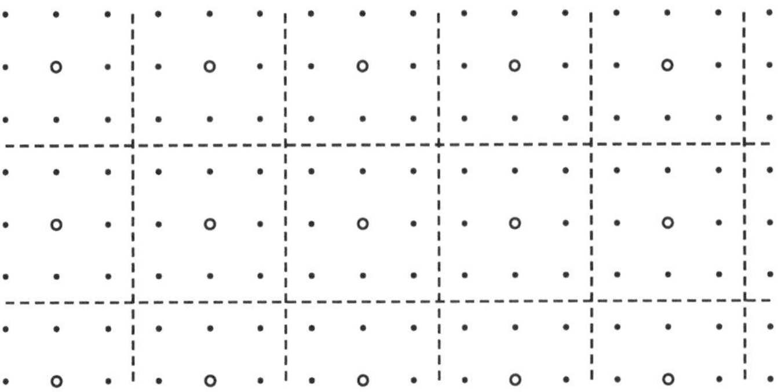

Fig. 2.4. Lattice of spacing a divided into cells, of linear dimension λa, containing λ^d spins.

space, y_K and y_L are constants. This means that $\mathfrak{r}(\sigma)$ acts as the single measure of the distance from the critical point and, under the sequence of transformations $\mathfrak{r}(\sigma) \to \lambda\mathfrak{r}(\sigma) \to \lambda^2\mathfrak{r}(\sigma)\cdots$, $\triangle K$ and L transform according to $\triangle K \to \lambda^{y_K}\triangle K \to \lambda^{2y_K}\triangle K\cdots$ and $L \to \lambda^{y_L}L \to \lambda^{2y_L}L\cdots$ respectively. Thus we have, in a neighbourhood of the critical point,

$$\mathfrak{r}(\sigma;\lambda^{y_L}L,\lambda^{y_K}\triangle K) = \lambda^{-1}\mathfrak{r}(\sigma;L,\triangle K)\,. \tag{2.7}$$

As we saw in Sect. 2.2, the critical point will correspond to non-smooth behaviour of the free energy of the system. It will, however, itself be finite non-zero and contain a smooth component. The same will be true of the potential ϕ and we suppose that we can make the separation

$$\phi = \phi_{\mathrm{smth}} + \phi_{\mathrm{sing}}\,, \tag{2.8}$$

where

$$\phi_{\mathrm{sing}} = \phi_{\mathrm{sing}}(L,\triangle K) \tag{2.9}$$

contains all the non-smooth behaviour of ϕ and $\phi_{\mathrm{sing}}(0,0) = 0$. If we are concerned only with the critical behaviour of the system, we can, in most cases (although see Sect. 2.5 below) neglect the smooth part of ϕ. Given that $\mathfrak{r}(\sigma)$ is the sole measure of the distance from the critical point, ϕ_{sing} can be given as a function of $\mathfrak{r}(\sigma)$. The free energy of the system is invariant under the transformation described above and the number of lattice sites is reduced from N to N/λ^d. Thus we have

$$\phi_{\mathrm{sing}}(\mathfrak{r}) = \lambda^{-d}\phi_{\mathrm{sing}}(\lambda^{-1}\mathfrak{r})\,, \tag{2.10}$$

or, equivalently,

$$\phi_{\mathrm{sing}}(\lambda^{y_L}L,\lambda^{y_K}\triangle K) = \lambda^{d}\phi_{\mathrm{sing}}(L,\triangle K)\,. \tag{2.11}$$

Since in the lattice of the cell centres $\tilde{\boldsymbol{r}}$ distances are measured in units of the new lattice spacing $\tilde{a} = \lambda a$, the rescaling for the site locations is given by the *dilatation transformation* $\tilde{\boldsymbol{r}} = \lambda^{-1}\boldsymbol{r}$ and the factor λ^{-d} in equation (2.10) is the Jacobian of the transformation $\boldsymbol{r} \to \tilde{\boldsymbol{r}}$. In Sect. 2.6 we show that the spin-spin correlation function satisfies the scaling form

$$\Gamma_2(\sigma(\lambda^{-1}\boldsymbol{r}_1), \sigma(\lambda^{-1}\boldsymbol{r}_2); \lambda^{-1}\mathfrak{r}) = \lambda^{2x_L}\Gamma_2(\sigma(\boldsymbol{r}_1), \sigma(\boldsymbol{r}_2); \mathfrak{r}), \qquad (2.12)$$

where $x_L = d - y_L$ is the *scaling dimension* of the spin operator $\sigma(\boldsymbol{r})$. When the system has translational and rotational symmetry with the spin-spin correlation function satisfying (1.73) we can set $\lambda = r = |\boldsymbol{r}_2 - \boldsymbol{r}_1|$ in (2.12) and, for sufficiently large r,

$$\Gamma_2(\sigma(\boldsymbol{r}_1), \sigma(\boldsymbol{r}_2); \mathfrak{r}) = \frac{f(\mathfrak{r}/r)}{r^{2x_L}}, \qquad (2.13)$$

for some unknown function f. The correlation length becomes the single measure of the dilatation scaling.

2.4 General Scaling Formulation

2.4.1 The Kadanoff Scaling Hypothesis

We now consider a system of the type described at the beginning of Sect. 2.2, with $n = n_{\mathrm{f}} + n_{\mathrm{e}} - 1$ independent couplings K_ℓ, $\ell = 1, 2, \ldots, n_{\mathrm{e}}$ and L_i, $i = 1, 2, \ldots, n_{\mathrm{f}} - 1$. Let there be a transition region $\mathcal{C}$ of dimension $n - s$ in the n-dimensional phase space of the couplings. We suppose that, close to $\mathcal{C}$, a set of curvilinear coordinates

$$\theta_j = \theta_j(L_i, K_\ell), \qquad j = 1, \ldots, n, \qquad (2.14)$$

called *scaling fields*, can be constructed. These are smooth functions of the couplings and in terms of them $\mathcal{C}$ is defined by

$$\theta_1 = \theta_2 = \cdots = \theta_s = 0. \qquad (2.15)$$

The variables $\theta_1, ..., \theta_s$ and $\theta_{s+1}, \ldots \theta_n$ are called *relevant* and *irrelevant* scaling fields respectively. The latter give coordinates for points in $\mathcal{C}$. As in the special case described in Sect. 2.3 it is supposed that the potential ϕ can be divided into smooth and singular[3] parts according to equation (2.8) with

$$\phi_{\mathrm{sing}} = \phi_{\mathrm{sing}}(\theta_1, \ldots, \theta_n). \qquad (2.16)$$

In this section a version of the scaling hypothesis is introduced, which is essentially that due to Kadanoff (1966), with a treatment which follows the work of Hankey and Stanley (1972). This procedure is sufficient for an analysis of power-law singularities. However, certain systems, in particular the Ising

[3] As we shall see later in this section, this analysis applies both to first-order transitions and to those for which some derivative of the potential diverges in the transition region. The term 'singular' is taken, therefore, to include all types of non-smooth behaviour, including finite discontinuities in derivatives of the potential.

model with $d = 2$ (see, for example, Volume 1, Chap. 8), exhibit logarithmic singularities. It has been shown by Nightingale and 'T Hooft (1974) that a slight generalization of the Kadanoff scaling hypothesis can be used to analyze these cases. This will be described in Sect. 2.5.

The *Kadanoff scaling hypothesis* is that there exists an origin and orientation for the scaling fields such that, in a region around $\mathcal{C}$, ϕ_{sing} satisfies the equation

$$\phi_{\text{sing}}(\lambda^{y_1}\theta_1, \ldots, \lambda^{y_n}\theta_n) = \lambda^d \phi_{\text{sing}}(\theta_1, \ldots, \theta_n), \tag{2.17}$$

for all real[4] $\lambda > 0$ and some set of *scaling exponents*[5] y_j, where d is the physical dimension of the system. The part of the potential ϕ_{sing} is a *generalized homogeneous function* and (2.17) is a generalization of (2.11). The scaling exponents fall into two subsets

$$\begin{aligned} y_j &> 0, \qquad & j &= 1, \ldots, s, \\ y_j &< 0, \qquad & j &= s+1, \ldots, n, \end{aligned} \tag{2.18}$$

and these are called *relevant exponents* and *irrelevant exponents* respectively like the corresponding scaling fields. A zero exponent is called *marginal.* We shall at this stage, for the sake of simplicity, assume that none of the exponents is marginal. The significance of marginal exponents in scaling theory is considered in Sect. 2.6. It will also be assumed that no exponent is complex. In practice, as will be seen in Sect. 6.7, this is not necessarily always the case, but situations arising from complex exponents are not difficult to interpret in particular examples. From equation (2.17)

$$\phi_{\text{sing}}(0, \ldots, 0) = 0 \tag{2.19}$$

and

$$\phi_{\text{sing}}(0, \ldots, 0, \theta_{s+1}, \ldots, \theta_n) = \lambda^{-d}\phi_{\text{sing}}(0, \ldots, 0, \lambda^{y_{s+1}}\theta_{s+1}, \ldots, \lambda^{y_n}\theta_n). \tag{2.20}$$

We know that ϕ_{sing} is finite throughout the phase space of couplings and, letting $\lambda \to \infty$ in (2.20), we see that

$$\phi_{\text{sing}}(0, \ldots, 0, \theta_{s+1}, \ldots, \theta_n) = 0 \tag{2.21}$$

throughout $\mathcal{C}$.

Let ψ be the function obtained by differentiating ϕ k_j times with respect to θ_j, for $j = 1, \ldots, n$. Then, from (2.17), assuming a division of ψ into smooth and singular parts as for ϕ,

---

4 The parameter λ introduced in Sect. 2.3 is an integer greater than unity. If, however, (2.17) were defined for $\lambda > 1$ it is not difficult to see that it would also apply for $1 \geq \lambda > 0$.

5 The term 'scaling exponent' is used to make the distinction with the critical exponents defined in Volume 1, Sect. 3.4 and Sect. 2.8 below. When there is no danger of confusion they will be referred to simply as exponents.

$$\psi_{\text{sing}}(\lambda^{y_1}\theta_1, \ldots, \lambda^{y_n}\theta_n) = \lambda^{d'}\psi_{\text{sing}}(\theta_1, \ldots, \theta_n), \tag{2.22}$$

where

$$d'(k_1, \ldots, k_n) = d - \sum_{j=1}^{n} k_j y_j . \tag{2.23}$$

Setting

$$\lambda = |\theta_m|^{-1/y_m} \tag{2.24}$$

in (2.17) and (2.22) gives

$$\begin{aligned} \phi_{\text{sing}}(\theta_1, \ldots, \theta_n) &= |\theta_m|^{\sigma_m}\phi_{\text{sing}}(g_1^{(m)}, \ldots, g_n^{(m)}), \\ \psi_{\text{sing}}(\theta_1, \ldots, \theta_n) &= |\theta_m|^{\omega_m}\psi_{\text{sing}}(g_1^{(m)}, \ldots, g_n^{(m)}), \end{aligned} \tag{2.25}$$

where

$$\sigma_m = \frac{d}{y_m}, \qquad \omega_m(k_1, \ldots, k_n) = \frac{d'(k_1, \ldots, k_n)}{y_m}, \qquad m = 1, \ldots, n \tag{2.26}$$

and

$$g_j^{(m)}(\theta_m, \theta_j) = \frac{\theta_j}{|\theta_m|^{\vartheta_j^{(m)}}}, \qquad m, j = 1, \ldots, n, \tag{2.27}$$

with

$$\vartheta_j^{(m)} = \frac{y_j}{y_m} = \frac{1}{\vartheta_m^{(j)}}, \qquad m, j = 1, \ldots, n. \tag{2.28}$$

From (2.23), (2.26) and (2.28)

$$\omega_m(k_1, \ldots, k_n) = \sigma_m - \sum_{j=1}^{n} k_j \vartheta_j^{(m)}. \tag{2.29}$$

The exponent $\vartheta_j^{(m)}$ gives the incremental change in the exponent ω_m of the function ψ_{sing} on differentiation with respect to θ_j. These quantities are therefore referred to as *gap exponents* . By extension it is convenient to refer to the quantities $g_j^{(m)}$ as *gap fields*. They satisfy the conditions

$$|g_m^{(\ell)}|^{\vartheta_j^{(m)}} |g_j^{(m)}| = |g_j^{(\ell)}|, \qquad m, j, \ell = 1, \ldots, n, \tag{2.30}$$

with $g_m^{(m)} = \pm 1$, according as $\theta_m \gtrless 0$. From (2.25) and (2.27)

$$\begin{aligned} \phi_{\text{sing}}(g_1^{(j)}, \ldots, g_n^{(j)}) &= |g_m^{(j)}|^{\sigma_m}\phi_{\text{sing}}(g_1^{(m)}, \ldots, g_n^{(m)}), \\ \psi_{\text{sing}}(g_1^{(j)}, \ldots, g_n^{(j)}) &= |g_m^{(j)}|^{\omega_m}\psi_{\text{sing}}(g_1^{(m)}, \ldots, g_n^{(m)}), \\ & j, m = 1, \ldots, n. \end{aligned} \tag{2.31}$$

For later reference it is useful to draw attention to the non-uniqueness of the choice of scaling fields. In particular the scaling hypothesis (2.17) and the analysis following from it continues to hold if a relevant scaling field θ_j is replaced by

$$\theta'_j = \theta_j - \varpi|\theta_m|^{\vartheta_j^{(m)}} , \tag{2.32}$$

for any other relevant scaling field θ_m and any positive constant ϖ (Example 2.1).

2.4.2 Approaches to the Transition Region

We now consider the asymptotic forms derived from (2.25) as $\mathcal{C}$ is approached ($\theta_j \to 0$, $j = 1, \ldots, s$) along some path $\mathcal{P}$. Suppose that

$$\lim_{\mathcal{P}} g_j^{(m)}(\theta_m, \theta_j) = \varpi_j^{(m)} . \tag{2.33}$$

It is not difficult to show (Example 2.1) that, for any path $\mathcal{P}$, it is always possible to find a relevant field θ_m such that $\varpi_j^{(m)}$ is finite for all $j = 1, \ldots, s$.[6] Conversely, as we see below, given a relevant field θ_m, there exists a part of the neighbourhood of $\mathcal{C}$ which contains paths for which $\varpi_j^{(m)}$ are finite for all $j = 1, \ldots, n$. A more difficult problem is to be sure that the amplitudes

$$\begin{aligned} &\lim_{\mathcal{P}} \phi_{\text{sing}}(g_1^{(m)}, \ldots, g_n^{(m)}) = \phi_{\text{sing}}(\varpi_1^{(m)}, \ldots, \varpi_n^{(m)}) , \\ &\lim_{\mathcal{P}} \psi_{\text{sing}}(g_1^{(m)}, \ldots, g_n^{(m)}) = \psi_{\text{sing}}(\varpi_1^{(m)}, \ldots, \varpi_n^{(m)}) , \end{aligned} \tag{2.34}$$

are finite *and* non-zero. If this is indeed the case then it follows from (2.25) that

$$\begin{aligned} &\phi_{\text{sing}}(\theta_1, \ldots, \theta_n) \simeq |\theta_m|^{\sigma_m} \phi_{\text{sing}}(\varpi_1^{(m)}, \ldots, \varpi_n^{(m)}) , \\ &\psi_{\text{sing}}(\theta_1, \ldots, \theta_n) \simeq |\theta_m|^{\omega_m} \psi_{\text{sing}}(\varpi_1^{(m)}, \ldots, \varpi_n^{(m)}) , \end{aligned} \tag{2.35}$$

as $\mathcal{C}$ is approached along $\mathcal{P}$. If $\varpi_j^{(m)} \neq 0$ for some j then, with the substitution $|\theta_m| = |\varpi_m^{(j)}||\theta_j|^{\vartheta_m^{(j)}}$ in (2.35) and using (2.31), the formulae (2.35) are recovered with m and j interchanged. For the moment we shall assume that the amplitudes in (2.35), given by (2.34) are finite and non-zero. The corrections to scaling arising from a breakdown in this assumption are discussed in Sect. 2.14. Since ϕ_{sing} is continuous the amplitude for this asymptotic form is invariant under change of sign of θ_m. For ψ_{sing} this is not in general the case.

The geometry of the path $\mathcal{P}$ close to $\mathcal{C}$ can be understood more clearly if we consider the family of surfaces

$$|\theta_j| = \varpi|\theta_m|^{\vartheta_j^{(m)}} , \tag{2.36}$$

[6] They are clearly zero for $j = s+1, \ldots, n$.

for all $0 < \varpi < \infty$. The members of this family are tangential to the surface $\theta_j = 0$ or $\theta_m = 0$ according as $y_j >, < y_m$. If $y_j = y_m$ the members of the family make angles greater than zero and less than $\pi/2$ with both surfaces. If $\varpi_j^{(m)} \neq 0$ the path $\mathcal{P}$ lies asymptotically on the member of the family with $\varpi = \varpi_j^{(m)}$ as $\mathcal{P}$ approaches $\mathcal{C}$. If $\varpi_j^{(m)} = 0$, for any ϖ and points sufficiently close to $\mathcal{C}$ the path $\mathcal{P}$ lies between that member of (2.36) and the surface $\theta_j = 0$. The family of surfaces (2.36) represent the border between one form of critical or coexistence behaviour and another, since they divide the phase space between paths on which the limiting value $\varpi_j^{(m)}$ of $g_j^{(m)}$ is finite or infinite. They are therefore referred to as *crossover surfaces* and in this context the gap exponent $\vartheta_j^{(m)}$ is called a *crossover exponent*. Looked at from a slightly different point of view the types of scaling behaviour can be classified by considering the different ranges of value of $\vartheta_j^{(m)}$ (see, for example, Pfeuty and Toulouse 1977, Chap. 3).

2.4.3 First-Order Transitions

Now the densities, which are the first-order partial derivatives of ϕ with respect to the couplings, can be expressed in terms of the first-order partial derivatives of ϕ with respect to the scaling fields. These are finite in the transition region and it therefore follows, from (2.23), (2.26) and (2.35), that

$$y_j \leq d\,, \qquad j = 1, \ldots, n\,. \tag{2.37}$$

In general we see from these same equations that, as $\mathcal{C}$ is approached along the path $\mathcal{P}$, ψ_{sing} is singular if

$$d < \sum_{j=1}^{n} k_j y_j \tag{2.38}$$

and zero if

$$d > \sum_{j=1}^{n} k_j y_j\,. \tag{2.39}$$

Consider now the possibility of there being a discontinuity in ψ, as $\mathcal{C}$ is approached along $\mathcal{P}$. This discontinuity must be in $\psi_{\text{sing}}(\varpi_1^{(m)}, \ldots, \varpi_n^{(m)})$, which must be finite and, on at least one side of the coexistence curve, non-zero. This in turn must give a discontinuity in $\psi_{\text{sing}}(\theta_1, \ldots, \theta_n)$. It follows from (2.35) that a necessary condition for this to be the case is that ω_m, and hence d' is zero; that is

$$d = \sum_{j=1}^{n} k_j y_j\,. \tag{2.40}$$

It must be emphasized that condition (2.40) is a necessary but not sufficient condition for the existence of a discontinuity in ψ. Indeed it is also the case when the function exhibits a logarithmic singularity (see Sect. 2.5). The most important case of (2.40) is $d = y_j$. It indicates the possibility of a first-order transition , that is a discontinuity in one or more of the densities (Nienhuis and Nauenberg 1975, Fisher and Berker 1982). In this case the 'evidence' for a discontinuity is rather stronger since the densities do not become infinite as $\mathcal{C}$ is approached, thereby excluding the possibility of a logarithmic singularity.

2.4.4 Effective Exponents

In most cases transition regions do not exist in isolation in phase space. They are normally connected to other transition regions, as in the case of the critical point and coexistence curve discussed in Sects. 2.8 and 2.9. This means that in the neighbourhood of $\mathcal{C}$ governed by the scaling hypothesis there will be singular behaviour apart from that exhibited on $\mathcal{C}$ itself. Suppose this exists in a transition region $\mathcal{C}'$ of dimension $n-s+1$. Remembering that the choice of scaling fields for $\mathcal{C}$ was not unique, we suppose that the scaling fields can be chosen (or subsequently modified; see Sect. 2.9) so that the transition region $\mathcal{C}'$ is given by

$$\theta_1 = \theta_2 = \ldots = \theta_{s-1} = 0 . \tag{2.41}$$

The problem now is to obtain the exponents, analogous to σ_m and ω_m, for the transition region $\mathcal{C}'$. One way to do this is to use a modification of the idea of *effective exponents* introduced by Riedel and Wegner (1974). Let

$$\begin{aligned} \Sigma_m(\theta_1,\ldots,\theta_n) &= \frac{\partial \ln \phi_{\text{sing}}}{\partial \ln |\theta_m|} , \\ \Omega_m(\theta_1,\ldots,\theta_n) &= \frac{\partial \ln \psi_{\text{sing}}}{\partial \ln |\theta_m|} . \end{aligned} \tag{2.42}$$

From (2.25),

$$\Sigma_m(\theta_1,\ldots,\theta_n) = \sigma_m - \sum_{j\neq m} \vartheta_j^{(m)} |g_j^{(m)}| \frac{\partial \ln \phi_{\text{sing}}(g_1^{(m)},\ldots,g_n^{(m)})}{\partial |g_j^{(m)}|} . \tag{2.43}$$

Let $\mathcal{P}$ be the path to $\mathcal{C}$ described above, but in this case suppose that $\varpi_j^{(m)} = 0$, for all $j \neq m$. It is clear, from (2.43), that

$$\lim_{\mathcal{P}} \Sigma_m(\theta_1,\ldots,\theta_n) = \sigma_m . \tag{2.44}$$

Consider now a slight variation of the path $\mathcal{P}$ to a new path $\mathcal{P}'$ which differs from $\mathcal{P}$ only in that $\theta_s \to \bar{\theta}_s \neq 0$. All the ϖs for the path $\mathcal{P}'$ are zero except $\varpi_s^{(m)}$, for which $g_s^{(m)} \to \infty$. The path $\mathcal{P}'$ is to a point on the transition surface $\mathcal{C}'$ specified by $\bar{\theta}_s$. The limiting behaviour of Σ_m will differ from σ_m by a contribution arising from the second term in (2.43). Suppose that as the path $\mathcal{P}'$ is followed to its final point

$$\phi_{\text{sing}}(g_1^{(m)},\ldots,g_n^{(m)}) \sim |g_s^{(m)}|^{\mu_s^{(m)}} . \tag{2.45}$$

Substituting from (2.45) into (2.43) we have

$$\lim_{\mathcal{P}'} \Sigma_m(\theta_1,\ldots,\theta_n) = \sigma_m - \vartheta_s^{(m)}\mu_s^{(m)} . \tag{2.46}$$

Thus we have the asymptotic form

$$\phi_{\text{sing}}(\theta_1,\ldots,\theta_n) \sim |\theta_m|^{\sigma_m - \vartheta_s^{(m)}\mu_s^{(m)}} , \tag{2.47}$$

as $\mathcal{C}'$ is approached along the path $\mathcal{P}'$. A similar analysis applies to the effective exponent Ω_m leading to an asymptotic form for ψ_{sing}.

2.5 Logarithmic Singularities

2.5.1 The Nightingale–'T Hooft Scaling Hypothesis

With the scaling fields $\theta_1,\ldots,\theta_n$, defined as in Sect. 2.4, the *Nightingale–'T Hooft scaling hypothesis* (Nightingale and 'T Hooft 1974) is that in a neighbourhood of the transition region $\mathcal{C}$

$$\phi(\lambda^{y_1}\theta_1,\ldots,\lambda^{y_n}\theta_n) = \lambda^d\phi(\theta_1,\ldots,\theta_n) - \bar{\phi}^{(\lambda)}(\theta_1,\ldots,\theta_n) , \tag{2.48}$$

for all real $\lambda > 1$, where the $\bar{\phi}^{(\lambda)}$ are a set of smooth functions of the scaling fields, parameterized by λ. This is essentially the scaling formula derived from the renormalization group (see Sect. 6.5), where $\bar{\phi}^{(\lambda)}$ is the term arising from the renormalization of the self-energy. There are two ways of understanding the relationship between the two scaling hypotheses (2.17) and (2.48). Given the decomposition (2.8) of ϕ into singular and smooth parts it can be seen that sufficiently close to the transition region $\mathcal{C}$ the equation relating the non-smooth contributions to (2.48) is (2.17). Alternatively by substituting from (2.8) into (2.17) and comparing with (2.48)

$$\bar{\phi}^{(\lambda)}(\theta_1,\ldots,\theta_n) = \lambda^d\phi_{\text{smth}}(\theta_1,\ldots,\theta_n) - \phi_{\text{smth}}(\lambda^{y_1}\theta_1,\ldots,\lambda^{y_n}\theta_n) . \tag{2.49}$$

Iterating equation (2.48) $\ell - 1$ times

$$\begin{aligned}\phi(\theta_1,\ldots,\theta_n) = {} & \lambda^{-\ell d}\phi(\lambda^{\ell y_1}\theta_1,\ldots,\lambda^{\ell y_n}\theta_n) \\ & + \lambda^{-d}\sum_{r=0}^{\ell-1}\lambda^{-rd}\bar{\phi}^{(\lambda)}(\lambda^{r y_1}\theta_1,\ldots,\lambda^{r y_n}\theta_n) .\end{aligned} \tag{2.50}$$

Since $\bar{\phi}^{(\lambda)}$ is a smooth function of its arguments, it has a power series expansion of the form

$$\bar{\phi}^{(\lambda)}(\theta_1,\ldots,\theta_n) = \sum_{\{q_j\}} B^{(\lambda)}(q_1,\ldots,q_n)\prod_{j=1}^{n}\frac{\theta_j^{q_j}}{q_j!} , \tag{2.51}$$

where the summation is over $q_j = 0, 1, 2, \ldots$, for all $j = 1, \ldots, n$. Substituting into (2.50) gives

$$\phi(\theta_1,\ldots,\theta_n) = \lambda^{-\ell d}\phi(\lambda^{\ell y_1}\theta_1,\ldots,\lambda^{\ell y_n}\theta_n) + \lambda^{-d}\sum_{\{q_j\}} B^{(\lambda)}(q_1,\ldots,q_n) \times \Lambda_\ell(\lambda; d'(q_1,\ldots,q_n)) \prod_{j=1}^{n} \frac{\theta_j^{q_j}}{q_j!}, \tag{2.52}$$

where

$$\Lambda_\ell(\lambda;x) = \sum_{r=0}^{\ell-1}\lambda^{-rx} = \begin{cases} \ell, & x=0, \\ \dfrac{1-\lambda^{-x\ell}}{1-\lambda^{-x}}, & x\neq 0, \end{cases} \tag{2.53}$$

and d' is as defined in (2.23). Throughout this section, for d' and $\omega_m = d'/y_m$, we shall adopt the notation of including the arguments of the quantity if they are the summation variables $q_1,\ldots,q_n$, but omitting them if they are the variables $k_1,\ldots,k_n$ used to define ψ. This function, which was introduced in Sect. 2.4, is obtained by differentiating ϕ k_j times with respect to θ_j for $j=1,\ldots,n$. From (2.52)

$$\psi(\theta_1,\ldots,\theta_n) = \lambda^{-\ell d'}\psi(\lambda^{\ell y_1}\theta_1,\ldots,\lambda^{\ell y_n}\theta_n) + \lambda^{-d}{\sum_{\{q_j\}}}' B^{(\lambda)}(q_1,\ldots,q_n) \times \Lambda_\ell(\lambda; d'(q_1,\ldots,q_n)) \prod_{j=1}^{n} \frac{\theta_j^{q_j-k_j}}{(q_j-k_j)!}, \tag{2.54}$$

where the summations are now restricted to $q_j = k_j, k_j+1,\ldots$ for $j = 1,\ldots,n$.

2.5.2 Constraints on Scaling

Having developed the formalism it is necessary to consider the limits of its applicability. The two forms of the scaling hypothesis (2.17) and (2.48) have been assumed to be valid in some neighbourhood of the transition region $\mathcal{C}$. This should be taken to mean that the part of ϕ_{sing}, in the case of (2.17), and ϕ and $\bar{\phi}^{(\lambda)}$, in the case of (2.48), which dominate as $\mathcal{C}$ is approached satisfy the scaling forms. This in itself implies no restriction on λ other than that contained in the hypotheses. However, to obtain the formula (2.50), it was necessary to substitute into (2.48) for a succession of points $(\lambda^{ry_1}\theta_1,\ldots,\lambda^{ry_n}\theta_n)$ for $r=1,2,\ldots,\ell-1$. As r increases, with $\lambda > 1$, the coordinate values of the relevant scaling fields grow and the trajectory of points moves away from $\mathcal{C}$. The need to remain within the neighbourhood of validity of the scaling hypothesis thus imposes a constraint which can be taken to be either an upper bound on λ or ℓ or a restriction on the neighbourhood containing $(\theta_1,\ldots,\theta_n)$. Suppose that the neighbourhood of validity of the scaling hypothesis (2.48) is given by

$$L(\theta_1,\ldots,\theta_s) \leq R, \tag{2.55}$$

where

$$L(\theta_1,\ldots,\theta_s) = \left[|\theta_1|^{2/y_1} + \ldots + |\theta_s|^{2/y_s}\right]^{\frac{1}{2}} \tag{2.56}$$

is the metric which measures distances from $\mathcal{C}$ and which scales according to the formula

$$L(\lambda^{y_1}\theta_1,\ldots,\lambda^{y_s}\theta_s) = \lambda L(\theta_1,\ldots,\theta_s)\,. \tag{2.57}$$

Then the iterative process remains in the scaling neighbourhood if

$$\lambda^{\ell-1} \le \frac{R}{L(\theta_1,\ldots,\theta_s)}\,. \tag{2.58}$$

This inequality is satisfied if we choose the value

$$\ell = \ell^* = \frac{1}{\ln(\lambda)}\ln\left[\frac{R}{L(\theta_1,\ldots,\theta_s)}\right], \tag{2.59}$$

assuming that λ is such that this is an integer. From (2.27), (2.56) and (2.59), for any relevant scaling field θ_m

$$\ell^* = -\frac{\ln|\theta_m|}{y_m\ln(\lambda)} + \frac{1}{\ln(\lambda)}\ln\left[\frac{R}{L(g_1^{(m)},\ldots,g_s^{(m)})}\right]. \tag{2.60}$$

2.5.3 Approaches to the Transition Region

Now consider the path $\mathcal{P}$ towards the transition region $\mathcal{C}$, described in Sect. 2.4, where the limiting values $\varpi_j^{(m)}$ of the gap fields, given by (2.33), are all finite or zero. Denoting by $\Lambda^*(\lambda;x)$ the value of $\Lambda_\ell(\lambda;x)$ when $\ell=\ell^*$, it follows from (2.53) and (2.60) that

$$\Lambda^*(\lambda;x) \simeq \begin{cases} A(\lambda;x)|\theta_m|^{x/y_m}, & x<0,\\ A(\lambda;0)\ln|\theta_m|, & x=0,\\ A(\lambda;x)\,, & x>0, \end{cases} \tag{2.61}$$

where

$$A(\lambda;x) = \begin{cases} -C_m^x/(1-\lambda^{-x})\,, & x<0,\\ -1/[y_m\ln(\lambda)]\,, & x=0,\\ 1/(1-\lambda^{-x}) & x>0, \end{cases} \tag{2.62}$$

and $C_m = L(\varpi_1^{(m)},\ldots,\varpi_s^{(m)})/R \neq 0$ (since $\varpi_m^{(m)} = 1$). We now need to derive asymptotic forms for (2.52) and (2.54) as $\mathcal{C}$ is approached along the path $\mathcal{P}$. From (2.60)

$$\lambda^{-\ell d}\phi(\lambda^{\ell y_1}\theta_1,\ldots,\lambda^{\ell y_n}\theta_n) \simeq |\theta_m|^{\sigma_m}C_m^d\phi(\varpi_1^{(m)}C_m^{-y_1},\ldots,\varpi_n^{(m)}C_m^{-y_m}), \tag{2.63}$$

$$\lambda^{-\ell d'}\psi(\lambda^{\ell y_1}\theta_1,\ldots,\lambda^{\ell y_n}\theta_n) \simeq |\theta_m|^{\omega_m}C_m^{d'}\psi(\varpi_1^{(m)}C_m^{-y_1},\ldots,\varpi_n^{(m)}C_m^{-y_m}),$$

where σ_m and ω_m are given by (2.26). These terms in (2.52) and (2.54) reproduce the power law behaviour obtained in (2.35). It is necessary to determine the asymptotic behaviour arising from the second terms on the right-hand sides of (2.52) and (2.54). For these equations the sums are divided into the sets of terms for which $q_1, \ldots, q_n$ satisfy the conditions: (1) $d'(q_1, \ldots, q_n) > 0$, (2) $d'(q_1, \ldots, q_n) = 0$ and (3) $d'(q_1, \ldots, q_n) < 0$, respectively. We then obtain

$$\begin{aligned}
&\lambda^{-d} \sum_{\{q_j\}} B^{(\lambda)}(q_1, \ldots, q_n) \Lambda^*(\lambda; d'(q_1, \ldots, q_n)) \prod_{j=1}^{n} \frac{\theta_j^{q_j}}{q_j!} \\
&\quad \simeq \lambda^{-d} |\theta_m|^{\sigma_m} \sum_{\{q_j\}}^{(1)} B^{(\lambda)}(q_1, \ldots, q_n) \\
&\quad \times A(\lambda; d'(q_1, \ldots, q_n)) |\theta_m|^{-\omega_m(q_1, \ldots, q_n)} \prod_{j=1}^{n} \frac{[\varpi_j^{(m)}]^{q_j}}{q_j!} \\
&\quad + \lambda^{-d} |\theta_m|^{\sigma_m} \ln |\theta_m| \sum_{\{q_j\}}^{(2)} B^{(\lambda)}(q_1, \ldots, q_n) A(\lambda; 0) \prod_{j=1}^{n} \frac{[\varpi_j^{(m)}]^{q_j}}{q_j!} \\
&\quad + \lambda^{-d} |\theta_m|^{\sigma_m} \sum_{\{q_j\}}^{(3)} B^{(\lambda)}(q_1, \ldots, q_n) A(\lambda; d'(q_1, \ldots, q_n)) \prod_{j=1}^{n} \frac{[\varpi_j^{(m)}]^{q_j}}{q_j!}
\end{aligned} \tag{2.64}$$

and

$$\begin{aligned}
&\lambda^{-d} \sum_{\{q_j\}}{}' B^{(\lambda)}(q_1, \ldots, q_n) \Lambda^*(\lambda; d'(q_1, \ldots, q_n)) \prod_{j=1}^{n} \frac{\theta_j^{q_j - k_j}}{(q_j - k_j)!} \\
&\quad \simeq \lambda^{-d} |\theta_m|^{\omega_m} \sum_{\{q_j\}}^{(1)\,'} B^{(\lambda)}(q_1, \ldots, q_n) \\
&\quad \times A(\lambda; d'(q_1, \ldots, q_n)) |\theta_m|^{-\omega_m(q_1, \ldots, q_n)} \prod_{j=1}^{n} \frac{[\varpi_j^{(m)}]^{q_j - k_j}}{(q_j - k_j)!} \\
&\quad + \lambda^{-d} |\theta_m|^{\omega_m} \ln |\theta_m| \sum_{\{q_j\}}^{(2)\,'} B^{(\lambda)}(q_1, \ldots, q_n) A(\lambda; 0) \prod_{j=1}^{n} \frac{[\varpi_j^{(m)}]^{q_j - k_j}}{(q_j - k_j)!} \\
&\quad + \lambda^{-d} |\theta_m|^{\omega_m} \sum_{\{q_j\}}^{(3)\,'} B^{(\lambda)}(q_1, \ldots, q_n) \\
&\quad \times A(\lambda; d'(q_1, \ldots, q_n)) \prod_{j=1}^{n} \frac{[\varpi_j^{(m)}]^{q_j - k_j}}{(q_j - k_j)!},
\end{aligned} \tag{2.65}$$

for (2.52) and (2.54) respectively. A term in the summations in (2.64) is zero if $q_j > 0$, for any j for which $\varpi_j^{(m)} = 0$. In particular to have a non-zero term q_j must be zero for all $j = s+1, \ldots, n$, since $\varpi_j^{(m)} = 0$ for all irrelevant fields. The same observation applies to (2.65), except that q_j must be equal to k_j rather than to zero. The residual terms in the summations (3) in (2.64) and (2.65) give contributions of the same power-law type as (2.63) and can therefore, in general, be ignored, except in the case when the amplitudes in (2.63) are not non-zero and finite, when they may provide the only terms with this asymptotic behaviour. The exponent of $|\theta_m|$ for a term in the summation (1) of (2.64) is positive and less that σ_m and since, for the corresponding summation in (2.65), $\omega_m > \omega_m(q_1, \ldots, q_s, 0, \ldots, 0) > 0$ this summation will be absent in the case where ψ has a power-law singularity as the path $\mathcal{P}$ approaches $\mathcal{C}$. The interesting and important terms in (2.64) and (2.65) are the summations (2). These will occur only when

$$d = \sum_{j=1}^{n} q_j y_j \tag{2.66}$$

has a solution, or solutions, for some set of non-negative integers $q_1, \ldots, q_n$. If we choose $k_1, \ldots, k_n$ to be such a solution, then ψ is such that (2.40) is satisfied and, from (2.54) and (2.65),

$$\psi(\theta_1, \ldots, \theta_n) \simeq \lambda^{-d} B^{(\lambda)}(k_1, \ldots, k_n) A(\lambda; 0) \ln |\theta_m| \,. \tag{2.67}$$

This particular derivative of ϕ exhibits a logarithmic singularity as $\mathcal{C}$ is approached along $\mathcal{P}$, as long as $B^{(\lambda)}(k_1, \ldots, k_n) \neq 0$. From a renormalization group point of view this divergence arises from the self-energy of the cells, whereas the more usual power-law singularities arise from the interaction between cells. The occurrence of logarithmic divergences in thermodynamic response functions are the first indication that the scaling ideas of Sect. 2.4 should be treated with some caution, and need modification in particular circumstances. Further modifications, arising from zero or singular amplitude factors are considered in Sect. 2.14.

2.6 Correlation Functions

2.6.1 Scaling Operators and Dimensions

Equation (1.57) gives the Hamiltonian in terms of the couplings and state operators. Now suppose that the formulae (2.14) are invertible near the transition region $\mathcal{C}$ to give

$$\begin{aligned} K_\ell &= K_\ell(\theta_1, \theta_2, \ldots, \theta_n)\,, & \ell &= 1, 2, \ldots, n_{\mathrm{e}}\,, \\ L_i &= L_i(\theta_1, \theta_2, \ldots, \theta_n)\,, & i &= 1, 2, \ldots, n_{\mathrm{f}} - 1\,. \end{aligned} \tag{2.68}$$

The division (2.8) of the potential ϕ into smooth and singular parts can now be achieved by dividing the Hamiltonian (1.57) into smooth and singular parts

$$\begin{aligned} \widehat{H}_{\text{smth}}(L_i, K_\ell; \boldsymbol{\sigma}) &= \widehat{H}(L_{ic}, K_{\ell c}; \boldsymbol{\sigma}), \\ \widehat{H}_{\text{sing}}(L_i, K_\ell; \boldsymbol{\sigma}) &= \widehat{H}(\triangle L_i, \triangle K_\ell; \boldsymbol{\sigma}), \end{aligned} \tag{2.69}$$

where

$$\begin{aligned} K_{\ell c} &= K_\ell(0, \dots, 0, \theta_{s+1}, \dots, \theta_n), & \ell &= 1, 2, \dots, n_e, \\ L_{ic} &= L_i(0, \dots, 0, \theta_{s+1}, \dots, \theta_n), & i &= 1, 2, \dots, n_f - 1 \end{aligned} \tag{2.70}$$

and

$$\begin{aligned} \triangle K_\ell &= K_\ell - K_{\ell c}, & \ell &= 1, 2, \dots, n_e, \\ \triangle L_{i\mathcal{C}} &= L_i - L_{ic}, & i &= 1, 2, \dots, n_f - 1. \end{aligned} \tag{2.71}$$

Since $\widehat{H}_{\text{sing}}(L_i, K_\ell; \boldsymbol{\sigma}) = 0$ throughout the transition region $\mathcal{C}$ in which the relevant scaling fields are zero, it can, close to $\mathcal{C}$, be expressed as a linear combination of the relevant scaling fields.

$$\widehat{H}_{\text{sing}}(L_i, K_\ell; \boldsymbol{\sigma}) = \sum_{\{\boldsymbol{r}\}} \sum_{j=1}^{s} \phi_j(\boldsymbol{r}) \theta_j, \tag{2.72}$$

where the coefficients $\phi_j(\boldsymbol{r})$, $j = 1, \dots, s$, which are in general functions of the irrelevant scaling fields, are called *scaling operators*. The development of Sect. 1.5 can be carried out for scaling fields and scaling operators just as well as for couplings and state operators. From (1.61) the connected correlation function of an arbitrary set of τ scaling operators satisfies the formula

$$\sum_{\{\boldsymbol{r}_1, \dots, \boldsymbol{r}_\tau\}} \Gamma_\tau(\phi_1(\boldsymbol{r}_1), \dots, \phi_\tau(\boldsymbol{r}_\tau)) = -N \frac{\partial^\tau \phi_{\text{sing}}}{\partial\theta_1 \cdots \partial\theta_\tau}. \tag{2.73}$$

Let this connected τ-point correlation function satisfied the scaling relation

$$\Gamma_\tau(\phi_1(\lambda^{-1}\boldsymbol{r}_1), \dots, \phi_\tau(\lambda^{-1}\boldsymbol{r}_\tau)) = \lambda^z \Gamma_\tau(\phi_1(\boldsymbol{r}_1), \dots, \phi_\tau(\boldsymbol{r}_\tau)), \tag{2.74}$$

where, for easy of presentation, we have omitted the dependence on the scaling fields and z is some as yet to be determined exponent. Applying the Fourier transformation (1.65) to (2.74) gives

$$\Gamma^*_\tau(\phi_1, \dots, \phi_\tau; \lambda\boldsymbol{k}_1, \dots, \lambda\boldsymbol{k}_\tau) = \lambda^{z-\tau d} \Gamma^*_\tau(\phi_1, \dots, \phi_\tau; \boldsymbol{k}_1, \dots, \boldsymbol{k}_\tau). \tag{2.75}$$

From (1.68),

$$\Gamma^*_\tau(\phi_1, \dots, \phi_\tau; \boldsymbol{0}, \dots, \boldsymbol{0}) = -\frac{1}{N} \frac{\partial^\tau \phi_{\text{sing}}}{\partial\theta_1 \cdots \partial\theta_\tau}, \tag{2.76}$$

so, when $\boldsymbol{k}_1 = \cdots = \boldsymbol{k}_\tau = \boldsymbol{0}$, (2.75) is the scaling form for the right-hand side of (2.76) which is given by (2.22) and (2.23). Thus we have

$$z = \sum_{j=1}^{\tau} x_j \,, \tag{2.77}$$

where

$$x_j = d - y_j \tag{2.78}$$

is the *scaling dimension* of the scaling operator ϕ_j. Equation (2.74) now takes the form

$$\Gamma_\tau(\phi_1(\lambda^{-1}\boldsymbol{r}_1), \ldots, \phi_\tau(\lambda^{-1}\boldsymbol{r}_\tau)) = \left[\prod_{j=1}^{\tau} \lambda^{x_j}\right] \Gamma_\tau(\phi_1(\boldsymbol{r}_1), \ldots, \phi_\tau(\boldsymbol{r}_\tau)) \,. \tag{2.79}$$

When the system has translational symmetry the τ-point correlation function takes the form (1.70) in terms of the relative lattice vectors $\bar{\boldsymbol{r}}_j = \boldsymbol{r}_{j+1} - \boldsymbol{r}_1$, $j = 1, \ldots, \tau - 1$ and (2.79) becomes

$$\begin{aligned} \Gamma_\tau(\phi_1(\boldsymbol{0}), \phi_2(\lambda^{-1}\bar{\boldsymbol{r}}_1), \ldots, \phi_\tau(\lambda^{-1}\bar{\boldsymbol{r}}_{\tau-1})) &= \left[\prod_{j=1}^{\tau} \lambda^{x_j}\right] \\ &\times \Gamma_\tau(\phi_1(\boldsymbol{0}), \phi_2(\bar{\boldsymbol{r}}_1), \ldots, \phi_\tau(\bar{\boldsymbol{r}}_{\tau-1})) \,. \end{aligned} \tag{2.80}$$

In the case where, for some ordering of the lattice site vectors, $|\bar{\boldsymbol{r}}_1| \sim |\bar{\boldsymbol{r}}_2| \sim \cdots \sim |\bar{\boldsymbol{r}}_{\tau-1}| \sim r$ we can make the substitution $\lambda = r$ in (2.80) to give

$$\begin{aligned} \Gamma_\tau(\phi_1(\boldsymbol{0}), \phi_2(\bar{\boldsymbol{r}}_1), \ldots, \phi_\tau(\bar{\boldsymbol{r}}_{\tau-1})) &= \left[\prod_{j=1}^{\tau} \frac{1}{r^{x_j}}\right] \\ &\times \Gamma_\tau(\phi_1(\boldsymbol{0}), \phi(\bar{\boldsymbol{r}}_1/r), \ldots, \phi_\tau(\bar{\boldsymbol{r}}_{\tau-1}/r)) \,. \end{aligned} \tag{2.81}$$

The asymptotic dependence of Γ_τ on r, in the limit of large r, on the transition surface is given by setting $\theta_1 = \cdots = \theta_s = 0$. As long as the amplitude factor on the right-hand side of (2.81) is non-zero and finite we have

$$\Gamma_\tau(\phi_1(\boldsymbol{0}), \phi_2(\bar{\boldsymbol{r}}_1), \ldots, \phi_\tau(\bar{\boldsymbol{r}}_{\tau-1})) \simeq G_\tau(\phi_1, \ldots, \phi_\tau) \prod_{j=1}^{\tau} \frac{1}{r^{x_j}} \,, \tag{2.82}$$

where $G_\tau(\phi_1, \ldots, \phi_\tau)$ is a constant dependent only on the choice of scaling operators. A similar expression

$$\bar{\Gamma}_\tau^*(\phi_1, \ldots, \phi_\tau; \bar{\boldsymbol{k}}_1, \ldots, \bar{\boldsymbol{k}}_{\tau-1}) \simeq G_\tau^*(\phi_1, \ldots, \phi_\tau) k^d \prod_{j=1}^{\tau} \frac{1}{k^{y_j}} \tag{2.83}$$

can be obtained for the Fourier transform taken with respect to the relative lattice vectors. It follows from (1.52), (1.53) and (1.56) that the total correlation function and fluctuation correlation function also satisfy similar scaling forms. In particular for the total correlation function

$$\langle \phi_1(\lambda^{-1}\boldsymbol{r}_1)\cdots\phi_\tau(\lambda^{-1}\boldsymbol{r}_\tau)\rangle = \left[\prod_{j=1}^{\tau}\lambda^{x_j}\right]\langle\phi_1(\boldsymbol{r}_1)\cdots\phi_\tau(\boldsymbol{r}_\tau)\rangle\,, \tag{2.84}$$

leading to the asymptotic form

$$\langle\phi_1(\boldsymbol{r}_1)\cdots\phi_\tau(\boldsymbol{r}_\tau)\rangle \simeq C_\tau(\phi_1,\ldots,\phi_\tau)\prod_{j=1}^{\tau}\frac{1}{r^{x_j}} \tag{2.85}$$

on the transition surface. In this context it is a common terminology to say (Kadanoff 1969a) that ϕ_j *scales as* $1/r^{x_j}$. This simply means that, in the expectation value of two or more of the scaling operators, ϕ_j contributes a scaling factor $1/r^{x_j}$.

2.6.2 Variable Scaling Exponents

Throughout the discussion in the previous sections we have assumed that the scaling exponents $y_1,\ldots,y_n$ are constants. There are, however, a number of models, the most well-known being the eight-vertex model (Chap. 5), in which the critical exponents are functions of one of the couplings (see Sects. 5.9 and 5.12). The way that this behaviour can be reconciled with the ideas of scaling has been explored by Kadanoff and Wegner (1971), using the *reduction hypothesis* of Kadanoff (1969a, 1969b). We shall now present a simplified version of their argument. Based on equation (2.82) with $\tau = 2$ and $\tau = 3$ we suppose that, when $|\bar{\boldsymbol{r}}_1| \sim |\bar{\boldsymbol{r}}_2|$,

$$\Gamma_3(\phi_1(\boldsymbol{0}),\phi_2(\bar{\boldsymbol{r}}_1),\phi_3(\bar{\boldsymbol{r}}_2)) \sim \frac{\Gamma_2(\phi_1(\boldsymbol{0}),\phi_2(\bar{\boldsymbol{r}}_1))}{|\bar{\boldsymbol{r}}_2|^{x_3}}\,. \tag{2.86}$$

From the formulae equivalent to (1.60) and (1.61) for scaling fields and operators

$$\frac{\partial\Gamma_2(\phi_1(\boldsymbol{0}),\phi_2(\bar{\boldsymbol{r}}_1))}{\partial\theta_3} = \sum_{\{\bar{\boldsymbol{r}}_2\}}\Gamma_3(\phi_1(\boldsymbol{0}),\phi_2(\bar{\boldsymbol{r}}_1),\phi_3(\bar{\boldsymbol{r}}_2))\,, \tag{2.87}$$

and using (A.20) to replace the summation by an integral from 0 to $r = |\bar{\boldsymbol{r}}_1|$ we have

$$\frac{\partial\Gamma_2(\phi_1(\boldsymbol{0}),\phi_2(\bar{\boldsymbol{r}}_1))}{\partial\theta_3} \sim \begin{cases} \dfrac{1}{r^{x_1+x_2-y_3}}\,, & \text{if } y_3 \neq 0,\\[2ex] \dfrac{\ln(r)}{r^{x_1+x_2}}\,, & \text{if } y_3 = 0. \end{cases} \tag{2.88}$$

The case for $y_3 \neq 0$ is easily understood on differentiating (2.81) for $\tau = 2$ with respect to θ_3. On the right-hand side θ_3 appears in the combination $r^{y_3}\theta_3$ and the leading term in the derivative of $\Gamma_2(\phi_1(\boldsymbol{0}),\phi_2(\bar{\boldsymbol{r}}_1/r))$ contributes a factor r^{y_3}. When $y_3 = 0$, ϕ_3 scales as $1/r^d$. It would appear that in this case differentiation of (2.81) leaves the scaling dependence on r unchanged. Suppose however that y_1 and y_2 are functions of θ_3. Then

$$\frac{\partial \Gamma_2(\phi_1(\mathbf{0}), \phi_2(\bar{\mathbf{r}}_1))}{\partial \theta_3} \sim \frac{\ln(r)}{r^{x_1+x_2}} \left(\frac{\partial y_1}{\partial \theta_3} + \frac{\partial y_2}{\partial \theta_3} \right) \tag{2.89}$$

and the required logarithmic dependence is obtained. A scaling field with zero exponent is marginal (see Sect. 2.4) and in renormalization group analysis is often associated with a surface or line of fixed points . We have now shown that, if such a field is present, it is not inconsistent with scaling theory that the other scaling exponents should be functions of that field. A marginal scaling field or coupling represents an 'underlying' parameter of the system. Scaling theory can proceed without taking it into account. Once the phase diagram is understood the additional dimension represented by a marginal parameter simply increases the dimension of the phase space, without affecting the type of phenomena observed. The application of this analysis to the eight-vertex model is considered in Sect. 5.12.

2.7 Densities and Response Functions

Before considering the application of scaling to some particular cases it is useful to relate the general ideas of Sect. 2.4 to the scaling properties of more familiar thermodynamic functions. The differential form (2.3) contains two types of densities: ρ_i, $i = 1, \ldots, n_{\mathrm{f}} - 1$ are associated with external fields and are quantities like magnetization density or particle density; $\mathfrak{u}_\ell$, $\ell = 1, \ldots, n_{\mathrm{e}}$ are the energy densities associated with each type of interparticle interaction. The linear sum (1.31) gives the total internal energy density and from (2.3)

$$\mathfrak{u}_\ell = -\frac{\partial \phi}{\partial K_\ell}, \qquad \rho_i = -\frac{\partial \phi}{\partial L_i}. \tag{2.90}$$

Let

$$\rho_{i\mathrm{c}} = -\frac{\partial \phi_{\mathrm{smth}}}{\partial L_i}, \qquad \mathfrak{u}_{\ell\mathrm{c}} = -\frac{\partial \phi_{\mathrm{smth}}}{\partial K_\ell}. \tag{2.91}$$

Then

$$\begin{aligned} \rho_i - \rho_{i\mathrm{c}} &= -\frac{\partial \phi_{\mathrm{sing}}}{\partial L_i} &= -\sum_{j=1}^{n} \frac{\partial \phi_{\mathrm{sing}}}{\partial \theta_j} \frac{\partial \theta_j}{\partial L_i}, \\ \mathfrak{u}_\ell - \mathfrak{u}_{\ell\mathrm{c}} &= -\frac{\partial \phi_{\mathrm{sing}}}{\partial K_\ell} &= -\sum_{j=1}^{n} \frac{\partial \phi_{\mathrm{sing}}}{\partial \theta_j} \frac{\partial \theta_j}{\partial K_\ell}. \end{aligned} \tag{2.92}$$

The scaling properties of the deviations of the densities from their transition values are thus determined by scaling properties of the first-order partial derivatives of ϕ_{sing} with respect to the scaling fields θ_j. If the transition region $\mathcal{C}$ is approached along a path $\mathcal{P}$ such that, according to the second of equations (2.34),

$$\frac{\partial \phi_{\mathrm{sing}}}{\partial \theta_j}(\varpi_1^{(m)}, \ldots, \varpi_n^{(m)}) \qquad \text{is finite and non-zero,} \tag{2.93}$$

it follows from (2.26), (2.29) and (2.35) that

$$\frac{\partial \phi_{\text{sing}}}{\partial \theta_j} \sim |\theta_m|^{(d-y_j)/y_m} . \tag{2.94}$$

The scaling behaviour in equations (2.92) is dominated by the term in the summation on the right-hand side which corresponds to the largest y_j for which the derivative of θ_j with respect to the appropriate coupling is non-zero on $\mathcal{C}$. We denote these exponents by y_i^{f} and y_ℓ^{e} respectively for the first and second of equations (2.92). Then, supposing that a path is chosen so that the finite non-zero amplitude condition (2.93) is satisfied,

$$\rho_i - \rho_{ic} \sim |\theta_m|^{(d-y_i^{\text{f}})/y_m} , \qquad \mathfrak{u}_\ell - \mathfrak{u}_{\ell c} \sim |\theta_m|^{(d-y_\ell^{\text{e}})/y_m} . \tag{2.95}$$

The response functions given by the second-order partial derivatives of ϕ with respect to the internal and external couplings can be analyzed in the same way. If again we assume that the appropriate amplitudes are finite and non-zero, the scaling forms can be seen to be

$$\frac{\partial^2 \phi}{\partial L_i \partial L_j} \sim |\theta_m|^{(d-y_i^{\text{f}}-y_j^{\text{f}})/y_m} , \tag{2.96}$$

$$\frac{\partial^2 \phi}{\partial K_\ell \partial K_j} \sim |\theta_m|^{(d-y_\ell^{\text{e}}-y_j^{\text{e}})/y_m} , \tag{2.97}$$

$$\frac{\partial^2 \phi}{\partial L_i \partial K_\ell} \sim |\theta_m|^{(d-y_i^{\text{f}}-y_\ell^{\text{e}})/y_m} . \tag{2.98}$$

2.8 Critical Point and Coexistence Curve

In Sect. 2.3 we discussed in an intuitive way the ideas of scaling as they relate to the critical point in the ferromagnetic Ising model. The case of a critical point, terminating a line of first-order transitions, is the situation which has received most attention both in discussions of scaling theory and more generally in the field of critical phenomena. The ferromagnetic Ising model has the virtue, from the point of view of simplicity, that the first-order transition and critical point lie on the zero-field axis, about which the phase diagram is symmetric. In this section we return to the case of a critical point, but here no intrinsic symmetry is assumed. This means that the discussion is applicable to both, fluid and magnetic models, these special cases being described in detail below.

The phase space of this system is a plane ($n = 2$) with couplings L and K. The potential $\phi = \phi(L, K)$ with

$$\mathrm{d}\phi = -\mathfrak{u}\mathrm{d}K - \rho\mathrm{d}L . \tag{2.99}$$

The first-order transition appears as a curve in the (L, K) plane terminating at the critical point (L_c, K_c) (Fig. 2.5). The phase transition makes its appearance in the (ρ, K) plane as a coexistence region bounded by two curves (Fig. 2.6)

$$\rho = \rho^{(\pm)}(K)\,, \tag{2.100}$$

where

$$\rho^{(+)}(K) > \rho^{(-)}(K), \qquad K > K_c\,, \tag{2.101}$$

$$\rho^{(+)}(K_c) = \rho^{(-)}(K_c) = \rho_c\,, \tag{2.102}$$

$$\left(\frac{d\rho^{(+)}}{dK}\right)_c = -\left(\frac{d\rho^{(-)}}{dK}\right)_c = \infty\,, \tag{2.103}$$

$$\left(\frac{d^2\rho^{(+)}}{dK^2}\right)_c = -\left(\frac{d^2\rho^{(-)}}{dK^2}\right)_c. \tag{2.104}$$

The equations (2.101)–(2.103) are necessary consequences of the definition of the critical point. Equation (2.104) is not a necessary condition but is true in most cases of critical points. The equation of state of the system in terms of the variables K, L and ρ is given by

$$\rho = -\left(\frac{\partial\phi}{\partial L}\right)_K\,, \tag{2.105}$$

yielding an equation of the form

$$L = L(\rho, K)\,, \tag{2.106}$$

which is smooth outside the coexistence region. The projection of the coexistence curve into the (L, K) plane, which is the line of first-order transitions shown in Fig. 2.5, is given, for $K \geq K_c$, by

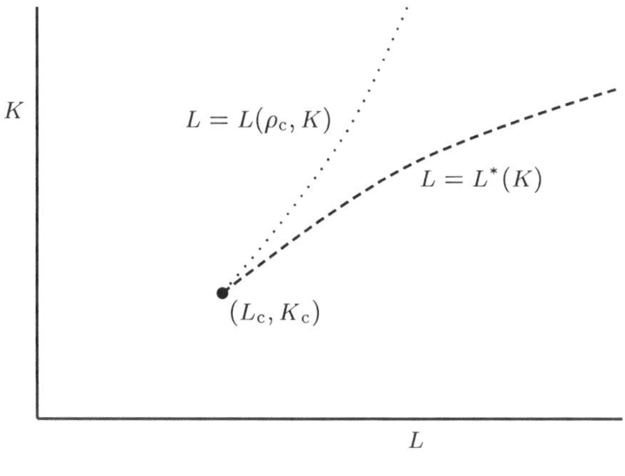

Fig. 2.5. A critical point (L_c, K_c) in the (L, K) plane. The first-order transition (coexistence curve) $L = L^*(K)$ is represented by a broken line and the critical isochore $L = L(\rho_c, K)$ by a dotted line. (The fact that these two curves have a common tangent at the critical point is established by scaling theory.)

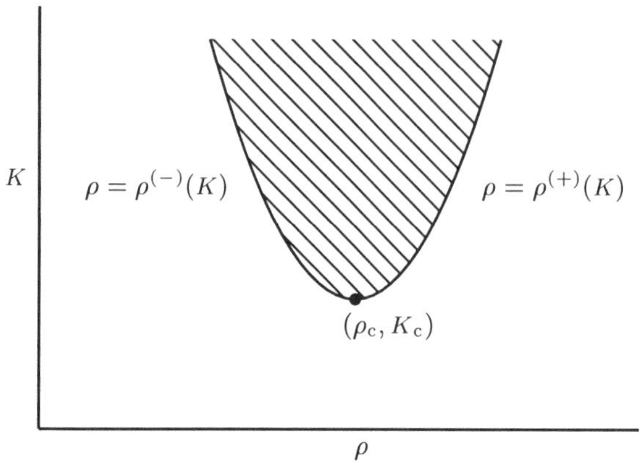

Fig. 2.6. The coexistence region (*shaded*) in the (ρ, K) plane.

$$L = L^*(K) = L(\rho^{(\pm)}(K), K) . \tag{2.107}$$

From (2.107),

$$\frac{dL^*}{dK} = \left(\frac{\partial L}{\partial K}\right)_\rho + \left(\frac{\partial L}{\partial \rho}\right)_K \frac{d\rho^{(\pm)}}{dK} . \tag{2.108}$$

The critical isochore $\rho = \rho_c$ makes the curve

$$L = L(\rho_c, K) \tag{2.109}$$

in the (L, K) plane passing through the critical point. The gradients of the critical isochore and coexistence curve differ at the critical point by the critical point value of the second term on the right-hand side of (2.108). This term contains an infinite factor at the critical point (see (2.103)). The fact that this term as a whole is zero will be established below using scaling theory. In terms of the scaling theory described in Sect. 2.4 the situation here is one in which we have two transition regions: the critical point of zero dimension, with $s = n = 2$ and two relevant scaling fields and the coexistence curve, with $s = n - 1 = 1$ and one relevant and one irrelevant scaling field.

2.8.1 Response Functions

The most commonly used critical exponents are those associated with the limiting properties of the density ρ and the three response functions

$$\varphi_T = T\left(\frac{\partial \rho}{\partial \xi}\right)_T , \tag{2.110}$$

$$c_\xi = k_B T\left(\frac{\partial s}{\partial T}\right)_\xi , \tag{2.111}$$

$$c_\rho = k_B T\left(\frac{\partial s}{\partial T}\right)_\rho , \tag{2.112}$$

where φ_T has the limiting properties of the compressibility of a fluid or the susceptibility of a magnet and c_ξ and c_ρ are the heat capacities at constant field and density respectively. These response functions must now be expressed in terms of the partial derivatives of ϕ with respect to the couplings. From the analysis of Sect. 1.4 with $n_e = 1$, $n = 2$ and $\eta = n - 1 = 1$

$$\varphi_T = -\frac{\partial^2 \phi}{\partial L^2}, \tag{2.113}$$

$$c_\xi = -k_B \left(K^2 \frac{\partial^2 \phi}{\partial K^2} + 2KL \frac{\partial^2 \phi}{\partial K \partial L} + L^2 \frac{\partial^2 \phi}{\partial L^2} \right). \tag{2.114}$$

For c_ρ the expression in terms of the derivatives of ϕ with respect to the couplings is rather more complicated. It is convenient to use the standard formula (see, for example, Volume 1, Chap. 1) relating the heat capacities at constant field and density. In our notation this is

$$c_\rho = c_\xi - k_B \frac{\alpha_\xi^2}{\varphi_T}, \tag{2.115}$$

where

$$\alpha_\xi = T \left(\frac{\partial \rho}{\partial T} \right)_\xi = L \frac{\partial^2 \phi}{\partial L^2} + K \frac{\partial^2 \phi}{\partial L \partial K} \tag{2.116}$$

has the limiting properties of the *coefficient of thermal expansion* .

2.8.2 Critical Exponents

In the standardized notation of critical phenomena (see, for example, Volume 1, Chap. 3) the critical exponents α, α', β, δ, γ and γ' are defined by

$$c_\rho \sim \begin{cases} (T - T_c)^{-\alpha}, & \text{along the critical isochore }, T > T_c, \\ (T_c - T)^{-\alpha'}, & \text{along the coexistence curve }, T < T_c, \end{cases} \tag{2.117}$$

$$\rho - \rho_c \sim (T_c - T)^{\beta}, \text{ along the coexistence curve}, T < T_c, \tag{2.118}$$

$$\varphi_T \sim \begin{cases} (T - T_c)^{-\gamma}, & \text{along the critical isochore}, T > T_c, \\ (T_c - T)^{-\gamma'}, & \text{along the coexistence curve}, T < T_c, \end{cases} \tag{2.119}$$

$$\xi - \xi_c \sim (\rho - \rho_c)|\rho - \rho_c|^{\delta - 1}, \text{ along the critical isotherm.} \tag{2.120}$$

It also convenient to define the exponents σ and σ' according to

$$c_\xi \sim \begin{cases} (T - T_c)^{-\sigma}, & \text{along the critical isochore}, T > T_c, \\ (T_c - T)^{-\sigma'}, & \text{along the coexistence curve}, T < T_c. \end{cases} \tag{2.121}$$

2.8.3 Exponent Inequalities

The heat capacities c_ρ and c_ξ in (2.115) are both positive. From this it follows that, if $\varphi_T > 0$, then c_ξ dominates both c_ρ and α_ξ^2/φ_T as $T \to T_c$. This means that

$$\sigma \geq \alpha\,, \qquad \sigma' \geq \alpha'\,, \tag{2.122}$$

$$\sigma' + 2\beta + \gamma' \geq 2\,. \tag{2.123}$$

The condition $\varphi_T > 0$ is true for a magnetic system and in this case the inequality (2.123) was first established by Rushbrooke (1963). The stronger condition

$$\alpha' + 2\beta + \gamma' \geq 2 \tag{2.124}$$

was obtained by Griffiths (1965) for both ferromagnetic and fluid systems using the convexity properties of the free energy. In fact, as we shall see in Sect. 2.9, for systems with the special symmetry of the ferromagnet $\alpha' = \sigma'$ and inequalities (2.123) and (2.124) become identical. Griffiths (1965) also derived a number of other inequalities. In particular

$$\gamma' \geq \beta(\delta - 1)\,. \tag{2.125}$$

2.9 Scaling for a Critical Point

From the discussion of Sect. 2.4 it is clear that there is a degree of latitude in the definition of the singular part of the thermodynamic potential. Given that it should contain all the non-smooth part of the thermodynamic potential and be zero in the transition region, it is convenient, for the critical point, to define

$$\phi_{\text{sing}}(\Delta L, \Delta K) = \phi(L, K) - \phi(L_c, K_c) + \rho_c \Delta L + \mathrm{u}_c \Delta K\,, \tag{2.126}$$

where $\Delta K = K - K_c$ and $\Delta L = L - L_c$, so that

$$\phi_{\text{sing}}(0,0) = 0\,, \tag{2.127}$$

$$\frac{\partial \phi_{\text{sing}}}{\partial \Delta L} = -\Delta\rho\,, \tag{2.128}$$

$$\frac{\partial \phi_{\text{sing}}}{\partial \Delta K} = -\Delta \mathrm{u}\,. \tag{2.129}$$

2.9.1 Scaling Fields for the Critical Point

In the light of Sect. 2.4, we see that (2.11) is the scaling hypothesis for a magnetic system using scaling fields $\Delta L = L$, ($L_c = 0$), and ΔK. The special feature of the magnetic system which is relevant here is that the first-order phase transition lies along the line $L = 0$. The scaling fields were chosen to pick out this special line. We now chose scaling fields for the general case which distinguish the one special direction at the critical point, namely the direction tangential to the coexistence curve. Assuming that the coexistence curve at the critical point is not parallel to the L-axis, we express $L^*(K)$ in the form

$$L^*(K) = L_c + a\Delta K + \bar{L}(K)\,, \tag{2.130}$$

where a is the (finite) slope of the coexistence curve at the critical point and

$$\bar{L}(K_c) = \left(\frac{d\bar{L}}{dK}\right)_c = 0\,, \qquad \bar{L}(K) \simeq b|\Delta K|^{\psi}\,, \tag{2.131}$$

b being zero if the coexistence curve is linear, and ψ being called the *shift exponent*. Orthogonal scaling fields are now defined by

$$\theta_1 = \Delta L - a\Delta K\,, \qquad \theta_2 = \Delta K + a\Delta L\,, \tag{2.132}$$

where $\theta_1 = 0$ gives the tangent to the coexistence curve. Along the coexistence curve $\Delta K \geq 0$ and, assuming that $0 \leq a < \infty$, with $0 \leq \bar{L}(K) < \infty$ for $\Delta K \geq 0$, the coexistence curve will leave the origin of the (θ_1, θ_2) plane and then satisfy the conditions $\theta_1 \geq 0$ and $\theta_2 > 0$. The scaling hypothesis now takes the form

$$\phi_{\text{sing}}(\lambda^{y_1}\theta_1, \lambda^{y_2}\theta_2) = \lambda^d \phi_{\text{sing}}(\theta_1, \theta_2)\,. \tag{2.133}$$

Following the procedure of Sect. 2.4 we take $\lambda = |\theta_1|^{-1/y_1}$, giving

$$\phi_{\text{sing}}(\theta_1, \theta_2) = |\theta_1|^{d/y_1} \phi_{\text{sing}}(\pm 1, g_2)\,, \tag{2.134}$$

and $\lambda = |\theta_2|^{-1/y_2}$, giving

$$\phi_{\text{sing}}(\theta_1, \theta_2) = |\theta_2|^{d/y_2} \phi_{\text{sing}}(g_1, \pm 1)\,. \tag{2.135}$$

For the sake of brevity, the gap exponent is denoted as $\vartheta = y_1/y_2$ and the gap fields as

$$g_1(\theta_1, \theta_2) = \theta_1/|\theta_2|^{\vartheta}\,, \qquad g_2(\theta_1, \theta_2) = \theta_2/|\theta_1|^{1/\vartheta}\,, \tag{2.136}$$

where $|g_1| = |g_2|^{-\vartheta}$. Both (2.134) and (2.135) are valid in a neighbourhood of the critical point and thus

$$\begin{aligned} \phi_{\text{sing}}(g_1, \pm 1) &= |g_1|^{d/y_1} \phi_{\text{sing}}(\pm 1, g_2)\,, \\ \phi_{\text{sing}}(\pm 1, g_2) &= |g_2|^{d/y_2} \phi_{\text{sing}}(g_1, \pm 1) \end{aligned} \tag{2.137}$$

(cf. (2.31)). Along the coexistence curve $\phi_{\text{sing}}(\theta_1,\theta_2)$ exhibits non-smooth behaviour; specifically one, or both, of the first-order partial derivatives is discontinuous. Since $\theta_2 \neq 0$ along the coexistence curve, apart from at the critical point, it follows from (2.135) that the coexistence curve corresponds to singular behaviour in $\phi_{\text{sing}}(g_1,+1)$. Supposing that, in a neighbourhood of the critical point, the only singular behaviour corresponds to the critical point and coexistence curve, $\phi_{\text{sing}}(g_1,+1)$ will be singular at only one value

$$g_1(\theta_1,\theta_2) = \varpi_0\,, \tag{2.138}$$

where $0 \leq \varpi_0 < \infty$ and (2.138) is the equation of the coexistence curve in a neighbourhood of the critical point. The constant will be zero only if $\bar{L}(K) \equiv 0$, when the coexistence curve lies along the $\theta_1 = 0$ axis. Except in this special case the coexistence curve is tangential to $\theta_1 = 0$ and it follows that

$$\vartheta > 1\,, \qquad y_1 > y_2\,. \tag{2.139}$$

The curves, defined by

$$|\theta_1| = \varpi|\theta_2|^{\vartheta} \tag{2.140}$$

for fixed finite non-zero ϖ, are tangential to the $\theta_1 = 0$ axis at the critical point, picking it out as a special direction, when (2.139) is satisfied. Experimental evidence indicates that, in all cases, there is one special direction at the critical point which is the direction tangential to the coexistence curve This supports the conclusion, even when the coexistence curve lies along $\theta_1 = 0$, that (2.139) must hold and that the curves (2.140), with non-zero, finite ϖ are tangential to $\theta_1 = 0$. The exponents y_1 and y_2 are referred to as the *strong* and *weak* exponents respectively.

Given that ϕ_{sing} is a continuous function we assume that the amplitudes

$$\lim_{g_2\to 0}\phi_{\text{sing}}(\pm 1, g_2) = A_{\infty}^{(\pm)}\,, \qquad \lim_{g_1\to 0}\phi_{\text{sing}}(g_1, \pm 1) = A_0^{(\pm)} \tag{2.141}$$

are finite and non-zero. (See Sect. 2.14 for a discussion of circumstances where this assumption breaks down.) Then , from (2.137),

$$\begin{aligned}\lim_{g_2\to\infty}\phi_{\text{sing}}(\pm 1, g_2) &\simeq |g_2|^{d/y_2} A_0^{(\pm)}\,,\\ \lim_{g_1\to\infty}\phi_{\text{sing}}(g_1, \pm 1) &\simeq |g_1|^{d/y_1} A_{\infty}^{(\pm)}\,.\end{aligned} \tag{2.142}$$

2.9.2 Approaches to the Critical Point

These can be divided into three classes:

(i) Paths on which $|g_1| \to \infty$ ($|g_2| \to 0$), including the axis $\theta_2 = 0$, for which, from (2.134) and (2.141), we have

$$\phi_{\text{sing}}(\theta_1,\theta_2) \simeq |\theta_1|^{d/y_1} A_{\infty}^{(\pm)}\,. \tag{2.143}$$

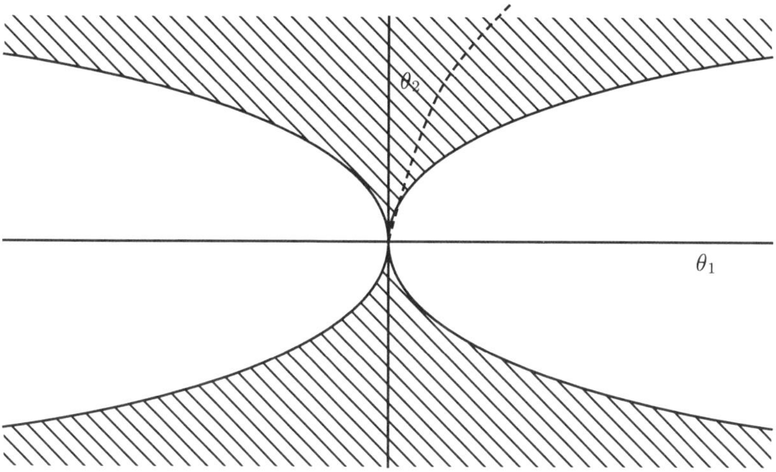

Fig. 2.7. The (θ_1, θ_2) plane in a neighbourhood of the critical point. The line $\theta_1 = 0, \theta_2 \geq 0$ is tangential to the coexistence curve and the crossover curves $|\theta_1| = \varpi|\theta_2|^{\vartheta}$ separate weak (*shaded*) and strong (*unshaded*) scaling approaches.

(ii) Paths on which $|g_1| \to 0$, including the axis $\theta_1 = 0$, for which, from (2.135) and (2.141), we have

$$\phi_{\text{sing}}(\theta_1, \theta_2) \simeq |\theta_2|^{d/y_2} A_0^{(\pm)} . \tag{2.144}$$

(iii) Paths on which $|g_1| \to \pi_1$, where π_1 is finite and non-zero, when

$$\phi_{\text{sing}}(\theta_1, \theta_2) \simeq |\theta_2|^{d/y_2} A_{\pi_1}^{(\pm)} , \tag{2.145}$$

with

$$\lim_{g_1 \to \pi_1} \phi_{\text{sing}}(g_1, \pm 1) = A_{\pi_1}^{(\pm)} \tag{2.146}$$

assumed to be finite and non-zero.

It is, of course, the case that an alternative to (2.145) could be given in terms of θ_1, when $\pi_1 \neq 0$. However, it is convenient to class cases (ii) and (iii) together with (2.145) and (2.146) applying for all finite values of π_1. All these approaches will be called *weak* as they differ only in the value of the amplitude $A_{\pi_1}^{(\pm)}$. Approaches in class (i) are called *strong*. Equation (2.140), for some finite non-zero ϖ gives two crossover curves in the (θ_1, θ_2) plane, which separate strong and weak scaling approaches (see Fig. 2.7).[7]

It is clear that an approach to the critical point along the coexistence curve itself is a weak approach with $\pi_1 = \varpi_0$. From (2.132)

[7] It will be seen from (2.140) that the actual value of ϖ, as long as it is finite and non-zero, is irrelevant to the asymptotic behaviour of ϕ_{sing}, representing only a multiplicative scaling of θ_1.

$$\theta_1\left(1+a\frac{\theta_2}{\theta_1}\right)=(1+a^2)\Delta L\,,\qquad \theta_2\left(1-a\frac{\theta_1}{\theta_2}\right)=(1+a^2)\Delta K \tag{2.147}$$

and since, by (2.140) with $\varpi=\varpi_0$, $\theta_1/\theta_2\to 0$ as the critical point is approached along the coexistence curve

$$\theta_1\simeq\varpi_0(1+a^2)^{\vartheta}|\Delta K|^{\vartheta}\,. \tag{2.148}$$

Then, from (2.130) and (2.132),

$$\bar{L}(K)\simeq\varpi_0(1+a^2)^{\vartheta}|\Delta K|^{\vartheta}\,, \tag{2.149}$$

showing that the shift exponent ψ, defined in (2.131), is equal to the gap exponent .

2.9.3 Experimental Variables

To link this analysis with experiment it is necessary to translate the asymptotic formulae, given above, into formulae expressed in thermodynamic variables. Two approaches to the critical point are of particular interest:

(a) The *critical isotherm* $T=T_c$, which in terms of the couplings, where $T=J/K$ with J constant, is the line $\Delta K=0$. From (2.132) and (2.136) $|g_1|=1/|a|^{\vartheta}|\Delta L|^{\vartheta-1}$. So, remembering that a is finite and possibly zero, it is clear that the critical isotherm gives a strong approach to the critical point.

(b) The *critical isochamp*, that is the line $\xi=\xi_c$, where ξ is the field associated with the coupling L by the formula $L=\xi/T$. It is straightforward to show that this is the line $K_c\Delta L=L_c\Delta K$ and this also corresponds to an approach to the critical point in a strong scaling direction unless

$$L_c=aK_c\,, \tag{2.150}$$

(including the case $a=0$, $L_c=0$).

For a weak scaling approach $\theta_1/\theta_2\to 0$ so, from (2.144) and (2.147),

$$\phi_{\text{sing}}(\theta_1,\theta_2)\sim|\Delta K|^{d/y_2}\,. \tag{2.151}$$

Strong scaling approaches include all those for which $\theta_2/\theta_1\to 0$ for which (2.143) can be replaced by

$$\phi_{\text{sing}}(\theta_1,\theta_2)\sim|\Delta L|^{d/y_1}\,. \tag{2.152}$$

They also include approaches for which θ_2/θ_1 remains finite. This affects the amplitude of the asymptotic form but (2.152) still applies. We should, however, bear in mind the case where $\theta_1/\theta_2\to 0$ with a curvature larger than that of the crossover curve. This is a somewhat pathological strong scaling approach. Equations (2.151) and (2.152) link the weak and strong exponents y_2 and y_1 with the couplings K and L and, as in Sect. 2.3, they could be denoted by y_K and y_L respectively. One may go further than this and link these exponents to thermodynamic variables. Since K is an internal coupling (see Sect. 1.4) the only physically variable parameter is the temperature T. The exponent y_2 can, therefore, be referred to as the *thermal exponent* often denoted as y_T. In a similar way, L is an external coupling, which can be changed by altering the field ξ. So y_1, can be referred to as the *field exponent* and denoted by y_ξ. We shall, however, in the remainder of this section continue to use the notation y_1 and y_2 for the exponents.

2.9.4 The Density and Response Functions

It is now necessary to express the density and the response functions in terms of the partial derivatives of ϕ_{sing} with respect to the scaling fields (see (2.92)). From (2.113), (2.114), (2.128) and (2.132)

$$\rho - \rho_c = \Delta\rho = -\left[\frac{\partial\phi_{\text{sing}}}{\partial\theta_1} + a\frac{\partial\phi_{\text{sing}}}{\partial\theta_2}\right], \tag{2.153}$$

$$\varphi_T = -\left[\frac{\partial^2\phi_{\text{sing}}}{\partial\theta_1^2} + 2a\frac{\partial^2\phi_{\text{sing}}}{\partial\theta_1\partial\theta_2} + a^2\frac{\partial^2\phi_{\text{sing}}}{\partial\theta_2^2}\right], \tag{2.154}$$

$$c_\xi = -k_B\left[\Theta_1{}^2\frac{\partial^2\phi_{\text{sing}}}{\partial\theta_1{}^2} + 2\Theta_1\Theta_2\frac{\partial^2\phi_{\text{sing}}}{\partial\theta_1\partial\theta_2} + \Theta_2{}^2\frac{\partial^2\phi_{\text{sing}}}{\partial\theta_2{}^2}\right], \tag{2.155}$$

where

$$\Theta_1(L,K) = L - aK\,, \qquad \Theta_2(L,K) = K + aL\,. \tag{2.156}$$

For c_ρ the expression (2.115) is used with

$$\alpha_\xi = \Theta_1\frac{\partial^2\phi_{\text{sing}}}{\partial\theta_1^2} + (a\Theta_1 + \Theta_2)\frac{\partial^2\phi_{\text{sing}}}{\partial\theta_1\partial\theta_2} + a\Theta_2\frac{\partial^2\phi_{\text{sing}}}{\partial\theta_2^2}\,. \tag{2.157}$$

Substituting from (2.154), (2.155) and (2.157) into (2.115)

$$c_\rho = -\frac{k_B}{\varphi_T}(a\Theta_1 - \Theta_2)^2\left[\left(\frac{\partial^2\phi_{\text{sing}}}{\partial\theta_1\partial\theta_2}\right)^2 - \frac{\partial^2\phi_{\text{sing}}}{\partial\theta_1^2}\frac{\partial^2\phi_{\text{sing}}}{\partial\theta_2^2}\right]. \tag{2.158}$$

2.9.5 Asymptotic Forms

To obtain asymptotic forms it is necessary to determine the dominant contributions in (2.153), (2.154), (2.155) and (2.158) using the reasoning of Sect. 2.7. Since $y_1 > y_2$, these correspond to the terms with the largest number of differentiations with respect to θ_1. On this basis we see that the terms in (2.153), (2.154) and (2.155) are ordered in descending dominance and those in (2.158) have equal weight. We can now repeat the reasoning between equations (2.133) and (2.146) for the partial derivatives of ϕ_{sing}, replacing d by the corresponding $d - k_1 y_1 - k_2 y_2$. For the weak scaling direction the formulae corresponding to (2.146), when $\pi_1 = \varpi_0$, that is across the coexistence curve, do not give an unambiguous limit, yielding possibly different values as $g_1 \to \varpi_0$ from the two sides. This means that the amplitudes for approaches along the two branches of the coexistence curve can be different. Except in the special case where one is zero and corrections to scaling need to be considered (see Sect. 2.14), this will not affect the exponent in the asymptotic form. Without further comment we shall now assume that all the appropriate amplitude factors given by the second of equations (2.34) are non-zero and use the arguments of Sect. 2.7. It follows from (2.153) and (2.154) that

$$\rho - \rho_c \sim \begin{cases} |\Delta L|^{(d-y_1)/y_1}, & \text{strong direction,} \\ |\Delta K|^{(d-y_1)/y_2}, & \text{weak direction,} \end{cases} \tag{2.159}$$

$$\varphi_T \sim \begin{cases} |\Delta L|^{(d-2y_1)/y_1}, & \text{strong direction,} \\ |\Delta K|^{(d-2y_1)/y_2}, & \text{weak direction.} \end{cases} \tag{2.160}$$

All the information necessary to consider the second term on the right-hand side of (2.108) is now available. Since this is the case of a weak direction approach to the critical point,

$$\frac{\mathrm{d}\rho^{(\pm)}}{\mathrm{d}K} \sim |\Delta K|^{-1+(d-y_1)/y_2}, \qquad \left(\frac{\partial L}{\partial \rho}\right)_K \sim |\Delta K|^{-(d-2y_1)/y_2},$$

giving

$$\left(\frac{\partial L}{\partial \rho}\right)_K \frac{\mathrm{d}\rho^{(\pm)}}{\mathrm{d}K} \sim |\Delta K|^{\vartheta - 1}. \tag{2.161}$$

As $\vartheta > 1$, this establishes the result that the critical isochore and coexistence curve share a common tangent (given by $\theta_1 = 0$) at the critical point.

2.9.6 Critical Exponents and Scaling Laws

Comparison of (2.159) and (2.160) with (2.118) and (2.119) gives

$$\begin{aligned} \beta &= (d - y_1)/y_2\,, \\ \delta &= y_1/(d - y_1)\,, \\ \gamma &= \gamma' = (2y_1 - d)/y_2\,. \end{aligned} \tag{2.162}$$

When we consider the asymptotic properties of c_ξ we need to take account of the possibility that $\Theta_1(L_c, K_c) = 0$, which is the condition given by (2.150). $\Theta_2(L_c, K_c) \neq 0$ if $\Theta_1(L_c, K_c) = 0$ and we have, from (2.155), two cases

$$c_\xi \sim \begin{cases} |\Delta L|^{(d-2y_1)/y_1}, & \text{strong direction,} \\ |\Delta K|^{(d-2y_1)/y_2}, & \text{weak direction,} \end{cases} \tag{2.163}$$

if $L_c \neq aK_c$ and

$$c_\xi \sim \begin{cases} |\Delta L|^{(d-2y_2)/y_1}, & \text{strong direction,} \\ |\Delta K|^{(d-2y_2)/y_2}, & \text{weak direction,} \end{cases} \tag{2.164}$$

if $L_c = aK_c$. Comparison of (2.163) and (2.164) with (2.121) gives

$$\sigma = \sigma' = \begin{cases} (2y_1 - d)/y_2 = \gamma, & L_c \neq aK_c, \\ (2y_2 - d)/y_2, & L_c = aK_c. \end{cases} \tag{2.165}$$

From (2.158) it can be seen that the asymptotic expressions for c_ρ will involve a combination of amplitude factors. Unless these cancel

$$c_\rho \sim \begin{cases} |\Delta L|^{(d-2y_2)/y_1}, & \text{strong direction,} \\ |\Delta K|^{(d-2y_2)/y_2}, & \text{weak direction.} \end{cases} \tag{2.166}$$

Comparing (2.166) with (2.117) we have

$$\alpha = \alpha' = (2y_2 - d)/y_2 . \tag{2.167}$$

It is now interesting to note that

$$\sigma = \sigma' = \begin{cases} \gamma, & L_c \neq aK_c, \\ \alpha, & L_c = aK_c. \end{cases} \tag{2.168}$$

From this result and the inequalities (2.122) it follows that, when $L_c \neq aK_c$, $\gamma \geq \alpha$, a result which can be established without the aid of scaling theory (Buckingham 1972).

The magnetic system, described in Sect. 2.3 has the first-order transition line lying along the zero-field axis. So $L_c = 0$, $a = 0$ and $\sigma = \alpha$. It is for this reason that, when defining critical exponents for fluids and magnets, it is common practice (Stanley 1987, Yeomans 1992) to use α for the heat capacity at constant density for the fluid and constant magnetic field for the magnet. We can now understand (see also Volume 1, Sect. 3.5) that the reason that the same exponent is appropriate to both cases is due to the special symmetry properties of the magnet, which also makes the analysis rather simpler (Example 2.2).

Equations (2.162) and (2.167) give formulae for the exponents α, β, γ and δ in terms of y_1 and y_2. They are, therefore, not independent and two relationships exist between them. These can be expressed in the form

$$\gamma = \beta(\delta - 1) , \tag{2.169}$$

called the *Widom scaling law* (Widom 1964) and

$$\alpha + 2\beta + \gamma = 2\,, \tag{2.170}$$

called the *Essam–Fisher scaling law* (Essam and Fisher 1963).

2.9.7 Scaling for the Coexistence Curve

Scaling behaviour across the coexistence curve can be analyzed using the method described in Sect. 2.4. To do this we need to modify the scaling field θ_1 to

$$\theta'_1 = \theta_1 - \varpi_0 |\theta_2|^{\vartheta} \tag{2.171}$$

with θ_2 unchanged. The gap fields are modified to

$$\begin{aligned} g'_1(\theta'_1, \theta_2) &= \theta'_1/|\theta_2|^{\vartheta} = g_1(\theta_1, \theta_2) - \varpi_0\,, \\ g'_2(\theta'_1, \theta_2) &= \theta_2/|\theta'_1|^{1/\vartheta} = g_2(\theta_1, \theta_2)/|1 - \varpi_0 g_1^{-1}(\theta_1, \theta_2)|^{1/\vartheta}\,. \end{aligned} \tag{2.172}$$

It is also necessary to modify the singular part of the thermodynamic potential to

$$\phi'_{\text{sing}}(\theta'_1, \theta_2) = \phi_{\text{sing}}(\theta'_1 + \varpi_0|\theta_2|^{\vartheta}, \theta_2) - \phi_{\text{sing}}(\varpi_0|\theta_2|^{\vartheta}, \theta_2)\,, \tag{2.173}$$

so that ϕ'_{sing} is zero not only at the critical point but on the coexistence curve. It is now straightforward to show that the scaling hypothesis (2.133) is satisfied with the new scaling fields in a neighbourhood of the critical point and all the subsequent analysis is valid except that the coexistence curve is given by $g'_1(\theta'_1, \theta_2) = 0$. From (2.42) and (2.43) we define the effective exponent

$$\Sigma_1(\theta'_1, \theta_2) = \frac{\partial \ln \phi'_{\text{sing}}}{\partial \ln |\theta'_1|} = \frac{d}{y_1} - \frac{y_2}{y_1}|g'_2| \frac{\partial \ln \phi'_{\text{sing}}(\pm 1, g'_2)}{\partial |g'_2|}\,. \tag{2.174}$$

As the critical point is approached $|g'_2| \to 0$ and $\Sigma_1 \to d/y_1$. As the coexistence curve is approached away from the critical point $|g'_2| \to \infty$. Suppose that

$$\phi'_{\text{sing}}(\pm 1, g'_2) \sim |g'_2|^{\mu_2}\,. \tag{2.175}$$

Then $\Sigma_1 \to d/y'_1$, where

$$y'_1 = \frac{d y_1}{d - y_2 \mu_2} \tag{2.176}$$

is the scaling exponent associated with θ'_1 on the coexistence curve.

2.10 Mean-Field Theory for the Ising Ferromagnet

The simplest example on which the scaling formulation of the previous section can be tested is the mean-field approximation to the ferromagnetic Ising model. This is described in detail in Volume 1, Sect. 3.1. The thermodynamic potential is given by

$$\phi(L,K) = \frac{1}{2}\ln(1-m^2) + \frac{Km^2}{2K_c} - \ln(2)\,, \tag{2.177}$$

where $K = J/T$, $L = \mathcal{H}/T$ and m, the magnetization per lattice site, is a root of the equation

$$L = \frac{1}{2}\ln\left(\frac{1+m}{1-m}\right) - \frac{Km}{K_c}\,. \tag{2.178}$$

For $K < K_c$ equation (2.178) has a single real root $m = m(L,K)$. For $K > K_c$ the equation exhibits the characteristic looping effect found in all mean-field approximations. The Maxwell equal-areas rule (see, for example, Volume 1, Chap. 1) implies the choice of the branch with m having the same sign as L and we have an implicit form for the magnetization as a function as L and K with

$$\lim_{L\to\pm 0} m(L,K) = \pm m_s(K)\,. \tag{2.179}$$

The magnetization has a discontinuity of $2m_s$ along the zero-field axis for $K > K_c$ with

$$\lim_{K\to K_c} m_s(K) = 0\,, \qquad \lim_{K\to\infty} m_s(K) = 1\,. \tag{2.180}$$

Substituting the computed form of m into (2.177) gives a function with a discontinuity in gradient along the zero-field axis terminating at the critical point $K = K_c$. This model represents a specific realization of the example of Kadanoff described in Sect. 2.3. The two relevant scaling fields in a neighbourhood of the critical point are $\theta_1 = L$ and $\theta_2 = \Delta K$. Near the critical point (2.177) can be approximated by

$$\phi(L,K) = \phi_{\rm sing}(\theta_1,\theta_2) - \ln(2)\,, \tag{2.181}$$

where

$$\phi_{\rm sing}(\theta_1,\theta_2) = \frac{\theta_2}{2K_c}m^2 - \frac{1}{4}m^4\,, \tag{2.182}$$

and, from (2.178), m is the unique root of

$$3\theta_1 + \frac{3\theta_2}{K_c}m - m^3 = 0 \tag{2.183}$$

with the same sign as θ_1. This is

$$m = \begin{cases} \pm(\theta_2/K_c)^{\frac{1}{2}}[\sqrt{3}(1-s^2)^{\frac{1}{2}}+s]\,, & \theta_2 > 0, \\ 2 \pm (|\theta_2|/K_c)^{\frac{1}{2}} s'\,, & \theta_2 < 0, \end{cases} \tag{2.184}$$

where the '±' corresponds to the sign of θ_1 and

$$\begin{aligned} s &= \sin\left\{\frac{1}{3}\arcsin\left[\left|\frac{3\theta_1}{2}\right|\left(\frac{K_c}{\theta_2}\right)^{\frac{3}{2}}\right]\right\}, \\ s' &= \sinh\left\{\frac{1}{3}\text{arcsinh}\left[\left|\frac{3\theta_1}{2}\right|\left(\frac{K_c}{|\theta_2|}\right)^{\frac{3}{2}}\right]\right\}. \end{aligned} \tag{2.185}$$

Substituting from (2.184) into (2.182) gives

$$\phi_{\text{sing}}(\theta_1, \theta_2) = \left(\frac{\theta_2}{K_c}\right)^2 \left[2s^4 - s^2 - 3 - 2\sqrt{3}s(1-s^2)^{\frac{3}{2}}\right] \tag{2.186}$$

for $\theta_2 > 0$, and

$$\phi_{\text{sing}}(\theta_1, \theta_2) = \left(\frac{\theta_2}{K_c}\right)^2 \left[2s'^2 - 4s'^4\right] \tag{2.187}$$

for $\theta_2 < 0$. It is clear that the form for ϕ_{sing} satisfies the scaling hypothesis (2.133) with

$$y_2 = y_T = \tfrac{1}{2}d\,, \qquad y_1 = y_{\mathcal{H}} = \tfrac{3}{4}d\,, \qquad \vartheta = \tfrac{3}{2} \tag{2.188}$$

(see Example 2.3). From (2.162) and (2.167) the mean-field critical exponent values $\alpha = 0$, $\beta = \frac{1}{2}$, $\gamma = 1$ and $\delta = 3$ are then obtained, satisfying the scaling laws (2.169) and (2.170). To consider scaling on the coexistence curve we need not change the scaling fields, as we did in Sect. 2.9, since the coexistence curve is given by $\theta_1 = 0$. However, in order for the singular part of the thermodynamic potential to be zero on the coexistence curve, we need to modify (2.186) by writing

$$\phi'_{\text{sing}}(\theta_1, \theta_2) = \phi_{\text{sing}}(\theta_1, \theta_2) + \frac{3}{4}\left(\frac{\theta_2}{K_c}\right)^2. \tag{2.189}$$

Then, from (2.185) and (2.189),

$$\phi'_{\text{sing}}(\pm 1, g_2) \simeq -\left(\frac{3|g_2|}{K_c}\right)^{\frac{1}{2}}. \tag{2.190}$$

Comparison with (2.175) and (2.176) shows that

$$\mu_2 = \tfrac{1}{2}\,, \qquad y'_1 = d\,. \tag{2.191}$$

This value for y'_1 is in accordance with the scaling criterion for a first-order transition discussed in Sect. 2.4.

2.11 Correlation Scaling at a Critical Point

In this section we consider scaling for the connected two-point correlation function $\Gamma_2(\mathfrak{q}(\boldsymbol{r}_1), \mathfrak{q}(\boldsymbol{r}_2))$ associated with the coupling L and defined by (1.54) with $\mathfrak{c}_1 = \mathfrak{c}_2 = \mathfrak{q}$. As before we assume a d-dimensional hypercubic lattice with periodic boundary conditions and we also assume translational invariance so that the correlation function is dependent on $\boldsymbol{r}_1$ and $\boldsymbol{r}_2$ through their difference $\bar{\boldsymbol{r}} = \boldsymbol{r}_2 - \boldsymbol{r}_1$. From equations (1.62) and (2.113),

$$\varphi_T = \frac{1}{N} \sum_{\{\boldsymbol{r}_1, \boldsymbol{r}_2\}} \Gamma_2(\mathfrak{q}(\boldsymbol{r}_1), \mathfrak{q}(\boldsymbol{r}_2)) = \sum_{\{\bar{\boldsymbol{r}}\}} \Gamma_2(\mathfrak{q}(\boldsymbol{0}), \mathfrak{q}(\bar{\boldsymbol{r}})) = \bar{\Gamma}_2^*(\mathfrak{q}; \boldsymbol{0}), \quad (2.192)$$

where $\bar{\Gamma}_2^*(\mathfrak{q}; \boldsymbol{0})$ is the zero wave vector component of the Fourier transform performed with respect to $\bar{\boldsymbol{r}}$. The correlation length is defined by (2.6) and it is easily seen that this can be expressed in the form

$$\mathfrak{r}^2(\mathfrak{q}) = -c(\mathfrak{d}) \frac{\nabla_{\bar{k}}^2 \bar{\Gamma}_2^*(\mathfrak{q}; \boldsymbol{0})}{\bar{\Gamma}_2^*(\mathfrak{q}; \boldsymbol{0})}. \quad (2.193)$$

In Sect. 2.6 we developed the scaling behaviour of correlation functions of scaling operators. To translate these results into the scaling behaviour of state operators like $\mathfrak{q}$ we must use the reasoning of Sect. 2.7 to determine the leading scaling behaviour associated with the coupling L. With the definitions (2.132) for the scaling fields we showed in Sect. 2.9 that L scales like θ_1. It follows that the two-point $\mathfrak{q}$ correlation function scales like the two-point ϕ_1 correlation function. Thus from (2.74), (2.75) and (2.78)

$$\Gamma_2(\lambda^{y_1}\theta_1, \lambda^{y_2}\theta_2; \mathfrak{q}(\boldsymbol{0}), \mathfrak{q}(\lambda^{-1}\bar{\boldsymbol{r}})) = \lambda^{2x_1}\Gamma_2(\theta_1, \theta_2; \mathfrak{q}(\boldsymbol{0}), \mathfrak{q}(\bar{\boldsymbol{r}})) \quad (2.194)$$

and

$$\bar{\Gamma}_2^*(\lambda^{y_1}\theta_1, \lambda^{y_2}\theta_2; \mathfrak{q}; \lambda\bar{\boldsymbol{k}}) = \lambda^{d-2y_1}\bar{\Gamma}_2^*(\theta_1, \theta_2; \mathfrak{q}; \bar{\boldsymbol{k}}), \quad (2.195)$$

where $x_1 = d - y_1$ is the scaling dimension of $\mathfrak{q}$. From (2.193), the scaling form for the correlation length is

$$\mathfrak{r}(\lambda^{y_1}\theta_1, \lambda^{y_2}\theta_2; \mathfrak{q}) = \lambda^{-1}\mathfrak{r}(\theta_1, \theta_2; \mathfrak{q}). \quad (2.196)$$

According to the standard notation the exponents ν and ν' are defined by

$$\mathfrak{r}(\mathfrak{q}) \sim \begin{cases} (T - T_c)^{-\nu}, & \text{along the critical isochore, } T > T_c, \\ (T_c - T)^{-\nu'}, & \text{along the coexistence curve, } T < T_c, \end{cases} \quad (2.197)$$

and thus we have

$$\nu = \nu' = 1/y_2. \quad (2.198)$$

This result and (2.167) yield the *Josephson hyper-scaling law*

$$d\nu = 2 - \alpha, \quad (2.199)$$

which was proved by Josephson (1967) as an inequality with '$\geq$' replacing '='. At the critical point, for large $|\boldsymbol{r}_2 - \boldsymbol{r}_1|$, the standard asymptotic form for the connected two-point correlation function is given in terms of the exponent η by

$$\Gamma_2(\mathfrak{q}(\boldsymbol{r}_1), \mathfrak{q}(\boldsymbol{r}_2)) \sim \frac{1}{|\boldsymbol{r}_2 - \boldsymbol{r}_1|^{d-2+\eta}} . \tag{2.200}$$

From (2.82) this means that

$$\eta = 2 - 2y_1 + d = 2 + 2x_1 - d , \tag{2.201}$$

if we assume that the appropriate amplitude factor is non-zero and finite. (For the ferromagnetic Ising model this is equivalent to assuming that the function $f(x)$ in (2.13) has a finite value as $x \to \infty$.) It now follows from (2.195) that, at the critical point, for small $|\bar{\boldsymbol{k}}|$,

$$\bar{\Gamma}_2^*(\mathfrak{q}; \bar{\boldsymbol{k}}) \sim |\bar{\boldsymbol{k}}|^{\eta-2} . \tag{2.202}$$

From (2.162), (2.198) and (2.201) we obtain the *Fisher scaling law*

$$(2 - \eta)\nu = \gamma , \tag{2.203}$$

which was proved by Fisher (1969) as an inequality with '$\geq$' replacing '=' .

From the definition of the gap exponent ϑ, given below equation (2.135), and (2.162)

$$\vartheta = \beta\delta . \tag{2.204}$$

From (2.140) and Fig. 2.7 it can be seen that the larger the value of ϑ the more of the (θ_1, θ_2) plane in a neighbourhood of the critical point corresponds to weak scaling approaches. We now see, from (2.204), that, for fixed δ, larger values of ϑ correspond to larger values of β, corresponding to a narrowing of the coexistence region in the (ρ, K) plane (see Fig. 2.6) and a corresponding increase in the part of the plane corresponding to strong scaling approaches. This apparently paradoxical situation is understood if we project the crossover curves into the (ρ, K) plane where they share a common tangent with the coexistence region. So in the (θ_1, θ_2) (or equivalently (L, K)) plane the region of weak scaling approaches is wedge shaped at the critical point, whereas in the (ρ, K) plane it is the region of strong scaling approaches which is wedge shaped. This interesting point was discussed by Griffiths and Wheeler (1970), who observe that weak scaling approaches to the critical point will be more easily achieved by adjustments of ρ and K, than by adjustments of L and K.

2.12 Tricritical Point

Sect. 2.2 contains a brief description of the kinds of critical phenomena which may be encountered in phase spaces of dimensions $n = 2$ and $n = 3$. We have described in Sects. 2.9 and 2.11 the application of scaling theory to the most well-known example of critical behaviour for $n = 2$, a critical point terminating a line of first-order transitions. We shall now consider the case of a tricritical point. In fact, as was indicated in Sect. 2.2, a tricritical point is normally a special case of a more general type of phase behaviour with a critical end-point and possibly a bicritical end-point (see Fig. 2.2). An adjustable parameter is needed to achieve a tricritical point. As a specific example we shall now amplify the description of the metamagnetic model introduced in Sect. 2.2. This is a generalization of the Ising ferromagnet on a d-dimensional hypercubic lattice, described in section 2.3. Suppose that one particular axis direction is selected and the $(d-1)$-dimensional hyperplanes orthogonal to this axis are labelled in sequence $\mathfrak{a}, \mathfrak{b}, \mathfrak{a}, \mathfrak{b}, \ldots$ to form a layered sublattice structure with two equivalent sublattices $\mathfrak{a}$ and $\mathfrak{b}$. The Hamiltonian (1.49) becomes

$$\begin{aligned}\hat{H}(\mathcal{H}_s, \mathcal{H}, J_1, J_2; \sigma(\boldsymbol{r})) = J_1 \sum_{\{\boldsymbol{r},\boldsymbol{r}'\}}^{(\mathfrak{a},\mathfrak{b})} \sigma(\boldsymbol{r})\sigma(\boldsymbol{r}') - J_2 \sum_{\boldsymbol{r},\boldsymbol{r}'\}}^{(\mathfrak{a},\mathfrak{a})} \sigma(\boldsymbol{r})\sigma(\boldsymbol{r}') \\ -J_2 \sum_{\{\boldsymbol{r},\boldsymbol{r}'\}}^{(\mathfrak{b},\mathfrak{b})} \sigma(\boldsymbol{r})\sigma(\boldsymbol{r}') - (\mathcal{H} + \mathcal{H}_s) \sum_{\{\boldsymbol{r}\}}^{(\mathfrak{a})} \sigma(\boldsymbol{r}) \\ -(\mathcal{H} - \mathcal{H}_s) \sum_{\{\boldsymbol{r}\}}^{(\mathfrak{b})} \sigma(\boldsymbol{r}) , \end{aligned} \tag{2.205}$$

where the nearest-neighbour pair and site summations are restricted as indicated by the superscripts and $J_1, J_2 > 0$. The interactions within sublattices are ferromagnetic and between sublattices are antiferromagnetic. The field $\mathcal{H}$ is uniform over the lattice, but $\mathcal{H}_s$ is a staggered field with a different sign on the two sublattices. The system can be thought of as having four independent fields T, $\mathcal{H}$, $\mathcal{H}_s$ and one of J_1 or J_2. Of these only T and $\mathcal{H}$ are the theoretical counterparts of experimentally controllable quantities. Mean-field calculations for this model (Volume 1, Sect. 4.4) and for an equivalent model, where J_1 and J_2 are taken to be nearest-neighbour and second-neighbour interactions (Kincaid and Cohen 1975), show that it is the variation of J_1 in proportion to J_2 which leads either to a critical end-point (together with a bicritical end-point) or a tricritical point. Translating the results of Volume 1, Sect. 4.4 to the case of a d-dimensional hypercubic lattice considered here a tricritical point will occur if $5J_1 < 3(d-1)J_2$. We suppose that the ratio of J_1 and J_2 is fixed within this range. The general structure of the phase diagram is known (Kincaid and Cohen 1975) to be of the form shown in Fig. 2.8. In the plane $\mathcal{H}_s = 0$ the phase diagram has the appearance of a line of first-order transitions $\mathcal{T}$ meeting a line of second-order transitions $\mathcal{C}$ at the point T. This point is the usual way that a tricritical point is seen in a two-dimensional phase space (see Fig. 2.1(b)). When the third dimension

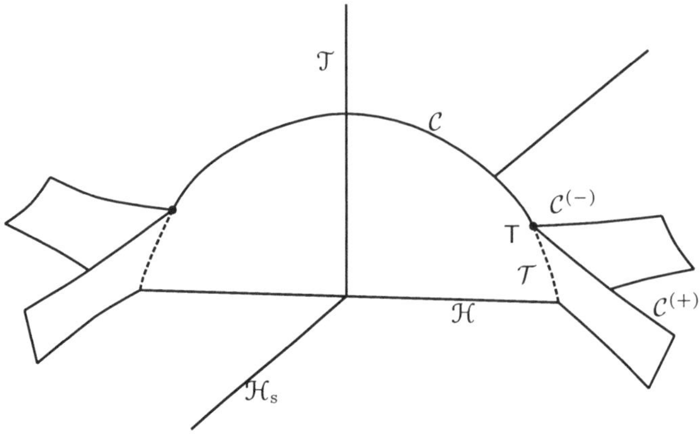

Fig. 2.8. Schematic phase diagram for a metamagnet in the space of the fields T, $\mathcal{H}$ and $\mathcal{H}_s$. The critical lines $\mathcal{C}$, $\mathcal{C}^{(+)}$, $\mathcal{C}^{(-)}$ and the triple line $\mathcal{T}$ meet at the tricritical point T. (Reprinted from Kincaid and Cohen (1975), p. 110, by permission of the publisher Elsevier Science.)

corresponding to $\mathcal{H}_s$ is included then this boundary line becomes the edge of a surface of first-order transitions. $\mathcal{T}$ is a triple line and $\mathcal{C}$ is a critical line. Two further critical lines $\mathcal{C}^{(+)}$ and $\mathcal{C}^{(-)}$, bounding critical surfaces symmetrically placed about the plane $\mathcal{H}_s = 0$ also meet at the tricritical point T.

For this system the potential is $\phi(L_s, L, K)$ with

$$d\phi = u dK - m dL - m_s dL_s\,, \tag{2.206}$$

where $K = J_1/T$, $L = \mathcal{H}/T$, $L_s = \mathcal{H}_s/T$, u is the internal energy per lattice site, m is the magnetization per lattice site and m_s measures the sublattice order.[8] The response functions relevant to our discussion are

$$\varphi_T = T\left(\frac{\partial m_s}{\partial \mathcal{H}_s}\right)_{T,\mathcal{H}} = -\frac{\partial^2\phi}{\partial L_s^2}\,, \tag{2.207}$$

$$\psi_T = T\left(\frac{\partial m}{\partial \mathcal{H}}\right)_{T,\mathcal{H}_s} = -\frac{\partial^2\phi}{\partial L^2} \tag{2.208}$$

and

[8] In this section the coupling K has the opposite sign to that in Sect. 2.8 since it is give in terms of the *antiferromagnetic* interaction between sublattices.

$$c_{\mathcal{H}} = k_{\mathrm{B}} T \left(\frac{\partial s}{\partial T} \right)_{\mathcal{H},\mathcal{H}_{\mathrm{s}}}$$

$$= -k_{\mathrm{B}} \left(K^2 \frac{\partial^2 \phi}{\partial K^2} + L^2 \frac{\partial^2 \phi}{\partial L^2} + L_{\mathrm{s}}^2 \frac{\partial^2 \phi}{\partial L_{\mathrm{s}}^2} \right.$$

$$\left. + 2KL \frac{\partial^2 \phi}{\partial K \partial L} + 2KL_{\mathrm{s}} \frac{\partial^2 \phi}{\partial K \partial L_{\mathrm{s}}} + 2LL_{\mathrm{s}} \frac{\partial^2 \phi}{\partial L \partial L_{\mathrm{s}}} \right) . \tag{2.209}$$

We shall assume that the critical line $\mathcal{C}$ and the triple line $\mathcal{T}$ are asymptotically parallel at the tricritical point T.[9] Let the curve given by $\mathcal{C}$ and $\mathcal{T}$ in the plane $\mathcal{H}_{\mathrm{s}} = 0$ have the equation $\mathcal{H} = \mathcal{H}^*(T)$ with the tricritical point T be given by the coordinates $T = T_{\mathrm{t}}$, $\mathcal{H} = \mathcal{H}_{\mathrm{t}} = \mathcal{H}^*(T_{\mathrm{t}})$. Consider now the point P on $\mathcal{C}$ with coordinates $T = T_{\mathrm{p}} (\geq T_{\mathrm{t}})$ and $\mathcal{H}_{\mathrm{p}} = \mathcal{H}^*(T_{\mathrm{p}})$. In the plane $\mathcal{P}_{\mathcal{H}}(T_{\mathrm{p}})$, defined by $\mathcal{H} = \mathcal{H}_{\mathrm{p}}$ (see Fig. 2.8), the phase behaviour with respect to the variables T and $\mathcal{H}_{\mathrm{s}}$ has the appearance of a first-order transition line along $\mathcal{H}_{\mathrm{s}} = 0$ terminating at the critical point $T = T_{\mathrm{p}}$. Exponents can, therefore, be defined as they were in equations (2.117)–(2.121), except that now we have the reflectional symmetry of a magnetic system with the exponents of c_ρ and c_ξ being equal and $\xi_c = \rho_c = 0$. Thus we define

$$c_{\mathcal{H}} \sim \begin{cases} (T - T_{\mathrm{p}})^{-\alpha_{\mathrm{P}}}, & \text{along } \mathcal{H}_{\mathrm{s}} = 0,\ \mathcal{H} = \mathcal{H}_{\mathrm{p}},\ T > T_{\mathrm{p}}, \\ (T_{\mathrm{p}} - T)^{-\alpha'_{\mathrm{P}}}, & \text{along } \mathcal{H}_{\mathrm{s}} = 0,\ \mathcal{H} = \mathcal{H}_{\mathrm{p}},\ T < T_{\mathrm{p}}, \end{cases} \tag{2.210}$$

$$m_{\mathrm{s}} \sim (T_{\mathrm{p}} - T)^{\beta_{\mathrm{P}}}, \quad \text{along } \mathcal{H}_{\mathrm{s}} = 0,\ \mathcal{H} = \mathcal{H}_{\mathrm{p}},\ T < T_{\mathrm{p}}, \tag{2.211}$$

$$\varphi_T \sim \begin{cases} (T - T_{\mathrm{p}})^{-\gamma_{\mathrm{P}}}, & \text{along } \mathcal{H}_{\mathrm{s}} = 0,\ \mathcal{H} = \mathcal{H}_{\mathrm{p}},\ T > T_{\mathrm{p}}, \\ (T_{\mathrm{p}} - T)^{-\gamma'_{\mathrm{P}}}, & \text{along } \mathcal{H}_{\mathrm{s}} = 0,\ \mathcal{H} = \mathcal{H}_{\mathrm{p}},\ T < T_{\mathrm{p}}, \end{cases} \tag{2.212}$$

$$\mathcal{H}_{\mathrm{s}} \sim m_{\mathrm{s}} |m_{\mathrm{s}}|^{\delta_{\mathrm{P}} - 1}, \quad \text{along } \mathcal{H} = \mathcal{H}_{\mathrm{p}},\ T = T_{\mathrm{p}}. \tag{2.213}$$

In a similar way in the plane $\mathcal{P}_T(\mathcal{H}_{\mathrm{p}})$, defined by $T = T_{\mathrm{p}}$, the phase pattern again has the appearance of a first-order line terminated by a critical point. The exponents are defined in a similar way to those of (2.210)–(2.213), except that the roles of T and $\mathcal{H}$ are reversed. We thus define

$$\psi_T \sim \begin{cases} (\mathcal{H} - \mathcal{H}_{\mathrm{p}})^{-\bar{\alpha}_{\mathrm{P}}}, & \text{along } \mathcal{H}_{\mathrm{s}} = 0,\ T = T_{\mathrm{p}},\ \mathcal{H} > \mathcal{H}_{\mathrm{p}}, \\ (\mathcal{H}_{\mathrm{p}} - \mathcal{H})^{-\bar{\alpha}'_{\mathrm{P}}}, & \text{along } \mathcal{H}_{\mathrm{s}} = 0,\ T = T_{\mathrm{p}},\ \mathcal{H} < \mathcal{H}_{\mathrm{p}}, \end{cases} \tag{2.214}$$

$$m_{\mathrm{s}} \sim (\mathcal{H}_{\mathrm{p}} - \mathcal{H})^{\bar{\beta}_{\mathrm{P}}}, \quad \text{along } \mathcal{H}_{\mathrm{s}} = 0,\ T = T_{\mathrm{p}},\ \mathcal{H} < \mathcal{H}_{\mathrm{p}}, \tag{2.215}$$

[9] This assumption is supported by mean-field calculations with the Hamiltonian (2.205) (see, for example, Volume 1, Sect. 4.4).

$$\varphi_T \sim \begin{cases} (\mathcal{H}-\mathcal{H}_\mathrm{p})^{-\bar{\gamma}_\mathrm{p}}\,, & \text{along } \mathcal{H}_\mathrm{s}=0,\ T=T_\mathrm{p},\ \mathcal{H}>\mathcal{H}_\mathrm{p}, \\ (\mathcal{H}_\mathrm{p}-\mathcal{H})^{-\bar{\gamma}'_\mathrm{p}} & \text{along } \mathcal{H}_\mathrm{s}=0,\ T=T_\mathrm{p},\ \mathcal{H}<\mathcal{H}_\mathrm{p}, \end{cases} \tag{2.216}$$

with equation (2.213) applying in both planes. The formulae (2.210)–(2.216) apply at all points on the critical curve $\mathcal{C}$ including the tricritical point, where the critical exponents will be labelled by the subscript 't' rather than 'p', (Griffiths 1973).

2.13 Scaling for a Tricritical Point

Following a procedure similar to that of Sect. 2.9, the singular part of the potential can be taken to be

$$\phi_\mathrm{sing}(L_\mathrm{s},\triangle L,\triangle K)=\phi(L_\mathrm{s},L,K)-\phi(0,L_\mathrm{t},K_\mathrm{t})+m_\mathrm{t}\triangle L-u_\mathrm{t}\triangle K\,, \tag{2.217}$$

where $\triangle K=K-K_\mathrm{t}$ and $\triangle L=L-L_\mathrm{t}$, so that

$$\phi_\mathrm{sing}(0,0,0)=0\,, \tag{2.218}$$

$$\frac{\partial\phi_\mathrm{sing}}{\partial\triangle L}=-\triangle m\,, \tag{2.219}$$

$$\frac{\partial\phi_\mathrm{sing}}{\partial\triangle K}=\triangle u\,. \tag{2.220}$$

2.13.1 Scaling Fields for the Tricritical Point

Unlike the situation described in Sect. 2.9, here the phase diagram has reflectional symmetry. The only other special direction at the tricritical point is the line of the curves $\mathcal{C}$ and $\mathcal{T}$. In terms of couplings, rather than fields, this curve can be taken to have the equation $L=L^*(K)$. Assuming that the curve $\mathcal{C}$ at the tricritical point is not parallel to the L-axis, we express $L^*(K)$ in the form

$$L^*(K)=L_\mathrm{t}+a\triangle K+\bar{L}(K)\,, \tag{2.221}$$

where a is the (finite) slope of $\mathcal{C}$ at the tricritical point and

$$\bar{L}(K_\mathrm{t})=\left(\frac{\mathrm{d}\bar{L}}{\mathrm{d}K}\right)_\mathrm{t}=0\,, \qquad \bar{L}(K)\simeq b|\triangle K|^{\psi_\mathrm{t}}\,, \tag{2.222}$$

where ψ_t is the *tricritical shift exponent* (cf. equations (2.130), (2.131)). We now define the orthogonal scaling fields

$$\theta_1=L_\mathrm{s}\,, \qquad \theta_2=\triangle L-a\triangle K\,, \qquad \theta_3=\triangle K+a\triangle L\,, \tag{2.223}$$

with the triad of orthogonal unit vectors

$$\hat{\boldsymbol{\theta}}_1 = (1,0,0)\,, \qquad \hat{\boldsymbol{\theta}}_2 = \frac{(0,1,-a)}{\sqrt{1+a^2}}\,, \qquad \hat{\boldsymbol{\theta}}_3 = \frac{(0,a,1)}{\sqrt{1+a^2}}\,. \tag{2.224}$$

Setting $\theta_1 = 0$ gives the plane of reflectional symmetry, with $\hat{\boldsymbol{\theta}}_3$ giving the direction of the tangent to $\mathcal{C}$ and $\mathcal{T}$ at the tricritical point. The scaling hypothesis now takes the form

$$\phi_{\text{sing}}(\lambda^{y_1}\theta_1, \lambda^{y_2}\theta_2, \lambda^{y_3}\theta_3) = \lambda^d \phi_{\text{sing}}(\theta_1, \theta_2, \theta_3)\,, \tag{2.225}$$

for $\lambda > 0$, (Riedel 1972, Hankey et al. 1972) and the general shapes of the critical curves in a neighbourhood of the tricritical point can be inferred from the relative magnitudes of y_1, y_2 and y_3. A consequence of (2.225) is that, if the thermodynamic potential has non-smooth behaviour of a particular type at $(\theta_1, \theta_2, \theta_3) = (\kappa_1, \kappa_2, \kappa_3)$, then it will also have the same non-smooth behaviour at

$$(\theta_1, \theta_2, \theta_3) = (\kappa_1\lambda^{y_1}, \kappa_2\lambda^{y_2}, \kappa_3\lambda^{y_3})\,. \tag{2.226}$$

It follows that the critical lines $\mathcal{C}$, $\mathcal{C}^{(+)}$ and $\mathcal{C}^{(-)}$ and the triple line $\mathcal{T}$ must all have parametric equations of the form (2.226), for some different sets of parameters $(\kappa_1, \kappa_2, \kappa_3)$ and $\lambda > 0$. We shall take the set of parameters for $\mathcal{C}$ to be $(\kappa_1, \kappa_2, \kappa_3)$, for $\mathcal{C}^{(\pm)}$ to be $(\kappa_1^{(\pm)}, \kappa_2^{(\pm)}, \kappa_3^{(\pm)})$ and for $\mathcal{T}$ to be (τ_1, τ_2, τ_3). It follows from the symmetry of the model and the choice of scaling fields that $\kappa_1 = \tau_1 = 0$, $\kappa_1^{(+)} = -\kappa_1^{(-)} > 0$, $\kappa_2^{(+)} = \kappa_2^{(-)}$ and $\kappa_3^{(+)} = \kappa_3^{(-)}$. It will be assumed that in each of these triplets not more than one member is zero. Otherwise, with the definitions (2.223) for the scaling fields, the corresponding curve would be linear and in the direction of one of the unit vectors (2.224). For $\mathcal{C}$ to approach the tricritical point in the direction of $\hat{\boldsymbol{\theta}}_3$ it must be the case that

$$\Delta_{\text{t}} = y_2/y_3 > 1\,. \tag{2.227}$$

It follows, by eliminating λ from equation (2.226), that the equation of $\mathcal{C}$ in a neighbourhood of the tricritical point is

$$\theta_2 = \varpi_0|\theta_3|^{\Delta_{\text{t}}}\,, \tag{2.228}$$

where $\varpi_0 = \kappa_2/|\kappa_3|^{\Delta_{\text{t}}}$. From equations (2.223)

$$\theta_2\left(1 + a\frac{\theta_3}{\theta_2}\right) = (1+a^2)\triangle L\,, \qquad \theta_3\left(1 - a\frac{\theta_2}{\theta_3}\right) = (1+a^2)\triangle K \tag{2.229}$$

and, by the same reasoning as that employed in Sect. 2.9, it follows that the shift exponent $\psi_{\text{t}} = \Delta_{\text{t}}$.

The way in which the curves $\mathcal{C}^{(\pm)}$ approach the tricritical point is determined by the magnitude of y_1 relative to y_2 and y_3 and whether one pair of $\kappa_2^{(\pm)}$ or $\kappa_3^{(\pm)}$ is zero. Given these various conditions the geometry of the curves can be described in terms of their projections onto the planes $\theta_j = 0$, $j = 1,2,3$. There are the following possibilities when the pairs $\kappa_2^{(\pm)}$ and $\kappa_3^{(\pm)}$ are non-zero:

I: $\underline{y_1 > y_2 > y_3}$. $\mathcal{C}^{(\pm)}$ approach the tricritical point tangentially to $\hat{\boldsymbol{\theta}}_3$, when they are projected into the planes $\theta_1 = 0$ or $\theta_2 = 0$ and tangentially to $\hat{\boldsymbol{\theta}}_2$, when projected into the plane $\theta_3 = 0$.

II: $\underline{y_1 = y_2 > y_3}$. This is the same as I, except that, when the curves are projected into the plane $\theta_3 = 0$, they approach the critical point at non-zero angles to both $\hat{\boldsymbol{\theta}}_1$ and $\hat{\boldsymbol{\theta}}_2$.

III: $\underline{y_2 > y_1 > y_3}$. Again, $\mathcal{C}^{(\pm)}$ approach the tricritical point tangentially to $\hat{\boldsymbol{\theta}}_3$, when they are projected into the planes $\theta_1 = 0$ or $\theta_2 = 0$. When they are projected into the plane $\theta_3 = 0$, the direction of approach to the tricritical point is along $\hat{\boldsymbol{\theta}}_1$.

IV: $\underline{y_2 > y_1 = y_3}$. The projections of $\mathcal{C}^{(\pm)}$ are tangential to $\hat{\boldsymbol{\theta}}_3$ in $\theta_1 = 0$ and to $\hat{\boldsymbol{\theta}}_1$ in $\theta_3 = 0$. In the plane $\theta_2 = 0$ the projections make non-zero angles to both $\hat{\boldsymbol{\theta}}_1$ and $\hat{\boldsymbol{\theta}}_3$.

V: $\underline{y_2 > y_3 > y_1}$. This is the same as IV except the projections into the plane $\theta_2 = 0$ are tangential to $\hat{\boldsymbol{\theta}}_1$.

This analysis is modified if either $\kappa_2^{(\pm)} = 0$ or $\kappa_3^{(\pm)} = 0$, when the geometry is described entirely by the shapes of the curves within the planes $\theta_2 = 0$ or $\theta_3 = 0$ respectively.

2.13.2 Tricritical Exponents and Scaling Laws

We now consider the asymptotic forms for the order parameter m_{s} and the response functions φ_T, ψ_T and $c_{\mathcal{H}}$ at the tricritical point. The exponents associated with these forms for different paths of approach are given by equations (2.210)–(2.216) (with ‘t’ replacing ‘p’). Following the procedure used for the general scaling formulation in Sect. 2.4 and for the critical point in Sect. 2.9 we take $\lambda = |\theta_1|^{-1/y_1}$, $\lambda = |\theta_2|^{-1/y_2}$ and $\lambda = |\theta_3|^{-1/y_3}$ in equation (2.225) to give respectively

$$\phi_{\mathrm{sing}}(\theta_1,\theta_2,\theta_3) = |\theta_1|^{d/y_1}\phi_{\mathrm{sing}}(\pm 1, g_2^{(1)}, g_3^{(1)}) \,, \tag{2.230}$$

$$\phi_{\mathrm{sing}}(\theta_1,\theta_2,\theta_3) = |\theta_2|^{d/y_2}\phi_{\mathrm{sing}}(g_1^{(2)}, \pm 1, g_3^{(2)}) \,, \tag{2.231}$$

$$\phi_{\mathrm{sing}}(\theta_1,\theta_2,\theta_3) = |\theta_3|^{d/y_3}\phi_{\mathrm{sing}}(g_1^{(3)}, g_2^{(3)}, \pm 1) \,, \tag{2.232}$$

respectively, where the (tricritical) gap fields $g_j^{(m)}(\theta_m,\theta_j)$, $m,j = 1,2,3$ are given by (2.27) and

$$\begin{aligned} &\vartheta_2^{(1)} = 1/\vartheta_{\mathrm{t}}\,, &&\vartheta_3^{(1)} = 1/\vartheta_{\mathrm{t}}\Delta_{\mathrm{t}}\,, \\ &\vartheta_1^{(2)} = \vartheta_{\mathrm{t}}\,, &&\vartheta_3^{(2)} = 1/\Delta_{\mathrm{t}}\,, \\ &\vartheta_1^{(3)} = \Delta_{\mathrm{t}}\vartheta_{\mathrm{t}}\,, &&\vartheta_2^{(3)} = \Delta_{\mathrm{t}}\,, \end{aligned} \tag{2.233}$$

where $\vartheta_{\mathrm{t}} = y_1/y_2$ is the *tricritical gap exponent.* It follows from (2.228) and (2.233) that in a neighbourhood of the tricritical point the equation of $\mathcal{C}$ in the plane $\theta_1 = 0$ is $g_2^{(3)}(\theta_2, \theta_3) = \varpi_0$. From (2.230), with $\theta_2 = \theta_3 = 0$, using similar reasoning to that used in Sect. 2.9, it follows that the exponent δ_{t}, defined in (2.213), is given by

$$\delta_{\mathrm{t}} = y_1/(d - y_1) \,. \tag{2.234}$$

A classification of the paths of approach to the tricritical point in the plane $\theta_1 = 0$ can now be made in a similar way to that for the critical point in Sect. 2.9. The two types of approach are:

(i) Paths on which $|g_3^{(2)}| \to 0$, including the direction of the vector $\hat{\boldsymbol{\theta}}_2$, for which, from (2.231),

$$\phi_{\mathrm{sing}}(0, \theta_2, \theta_3) \simeq |\theta_2|^{d/y_2} A_\infty^{(\pm)} \,, \tag{2.235}$$

where

$$\lim_{g_3^{(2)} \to 0} \phi_{\mathrm{sing}}(0, \pm 1, g_3^{(2)}) = A_\infty^{(\pm)} \tag{2.236}$$

is assumed to be finite and non-zero.

(ii) Paths on which $|g_2^{(3)}| \to \pi_2$, where π_2 is finite (possibly zero), when

$$\phi_{\mathrm{sing}}(0, \theta_2, \theta_3) \simeq |\theta_3|^{d/y_3} A_{\pi_2}^{(\pm)} \,, \tag{2.237}$$

where

$$\lim_{g_2^{(3)} \to \pi_2} \phi_{\mathrm{sing}}(0, g_2^{(3)}, \pm 1) = A_{\pi_2}^{(\pm)} \tag{2.238}$$

is assumed to be finite and non-zero.

Since $\Delta_{\mathrm{t}} > 1$, all curves in class (ii), including $\mathcal{C}$, are tangential to the direction of the unit vector $\hat{\boldsymbol{\theta}}_3$. Some curves in class (i) are also tangential to $\hat{\boldsymbol{\theta}}_3$, but all curves not tangential to this vector are certainly in class (i). The asymptotic forms in equations (2.210)–(2.212) are along the line $\mathcal{H} = \mathcal{H}_{\mathrm{t}}$. It follows from (2.223) that unless $\mathcal{H}_{\mathrm{t}} = aJ_1$ or $-J_1/a$, $\theta_2 \sim |T - T_{\mathrm{t}}|$ and $\theta_3 \sim |T - T_{\mathrm{t}}|$ as the tricritical point is approached along this line. It therefore falls into class (i) with

$$\phi_{\mathrm{sing}} \sim |T - T_{\mathrm{t}}|^{d/y_2} \,, \qquad \text{along } \mathcal{H}_{\mathrm{s}} = 0,\, \mathcal{H} = \mathcal{H}_{\mathrm{t}},\, T \gtrless T_{\mathrm{t}}. \tag{2.239}$$

The asymptotic forms in equations (2.214)–(2.216) are along the line $T = T_{\mathrm{t}}$. It follows from (2.223) that $\theta_2 \sim |\mathcal{H} - \mathcal{H}_{\mathrm{t}}|$ and $\theta_3 \sim |\mathcal{H} - \mathcal{H}_{\mathrm{t}}|$, unless $a = 0$, as the tricritical point is approached along this line. It therefore falls into class (i) with

$$\phi_{\mathrm{sing}} \sim |\mathcal{H} - \mathcal{H}_{\mathrm{t}}|^{d/y_2} \,, \qquad \text{along } \mathcal{H}_{\mathrm{s}} = 0,\, T = T_{\mathrm{t}},\, \mathcal{H} \gtrless \mathcal{H}_{\mathrm{t}}. \tag{2.240}$$

Following similar reasoning to that used in Sect. 2.9 the critical exponents defined in equations (2.210)–(2.212), (2.214)–(2.216) can now be obtained as functions of y_1 and y_2. The formulae obtained are

$$\alpha_t = \alpha_t' = \bar{\alpha}_t = \bar{\alpha}_t' = (2y_2 - d)/y_2\,, \tag{2.241}$$

$$\beta_t = \bar{\beta}_t = (d - y_1)/y_2\,, \tag{2.242}$$

$$\gamma_t = \gamma_t' = \bar{\gamma}_t = \bar{\gamma}_t' = (2y_1 - d)/y_2\,. \tag{2.243}$$

From these equations and (2.234) the tricritical exponents can be show to satisfy

$$\gamma_t = \beta_t(\delta_t - 1)\,, \tag{2.244}$$

$$2 = \alpha_t + 2\beta_t + \gamma_t\,, \tag{2.245}$$

which are respectively the tricritical forms of the Widom and Essam–Fisher scaling laws (see equations (2.169) and (2.170)).

2.13.3 Connected Transition Regions

In terms of the scaling formulation of Sect. 2.4, the tricritical point is a transition region connected to seven other transition regions; the critical curves $\mathcal{C}$, $\mathcal{C}^{(+)}$ and $\mathcal{C}^{(-)}$, the triple curve $\mathcal{T}$ and the three first-order surfaces. To construct a scaling hypothesis for any one of these we need to define appropriate scaling fields such that the relevant fields are zero for the region in question. We shall confine our attention to the critical curve $\mathcal{C}$. Following a similar procedure to that adopted for the coexistence curve in Sect. 2.9 we retain the scaling fields θ_1 and θ_3 of (2.223) and replace θ_2 by

$$\theta_2' = \theta_2 - \varpi_0|\theta_3|^{\Delta_t}\,. \tag{2.246}$$

The gap fields, involving θ_2, and the singular part of the thermodynamic potential are modified to

$$g'^{(1)}_2(\theta_1, \theta_2') = \theta_2'/|\theta_1|^{1/\vartheta_t} = g_2^{(1)} - \varpi_0|g_3^{(1)}|^{\Delta_t}\,, \tag{2.247}$$

$$g'^{(3)}_2(\theta_2', \theta_3) = \theta_2'/|\theta_3|^{\Delta_t} = g_2^{(3)} - \varpi_0\,, \tag{2.248}$$

$$g'^{(2)}_1(\theta_1, \theta_2') = \theta_1/|\theta_2'|^{\vartheta_t} = g_1^{(2)}/|1 - \varpi_0[g_2^{(3)}]^{-1}|^{\vartheta_t}\,, \tag{2.249}$$

$$g'^{(2)}_3(\theta_2', \theta_3) = \theta_3/|\theta_2'|^{1/\Delta_t} = g_3^{(2)}/|1 - \varpi_0[g_2^{(3)}]^{-1}|^{1/\Delta_t}\,, \tag{2.250}$$

$$\begin{aligned}\phi'_{\text{sing}}(\theta_1, \theta_2', \theta_3) = {}& \phi_{\text{sing}}(\theta_1, \theta_2' + \varpi_0|\theta_3|^{\Delta_t}, \theta_3) \\ & - \phi_{\text{sing}}(\theta_1, \varpi_0|\theta_3|^{\Delta_t}, \theta_3)\,,\end{aligned} \tag{2.251}$$

so that ϕ'_{sing} is zero not only at the tricritical point but on the critical curve. The scaling hypothesis (2.225) is satisfied with the new scaling fields in a neighbourhood of the tricritical point and all the subsequent analysis is valid except that the coexistence curve is given by $g'^{(3)}_2(\theta'_2, \theta_3) = 0$. From (2.42) and (2.43) we define the effective exponents

$$\Sigma_1(\theta_1, \theta'_2, \theta_3) = \frac{\partial \ln \phi'_{\text{sing}}}{\partial \ln |\theta_1|}$$

$$= \frac{d}{y_1} - \frac{y_2}{y_1}|g'^{(1)}_2| \frac{\partial \ln \phi'_{\text{sing}}(\pm 1, g'^{(1)}_2, g^{(1)}_3)}{\partial |g'^{(1)}_2|}$$

$$- \frac{y_3}{y_1}|g^{(1)}_3| \frac{\partial \ln \phi'_{\text{sing}}(\pm 1, g'^{(1)}_2, g^{(1)}_3)}{\partial |g^{(1)}_3|}, \tag{2.252}$$

$$\Sigma_2(\theta_1, \theta'_2, \theta_3) = \frac{\partial \ln \phi'_{\text{sing}}}{\partial \ln |\theta'_2|}$$

$$= \frac{d}{y_2} - \frac{y_1}{y_2}|g'^{(2)}_1| \frac{\partial \ln \phi'_{\text{sing}}(g'^{(2)}_1, \pm 1, g'^{(2)}_3)}{\partial |g'^{(2)}_1|}$$

$$- \frac{y_3}{y_2}|g'^{(2)}_3| \frac{\partial \ln \phi'_{\text{sing}}(g'^{(2)}_1, \pm 1, g'^{(2)}_3)}{\partial |g'^{(2)}_3|}. \tag{2.253}$$

Along the line of approach to the tricritical point $\mathcal{H} = \mathcal{H}_{\text{t}}$, $T = T_{\text{t}}$, $g'^{(1)}_2 = g^{(1)}_3 = 0$ and $\Sigma_1 = d/y_1$. As the critical curve is approached in a direction perpendicular to the plane $\theta_1 = 0$, $g'^{(1)}_2 = 0$ but $|g^{(1)}_3| \to \infty$. Suppose that

$$\phi'_{\text{sing}}(\pm 1, 0, g^{(1)}_3) \sim |g^{(1)}_3|^{\mu^{(1)}_3}. \tag{2.254}$$

Then $\Sigma_1 \to d/y'_1$, where

$$y'_1 = \frac{d\, y_1}{d - y_3 \mu^{(1)}_3} \tag{2.255}$$

is the scaling exponent associated with θ_1 on the critical curve. In the plane $\theta_1 = 0$, $g'^{(2)}_1 = 0$. As the tricritical point is approached along a class (i) path in the plane, $|g'^{(2)}_3| \to 0$ and $\Sigma_2 \to d/y_2$. As the critical curve is approached away from the tricritical point $|g'^{(2)}_3| \to \infty$. Suppose that

$$\phi'_{\text{sing}}(0, \pm 1, g'^{(2)}_3) \sim |g'^{(2)}_3|^{\mu^{(2)}_3}. \tag{2.256}$$

Then $\Sigma_2 \to d/y'_2$, where

$$y'_2 = \frac{d\, y_2}{d - y_3 \mu^{(2)}_3} \tag{2.257}$$

is the scaling exponent associated with θ_2' on the critical curve. In formulating the asymptotic expressions (2.210)–(2.216) for the critical curve, we allowed the possibility that the exponents vary along the curve. We now see that this will be the case only if $\mu_3^{(1)}$ or $\mu_3^{(2)}$ is dependent on the point on the critical curve to which convergence is taken. In general this will not be so. In the language of renormalization group theory (see Chap. 6), the whole of the critical curve will be controlled by a single fixed point leading to exponents which are the same throughout the length of the curve. These exponents, which we simply denote by α, β, γ and δ will be given by the formulae (2.234), (2.241)–(2.243) with y_1 replaced by y_1' and y_2 by y_2'. They will, therefore, satisfy the Widom and Essam–Fisher scaling laws (2.169) and (2.170). The critical and tricritical exponents are, from (2.234), (2.241)–(2.243), (2.255) and (2.257), related by

$$\alpha = \alpha_t + \frac{\mu_3^{(2)}}{\Delta_t}, \tag{2.258}$$

$$\beta = \left(\beta_t - \frac{\mu_3^{(1)}}{\Delta_t}\right)\left(\frac{d - y_3\mu_3^{(2)}}{d - y_3\mu_3^{(1)}}\right), \tag{2.259}$$

$$\gamma = \left(\gamma_t + \frac{\mu_3^{(1)}}{\Delta_t}\right)\left(\frac{d - y_3\mu_3^{(2)}}{d - y_3\mu_3^{(1)}}\right), \tag{2.260}$$

$$\delta = \delta_t\left(\frac{d - y_3\Delta_t\vartheta_t}{d - y_3\Delta_t\vartheta_t - y_3\mu_3^{(1)}}\right). \tag{2.261}$$

A similar analysis applies to the critical curves $\mathcal{C}^{(+)}$ and $\mathcal{C}^{(-)}$, although, since these curves do not lie in the plane $\theta_1 = 0$, the choice of appropriate scaling fields is more complicated. The same reasoning can also be applied to the triple curve $\mathcal{T}$. In that case we expect the corresponding exponents $\mu_3^{(1)}$ and $\mu_3^{(2)}$ to be such as to give the value d for both y_1' and y_2'. This is because across $\mathcal{T}$ there is a first-order transition both within and orthogonal to the plane $\theta_1 = 0$. Two factors affect the crossover from critical behaviour along the curve $\mathcal{C}$, with exponents α, β, γ and δ to tricritical behaviour at the point T, with exponents α_t, β_t, γ_t and δ_t. The first of these is the change in limiting behaviour of the gap fields $g_3^{(1)}$ and $g'^{(2)}_3$. The effect of this change is in turn mediated by the exponents $\mu_3^{(1)}$ and $\mu_3^{(2)}$ respectively. The other parameter which affects the formulae (2.258)–(2.261) is y_3 (and consequently Δ_t). In the limit when y_3 becomes less relevant ($y_3 \to 0$, $\Delta_t \to \infty$) the critical and tricritical exponents approach the same values. It can also be seen, from (2.228), that, in this limit the curve $\mathcal{C}$ becomes flat at T.

2.14 Corrections to Scaling

In Sect. 2.5 we considered one case where the simple power law asymptotic behaviour of equations (2.35) broke down. This arose from the exponent ω_m being zero and led to a logarithmic divergence in the corresponding derivative ψ of ϕ. In this section modifications to scaling which arise from the amplitude factors are considered. For simplicity we concentrate on the case $n = 3$ with a transition region $\mathcal{C}$ with $s = 2$. That is to say, we have a system with two relevant fields θ_1 and θ_2 and one irrelevant field θ_3. From (2.25)–(2.27)

$$\phi_{\text{sing}}(\theta_1, \theta_2, \theta_3) = |\theta_1|^{\sigma_1} \phi_{\text{sing}} \left(\pm 1, \frac{\theta_2}{|\theta_1|^{\vartheta_2^{(1)}}}, \frac{\theta_3}{|\theta_1|^{\vartheta_3^{(1)}}} \right), \tag{2.262}$$

$$\phi_{\text{sing}}(\theta_1, \theta_2, \theta_3) = |\theta_2|^{\sigma_2} \phi_{\text{sing}} \left(\frac{\theta_1}{|\theta_2|^{\vartheta_1^{(2)}}}, \pm 1, \frac{\theta_3}{|\theta_2|^{\vartheta_3^{(2)}}} \right). \tag{2.263}$$

In Sect. 2.4 we considered the asymptotic forms which can be derived from (2.262) and (2.263). In each case it is necessary to choose a path of approach to the transition region so that the arguments of the functions on the right-hand sides tend to finite values. Even then, in order to obtain the simple form of scaling described in Sect. 2.4 and utilized for a critical point and tricritical point in Sects. 2.9 and 2.13 respectively, it is necessary that the amplitude factors are finite and non-zero. We begin by considering the functions

$$f^{(\pm)}(u, v) = \phi_{\text{sing}}(\pm 1, u, v), \tag{2.264}$$

which appear on the right-hand side of (2.262). The limit $u \to 0$ with v small and non-zero can be achieved by taking $\theta_2 \to 0$ with θ_1 and θ_3 non-zero. Excluding the possibility that $\theta_2 = 0$ is some transition region connected to $\mathcal{C}$,[10] then $f^{(\pm)}(u, v)$ have power-series expansions of the form

$$f^{(\pm)}(u, v) = w_0^{(\pm)}(v) + u w_1^{(\pm)}(v) + \cdots. \tag{2.265}$$

If $w_0^{(\pm)}(0)$ is finite and non-zero then the power-law behaviour is that predicted by simple scaling theory, although the magnitude of $w_1^{(\pm)}(0)$ will affect the range over which the leading scaling behaviour can be observed. If $w_0^{(\pm)}(0) = 0$ there is a breakdown of simple scaling and the power-law predictions rest on the first non-zero term in the expansion (2.265). If this is given by $w_1^{(\pm)}(0)$ being finite and non-zero, then the singular behaviour is not only weaker but will depend on the next layer of detail of the path of approach $\mathcal{P}$. A possible path $\mathcal{P}$ consistent with the condition $u \to 0$ is $\theta_2 \sim |\theta_1|^{\tau}$, where $\tau > \vartheta_2^{(1)}$. Then from (2.262)

$$\phi_{\text{sing}}(\theta_1, \theta_2, \theta_3) \sim |\theta_1|^{\sigma_1 + \tau - \vartheta_2^{(1)}}, \tag{2.266}$$

[10] The situation of connected transition regions has been considered in Sect. 2.4.

which is a weaker power-law behaviour. The same analysis can be applied to (2.263).

We now consider the possibility of $f^{(\pm)}(0,0)$ being infinite. It is clear that, for an isolated transition region $f^{(\pm)}(0,v)$ cannot be infinite and thus, any singular behaviour must be arise from $w_0^{(\pm)}(v)$, and therefore involve the irrelevant variable θ_3. That this is a possibility was pointed out by Fisher (1983), who classified irrelevant variables as *dangerous* or *harmless* according as they do or do not yield such divergent behaviour. Suppose that

$$w_0^{(\pm)}(v) \sim |v|^{-\tau}, \tag{2.267}$$

where $\tau > 0$. Then, from (2.262)–(2.265),

$$\phi_{\text{sing}}(\theta_1, \theta_2, \theta_3) \sim |\theta_1|^{\bar{\sigma}_1}, \tag{2.268}$$

where

$$\bar{\sigma}_1 = \frac{\bar{d}}{y_1} \tag{2.269}$$

and

$$\bar{d} = d - \tau|y_3|. \tag{2.270}$$

The effect of the dangerous irrelevant variable θ_3 can be view as an effective reduction in the dimension of the system, from d to $\bar{d}$ provided that all paths to the transition region produce the same change. This means that we must be satisfied that a parallel analysis of (2.263) yields the same formula (2.270). In Sect. 3.3.2 the significance of the presence of a dangerous irrelevant variable for the interpretation of Gaussian critical exponents in relation to scaling theory is discussed.

2.15 Scaling and Universality

The universality hypothesis, which is described in Sect. 2.3.2, and the scaling hypotheses of Sects. 2.4.1 and 2.5.1 both apply to a particular transition region rather than to a physical system or theoretical model as a whole. Different transition regions of the same system may be in different universality classes and may have different scaling functions and exponents. Discussion of the relationship between scaling and universality is made easier if we take a particular type of transition region. The simplest example to choose is a critical point with power-law scaling. In the weak scaling direction, from (2.140), $\theta_1/\theta_2 \to 0$ as the critical point is approached and thus, from (2.132) and (2.147),

$$\theta_1 = \Delta L - a\Delta K\,, \qquad \theta_2 \simeq (1+a^2)\Delta K\,. \tag{2.271}$$

From equations (2.135), (2.136), (2.162), (2.167) and (2.204), the forms for the singular part of the thermodynamic potential, the density difference from its critical value and the response function φ_T can be expressed as

$$\phi_{\text{sing}} = A_0 |\Delta K|^{2-\alpha} W_0^{(\pm)} \left(C \frac{[\Delta L - a\Delta K]}{|\Delta K|^{\beta\delta}} \right), \tag{2.272}$$

$$\Delta\rho = -\frac{\partial \phi_{\text{sing}}}{\partial \Delta L} = A_1^{(\pm)} |\Delta K|^{\beta} W_1^{(\pm)} \left(C \frac{[\Delta L - a\Delta K]}{|\Delta K|^{\beta\delta}} \right), \tag{2.273}$$

$$\varphi_T = -\frac{\partial^2 \phi_{\text{sing}}}{\partial \Delta L^2} = A_2^{(\pm)} |\Delta K|^{-\gamma} W_2^{(\pm)} \left(C \frac{[\Delta L - a\Delta K]}{|\Delta K|^{\beta\delta}} \right), \tag{2.274}$$

where C, A_0, $A_1^{(\pm)}$ and $A_2^{(\pm)}$ are constants and

$$A_s^{(\pm)} W_s^{(\pm)}(x) = -C^s A_0 \frac{\mathrm{d}^s W_0^{(\pm)}(x)}{\mathrm{d}x^s}, \qquad s = 1, 2. \tag{2.275}$$

By imposing the conditions

$$W_0^{(\pm)}(0) = W_1^{(\pm)}(0) = W_2^{(\pm)}(0) = 1, \tag{2.276}$$

we ensure that A_0, $A_1^{(\pm)}$ and $A_2^{(\pm)}$ are the critical amplitudes with

$$\begin{aligned} A_s^{(\pm)} &= C^s A_0 Q_s^{(\pm)}, \\ Q_s^{(\pm)} &= -\left(\frac{\mathrm{d}^s W_0^{(\pm)}(x)}{\mathrm{d}x^s} \right)_{x=0} \qquad s = 1, 2. \end{aligned} \tag{2.277}$$

Each of the equations (2.272)–(2.274) represents a family of functions in two variables, $(\phi_{\text{sing}}, \Delta L)$, $(\Delta\rho, \Delta L)$, and $(\varphi_T, \Delta L)$ respectively, parameterized by ΔK. The scaling hypothesis implication of the form of these equations is that each family of isotherms (ΔK constant)'collapses' onto one curve when plotted in the space of $(\Delta L - a\Delta K)/|\Delta K|^{\beta\delta}$ against $\phi_{\text{sing}}/|\Delta K|^{2-\alpha}$, $\Delta\rho/|\Delta K|^{\beta}$ and $\varphi_T|\Delta K|^{\gamma}$ respectively. Equation (2.273) is the equation of state. For a ferromagnetic critical point $a = 0$, $\Delta L = \mathcal{H}/T \simeq \mathcal{H}/T_c$, $\Delta K = K - K_c \simeq= J(T_c - T)/T_c^2$ and $\Delta\rho = m$, the magnetization per lattice site. The mean-field equations (2.184) and (2.185) yield an expression like (2.273) with $\beta = \frac{1}{2}$, $\beta\delta = \frac{3}{2}$. The validity of an equation of the form (2.273) has been investigated for nickel by Arrott and Noakes (1967), Kouvel and Rodbell (1967) and Kouvel and Comly (1968), for CrO_2 by Kouvel and Rodbell (1967) and for $CrBr_3$ by Ho and Litster (1969). Similar tests for a number of fluid systems have been made by Green et al. (1967). In all cases the experimental data gives strong support for the scaling hypothesis. Although the critical exponent equalities $\alpha = \alpha'$ and $\gamma = \gamma'$ and the scaling laws (2.169), (2.170) are analytic consequences of (2.272)–(2.274), they can be used as a direct test of scaling for theoretical models and experimental data. The well-known mean-field values for the purely thermodynamic exponents (see, for example, Volume 1, Chap. 3) are $\alpha = \alpha' = 0$, $\beta = \frac{1}{2}$, $\gamma = \gamma' = 1$ and $\delta = 3$. These values, which satisfy the Widom scaling law

(2.169) and the Essam–Fisher scaling law (2.170), are those which arise for systems of any dimension when the free energy is assumed to be smooth in a neighbourhood of the critical point. In particular they follow from Landau theory, as we shall see in Chap. 3. For the two-dimensional Ising model (see, for example, Volume 1, Chap. 8 or Baxter 1982a) $\alpha = \alpha' = 0$, $\beta = \frac{1}{8}$ and $\gamma = \gamma' = \frac{7}{4}$, where the value of α corresponds to a logarithmic singularity. These values satisfy the Essam–Fisher scaling law (2.170). If the Widom scaling law (2.169) is assumed then $\delta = 15$, which is the generally accepted value (Baxter 1982a). Ho and Litster (1969) obtained the values $\gamma = \gamma' = 1.215 \pm 0.015$, $\beta = 0.368 \pm 0.005$ and $\delta = 4.28 \pm 0.1$ for $CrBr_3$, which yields $\gamma' - \beta(\delta - 1) = 0.00746 \pm 0.0682$.

The universality hypothesis was presented in Sect. 2.3.2 as a list of criteria for dividing critical problems into universality classes. This hypothesis can be tested when it is clear what is understood by the critical properties of two problems being 'the same'. This is usually taken to mean that the problems have the same values for critical exponents and the same scaling functions to within metric factors (Fisher 1983). For equations (2.272)–(2.274), this means that the functions $W_s^{(\pm)}(x)$, $s = 0, 1, 2$ and the exponents α, β, γ and δ are the same for all problems in the same universality class but the constants C, $A_s^{(\pm)}$, $s = 0, 1, 2$ may be different. As indicated above the work of Green et al. (1967) represents a test of scaling. It can also be taken as a test of universality, which is clearly supported by their results which show the collapse of data points, for the branches of the coexistence curves of a number of fluids, onto a single universal curve.

So far in this section we have concentrated the discussion on thermodynamic functions and variables. Now we consider the correlation function and correlation length for the statistical mechanical variable $\mathfrak{q}(\boldsymbol{r})$ associated with the coupling L. Let $\Delta\mathfrak{q}(\boldsymbol{r}) = \mathfrak{q}(\boldsymbol{r}) - \rho_c$. Then

$$\langle \Delta\mathfrak{q}(\boldsymbol{r})\rangle = \langle \mathfrak{q}(\boldsymbol{r})\rangle - \rho_c = \rho - \rho_c = \Delta\rho\,, \tag{2.278}$$

$$\langle \Delta\mathfrak{q}(\boldsymbol{r}_1)\Delta\mathfrak{q}(\boldsymbol{r}_2)\rangle = \langle \mathfrak{q}(\boldsymbol{r}_1)\mathfrak{q}(\boldsymbol{r}_2)\rangle - 2\rho\rho_c + \rho_c^2 \tag{2.279}$$

and, from (1.54) with $\mathfrak{c}_1 = \mathfrak{c}_2 = \mathfrak{q}$,

$$\Gamma_2(\mathfrak{q}(\boldsymbol{r}_1), \mathfrak{q}(\boldsymbol{r}_2)) = \langle \Delta\mathfrak{q}(\boldsymbol{r}_1)\Delta\mathfrak{q}(\boldsymbol{r}_2)\rangle - \langle \Delta\mathfrak{q}(\boldsymbol{r}_1)\rangle\langle \Delta\mathfrak{q}(\boldsymbol{r}_2)\rangle\,. \tag{2.280}$$

As in Sect. 2.11 we assume that the system is translational invariant so that the connected two-point correlation function is dependent on the single spatial variable $\bar{\boldsymbol{r}} = \boldsymbol{r}_2 - \boldsymbol{r}_1$. Then, from (2.194), (2.195) and (2.271),

$$\Gamma_2(\mathfrak{q}(\boldsymbol{0}), \mathfrak{q}(\bar{\boldsymbol{r}})) = B^{(\pm)}|\Delta K|^{d\nu-\gamma} X^{(\pm)}\left(C\,\frac{[\Delta L - a\Delta K]}{|\Delta K|^{\beta\delta}}, \frac{\bar{r}|\Delta K|^{\nu}}{R_0}\right), \tag{2.281}$$

$$\Gamma_2^*(\mathfrak{q};\bar{\boldsymbol{k}}) = B^{(\pm)}(R_0^d/v_0)|\Delta K|^{-\gamma} X^{*(\pm)}\left(C\,\frac{[\Delta L - a\Delta K]}{|\Delta K|^{\beta\delta}}, \frac{R_0\bar{\boldsymbol{k}}}{|\Delta K|^{\nu}}\right), \tag{2.282}$$

where v_0 is the volume per lattice site and

$$X^{*(\pm)}(x;\bar{\boldsymbol{k}}) = \sum_{\{\bar{\boldsymbol{r}}\}} X^{(\pm)}(x;\bar{\boldsymbol{r}}) \exp(-\mathrm{i}\bar{\boldsymbol{k}} \boldsymbol{.} \bar{\boldsymbol{r}})\,. \tag{2.283}$$

That the metric factor C in (2.281) and (2.282) is the same as that in (2.272)–(2.274) follows from (2.192), from which it can also be seen that

$$B^{(\pm)} = A_2^{(\pm)} v_0/R_0^d\,, \tag{2.284}$$

$$X^{*(\pm)}(x;\mathbf{0}) = W_2^{(\pm)}(x)\,. \tag{2.285}$$

From (2.193),

$$\mathfrak{r}(\mathfrak{q}) = R_0 Q_3^{(\pm)}|\Delta K|^{-\nu} Y^{(\pm)}\left(C\,\frac{[\Delta L - a\Delta K]}{|\Delta K|^{\beta\delta}}\right), \tag{2.286}$$

where

$$Q_3^{(\pm)}\, Y^{(\pm)}(x) = \sqrt{-c(d)\frac{\nabla_{\bar{k}}^2 X^{*(\pm)}(x,\bar{\mathbf{0}})}{W_2^{(\pm)}(x)}}\,. \tag{2.287}$$

By imposing the condition

$$Y^{(\pm)}(0) = 1\,, \tag{2.288}$$

we have

$$Q_3^{(\pm)} = \sqrt{-c(d)\nabla_{\bar{k}}^2 X^{*(\pm)}(0,\bar{\mathbf{0}})}\,. \tag{2.289}$$

The functions $X^{(\pm)}(x;\bar{\boldsymbol{r}})$ are supposed to be universal making $X^{*(\pm)}(x;\bar{\boldsymbol{k}})$ and $Y^{(\pm)}(x)$ also universal. It thus follows, from (2.289), that $Q_3^{(\pm)}$, like $Q_1^{(\pm)}$ and $Q_2^{(\pm)}$ defined in (2.277), are universal constants.

It would seem that in each of the regions $\Delta K >, < 0$ there are three independent non-universal metric parameters C, A_0 and R_0. It has, however, been argued by Stauffer et al. (1972) that, at the critical point, the singular part of dimensionless free energy in a region of the dimensions of the correlation length is a universal constant. In our notation this quantity is the critical point value of $\mathfrak{r}^d(\mathfrak{q})\phi_{\text{sing}}/v_0$. From (2.272), (2.276), (2.286), (2.288) and (2.289)

$$\frac{\mathfrak{r}^d(\mathfrak{q})\phi_{\text{sing}}}{v_0} \sim A_0 R_0^d v_0^{-1}|\Delta K|^{2-\alpha-d\nu}[Q_3^{(\pm)}]^d \tag{2.290}$$

at the critical point. When the Josephson hyper-scaling law (2.199) is satisfied the right-hand side of (2.290) is a constant and the proposition, referred to as the *two scale factor* or *hyper-universality hypothesis*, is that the quantity $A_0 R_0^d / v_0$ is universal. In Sect. 2.14 we discussed the idea of dangerous irrelevant variables. In Sect. 3.3.2 these ideas will be amplified in relation to the Gaussian critical exponents to show that the hyper-scaling law is satisfied with the 'real' physical dimension d when $d \leq d_{\mathrm{u}}$, the upper borderline dimension, which is four for a critical point. For $d > d_{\mathrm{u}}$ the hyper-scaling law is satisfied (at least for the Gaussian exponents) when d is replaced by $\bar{d}$, the dimension reduced by the presence of a dangerous irrelevant variable (see equation (2.270)). The hyper-universality hypothesis is thus expected to be true for $d \leq 4$. This is supported by the results for theoretical models and experimental systems given by Stauffer et al. (1972). Later work (Ferer et al. 1973) shows good agreement for systems, with different lattice structures, which on other grounds are taken to be in the same universality class, and Aharony (1974) used a renormalization group ϵ-expansion ($\epsilon = 4 - d$) to establish its validity to order ϵ^2.

The argument of Privman and Fisher (1984) for the truth of the hyper-universality hypothesis is also based on renormalization group ideas. They conclude that $\langle \Delta \mathfrak{q}(\boldsymbol{r}_1) \Delta \mathfrak{q}(\boldsymbol{r}_2) \rangle$ and $\Gamma_2(\mathfrak{q}; \bar{\boldsymbol{r}})$ have scaling forms which may differ in their universal scaling functions, but will have the same non-universal metric factors. Thus

$$\langle \Delta \mathfrak{q}(\boldsymbol{r}_1) \Delta \mathfrak{q}(\boldsymbol{r}_2) \rangle = B^{(\pm)} |\Delta K|^{d\nu - \gamma} Z^{(\pm)} \left(C \frac{[\Delta L - a \Delta K]}{|\Delta K|^{\beta\delta}}, \frac{\bar{\boldsymbol{r}} |\Delta K|^{\nu}}{R_0} \right), \tag{2.291}$$

and, taking the limit $|\bar{\boldsymbol{r}}| \to \infty$, it follows that

$$(\Delta \rho)^2 = B^{(\pm)} |\Delta K|^{d\nu - \gamma} Z^{(\pm)} \left(C \frac{[\Delta L - a \Delta K]}{|\Delta K|^{\beta\delta}}, \infty \right). \tag{2.292}$$

Comparing equations (2.273) and (2.292) gives

$$B^{(\pm)} = [A_1^{(\pm)}]^2 \tag{2.293}$$

and, from (2.277) and (2.284),

$$A_0 R_0^d / v_0 = Q_2^{(\pm)} / [Q_1^{(\pm)}]^2 . \tag{2.294}$$

In Sect. 3.3.2 we compute $A_0 R_0^d / v_0$ in the Gaussian approximation.

The universality hypothesis for the equation of state (2.273) is rather similar to the law of corresponding states (see Sect. 2.3.1). One difference is that now the universal scaling function is not analytic in the thermodynamic variables but in non-analytic combinations of them. It also contains non-universal metric parameters which are not simply the critical point parameters. Another important difference is that (2.273) applies to the locality of the critical point rather than to the system as a whole. In line with the quote from Levelt Sengers (1974), one may nevertheless regard the universality hypothesis as a result of the same 'unifying' impulse which led to the earlier formulation of the law of corresponding states.

2.16 Finite-Size Scaling

All the discussion in this chapter until now has assumed that the system of interest is one for which the thermodynamic limit has been taken. For the case of a one-component lattice gas the thermodynamic limit is described in greater detail in Sect. 4.2. For present purposes we suppose that we have a d-dimensional lattice system $\mathcal{N}$ of N sites, with N_ℓ lattice sites in the direction of the ℓ-th dimension of the lattice, so that

$$N = \prod_{\ell=1}^{d} N_\ell \,. \tag{2.295}$$

The thermodynamic limit when $\mathcal{N} \to \mathcal{L}$, the corresponding d-dimensional infinite system, is achieved by taking $N_\ell \to \infty$ simultaneously for every $\ell = 1, 2, \ldots, d$, in such a way that the van Hove criterion (4.12) is satisfied. Consider now the case where the thermodynamic limit is taken only for the variables N_ℓ, $\ell \le \eth < d$. This is a system which is infinite in $\eth$ dimensions and finite in $d - \eth$ dimensions. The parameter

$$\aleph = \Big[\prod_{\ell=\eth+1}^{d} N_\ell \Big]^{1/(d-\eth)} \tag{2.296}$$

can be understood as the *thickness* of the system. We denote this system by $\mathcal{L}_\eth(\aleph)$. The thermodynamic limit $\mathcal{L}_\eth(\aleph) \to \mathcal{L}$ now means that $\aleph \to \infty$ with the *shape ratios* $N_\ell/\aleph$, $\ell = \eth + 1, \ldots, d$ kept constant.

Two distinguishable phenomena are of interest for $\mathcal{L}_\eth(\aleph)$, surface effects and finite-size effects. The former, which are present only in the absence of periodic boundary conditions, concern the effect on thermodynamic and statistical mechanical properties of the presence of a surface, including the occurrence of surface phase transitions.[11] This is not, however, the subject of this section which will be solely concerned with the changes to bulk scaling which arise from the finite value of $\aleph$. This subject was initiated by Fisher (1971) and Fisher and Barber (1972).[12] We have seen in Sect. 2.9.7 that bulk scaling ideas can, with suitable modification, be applied to first-order transitions. Similar developments can also be made to finite-size scaling methods (Fisher and Berker 1982, Privman and Fisher 1983, Binder and Landau 1984, Fisher

[11] For a review, see Binder (1983).

[12] For a review see Barber (1983) and, for a collection of papers on finite-size scaling, Cardy (1988).

and Privman 1985). We shall, however, in this section restrict our attention to continuous transitions.

If $\mathfrak{d} = 0$ the system $\mathcal{L}_0(\aleph) = \mathcal{N}$ is completely finite. In terms of some suitable variable like the fugacity (see Sect. 4.1) the partition function will be a polynomial with positive coefficients and no phase transitions can occur. In spite of this such systems play an important role as approximations to infinite systems in computer simulation methods and in real-space renormalization group finite-lattice methods (see Sect. 6.11). When $\mathfrak{d} > 0$, $\mathcal{L}_{\mathfrak{d}}(\aleph)$ can be regarded as a perhaps somewhat complicated infinite lattice system of $\mathfrak{d}$ dimensions and the type of phase transition behaviour will be characterized by $\mathfrak{d}$ rather than d. For $\mathfrak{d} = 1$ and short-range interactions transfer matrix methods of the type described in Sect. 4.9 can be applied and, as will be shown in that section, no phase transitions can occur at non-zero temperature. The correlation length may, however, (for reasonably larger values of $\aleph$) attain a steep maximum at some values of the couplings. These are termed *pseudocritical values*, (Fisher 1971, Fisher and Ferdinand 1967). Many authors (Ree and Chesnut 1966, Runnels and Combs 1966, Lavis 1976) have used such maxima and similar maxima in response functions as indicating the approximate location of the phase transition in the corresponding infinite system. The pseudocritical values of the couplings will be functions of the thickness $\aleph$ of the system. When $\mathfrak{d} > 1$ the maxima associated with pseudocritical coupling values will be replaced by genuine critical values. These will be dependent on both $\aleph$ and $\mathfrak{d}$, although, according to the discussion of universality in Sects. 2.3 and 2.15, the corresponding exponents will be dependent solely on $\mathfrak{d}$.

2.16.1 The Finite-Size Scaling Field

Consider the situation described in Sect. 2.4 of a system with n couplings and a transition region $\mathcal{C}$ of dimension $n - s$, defined in terms of the scaling fields $\theta_1, \theta_2, \ldots, \theta_n$ by (2.15). The basic assumption of finite-size scaling theory is that, for any specified boundary conditions, the variable $\theta_{n+1} = 1/\aleph$ can be treated as another scaling field. The defining condition (2.15) for $\mathcal{C}$ must then be augmented by

$$\theta_{n+1} = 0\,. \tag{2.297}$$

It follows from the block-spin idea of Kadanoff, described in Sect. 2.3, that the scaling exponent y_{n+1} associated with the finite-size scaling field θ_{n+1} is unity. It is therefore a relevant scaling field. In the case $\mathfrak{d} > 1$ there is a transition region $\mathcal{C}_{\mathfrak{d}}(\aleph)$ for $\mathcal{L}_{\mathfrak{d}}(\aleph)$, which can be regarded as the projection of $\mathcal{C}$ into the extended phase space. Since $\mathfrak{d} \neq d$, the critical exponents for this projection will have different critical exponents from those of $\mathcal{C}$, which is consistent with θ_{n+1} being a relevant scaling field.

For a d-dimensionally infinite system, with $d > 1$, the singular part of the thermodynamic potential ϕ can be identified unambiguously. For a d-

dimensional system with $\mathfrak{d} < 2$ this is not the case. To cover both situations Privman and Fisher (1984) proposed that $\phi_{\text{sing}}(K_1, \ldots, L_1, \ldots, \aleph)$, the singular part of the thermodynamic potential of a system of thickness $\aleph$, is *defined* by

$$\phi_{\text{sing}}(K_1, \ldots, L_1, \ldots, \aleph) = \phi(K_1, \ldots, L_1, \ldots, \aleph) - \phi_{\text{smth}}(K_1, \ldots, L_1, \ldots), \tag{2.298}$$

where the second term on the right is the smooth component of the thermodynamic potential for the $\aleph$-infinite system. For this singular part of ϕ, and ψ obtained by differentiating it k_j times with respect to θ_j, for $j = 1, 2, \ldots, n$, (2.17) and (2.22) must be modified to give

$$\phi_{\text{sing}}(\lambda^{y_1}\theta_1, \ldots, \lambda^{y_n}\theta_n, \lambda\theta_{n+1}) = \lambda^d \phi_{\text{sing}}(\theta_1, \ldots, \theta_n, \theta_{n+1}), \tag{2.299}$$

$$\psi_{\text{sing}}(\lambda^{y_1}\theta_1, \ldots, \lambda^{y_n}\theta_n, \lambda\theta_{n+1}) = \lambda^{d'} \psi_{\text{sing}}(\theta_1, \ldots, \theta_n, \theta_{n+1}), \tag{2.300}$$

where d' is given by (2.23). In a similar way the formula (2.196) for the correlation length generalizes to

$$\mathfrak{r}(\mathfrak{q}; \lambda^{y_1}\theta_1, \ldots, \lambda^{y_n}\theta_n, \lambda\theta_{n+1}) = \lambda^{-1}\mathfrak{r}(\mathfrak{q}; \theta_1, \ldots, \theta_n, \theta_{n+1}). \tag{2.301}$$

Substituting $\lambda = 1/\theta_{n+1} = \aleph$ in equations (2.299)–(2.301) gives

$$\phi_{\text{sing}}(\theta_1, \ldots, \theta_{n+1}) = \aleph^{-d}\phi_{\text{sing}}(\theta_1\aleph^{y_1}, \ldots, \theta_n\aleph^{y_n}, 1), \tag{2.302}$$

$$\psi_{\text{sing}}(\theta_1, \ldots, \theta_{n+1}) = \aleph^{-d'}\psi_{\text{sing}}(\theta_1\aleph^{y_1}, \ldots, \theta_n\aleph^{y_n}, 1), \tag{2.303}$$

$$\mathfrak{r}(\mathfrak{q}; \theta_1, \ldots, \theta_{n+1}) = \aleph\, \mathfrak{r}(\mathfrak{q}; \theta_1\aleph^{y_1}, \ldots, \theta_n\aleph^{y_n}, 1) \tag{2.304}$$

and it is also useful to note that

$$\frac{\partial \mathfrak{r}}{\partial \theta_m}(\mathfrak{q}; \theta_1, \ldots, \theta_{n+1}) = \aleph^{1-y_m}\frac{\partial \mathfrak{r}}{\partial \theta_m}(\mathfrak{q}; \theta_1\aleph^{y_1}, \ldots, \theta_n\aleph^{y_n}, 1), \quad m = 1, 2, \ldots, n. \tag{2.305}$$

Equations (2.303) and (2.305) encapsulate the important practical value of finite-size scaling. If the transition region $\mathcal{C}$ is approached along any path for which the function on the right-hand side has a finite non-zero limit then from (2.23)

$$\lim_{\aleph \to \infty} \frac{\ln(\psi_{\text{sing}})}{\ln(\aleph)} = -d'(k_1, \ldots, k_n) = \sum_{j=1}^{n} k_j y_j - d, \tag{2.306}$$

$$\lim_{\aleph \to \infty} \frac{\ln\left(\dfrac{\partial \mathfrak{r}}{\partial \theta_m}\right)}{\ln(\aleph)} = 1 - y_m, \qquad m = 1, 2, \ldots, n. \tag{2.307}$$

From the second of equations (2.26) it can be seen that the variable $d'(k_1, \ldots, k_n)$ is the numerator of the critical exponent $\omega_m(k_1, \ldots, k_n)$ associated with the response function (or density) ψ as $\mathcal{C}$ is approached along a path $\mathcal{P}$ which satisfies (2.33) with finite values of $\varpi_j^{(m)}$, $j = 1, 2, \ldots, s$. The corresponding denominator y_m appears in formula (2.307). This means that the critical exponents for the infinite system $\mathcal{L}$ can be obtained approximately from investigating systems of finite thickness. Just as in the case of scaling for infinite systems modifications to this analysis are needed for the case of logarithmic singularities (Sect. 2.5) and when zero amplitude factors imply the need for considering corrections to scaling (Sect. 2.14).

2.16.2 The Shift and Rounding Exponents

We now investigate in more detail the relationship between the transition region $\mathcal{C}$ for the d-dimensionally infinite system and the transition region $\mathcal{C}_\eth(\aleph)$ for the system of thickness $\aleph$. For simplicity we confine our attention to the case of a critical point, with two couplings L and K, where the approach to the critical point is in the weak scaling direction. For the infinite system $\mathcal{L}$ the scaling fields θ_1 and θ_2 are given by equations (2.271) and from (2.274) the response function φ_T is given by

$$\varphi_T = -\frac{\partial^2 \phi_{\text{sing}}}{\partial \theta_1^2} \,. \tag{2.308}$$

Thus, from (2.23) and (2.303),

$$\varphi_T(\theta_1, \theta_2, \theta_3) = \aleph^{2y_1 - d} \varphi_T(\theta_1 \aleph^{y_1}, \theta_2 \aleph^{y_2}, 1) \,. \tag{2.309}$$

When $\eth > 1$, φ_T will exhibit some kind of singular behaviour in the transition region $\mathcal{C}_\eth(\aleph)$. Since $\aleph$ is finite and non-zero on $\mathcal{C}_\eth(\aleph)$ it follows that this transition region for systems of finite width must be given by some finite values ϖ_j for the arguments $\theta_j \aleph^{y_j}$, $j = 1, 2$ of the function on the right-hand side of (2.309). The transition region is, therefore, given by

$$[L_{\text{c}}(\eth; \aleph) - L_{\text{c}}] - a[K_{\text{c}}(\eth; \aleph) - K_{\text{c}}] \simeq \varpi_1 \aleph^{-y_1} \,, \tag{2.310}$$

$$K_{\text{c}}(\eth; \aleph) - K_{\text{c}} \simeq (1 + a^2)^{-1} \varpi_2 \aleph^{-y_2} \,, \tag{2.311}$$

where $L_{\text{c}}(\eth; \aleph)$ and $K_{\text{c}}(\eth; \aleph)$ are the critical values of L and K for the system of dimension $\eth$ and thickness $\aleph$. In this context the exponent y_2 which appears on the right-hand side of (2.311) is called the *shift exponent* (Fisher and Ferdinand 1967, Fisher 1971) and is usually denoted by λ. According to (2.198)

$$\lambda = y_2 = 1/\nu \,. \tag{2.312}$$

In the case $\eth = 1$, $L_{\text{c}}(\eth; \aleph)$ and $K_{\text{c}}(\eth; \aleph)$ are replaced by $L_{\max}(\eth; \aleph)$ and $K_{\max}(\eth; \aleph)$, the values where φ_T attains a maximum. A second value of K which is of interest (Fisher and Ferdinand 1967) is the *rounding value* $K^*(\eth; \aleph)$ at which φ_T for $\mathcal{L}_\eth(\aleph)$ differs significantly from its value for $\mathcal{L}$. A *rounding exponent* w can then be defined by[13]

[13] The rounding exponent is usually denoted by θ. We have used w to avoid confusion with the use of θ for scaling fields.

$$|K^*(\mathfrak{d};\aleph) - K_c| \sim \aleph^{-w}, \tag{2.313}$$

If, following Fisher and Ferdinand (1967), we suppose that this will occur when $\mathfrak{r}(\mathfrak{q}) \sim \aleph$ then, from (2.197) and (2.313),

$$w = 1/\nu\,. \tag{2.314}$$

The relationship (2.312) between the shift exponent and ν is a consequence of the finite-size scaling assumption (2.299) for the singular part of the free energy density as long as the coefficient ϖ_2 in (2.311) is non-zero. If we include the possibility that $\varpi_2 = 0$, then corrections to scaling lead to

$$\lambda \geq 1/\nu\,. \tag{2.315}$$

For the two-dimensional Ising model (Ferdinand and Fisher 1969, Au-Yang and Fisher 1975) $\lambda = w = 1/\nu = 1$. Fisher and Ferdinand (1967) have, however, argued for the possibility that $\lambda > 1/\nu$ in three dimensions and the results of Barber and Fisher (1973a, 1973b) for the spherical model and ideal Bose gas respectively in d dimensions with layer geometries and open boundary conditions gave $\lambda = 1 > 1/\nu$. This was later understood to be a pathological result arising from constraints which give an effective long-range interaction (Brézin 1983) and (2.312) is now thought to be generally true (Barber 1983).

As indicated above when $\mathfrak{d} > 1$ the transition regions $\mathcal{C}$ and $\mathcal{C}_{\mathfrak{d}}(\aleph)$ are distinct but connected in phase space. In this sense the situation is somewhat similar to the relationship between the first-order line and the critical point in the ferromagnetic Ising model. A general situation of this kind is analyzed using effective exponents at the end of Sect. 2.4 and applied to the critical point and coexistence curve in Sect. 2.9. If, in (2.299), we set λ according to (2.24) then we obtain the modification

$$\phi_{\text{sing}}(\theta_1,\ldots,\theta_{n+1}) = |\theta_m|^{d/y_m}\phi_{\text{sing}}(g_1^{(m)},\ldots,g_{n+1}^{(m)}) \tag{2.316}$$

of the first of equations (2.25), where $g_j^{(m)}$, $j = 1,2,\ldots,n$ are given by equations (2.27) and (2.28) and

$$g_{n+1}^{(m)} = \frac{\theta_{n+1}}{|\theta_m|^{1/y_m}}\,. \tag{2.317}$$

For an approach to the transition region $\mathcal{C}$ all the scaling fields tend to zero and the asymptotic behaviour of ϕ_{sing} is given by the leading factor on the right of (2.317) if all arguments $g_j^{(m)}$, $j = 1,\ldots,n+1$ attain finite values with a finite non-zero value for the function. If, however, the approach is to $\mathcal{C}_{\mathfrak{d}}(\aleph)$, $\aleph$ remains finite as $g_{n+1}^{(m)} \to \infty$. If then

$$\phi_{\text{sing}}(g_1^{(m)},\ldots,g_{n+1}^{(m)}) \sim |g_{n+1}^{(m)}|^{\mu_{n+1}^{(m)}}, \tag{2.318}$$

it follows, from (2.316), that

$$\phi_{\text{sing}}(\theta_1,\ldots,\theta_{n+1}) \sim |\theta_m|^{(d-\mu_{n+1}^{(m)})/y_m}\,. \tag{2.319}$$

For consistency in the scaling theories for $\mathcal{L}$ and $\mathcal{L}_{\mathfrak{d}}(\aleph)$ we must have

$$\mathfrak{d} = d - \mu_{n+1}^{(m)} . \tag{2.320}$$

2.16.3 Universality and Finite-Size Scaling

As in Sect. 2.15 we concentrate on the case of a critical point. From (2.299) and (2.301), the scaling forms (2.272) and (2.286) are replaced by

$$\phi_{\text{sing}} = A_0|\Delta K|^{2-\alpha}\mathring{W}_0^{(\pm)}\left(C\,\frac{[\Delta L - a\Delta K]}{|\Delta K|^{\beta\delta}}, \frac{D}{|\Delta K|^{\nu}\aleph}\right) , \tag{2.321}$$

$$\mathfrak{r}(\mathfrak{q}) = R_0 Q_3^{(\pm)}|\Delta K|^{-\nu}\mathring{Y}^{(\pm)}\left(C\,\frac{[\Delta L - a\Delta K]}{|\Delta K|^{\beta\delta}}, \frac{D}{|\Delta K|^{\nu}\aleph}\right) , \tag{2.322}$$

with the bulk forms being recovered, in the limit $\aleph \to \infty$, with

$$\mathring{W}_0^{(\pm)}(x,0) = W_0^{(\pm)}(x) , \qquad \mathring{Y}^{(\pm)}(x,0) = Y^{(\pm)}(x) . \tag{2.323}$$

When the Josephson hyper-scaling law (2.199) is valid, (2.321) can be rescaled.

$$\begin{aligned}\phi_{\text{sing}} &= \frac{A_0 D^d}{\aleph^d}\left[\frac{|\Delta K|\aleph^{\frac{1}{\nu}}}{D^{\frac{1}{\nu}}}\right]^{2-\alpha} \\ &\times \mathring{W}_0^{(\pm)}\left(\left[\frac{C}{D^{\frac{\beta\delta}{\nu}}}\right][(\Delta L - a\Delta K)\aleph^{\frac{\beta\delta}{\nu}}]\left[\frac{|\Delta K|\aleph^{\frac{1}{\nu}}}{D^{\frac{1}{\nu}}}\right]^{-\beta\delta}, \left[\frac{|\Delta K|\aleph^{\frac{1}{\nu}}}{D^{\frac{1}{\nu}}}\right]^{-\nu}\right) ,\end{aligned} \tag{2.324}$$

gives

$$\phi_{\text{sing}} = A_0 D^d \aleph^{-d} U(C'[\Delta L - a\Delta K]\aleph^{\frac{\beta\delta}{\nu}}, D'\Delta K\aleph^{\frac{1}{\nu}}) , \tag{2.325}$$

where

$$U(x, \pm y) = y^{2-\alpha}\mathring{W}_0^{(\pm)}(xy^{-\beta\delta}, y^{-\nu}) , \tag{2.326}$$

$$C' = C/D^{\frac{\beta\delta}{\nu}} , \qquad D' = 1/D^{\frac{1}{\nu}} . \tag{2.327}$$

In a similar way (2.322) can be re-expressed in the form

$$\mathfrak{r}(\mathfrak{q}) = \frac{R_0 Q_3^{(\pm)}}{D}\aleph V(C'[\Delta L - a\Delta K]\aleph^{\frac{\beta\delta}{\nu}}, D'\Delta K\aleph^{\frac{1}{\nu}}) , \tag{2.328}$$

where

$$V(x, \pm y) = y^{-\nu}\mathring{Y}^{(\pm)}(xy^{-\beta\delta}, y^{-\nu}) . \tag{2.329}$$

Equations (2.325) and (2.328) could be derived directly from (2.302) and (2.304) except that then no relationship would be established with the metric factors in (2.321) and (2.322).

As in the case of the d-dimensionally infinite system discussed in Sect. 2.15 the scaling functions $\mathring{W}_0^{(\pm)}(x,y)$, $\mathring{Y}^{(\pm)}(x,y)$, $U(x,y)$ and $V(x,y)$ are supposed to be universal with non-universal metric factors A_0, C, D and R_0. However, we can in this case go further if we adopt a standard assumption of finite-size scaling theory (Barber 1983) that all lengths which diverge at bulk criticality should scale like the bulk correlation length. This means that the second argument in (2.321) and (2.322) $D/|\Delta K|^{\nu}\aleph \simeq \mathfrak{r}(\mathfrak{q})/v_0^{\frac{1}{d}}\aleph$. From (2.286) $D \simeq R_0 Q_3^{(\pm)}/v_0^{\frac{1}{d}}$. Substituting into (2.325) and (2.328) and using the hyper-universality formula (2.294) gives

$$\phi_{\text{sing}} = E_0 \aleph^{-d} U(C'[\Delta L - a\Delta K]\aleph^{\frac{\beta\delta}{\nu}}, D'\Delta K \aleph^{\frac{1}{\nu}})\,, \tag{2.330}$$

$$\mathfrak{r}(\mathfrak{q}) = v_0^{\frac{1}{d}} \aleph V(C'[\Delta L - a\Delta K]\aleph^{\frac{\beta\delta}{\nu}}, D'\Delta K \aleph^{\frac{1}{\nu}}), \tag{2.331}$$

where

$$E_0 = A_0 D^d = Q_2^{(\pm)}[Q_3^{(\pm)}]^d/[Q_1^{(\pm)}]^2\,. \tag{2.332}$$

That the scaling forms (2.330) and (2.331) have leading universal amplitude factors is a result derived by Privman and Fisher (1984). At the bulk critical point $\Delta L = \Delta K = 0$ (2.330) and (2.331) becomes

$$\phi_{\text{sing}} = \aleph^{-d} E_0 U(0,0)\,, \tag{2.333}$$

$$\mathfrak{r}(\mathfrak{q})/v_0^{\frac{1}{d}} = \aleph V(0,0)\,. \tag{2.334}$$

As we shall see in Sect. 2.17 the value of the universal constants $V(0,0)$ and $E_0 U(0,0)$ can in the case $d = 2$, $\mathfrak{d} = 1$ be derived from conformal invariance.

2.17 Conformal Invariance

As indicated in the introduction to this chapter the application of conformal invariance to critical phenomena began with the work of Polyakov (1970). Since then it has developed into a powerful tool which provides results beyond those given by scaling theory. In this section we indicate briefly some of the main features of the theory and prove a famous result due to Cardy (1984), which establishes the value of the universal constant $V(0,0)$ of equation (2.334) in two dimensions.[14]

[14] For more detailed discussions of conformal invariance the reader is referred to the review of Cardy (1987) and the book of Christe and Henkel (1993).

2.17.1 From Scaling to the Conformal Group

In Sect. 1.5 we considered simplifications which can arise in lattice systems when translational and rotational invariance are present. Such global properties arise from the nature of the Hamiltonian and their presence may be qualified by the existence of a sublattice structure or anisotropic interactions.

The scaling transformation which, for scaling fields and operators is characterized by

$$\begin{array}{llll} \boldsymbol{r} & \to \tilde{\boldsymbol{r}} & = & \lambda^{-1}\boldsymbol{r}\,, \\ \theta_j & \to \tilde{\theta}_j & = & \lambda^{y_j}\theta_j\,, \\ \phi_j(\boldsymbol{r}) & \to \phi_j(\tilde{\boldsymbol{r}}) & = & \lambda^{x_j}\phi_j(\boldsymbol{r})\,, \end{array} \tag{2.335}$$

where, as in (2.78), the scaling exponent y_j of θ_j and the scaling dimension x_j of ϕ_j are related by $x_j = d - y_j$, is of a rather different type. Applying only as an asymptotic property near to critical regions, its motivation is given an intuitive basis in Sect. 2.3 and a more mathematical foundation is Sects. 2.4–2.6. The salient feature for present purposes is the transformation property of the singular part (2.72) of the Hamiltonian. This is given by

$$\begin{aligned} \widehat{H}_{\text{sing}}(\theta_1,\dots,\theta_n;\phi_1(\boldsymbol{r}_1),\dots,\phi_s(\boldsymbol{r}_s);\boldsymbol{\sigma}) &= \sum_{j=1}^{s}\theta_j\sum_{\{\boldsymbol{r}\}}\phi_j(\boldsymbol{r}) \\ \to \sum_{j=1}^{s}\tilde{\theta}_j\sum_{\{\tilde{\boldsymbol{r}}\}}\phi_j(\tilde{\boldsymbol{r}}) &= \sum_{j=1}^{s}\lambda^{y_j}\theta_j\lambda^{-d}\sum_{\{\boldsymbol{r}\}}\phi_j(\lambda^{-1}\boldsymbol{r}) \\ &= \sum_{j=1}^{s}\theta_j\lambda^{x_j+y_j-d}\sum_{\{\boldsymbol{r}\}}\phi_j(\boldsymbol{r}) \\ &= \widehat{H}_{\text{sing}}(\theta_1,\dots,\theta_n;\phi_1(\boldsymbol{r}_1),\dots,\phi_s(\boldsymbol{r}_s);\boldsymbol{\sigma})\,, \end{aligned} \tag{2.336}$$

where λ^{-d} enters into this analysis as the Jacobian of the transformation $\boldsymbol{r}_j \to \lambda^{-1}\boldsymbol{r}_j$.

As we have seen in Sects. 2.9 and 2.13, the scaling theory outlined here yields the scaling laws for critical and tricritical points. Stronger assumptions are needed to give the possibility of extracting further information. Suppose that the elements of the scaling transformation (2.335) are 'localized' to give

$$\begin{array}{llll} \boldsymbol{r} & \to \tilde{\boldsymbol{r}} & = & [\lambda(\boldsymbol{r})]^{-1}\boldsymbol{r}\,, \\ \theta_j & \to \tilde{\theta}_j & = & [\lambda(\boldsymbol{r})]^{y_j}\theta_j\,, \\ \phi_j(\boldsymbol{r}) & \to \phi_j(\tilde{\boldsymbol{r}}) & = & [\lambda(\boldsymbol{r})]^{x_j}\phi_j(\boldsymbol{r})\,. \end{array} \tag{2.337}$$

This modification to scaling, which can be justified with the same kind of argument that was used in Sect. 2.3, will leave (2.335) intact with $\lambda(\boldsymbol{r})$ replacing λ as long as the spatial transformation in the first line of (2.337) has the Jacobian $[\lambda(\boldsymbol{r})]^{-d}$. It is not difficult to confirm that this will be the case if either $\lambda(\boldsymbol{r})$ is a constant, which is the usual dilatation transformation of scaling, or $\lambda(\boldsymbol{r}) = 1/|\boldsymbol{r}|^2$, which is the inversion transformation. As we describe in Appendix A.2 the group generated by translations, rotations, dilatations and inversions is the *conformal group.*

2.17.2 Correlation Functions for $d \geq 2$

Given that the transformation (2.337) belongs to the conformal group, the analysis of Sect. 2.6 leading to the scaling form (2.84) for the total correlation function is not affected by the Jacobian having a spatial dependence and thus we now have

$$\langle \phi_1(\tilde{\boldsymbol{r}}_1) \cdots \phi_\tau(\tilde{\boldsymbol{r}}_\tau)\rangle = \left\{ \prod_{j=1}^{\tau} [\lambda(\boldsymbol{r}_j)]^{x_j} \right\} \langle \phi_1(\boldsymbol{r}_1) \cdots \phi_\tau(\boldsymbol{r}_\tau)\rangle \,. \tag{2.338}$$

In the case of the two-point correlation function, rotation and translation can be used to interchange $\boldsymbol{r}_1$ and $\boldsymbol{r}_2$. Thus we have

$$\begin{aligned} \langle \phi_1(\tilde{\boldsymbol{r}}_1)\phi_2(\tilde{\boldsymbol{r}}_2)\rangle &= [\lambda(\boldsymbol{r}_1)]^{x_1}[\lambda(\boldsymbol{r}_2)]^{x_2}\langle \phi_1(\boldsymbol{r}_1)\phi_2(\boldsymbol{r}_2)\rangle \,, \\ \langle \phi_1(\tilde{\boldsymbol{r}}_2)\phi_2(\tilde{\boldsymbol{r}}_1)\rangle &= [\lambda(\boldsymbol{r}_2)]^{x_1}[\lambda(\boldsymbol{r}_1)]^{x_2}\langle \phi_1(\boldsymbol{r}_2)\phi_2(\boldsymbol{r}_1)\rangle \end{aligned} \tag{2.339}$$

with

$$\begin{aligned} \langle \phi_1(\tilde{\boldsymbol{r}}_1)\phi_2(\tilde{\boldsymbol{r}}_2)\rangle &= \langle \phi_1(\tilde{\boldsymbol{r}}_2)\phi_2(\tilde{\boldsymbol{r}}_1)\rangle \,, \\ \langle \phi_1(\boldsymbol{r}_1)\phi_2(\boldsymbol{r}_2)\rangle &= \langle \phi_1(\boldsymbol{r}_2)\phi_2(\boldsymbol{r}_1)\rangle \,. \end{aligned} \tag{2.340}$$

For a non-zero two-point correlation function it would appear that these results must be inconsistent unless $x_1 = x_2$, leading to the conclusion that the two-point total correlation function of two operators with different scaling dimensions is zero. In fact it is clear that this conclusion must be modified to take into account two-point correlation functions like

$$\langle \phi(\boldsymbol{r}_1)\nabla_{\boldsymbol{r}_2}\phi(\boldsymbol{r}_2)\rangle = \nabla_{\boldsymbol{r}_2}\langle \phi(\boldsymbol{r}_1)\phi(\boldsymbol{r}_2)\rangle \,, \tag{2.341}$$

which will not, in general, be zero in spite of the fact that the scaling dimensions differ by two. This argument has been analyzed by Schäfer (1976) (see also, Cardy 1987) who concluded that

$$\langle \phi_1(\boldsymbol{r}_1)\phi_2(\boldsymbol{r}_2)\rangle = 0 \,, \qquad \text{if } x_1 - x_2 \neq \text{integer.} \tag{2.342}$$

When x_1 and x_2 differ by an integer, the two-point correlation is non-zero in some, but not all, cases (Schäfer 1976). In particular the formula (2.85) is exact within the context of conformal invariance for $\tau = 2$.

$$\langle \phi(\boldsymbol{r}_1)\phi(\boldsymbol{r}_2)\rangle = \frac{C_2(\phi,\phi)}{|\boldsymbol{r}_2 - \boldsymbol{r}_1|^{2x}} \,. \tag{2.343}$$

In the case of the three-point correlation function, the operations of the conformal group can be used to map three arbitrary points $\boldsymbol{r}_1$, $\boldsymbol{r}_2$ and $\boldsymbol{r}_3$ into any three chosen points $\boldsymbol{\alpha}$, $\boldsymbol{\beta}$, γ It has been shown by Polyakov (1970) that

$$\langle \phi_1(\boldsymbol{r}_1)\phi_2(\boldsymbol{r}_2)\phi_3(\boldsymbol{r}_3)\rangle = \frac{C_3(\phi_1,\phi_2,\phi_3)}{|\boldsymbol{r}_1-\boldsymbol{r}_2|^{x_1+x_2-x_3}|\boldsymbol{r}_2-\boldsymbol{r}_3|^{x_2+x_3-x_1}|\boldsymbol{r}_3-\boldsymbol{r}_1|^{x_3+x_1-x_2}}, \quad (2.344)$$

where $C_3(\phi_1,\phi_2,\phi_3)$ is a constant dependent only on the choice of scaling operators. A expression can also be obtained for the four-point correlation function, although since it is not possible to map four arbitrary points into four prescribed points with the operations of the conformal group an arbitrary function is involved (Polyakov 1970).

2.17.3 Universal Amplitudes for $d = 2$

We now consider the square lattice with lattice spacing a. In two dimensions the conformal group is infinite dimensional and consists, for $\boldsymbol{r} = a\,(\mathsf{x},\mathsf{y})$, of transformations given by analytic functions of the complex variable $\mathsf{z} = \mathsf{x}+i\mathsf{y}$ (see Appendix A.2). For the transformation $\mathsf{z} \to \mathsf{Z}(\mathsf{z})$ the Jacobian is $|\mathsf{Z}'(\mathsf{z})|^2$. So, for the two-point correlation function of the scaling operator ϕ_1, (2.338) gives

$$\langle \phi(\mathsf{z}_1)\phi(\mathsf{z}_2)\rangle = |\mathsf{Z}'(\mathsf{z}_1)|^x|\mathsf{Z}'(\mathsf{z}_2)|^x\langle\phi(\mathsf{Z}(\mathsf{z}_1))\phi(\mathsf{Z}(\mathsf{z}_2))\rangle\,. \quad (2.345)$$

The transformation given by

$$\mathsf{Z}(\mathsf{z}) = \frac{\aleph}{2\pi}\ln(\mathsf{z})\,, \quad (2.346)$$

is analytic for all $\mathsf{z} \neq 0$ and maps the z-plane onto the strip $0 \le \tilde{\mathsf{y}} < \aleph$. Thus, from (2.343), (2.345) and (2.346), the connected pair correlation function on the strip of width $\aleph$ lattice sites with periodic boundary conditions is

$$\langle \phi(\tilde{\boldsymbol{r}})\phi(\tilde{\boldsymbol{r}}_2)\rangle = \frac{C_2(\phi,\phi)(2\pi/\aleph a)^{2x}}{\left\{2\cosh\left[\frac{2\pi(\tilde{\mathsf{x}}_2-\tilde{\mathsf{x}}_1)}{\aleph}\right] - 2\cos\left[\frac{2\pi(\tilde{\mathsf{y}}_2-\tilde{\mathsf{y}}_1)}{\aleph}\right]\right\}^x}, \quad (2.347)$$

where $\tilde{\boldsymbol{r}} = a(\tilde{\mathsf{x}},\tilde{\mathsf{y}})$. When $|\tilde{\mathsf{z}}_2 - \tilde{\mathsf{z}}_1| \ll \aleph$,

$$\langle \phi(\tilde{\boldsymbol{r}})\phi(\tilde{\boldsymbol{r}}_2)\rangle \simeq \frac{C_2(\phi,\phi)}{|\tilde{\boldsymbol{r}}_1-\tilde{\boldsymbol{r}}_2|^{2x}}, \quad (2.348)$$

but, when $|\tilde{\mathsf{x}}_2 - \tilde{\mathsf{x}}_1| \gg \aleph$,

$$\langle \phi(\tilde{\boldsymbol{r}})\phi(\tilde{\boldsymbol{r}}_2)\rangle \simeq C_2(\phi,\phi)\left(\frac{2\pi}{\aleph a}\right)^{2x}\exp\left[-\frac{2\pi x(\tilde{\mathsf{x}}_2-\tilde{\mathsf{x}}_1)}{\aleph}\right]. \quad (2.349)$$

We now consider the case of finite-size scaling at a critical point discussed in Sect. 2.16. If the leading behaviour of the state operator $\mathfrak{q}$ is given by the scaling operator ϕ with $\langle\phi(\boldsymbol{r})\rangle = 0$ at the critical point, (2.349) can be substituted into (2.6) (with σ replaced by $\mathfrak{q}$) to give the correlation length $\mathfrak{r}(\mathfrak{q})$. In the limit of small lattice spacing the sums become integrals and

$$\mathfrak{r}^2(\mathfrak{q}) = \frac{1}{2}a^2 \frac{\displaystyle\int_0^\infty r^2 \exp\left(-\frac{2\pi x r}{\aleph}\right) \mathrm{d}r}{\displaystyle\int_0^\infty \exp\left(-\frac{2\pi x r}{\aleph}\right) \mathrm{d}r}, \tag{2.350}$$

giving

$$\mathfrak{r}(\mathfrak{q}) = \frac{\aleph a}{2\pi x}. \tag{2.351}$$

This result was first obtained by Cardy (1984). Substituting into (2.334) with $v_0^{\frac{1}{d}} = a$ gives

$$V(0,0) = \frac{1}{2\pi x}. \tag{2.352}$$

The detailed analysis which leads to the determination of the critical amplitude $E_0 U(0,0)$ in (2.333), for $d = 2$, is beyond the scope of this book. We simply state the result, obtained by Blöte et al. (1986) and Affleck (1986), that

$$E_0 U(0,0) = -\frac{1}{6}\pi \mathrm{c}, \tag{2.353}$$

where c is the *conformal anomaly number* defined by $\frac{1}{2}\mathrm{c}$ being the coefficient of the leading term of the operator product expansion of the stress tensor with itself.[15] From this it can be shown that c is the central charge in the Virasoro algebra of the generators of conformal transformations. For models where the transfer matrix (see Sect. 4.9) is symmetric (or can be made symmetric at criticality) the Virasoro algebra is unitary and, for $\mathrm{c} < 1$, the only allowed values are given by

$$\mathrm{c} = 1 - \frac{6}{m(m+1)}, \qquad m = 3, 4, 5, \ldots. \tag{2.354}$$

If a particular model can be identified with a value of m and c the scaling dimensions of its primary scalar operators, indexed by i and j with $1 \le j \le i \le m-1$, are given by the *Kac formula*

$$x_{ij}(m) = \frac{[i(m+1) - jm]^2 - 1}{2m(m+1)}. \tag{2.355}$$

The (critical) Ising model has been identified with the case $m = 3$, the tricritical Ising model with $m = 4$ and the 3-state Potts model with $m = 5$ For the Ising model the three allowed pairs of values for the parameters are:

$i = j = 1$ *giving* $x_{11} = 0$. This corresponds to an operator present in all models. It is similar to the trivial coupling in renormalization theory (Sect. 5.1).

[15] For a full explanation of this and all the following discussion see Cardy (1987).

$i = 2,\ j = 1$ *giving* $x_{21} = 1$. This gives the scaling exponent $y_{21} = d - x_{21} = 1$. By identifying y_{21} with the scaling exponent y_2 of Sect. 2.9 and using (2.167) we obtain the value $\alpha = 0$, in agreement with the exact result (5.126). Since by the fluctuation-response function relation (1.62) the heat capacity is associated with the energy-energy correlation function and we are able to identify the operator as the energy density.

$i = j = 2$ *giving* $x_{22} = \frac{1}{8}$. This gives the scaling exponent $y_{22} = d - x_{22} = \frac{15}{8}$. By identify y_{22} with the exponent y_1 of Sect. 2.9 and using (2.162) we obtain the value $\gamma = \frac{7}{4}$ in agreement with the exact result (5.126). In this case the operator is the magnetization density.

In the cases of the tricritical Ising model and 3-state Potts model there are a larger number of allowed parameter values. It is only with the use of additional information that it becomes possible to identify the physical nature of the operators. Thus, for example, for the 3-state Potts model the known values $\alpha = \frac{1}{3}$ and $\gamma = \frac{13}{9}$ (see, for example, Volume 1, Sect. 10.14) allows the identification of the cases $i = 2,\ j = 1\ x_{21} = \frac{4}{5}$ and $i = j = 3,\ x_{33} = \frac{2}{15}$ with the energy density and magnetization density respectively.

Examples

2.1 Let θ_j and θ_m be two relevant scaling fields as defined for the Kadanoff scaling hypothesis (2.17). Show that

(a) With θ'_j defined by (2.32)

$$\lambda^{y_j}\theta'_j = \lambda^{y_j}\theta_j - \varpi|\lambda^{y_m}\theta_m|^{\vartheta_j^{(m)}} .$$

This means that the change of variable $\theta_j \to \theta'_j$ can be made in (2.17) without affecting the exponents.

(b) With $\varpi_j^{(m)}$ defined by (2.33), show that, if $\varpi_j^{(m)}$ is finite and $\varpi_\ell^{(m)}$ is infinite, for y_ℓ relevant, then $\varpi_j^{(\ell)}$ and $\varpi_m^{(\ell)}$ are finite. Use this result to complete the argument showing that it is possible to chose a relevant scaling field θ_ℓ so that $\varpi_j^{(\ell)}$ is finite for all $j = 1, 2, \ldots, s$.

2.2 Consider a magnetic system with independent fields $\mathcal{H}$ and T, which has a critical point on the zero-field axis at $T = T_c$ terminating a line of first-order transitions along $\mathcal{H} = 0$ for $T < T_c$. In this case the scaling fields can be defined, as in Sect. 2.10, by

$$\theta_1 = L\,, \qquad \theta_2 = \Delta K = K - K_c\,.$$

Work through the development of scaling theory for this case (assuming that $y_1 > y_2$), showing that $\sigma = \alpha$ and proving the scaling laws (2.169) and (2.170).

2.3 The 'obvious' way to obtain the exponent expressions (2.188) is to use equations (2.185)–(2.187) together with the scaling hypothesis (2.133). Alternatively, the formula (2.182) can be treated as a generalized homogeneous function $\phi_{\text{sing}}(\theta_2, m)$ with

$$\phi_{\text{sing}}(\lambda^{y_2}\theta_2, \lambda^z m) = \lambda^d \phi_{\text{sing}}(\theta_2, m).$$

This gives both y_2 and z in terms of d. The form of y_1 in terms of d can then be obtained from (2.183). Show, by using this method, that if

$$\phi_{\text{sing}}(\theta_2, m) = \frac{1}{2}m^2 \left(\frac{\theta_2}{2K_c} - \frac{1}{6}m^2(3 - 2\alpha)\right) \left(\frac{1}{6}m^2 - \frac{\theta_2}{K_c}\right)^{-\alpha},$$

$$0 = \theta_1 + m \left(\frac{\theta_2}{K_c} - \frac{1}{6}m^2(2 - \alpha)\right) \left(\frac{1}{6}m^2 - \frac{\theta_2}{K_c}\right)^{-\alpha},$$

with $\alpha \leq 0$, then $\beta = \frac{1}{2}$, $\gamma = 1 - \alpha$ and $\delta = 3 - 2\alpha$.

3. Landau and Landau–Ginzburg Theory

3.1 The Ferromagnetic Ising Model

As an introduction to the methods of Landau theory we consider the simplest version of the Ising model ferromagnet described in Sect. 2.3 with Hamiltonian given by (1.49). The initial step in the development of Landau theory is essentially the same as in any other version of mean-field theory (see, for example, Volume 1, Sect. 3.1). This leads (Example 3.1) to the replacement of the Hamiltonian (1.49) by

$$\hat{H}(\mathcal{H}, J; \hat{m}) = -N\left(\tfrac{1}{2}zJ\hat{m}^2 + \mathcal{H}\hat{m}\right), \tag{3.1}$$

where $\hat{m} = \widehat{\mathcal{M}}/N$, with $\widehat{\mathcal{M}}$ given by (1.46), and z is the coordination number of the lattice. The partition function Z, which is a function of $L = \mathcal{H}/T$ and $K = J/T$, is

$$Z(L, K) = \sum_{\{\sigma(\boldsymbol{r})\}} \exp[-\widehat{H}(\mathcal{H}, J; \hat{m})/T]. \tag{3.2}$$

We define the *Landau partition function*

$$Z_{\mathrm{L}}(L, K; m) = \sum_{\{\sigma(\boldsymbol{r})\}} \delta^{\mathrm{Kr}}(m - \hat{m}) \exp[-\widehat{H}(\mathcal{H}, J; \hat{m})/T]. \tag{3.3}$$

This contains the subset of the terms of Z, for which the spin states are restricted by the condition $\hat{m} = m$. It is, of course, the case that

$$Z(L, K) = \sum_{\{m\}} Z_{\mathrm{L}}(L, K; m), \tag{3.4}$$

with the summation over all allowed values of m. The next assumption of Landau theory, again following the usual development of mean-field theory, is to suppose that the sum in (3.4) is dominated by its largest term. We then define the *Landau potential*

$$\phi_{\mathrm{L}}(L, K; m) = -\frac{1}{N} \ln Z_{\mathrm{L}}(L, K; m). \tag{3.5}$$

It follows from our assumptions that the thermodynamic potential $\phi(L, K)$ is given by minimizing $\phi_{\mathrm{L}}(L, K; m)$ with respect to m.

We now consider the possible form of ϕ_{L} consistent with the symmetry properties of the Hamiltonian (3.1). It is clear that, in the absence of a magnetic field ($L = 0$), the Hamiltonian is invariant under inversion of all the spins on the lattice. If, on the other hand, the magnetic field is non-zero the inversion of all the spins *and* the magnetic field leaves the Hamiltonian invariant. These symmetry properties of the Hamiltonian, when operating on the spins, or $\hat{m}$, are carried through to ϕ_{L}, when operating on m. Representing spin inversion by the operator $\mathrm{Q}^{(m)}$ and field inversion by $\mathrm{Q}^{(L)}$ we have

$$\begin{aligned} \mathrm{Q}^{(m)}\phi_{\mathrm{L}}(L,K;m) &= \phi_{\mathrm{L}}(L,K;-m)\,, \\ \mathrm{Q}^{(L)}\phi_{\mathrm{L}}(L,K;m) &= \phi_{\mathrm{L}}(-L,K;m)\,, \end{aligned} \tag{3.6}$$

and the symmetry properties, described above, are then given by

$$\begin{aligned} \mathrm{Q}^{(m)}\phi_{\mathrm{L}}(0,K;m) &= \phi_{\mathrm{L}}(0,K;m)\,, \\ \mathrm{Q}^{(L)}\mathrm{Q}^{(m)}\phi_{\mathrm{L}}(L,K;m) &= \phi_{\mathrm{L}}(L,K;m)\,. \end{aligned} \tag{3.7}$$

The form of ϕ_{L} must be such that L and m appear only in combinations which satisfy the invariance conditions (3.7). For $L = 0$, ϕ_{L} can contain only even powers of m. If $L \neq 0$, it would appear from the symmetry properties that all terms of the form $L^i m^j$, where $i + j$ is an even integer, can occur. It must, however, be remembered that m is given by the equilibrium condition

$$\frac{\partial \phi_{\mathrm{L}}}{\partial m} = 0 \tag{3.8}$$

and, from (2.105) (where ρ is now m),

$$\begin{aligned} m &= -\left(\frac{\partial \phi_{\mathrm{L}}}{\partial L}\right)_K \\ &= -\left(\frac{\partial \phi_{\mathrm{L}}}{\partial L}\right)_{K,m} - \frac{\partial \phi_{\mathrm{L}}}{\partial m}\left(\frac{\partial m}{\partial L}\right)_K \\ &= -\left(\frac{\partial \phi_{\mathrm{L}}}{\partial L}\right)_{K,m}. \end{aligned} \tag{3.9}$$

The only L-dependent term in ϕ_{L} must, therefore, be of the form $-Lm$ and

$$\phi_{\mathrm{L}}(L,K;m) = -Lm + B_0(K) + \sum_{i=1}^{\infty} B_{2i}(K)m^{2i}\,. \tag{3.10}$$

It should be noted that the only use we have made of the Ising model ferromagnet Hamiltonian (3.1) in the derivation of (3.10) is to ensure that the Landau potential (3.10) satisfies the correct symmetry properties. The formula (3.10) is, therefore, more general than (3.1). Other terms with the same symmetry (e.g. a quartic term in $\hat{m}$) could be added to (3.10) without affecting the form of the Landau potential.

3.2 Landau Theory for a Critical Point

The expansion (3.10) is capable of producing all manner of phase behaviour depending on the relative magnitudes of the coefficients $B_{2i}(K)$ and the form of their dependence on K. The simplest case of any interest is when the expansion is dominated by its leading terms, such that

$$\phi_{\mathrm{L}}(L, K; m) \simeq -Lm + B_0(K) + B_2(K)m^2 + B_4(K)m^4. \tag{3.11}$$

The equilibrium conditions for minimizing ϕ_{L}, given by (3.11) as a function of m, are then

$$2B_2(K)m + 4B_4(K)m^3 - L = 0\,, \tag{3.12}$$

$$2B_2(K) + 12B_4(K)m^2 > 0\,. \tag{3.13}$$

Equations (3.11)–(3.13), with explicit expressions for $B_2(K)$ and $B_4(K)$, are equivalent to those considered in Sect. 2.10. The analysis could, therefore, proceed in the same way as it did in that section. The cubic equation (3.12) could be solved and the appropriate root substituted in (3.11). It is, however, illuminating to pursue a slightly different approach. Consider first the zero-field case ($L = 0$). Equation (3.11) has two solutions $m = 0$ and

$$m^2 = -\frac{B_2(K)}{2B_4(K)}\,. \tag{3.14}$$

A magnetized state can occur only if $B_2(K)$ and $B_4(K)$ have opposite signs and substituting (3.14) into (3.13) it follows that a stable equilibrium solution of this kind will need $B_2(K) < 0$ and $B_4(K) > 0$. On the other hand a non-magnetized state is stable if $B_2(K) > 0$. Thus, for a magnetized state to develop with zero field at $K = K_{\mathrm{c}}$, this must the value at which $B_2(K)$ changes sign. In order for the model to represent the physical situation of ferromagnetism, the non-magnetized state must occur at high temperature (low K), with the onset of magnetism as K is increased. The simplest, although not unique way, to achieve this is to suppose that, to leading order in $\Delta K = K - K_{\mathrm{c}}$,

$$B_2(K) \simeq -A_2\Delta K\,, \qquad B_4(K) \simeq A_4\,, \tag{3.15}$$

where A_2 and A_4 are positive constants and curves of the general form of $\phi_{\mathrm{L}}(0, K; m)$ plotted against m for different values of K are shown in Fig. 3.1.

From equation (3.12) together with the conditions (3.15) it is simple to derive the classical (mean-field) values for the critical exponents β, γ and δ and the value for α follows if it is assumed that $B_0(K)$ is a smooth function of K in a neighbourhood of the critical point (see, for example, Landau and Lifschitz 1980). The terms Landau theory and mean-field theory are sometimes used as synonyms. We shall, however, restrict the name 'Landau theory' to formulations for which the thermodynamic potential is given near a critical or coexistence region as a power series, determined only by a particular group

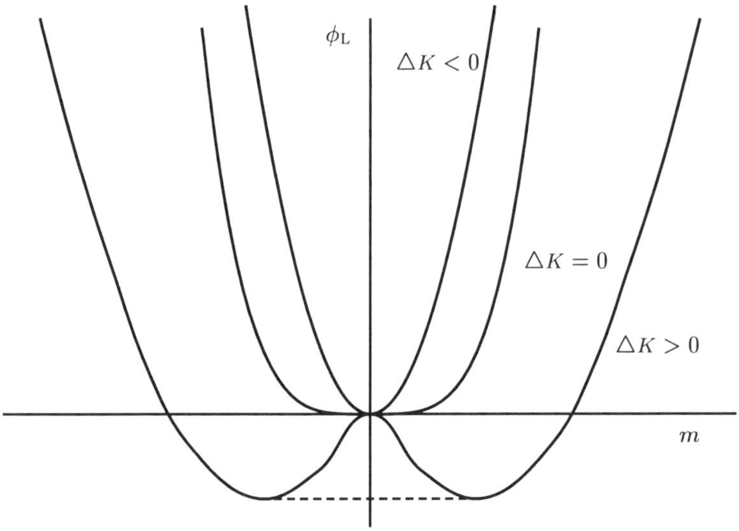

Fig. 3.1. Curves of $\phi_L(0, K; m)$ plotted against m for a temperatures above, at and below the critical point, ($\Delta K < 0$, $\Delta K = 0$ and $\Delta K > 0$ respectively).

of symmetry operations. 'Mean-field theory' will be used, as in Volume 1, to denote approximation methods applied to particular models, in which such a power series is obtained as well as the phase diagram.

Landau theory and scaling theory can both be seen as methods which extract information about a whole class of physical systems, based on a minimum of assumptions about their physical structure. The difference between the two is, of course, that, while scaling theory is capable of determining only scaling law relationships between critical exponents, Landau theory allows actual values for the exponents to be obtained. It is an illuminating exercise to apply the procedures of scaling theory to the Landau potential (3.11), in order to obtain the critical exponents. The scaling fields in this case are, as for the example of Kadanoff (1966) described in Sect. 2.3 and for the mean-field calculation of Sect. 2.10, $\theta_1 = L$ and $\theta_2 = \Delta K$. Let the smooth part of ϕ_L be identified as $B_0(K)$. Then

$$\phi_{\text{sing}}(\theta_1, \theta_2; m) = -\theta_1 m - A_2\theta_2 m^2 + A_4 m^4. \tag{3.16}$$

The method of Example 2.3 can now be used, supposing that the homogeneity condition

$$\phi_{\text{sing}}(\lambda^{y_1}\theta_1, \lambda^{y_2}\theta_2; \lambda^z m) = \lambda^d \phi_{\text{sing}}(\theta_1, \theta_2; m) \tag{3.17}$$

is satisfied. It follows from (3.16) and (3.17) that

$$y_1 = \tfrac{3}{4}d, \qquad y_2 = \tfrac{1}{2}d, \qquad z = \tfrac{1}{4}d. \tag{3.18}$$

The exponent z for the magnetization satisfies, as it must (see equation (2.23)), the condition $z = d - y_1$. Substituting into (2.162) and (2.167) we obtain $\alpha = 0$, $\beta = \frac{1}{2}$, $\gamma = 1$ and $\delta = 3$. The exponent γ applies to the divergence of the susceptibility as the critical point is approached from above and below along the zero-field axis. However, it can be easily seen (Example 3.2) that these directions yield different amplitudes. The value $\alpha = 0$ indicates that, in Landau theory, there is no divergence, at the critical point, in the heat capacity $c_{\mathcal{H}}$ at constant field along the zero-field axis. However, there is a discontinuity at the critical point. From (2.155)

$$c_{\mathcal{H}} = -k_{\mathrm{B}} K^2 \frac{\partial^2 \phi}{\partial K^2} \tag{3.19}$$

and, from (3.11), (3.14) and (3.15) the discontinuity in $c_{\mathcal{H}}$ is

$$\Delta c_{\mathcal{H}} = \frac{k_{\mathrm{B}} K_{\mathrm{c}}^2 A_2^2}{2A_4} . \tag{3.20}$$

In Sects. 3.4–3.6 Landau theory is applied to some other models. Before that a treatment of the critical point which incorporates harmonic fluctuations is considered.

3.3 Landau–Ginzberg Theory for a Critical Point

A characteristic of Landau theory, as described in Sects. 3.1 and 3.2, or indeed of any form of classical mean-field theory, is that all reference to spatially fluctuating variables is eliminated. This prevents an investigation of the critical properties of any quantities, like the correlation length and pair correlation function, which depend on fluctuations. In this section a generalization of Landau theory, due to Ginzburg and Landau (1950), is introduced, in which the partition function is expressed as a functional integral. This procedure is considered in detail using the *Gaussian approximation*, in which only harmonic fluctuations about mean-field values are retained. This is the simplest way to include some fluctuations and, in the context of quantum field theory, it is equivalent to the 'tree approximation' (see, for example, Amit 1978).[1]

Let $\mathcal{V}$ be a d-dimensional hypercube of volume R^d. In terms of some reference volume v_0, $V = R^d/v_0$ is the dimensionless volume of the system. At every point $\boldsymbol{r}$ of $\mathcal{V}$ an n-dimensional vector field $\boldsymbol{\sigma}(\boldsymbol{r})$ is defined, in terms of which the statistical mechanical magnetization density $\hat{\boldsymbol{m}}$ is given by

$$\hat{\boldsymbol{m}} = \frac{1}{R^d} \int_{\mathcal{V}} \boldsymbol{\sigma}(\boldsymbol{r}) \mathrm{d}\mathcal{V} , \tag{3.21}$$

(cf. (1.46)). The vector field $\boldsymbol{\sigma}(\boldsymbol{r})$ is the *local* magnetization density with conjugate vector coupling $\boldsymbol{L} = \boldsymbol{\mathcal{H}}/T$, where $\boldsymbol{\mathcal{H}}$ is an n-dimensional magnetic field. We now define the local free energy density or potential to the fourth degree in $|\boldsymbol{\sigma}(\boldsymbol{r})|$ by

[1] For a more detailed treatment of generalizations to Landau theory the reader is referred to Luban (1976).

$$\hat{\phi}_{\mathrm{L}}(\boldsymbol{L},K;\boldsymbol{\sigma}(\boldsymbol{r})) \simeq -\boldsymbol{L}.\boldsymbol{\sigma}(\boldsymbol{r}) + B_0(K) + B_2(K)|\boldsymbol{\sigma}(\boldsymbol{r})|^2 + B_4(K)|\boldsymbol{\sigma}(\boldsymbol{r})|^4 + C(K)\sum_{j=1}^{n}|\boldsymbol{\nabla}\sigma_j(\boldsymbol{r})|^2\,, \tag{3.22}$$

(cf. (3.11)). Then the thermodynamic potential is

$$\phi(\boldsymbol{L},K) = -\frac{1}{V}\ln Z(\boldsymbol{L},K)\,, \tag{3.23}$$

where the partition function is given by the functional integral

$$Z(\boldsymbol{L},K) = \int_{\Sigma}\exp\left[-\frac{1}{v_0}\int_{\mathcal{V}}\hat{\phi}_{\mathrm{L}}(\boldsymbol{L},K;\boldsymbol{\sigma}(\boldsymbol{r}))\mathrm{d}\mathcal{V}\right]\mathrm{d}\Sigma \tag{3.24}$$

and Σ is the space of values of $\boldsymbol{\sigma}$.

Consider the case $n = 1$ (i.e. assume a scalar magnetization and magnetic field) and select the form of σ which minimizes $\hat{\phi}_{\mathrm{L}}$. For $C(K) > 0$ this will occur when σ is uniform over $\mathcal{V}$ and, assuming that this distribution dominates the functional integral in (3.24) the form of the Landau potential (3.11) is retrieved.

In general suppose that the magnetic field $\mathcal{H}$ is in the direction of the first component σ_1 of $\boldsymbol{\sigma}$. The first term in (3.22) is now replaced by $-L\sigma_1$. The first component of $\boldsymbol{\sigma}$ is referred to as *longitudinal*, the other components as *transverse*. Let $\mathring{\boldsymbol{\sigma}}$ be defined according to the high-temperature minimization conditions for $\hat{\phi}_{\mathrm{L}}$, when the distribution is uniform. Thus

$$\begin{aligned} 0 &= 2B_2(K)\mathring{\sigma}_1 + 4B_4(K)\mathring{\sigma}_1^3 - L\,, \\ 0 &= \mathring{\sigma}_j\,, \qquad j \neq 1\,. \end{aligned} \tag{3.25}$$

3.3.1 The Gaussian Approximation

This involves retaining only harmonic fluctuations around the uniform values given by (3.25). With

$$\sigma'_j(\boldsymbol{r}) = \sigma_j(\boldsymbol{r}) - \mathring{\sigma}_j\,, \qquad j = 1,\dots,n\,, \tag{3.26}$$

we have

$$\hat{\phi}_{\mathrm{L}}(\boldsymbol{L},K;\boldsymbol{\sigma}(\boldsymbol{r})) \simeq \hat{\phi}_{\mathrm{L}}(\boldsymbol{L},K;\mathring{\boldsymbol{\sigma}}) + 4B_4(K)\mathring{\sigma}_1^2{\sigma'_1}^2(\boldsymbol{r}) + D(K)|\boldsymbol{\sigma}'(\boldsymbol{r})|^2 + C(K)\sum_{j=1}^{n}|\boldsymbol{\nabla}\sigma'_j(\boldsymbol{r})|^2, \tag{3.27}$$

where

$$D(K) = B_2(K) + 2B_4(K)\mathring{\sigma}_1^2\,. \tag{3.28}$$

With periodic boundary conditions on the hypercube $\mathcal{V}$, Fourier transforms are defined (see equations (A.8)and (A.9)) by

$$\psi_j(\boldsymbol{k}) = \frac{1}{R^d}\int_{\mathcal{V}} \sigma'_j(\boldsymbol{r})\exp(-\mathrm{i}\boldsymbol{k}\cdot\boldsymbol{r})\mathrm{d}\mathcal{V}\,, \tag{3.29}$$

with

$$\sigma'_j(\boldsymbol{r}) = \sum_{\{\boldsymbol{k}\}} \psi_j(\boldsymbol{k})\exp(\mathrm{i}\boldsymbol{k}\cdot\boldsymbol{r})\,, \tag{3.30}$$

where the wave vectors $\boldsymbol{k} = 2\pi(m_1, m_2, \ldots, m_d)/R$, with $m_i = 0, \pm 1, \pm 2, \ldots$. It can be seen from (3.30) that $\psi_j(\boldsymbol{k})$ corresponds to the amplitude fluctuation of $\sigma_j(\boldsymbol{r})$ with wave vector $\boldsymbol{k}$. The larger the magnitude of $\boldsymbol{k}$ the shorter the wave length of the fluctuation. The coefficients $\psi_1(\boldsymbol{k})$ represent fluctuations in the direction of the external field, called *longitudinal fluctuations*; $\psi_j(\boldsymbol{k})$, for $j \neq 1$, correspond to *transverse fluctuations*. Since all the variables $\sigma'_j(\boldsymbol{k})$ are real, $\psi_j(-\boldsymbol{k})$ is the conjugate complex of $\psi_j(\boldsymbol{k})$, with $\psi_j(\boldsymbol{0})$ real. From (3.21), (3.26) and (3.29)

$$\hat{\boldsymbol{m}} = \mathring{\boldsymbol{\sigma}} + \boldsymbol{\psi}(\boldsymbol{0})\,, \tag{3.31}$$

where $\boldsymbol{\psi}(\boldsymbol{0})$ is the n-dimensional vector with components $\psi_j(\boldsymbol{0})$. This term in (3.31) represents the fluctuations in magnetization per unit volume. For $\boldsymbol{k} \neq \boldsymbol{0}$ let

$$\psi_j(\pm\boldsymbol{k}) = \psi_j^{(\Re)}(\boldsymbol{k}) \pm \mathrm{i}\psi_j^{(\Im)}(\boldsymbol{k})\,. \tag{3.32}$$

From (3.27) and (3.30),

$$\begin{aligned}\frac{1}{R^d}\int_{\mathcal{V}} \hat{\phi}_{\mathrm{L}}(\boldsymbol{L}, K; \boldsymbol{\sigma}(\boldsymbol{r}))\mathrm{d}\mathcal{V} &= \hat{\phi}_{\mathrm{L}}(\boldsymbol{L}, K; \mathring{\boldsymbol{\sigma}}) + 4B_4(K)\mathring{\sigma}_1^2\psi_1^2(\boldsymbol{0}) + D(K)|\boldsymbol{\psi}(\boldsymbol{0})|^2 \\ &+ 2\sum_{\{\boldsymbol{k}>\boldsymbol{0}\}} [4B_4(K)\mathring{\sigma}_1^2 + D(K) + C(K)k^2][\psi_1^{(\Re)2}(\boldsymbol{k}) + \psi_1^{(\Im)2}(\boldsymbol{k})] \\ &+ 2\sum_{j=2}^{n}\sum_{\{\boldsymbol{k}>\boldsymbol{0}\}} [D(K) + C(K)k^2][\psi_j^{(\Re)2}(\boldsymbol{k}) + \psi_j^{(\Im)2}(\boldsymbol{k})]\,,\end{aligned} \tag{3.33}$$

where '$\boldsymbol{k} > \boldsymbol{0}$' means all the wave vectors in half the $\boldsymbol{k}$-vector space excluding $\boldsymbol{k} = \boldsymbol{0}$. The partition function is now given by substituting from (3.33) into (3.24). The functional integral over the space Σ is replaced by integrals over the variables $\psi_j(\boldsymbol{0})$, $\psi_j^{(\Re)}(\boldsymbol{k})$ and $\psi_j^{(\Im)}(\boldsymbol{k})$, for $j = 1, \ldots, n$, which we take to extend over the range $(-\infty, \infty)$. This gives

$$Z(\boldsymbol{L}, K) = \exp[-V\hat{\phi}_{\mathrm{L}}(\boldsymbol{L}, K; \mathring{\boldsymbol{\sigma}})]\prod_{j=1}^{n}\left[\Psi_j(0)\prod_{\{\boldsymbol{k}>\boldsymbol{0}\}}\Psi_j(\boldsymbol{k})\right], \tag{3.34}$$

where

$$\Psi_j(\boldsymbol{0}) = \begin{cases}\displaystyle\int_{-\infty}^{\infty} \mathrm{d}\psi_1(\boldsymbol{0})\exp\{-V[4B_4(K)\mathring{\sigma}_1^2 + D(K)]\psi_1^2(\boldsymbol{0})\}\,, & j = 1, \\ \displaystyle\int_{-\infty}^{\infty} \mathrm{d}\psi_j(\boldsymbol{0})\exp\{-VD(K)\psi_j^2(\boldsymbol{0})\}\,, & j \neq 1,\end{cases} \tag{3.35}$$

and, for $\boldsymbol{k} \neq \boldsymbol{0}$,

$$\Psi_j(\boldsymbol{k}) = \begin{cases} \displaystyle\int_{-\infty}^{\infty} \mathrm{d}\psi_1^{(\Re)}(\boldsymbol{k}) \int_{-\infty}^{\infty} \mathrm{d}\psi_1^{(\Im)}(\boldsymbol{k}) \exp\{-V[4B_4(\kappa)\sigma_1^{\star 2} + D(\kappa) \\ \qquad + C(\kappa)k^2][\psi_1^{(\Re)2}(\boldsymbol{k}) + \psi_1^{(\Im)2}(\boldsymbol{k})]\}\,, \qquad j = 1, \\ \displaystyle\int_{-\infty}^{\infty} \mathrm{d}\psi_j^{(\Re)}(\boldsymbol{k}) \int_{-\infty}^{\infty} \mathrm{d}\psi_j^{(\Im)}(\boldsymbol{k}) \exp\{-V[D(\kappa) \\ \qquad + C(\kappa)k^2][\psi_j^{(\Re)2}(\boldsymbol{k}) + \psi_j^{(\Im)2}(\boldsymbol{k})]\}\,, \qquad j \neq 1. \end{cases} \tag{3.36}$$

The variables $\psi_j(\boldsymbol{0})$, $\psi_j^{(\Re)}(\boldsymbol{k})$ and $\psi_j^{(\Im)}(\boldsymbol{k})$, for all j and $\boldsymbol{k}$ are normally distributed and uncorrelated with

$$\langle \psi_j(\boldsymbol{k}) \rangle = 0\,, \tag{3.37}$$

$$\langle |\psi_j(\boldsymbol{k})|^2 \rangle = \begin{cases} g_{\|}(\boldsymbol{L}, \kappa; k)\,, & j = 1, \\ g_{\perp}(\boldsymbol{L}, \kappa; k)\,, & j \neq 1, \end{cases} \tag{3.38}$$

where

$$\begin{aligned} g_{\|}(\boldsymbol{L}, \kappa; k) &= \{2V[4B_4(\kappa)\mathring{\sigma}_1^2 + D(\kappa) + C(\kappa)k^2]\}^{-1}\,, \\ g_{\perp}(\boldsymbol{L}, \kappa; k) &= \{2V[D(\kappa) + C(\kappa)k^2]\}^{-1}\,. \end{aligned} \tag{3.39}$$

From equations (3.25), (3.30), (3.31) and (3.37)

$$\begin{aligned} \langle \boldsymbol{\sigma}'(\boldsymbol{r}) \rangle &= 0\,, \\ \langle \boldsymbol{\sigma}(\boldsymbol{r}) \rangle &= \langle \mathring{\boldsymbol{\sigma}} \rangle = \mathring{\boldsymbol{\sigma}}\,, \end{aligned} \tag{3.40}$$

$$\boldsymbol{m} = \langle \hat{\boldsymbol{m}} \rangle = \mathring{\boldsymbol{\sigma}} = \mathring{\sigma}_1 \hat{\boldsymbol{h}}\,, \tag{3.41}$$

where $\hat{\boldsymbol{h}}$ is the unit vector in the direction of the external magnetic field. It follows that the thermodynamic magnetization per unit volume is the same as that per lattice site in Landau theory (cf. equations (3.12) and (3.25)).

Formula (2.5) for the pair correlation function applies to the continuum, as well as to a lattice system, and from (3.30) using the notion introduced in (1.73) for a system with translational and rotational invariance,

$$\begin{aligned} \Gamma_2(\sigma; |\boldsymbol{r} - \boldsymbol{r}'|) &= \langle \boldsymbol{\sigma}'(\boldsymbol{r}) . \boldsymbol{\sigma}'(\boldsymbol{r}') \rangle \\ &= \sum_{j=1}^{n} \sum_{\{\boldsymbol{k}\}} \langle |\psi_j(\boldsymbol{k})|^2 \rangle \exp[\mathrm{i}\boldsymbol{k} \centerdot (\boldsymbol{r} - \boldsymbol{r}')]\,. \end{aligned} \tag{3.42}$$

With $r = |\boldsymbol{r} - \boldsymbol{r}'|$, the Fourier transform $\Gamma_2^*(\sigma; k)$ of $\Gamma_2(\sigma; r)$ is defined by the continuum form of (1.65) (see (3.29)) and thus, from (3.39) and (3.42),

$$\Gamma_2^*(\sigma;k) = \sum_{j=1}^{n} \langle |\psi_j(\boldsymbol{k})|^2 \rangle = g_{\|}(\boldsymbol{L}, K; k) + (n-1) g_{\perp}(\boldsymbol{L}, K; k) . \tag{3.43}$$

From (2.193),[2] (3.39) and (3.43) the correlation length $\mathfrak{r}$ is given by

$$\mathfrak{r}^2 = 4c(d) d V C(K) \frac{g_{\|}^2(\boldsymbol{L}, K; 0) + (n-1) g_{\perp}^2(\boldsymbol{L}, K; 0)}{g_{\|}(\boldsymbol{L}, K; 0) + (n-1) g_{\perp}(\boldsymbol{L}, K; 0)} . \tag{3.44}$$

To evaluate the thermodynamic potential, given by (3.23) and (3.34), it is necessary to insert explicit expressions for the Gaussian integrals (3.35) and (3.36). From (3.38) and (3.39) this gives

$$\begin{aligned} \phi(\boldsymbol{L}, K) &= \hat{\phi}_{\mathrm{L}}(\boldsymbol{L}, K; \mathring{\boldsymbol{\sigma}}) \\ &\quad - \frac{1}{2V} \sum_{\{\boldsymbol{k}\}} \{ \ln[2\pi g_{\|}(\boldsymbol{L}, K; k)] + (n-1) \ln[2\pi g_{\perp}(\boldsymbol{L}, K; k)] \} . \end{aligned} \tag{3.45}$$

3.3.2 Gaussian Critical Exponents

Adopting the forms for $B_2(K)$ and $B_4(K)$ given by (3.15), it can be shown from (3.25) and (3.41) that the exponents β, γ and δ take their Landau (mean-field) values $\beta = \frac{1}{2}$, $\gamma = 1$, $\delta = 3$. We now consider the situation along the zero-field axis, $\boldsymbol{L} = \boldsymbol{0}$, and suppose that, to leading order in $\triangle K = K - K_{\mathrm{c}}$,

$$C(K) \simeq C_0 > 0 . \tag{3.46}$$

From (3.15) and (3.39),

$$g_{\|}(\boldsymbol{0}, K; k) = g_{\perp}(\boldsymbol{0}, K; k) = [2V(-A_2 \triangle K + C_0 k^2)]^{-1} \tag{3.47}$$

for $K < K_{\mathrm{c}}$, when $\mathring{\sigma}_1 = 0$, and

$$\begin{aligned} g_{\|}(\boldsymbol{0}, K; k) &= [2V(2A_2 \triangle K + C_0 k^2)]^{-1} , \\ g_{\perp}(\boldsymbol{0}, K; k) &= (2V C_0 k^2)^{-1} \end{aligned} \tag{3.48}$$

for $K > K_{\mathrm{c}}$, when, from (3.14) and (3.41), $\mathring{\sigma}_1^2 = A_2 \triangle K / (2A_4)$. Thus, from (3.44), it can be seen that throughout the magnetized, low-temperature, phase the correlation length is infinite when $n \neq 1$. This behaviour arises from the transverse singular amplitude of the long-wavelength contribution to the pair correlation function. The magnetized phase is not well-defined in the Gaussian approximation. We shall, for the rest of this section, concentrate on the high-temperature phase. From (3.44) and (3.47)

$$\mathfrak{r} = \left[\frac{2c(d) d C_0}{A_2 (K_{\mathrm{c}} - K)} \right]^{\frac{1}{2}} = \left[\frac{C_0}{A_2 (K_{\mathrm{c}} - K)} \right]^{\frac{1}{2}} , \tag{3.49}$$

[2] In this case $\mathfrak{d} = d$, the thermodynamic limit is taken in each of the d dimensions.

with $c(d) = 1/(2d)$, (a choice which is justified below). From (2.197) this gives the value for the exponent $\nu = \frac{1}{2}$. From (3.43), (3.47) and (3.49),

$$\Gamma_2^*(\sigma;k) = \frac{n v_0}{2VC_0(k^2 + \mathfrak{r}^{-2})} \tag{3.50}$$

and thus

$$\Gamma_2^*(\sigma;k) = \frac{n v_0}{2VC_0 k^2} \tag{3.51}$$

at the critical point. Comparing this formula with the asymptotic form (2.202) yields the result $\eta = 0$. The form of the Fourier transform (3.50) is that given by (A.24). In the thermodynamic limit, $\Gamma_2(\sigma;r)$ is given, for $d < 5$ and $\mathfrak{r}$ finite, from (A.30), by

$$\Gamma_2(\sigma;r) = \frac{n v_0}{2C_0(2\pi)^{\frac{d}{2}}}(r\mathfrak{r})^{1-\frac{d}{2}}\mathsf{K}_{\frac{d}{2}-1}(r/\mathfrak{r}), \tag{3.52}$$

where $\mathsf{K}_\nu(z)$ is the Bessel function of complex argument. Substituting from (3.52) into the continuum form of (2.6) and using the formula (A.20) for the surface area of a d-dimensional hypersphere, gives

$$\mathfrak{r}^2 = c(d)\frac{\displaystyle\int_0^\infty r^{\frac{d}{2}+2}\mathsf{K}_{\frac{d}{2}-1}(r/\mathfrak{r})\mathrm{d}r}{\displaystyle\int_0^\infty r^{\frac{d}{2}}\mathsf{K}_{\frac{d}{2}-1}(r/\mathfrak{r})\mathrm{d}r}. \tag{3.53}$$

The integrals in (3.53) are of a standard form (see, for example, Gradshteyn and Ryzhik 1980, p. 684) yielding the result $c(d) = 1/(2d)$. In the limit $r \to \infty$, from (A.32),

$$\Gamma_2(\sigma;r) \simeq \frac{n v_0}{2^{\frac{d}{2}+\frac{3}{2}}(\pi r\mathfrak{r})^{\frac{d}{2}-\frac{3}{2}}}\frac{\exp(-r/\mathfrak{r})}{r}. \tag{3.54}$$

Equation (3.54) is exact for $d = 1$ and $d = 3$, when it includes the limiting behaviour as $\mathfrak{r} \to \infty$. From (A.33) and (A.34),

$$\begin{aligned} \Gamma_2(\sigma;r) &\simeq \frac{n v_0}{2\pi C_0}\ln\left(\frac{r}{2\mathfrak{r}}\right), && \text{if } d = 2,\\ \Gamma_2(\sigma;r) &= \frac{n v_0}{8\pi^2 C_0}\frac{1}{r^2}, && \text{if } d = 4, \end{aligned} \tag{3.55}$$

in the limit $\mathfrak{r} \to \infty$. For $d \geq 5$ the integral for $\Gamma_2(\sigma;r)$ diverges at the infinite-frequency limit. From the point of view of critical phenomena it is the long-wavelength, low-frequency, fluctuations which are important. This short-wavelength, high-frequency, divergence is dealt with by supposing that the system has a minimum resolution length r_0. This means that fluctuations of frequency greater than the cut off value $k_0 = 2\pi/r_0$ are dampened by the system. Introducing k_0 as the upper limit of integration for $d \geq 5$ we have, from (A.38),

$$\Gamma_2(\sigma;r) \simeq \frac{nv_0}{2C_0 r^{d-2}(2\pi)^{\frac{d}{2}}}\left\{k_0^{\frac{d}{2}-4} - \left(\frac{2}{\pi}\right)^{\frac{1}{2}} k_0^{\frac{d}{2}-\frac{5}{2}}\cos\left[k_0 - \frac{\pi}{4}(d+1)\right]\right\}, \tag{3.56}$$

in the limit $k_0 \to \infty$. From (3.54), (3.55) and (3.56),

$$\Gamma_2(\sigma;r) \sim \frac{1}{r^{d-2}}, \tag{3.57}$$

for $d \geq 3$. Comparison with (2.200) again shows that $\eta = 0$, but now we have additional information about the cases $d = 1$ and $d = 2$. In the limit $\mathfrak{r} \to \infty$, from (3.54) and (3.55) respectively,

$$\Gamma_2(\sigma;r) \sim \begin{cases} \mathfrak{r}\exp(-r/\mathfrak{r}), & d = 1, \\ \ln(r/\mathfrak{r}), & d = 2. \end{cases} \tag{3.58}$$

The Landau values for the exponents β, γ and δ satisfy the Widom scaling law (2.169) and these together with the values, derived above for η and ν satisfy the Fisher scaling law (2.203).

It is interesting to see from (3.52) that, according to this analysis for $d < 5$, the correlation function can be expressed in the form

$$\Gamma_2(\sigma;r) = \frac{f(\mathfrak{r}/r)}{r^{d-2-\eta}}, \tag{3.59}$$

given by (2.12), (2.13), (2.194) and (2.201), where in this case $\eta = 0$ and

$$f(x) = \frac{nv_0 x^{\frac{d}{2}-1}}{2C_0(2\pi)^{\frac{d}{2}}} \mathsf{K}_{\frac{d}{2}-1}(x). \tag{3.60}$$

It is now necessary to determine the exponent α. Again we consider the thermodynamic limit. The sum in (3.45) becomes an integral with the wave vectors forming an effective continuum with $\mathcal{S}_d(k)V(2\pi)^{-d}\mathrm{d}k$ states in the hypershell $k < |\boldsymbol{k}| < k+\mathrm{d}k$, where $\mathcal{S}_d(k)$ is the surface area of a d-dimensional hypersphere of radius k, given by (A.20). Introducing the high-frequency cut off k_0, from (3.47), (3.45),

$$\begin{aligned}\phi(\mathbf{0},K) &= \hat{\phi}_{\mathrm{L}}(\mathbf{0},K;\mathbf{0}) \\ &\quad + \frac{nv_0}{\Gamma(d/2)2^d\pi^{d/2}}\int_0^{k_0} k^{d-1}\ln\left\{\frac{V[A_2(K_c-K)+C_0k^2]}{\pi}\right\}\mathrm{d}k,\end{aligned} \tag{3.61}$$

for $K < K_c$. The heat capacity $c_{\mathcal{H}}$ at constant field along the zero-field axis is given by (3.19). The first term on the right-hand side of (3.61) is the contribution arising from Landau theory and gives the discontinuity $\Delta c_{\mathcal{H}}$ of (3.20). In differentiating the integral in (3.61) only the most potentially singular contribution is retained to give

$$c_{\mathcal{H}} \simeq \frac{k_{\mathrm{B}} n K_c^2 v_0 A_2^2}{\Gamma(d/2)2^d\pi^{d/2}}\int_0^{k_0}\frac{k^{d-1}\mathrm{d}k}{[A_2(K_c-K)+C_0k^2]^2}. \tag{3.62}$$

Transforming to the variable

$$\kappa(k) = k\mathfrak{r} = k\left[\frac{C_0}{A_2(K_\mathrm{c} - K)}\right]^{\frac{1}{2}} \tag{3.63}$$

with $\kappa_0 = \kappa(k_0)$ gives

$$c_{\mathcal{H}} \simeq \frac{k_\mathrm{B} n K_\mathrm{c}^2 v_0}{\Gamma(d/2) 2^d \pi^{d/2}} \left(\frac{A_2}{C_0}\right)^2 \left[\frac{C_0}{A_2(K_\mathrm{c} - K)}\right]^{\frac{4-d}{2}} \int_0^{\kappa_0} \frac{\kappa^{d-1}\mathrm{d}\kappa}{(1+\kappa^2)^2} . \tag{3.64}$$

A singularity in $c_{\mathcal{H}}$, at the critical point can arise, either from the integral, or, when $d < 4$, from the prefactor. The integral is bounded above by its limiting value, given by taking $\kappa_0 \to \infty$. This is

$$\int_0^{\infty} \frac{\kappa^{d-1}\mathrm{d}\kappa}{(1+\kappa^2)^2} = \Gamma\left(\tfrac{1}{2}d\right)\Gamma\left(2 - \tfrac{1}{2}d\right) . \tag{3.65}$$

The gamma function is singular at $0, -1, -2, \ldots$ and the integral works as a finite upper bound only when $d < 4$. In this case the prefactor diverges giving $\alpha = (4-d)/2$. When $d = 4$ the prefactor is absent and any singular behaviour must come from the integral. Since the integrand is zero at the lower limit, for K sufficiently close to K_c,

$$\int_0^{\kappa_0} \frac{\kappa^3\mathrm{d}\kappa}{(1+\kappa^2)^2} \sim \int_\epsilon^{\kappa_0} \frac{\mathrm{d}\kappa}{\kappa} \sim -\tfrac{1}{2}\ln(K_\mathrm{c} - K) . \tag{3.66}$$

The heat capacity has a logarithmic divergence at the critical point. For $d > 4$ the prefactor does not diverge and, as in the case of $d = 4$, any singularity must arise from the integral in a neighbourhood of its upper limit. But

$$\int_0^{\kappa_0} \frac{\kappa^{d-1}\mathrm{d}\kappa}{(1+\kappa^2)^2} \sim \int_\epsilon^{\kappa_0} \kappa^{d-5}\mathrm{d}\kappa = \frac{\kappa_0^{d-4} - \epsilon^{d-4}}{d-4} . \tag{3.67}$$

The heat capacity does not diverge, but there is a discontinuity from the Landau term. The exponent $\alpha = 0$.

The results of the combination of Landau theory and the Gaussian approximation applied to Landau–Ginzberg theory can now be summarized as follows:

- *When* $d \geq 4$, the values for the critical exponents are

$$\alpha = 0 , \qquad \beta = \tfrac{1}{2} , \qquad \delta = 3 , \tag{3.68}$$

$$\gamma = 1 , \qquad \eta = 0 , \qquad \nu = \tfrac{1}{2} . \tag{3.69}$$

  These values satisfy the three scaling laws (2.169) (Widom), (2.170) (Essam–Fisher) and (2.203) (Fisher), but not the Josephson hyperscaling law (2.199), except at $d = 4$.

- *When* $d < 4$, η and ν are still given by (3.69) and to satisfy the Fisher scaling law (2.203) we need also to retain the value of γ given in (3.69). Since we have shown that now

$$\alpha = \frac{4-d}{2}, \tag{3.70}$$

the Josephson hyper-scaling law (2.199) is satisfied. However, the Widom and Essam–Fisher scaling laws are not satisfied by α from (3.70) and β and δ from (3.68). To satisfy these two laws we need the new expressions

$$\beta = \frac{d-2}{4}, \qquad \delta = \frac{d+2}{d-2}. \tag{3.71}$$

The mean-field values for β and δ were derived from Landau theory in Sect. 3.2 and are intended to be true for all values of d. The analysis, carries over to the Landau–Ginzberg theory and this seems, therefore, to contradict the validity of the values (3.71) (or alternatively of scaling theory) unless a reconciliation can be achieved. The key to this can be found in the treatment of Landau–Ginzberg using the renormalization group (see, for example, Ma 1976a). In this treatment the parameters A_2 and A_4 of equation (3.16) are also treated as renormalizable (rescalable) parameters. If we modify the treatment of (3.16) by including exponents u_2 and u_4 for A_2 and A_4 respectively then it follows that

$$z = d - y_1\,, \qquad u_2 = 2y_1 - y_2 - d\,, \qquad u_4 = 4y_1 - 3d\,. \tag{3.72}$$

It is not possible to obtain unique exponent values from these expressions. They are, however, satisfied by

$$y_2 = 2\,, \qquad y_1 = \tfrac{1}{2}d + 1\,, \qquad u_2 = 0\,, \qquad u_4 = 4 - d \tag{3.73}$$

and these are the values associated with the *Gaussian fixed point* of renormalization theory (Ma 1976a). From (2.162) and (2.167) the values of $\gamma = 1$ and α, β and δ given by (3.70) and (3.71) are then obtained. These values are called the *Gaussian critical exponents*. In this new context we can view the derivation of the exponent values (3.18) as based on the tacit assumption that both A_2 and A_4 are marginal parameters ($u_2 = u_4 = 0$). The modified derivation leading to (3.73) still leaves A_2 marginal. Indeed, if this were not the case, the derivation of $\nu = \frac{1}{2}$ from (3.49) would not be valid. However, for $d < 4$, A_4 is relevant, for $d = 4$ it is marginal and for $d > 4$ it is irrelevant. A discussion of the case $d < 4$, for which the Gaussian fixed point is unstable requires the full renormalization group analysis. The case of interest here is $d > 4$. Taking the singular part of the thermodynamic potential (3.16) on the zero-field axis ($\theta_1 = 0$) and substituting from (3.14) and (3.15) we have

$$\phi_{\text{sing}}(0, \theta_2, A_4) = \frac{3A_2^2\theta_2^2}{4A_4}. \tag{3.74}$$

Using the homogeneity condition

$$\phi_{\text{sing}}(0, \lambda^{y_2}\theta_2, \lambda^{u_4} A_4) = \lambda^d \phi_{\text{sing}}(0, \theta_2, A_4) \tag{3.75}$$

and substituting $\lambda = |\theta_2|^{-1/y_2}$ we have

$$\phi_{\text{sing}}(0, \theta_2, A_4) = |\theta_2|^{d/y_2} w_0^{(\pm)}(v), \tag{3.76}$$

where $v = A_4/|\theta_2|^{u_4/y_2}$ and

$$w_0^{(\pm)}(v) = \frac{3A_2^2}{4v} \tag{3.77}$$

is a particular realization of the function appearing in equation (2.265). It follows that, in terms of the discussion in Sect. 2.14, A_4 is a dangerous irrelevant variable, with $\tau = 1$ and, from (2.270), an effective change of dimension in the system to

$$\bar{d} = d - |d - 4| = 4\,. \tag{3.78}$$

Using $\bar{d}$ in place of d in equations (3.70) and (3.71) gives the mean-field exponents (3.68). We have, therefore, obtained a consistent scaling analysis using the Gaussian exponents, with the proviso that, for $d > 4$, account is taken of the dangerous irrelevant variable A_4, which leads to the effective dimension of the system being fixed at the value four.[3] For any form of critical behaviour the *upper borderline dimension* d_{u} is defined to be that above which the critical exponents are those of mean-field theory (Fisher 1983).[4] We have shown that for a simple critical point $d_{\text{u}} = 4$.

Although the existence of a singularity in the heat capacity at the critical point, superimposed upon a discontinuity, is dependent only on the dimension d, the ability to detect the singularity experimentally will depend on the range of temperature over which the singularity dominates the discontinuity. This in turn will depend on the relative size of the parameters A_2, A_4 and C_0 and on the dimension n of the order parameter. The *width of the critical region* $\mathcal{W}_{\text{c}}$ can be defined by the condition that

$$\text{msp}\{c_{\mathcal{H}}\} \geq \Delta c_{\mathcal{H}} \qquad \text{if and only if}$$

$$\left|1 - \frac{T}{T_{\text{c}}}\right| \leq \mathcal{W}_{\text{c}}\,, \tag{3.79}$$

where "msp" stands for 'most singular part'. For $T > T_{\text{c}}$ the most singular part of $c_{\mathcal{H}}$ is given by substituting from (3.65) into (3.64) and $\Delta c_{\mathcal{H}}$ is given by (3.20). This yields the result

[3] The discussion presented here is not quite complete. As pointed out in Sect. 2.14 it is necessary to consider other approaches to the critical point to determine whether the change of effective dimension can be consistently applied. This involves consideration of the case $\theta_1 \neq 0$, when the magnetization is given, as in Sect. 2.10, by a root of a cubic equation.

[4] The *lower borderline dimension* is defined to be that below which no phase transition occurs. For the Ising model ferromagnet this is two.

$$\mathcal{W}_\mathrm{c} = \frac{1}{K_\mathrm{c} A_2}\left[\frac{n v_0 \Gamma\left(2-\frac{1}{2}d\right) A_4}{2^{d-1}(C_0 \pi)^{\frac{d}{2}}}\right]^{2/(4-d)} . \tag{3.80}$$

Equation (3.80) is a form of the *Ginzburg criterion*, (Ginzburg 1960). The singularity in $c_\mathcal{H}$ is caused by fluctuations and the critical region is the range of temperature around the critical point within which fluctuations are important. The motivation for Ginzburg's work was to understand why a finite discontinuity was seen at some critical points, like those in superconductors, whereas in others, like the λ point in liquid helium, a singularity was observed in addition to the discontinuity. On the basis of (3.80) it can now be seen that, when $\mathcal{W}_\mathrm{c}$ is too small, the critical region cannot be resolved experimentally and the singularity will not be seen. Estimates of $\mathcal{W}_\mathrm{c}$ for various systems are given by Ginzburg (1960). He concludes that, for superconductors $\mathcal{W}_\mathrm{c} \sim 3 \times 10^{-16}$, whereas in helium $\mathcal{W}_\mathrm{c} \sim 0.3$.

3.4 The Equilibrium Dilute Ising Model

Consider now a generalization of the ferromagnetic Ising model in which the variable $\sigma(\boldsymbol{r})$ at any lattice site $\boldsymbol{r}$ can take the three values ± 1 and 0. The Hamiltonian is extended from the form (1.49) to

$$\widehat{H}(\mathcal{H}, \mu, J, \varepsilon; \sigma(\boldsymbol{r})) = -J \sum_{\{\boldsymbol{r},\boldsymbol{r}'\}}^{(\text{n.n.})} \sigma(\boldsymbol{r})\sigma(\boldsymbol{r}') - \varepsilon \sum_{\{\boldsymbol{r},\boldsymbol{r}'\}}^{(\text{n.n.})} \sigma^2(\boldsymbol{r})\sigma^2(\boldsymbol{r}')$$

$$- \mathcal{H} \sum_{\{\boldsymbol{r}\}} \sigma(\boldsymbol{r}) - \mu \sum_{\{\boldsymbol{r}\}} \sigma^2(\boldsymbol{r}) . \tag{3.81}$$

There are now two densities m, conjugate to $\mathcal{H}$ and defined as in Sect. 3.1, and ρ conjugate to μ and defined by

$$\rho = \langle \hat{\rho} \rangle , \qquad \hat{\rho} = \frac{1}{N} \sum_{\{\boldsymbol{r}\}} \sigma^2(\boldsymbol{r}) . \tag{3.82}$$

A common terminology for the 'standard' Ising model of Sect. 3.1 and this extended model is to refer to them respectively as the *spin-*$\frac{1}{2}$ and *spin-1* Ising models, by analogy between the values of the site variables and that of a quantum spin variable. The current model is also called the *Blume–Emery–Griffiths model* in reference to the work of Blume, Emery and Griffiths (1971). The model has many applications (see Volume 1, Sect. 6.6), including a lattice gas of magnetic particles, where $\sigma(\boldsymbol{r}) = 0$ signifies a vacant site and μ the chemical potential of a particle, binary mixtures with one ferromagnetic component and ternary mixtures.

By the procedure used in Sect. 3.1 for the spin-$\frac{1}{2}$ ferromagnetic Ising model (see Example 3.1), the mean-field form for the Hamiltonian is

$$\hat{H}(\mathcal{H},\mu,J,\varepsilon;\hat{m},\hat{\rho}) = -N\left(\tfrac{1}{2}zJ\hat{m}^2 + \tfrac{1}{2}z\varepsilon\hat{\rho}^2 + \mathcal{H}\hat{m} + \mu\hat{\rho}\right) . \tag{3.83}$$

The argument leading to the Laudau potential ϕ_L proceeds as described in Sect. 3.1 except that now in the formula corresponding to (3.3) two constraints are applied, $\hat{m} = m$ and $\hat{\rho} = \rho$. The final potential is a function of the two densities m and ρ and four couplings, $K_1 = J/T$, $K_2 = \varepsilon/T$, $L_1 = \mathcal{H}/T$ and $L_2 = \mu/T$. For the spin-$\frac{1}{2}$ Ising model we saw that, when $L_1 = 0$, ϕ_L is invariant under the operator $\mathrm{Q}^{(m)}$, which inverts all the spins on the lattice, changing the sign of m. The symmetry group, with $L_1 = 0$, therefore contains two elements $\mathrm{Q}^{(m)}$ and the identity element I. This is the permutation group $\mathcal{S}_2$. In the terminology of group representation theory (see Appendix A.3) m is a one-dimensional antisymmetric representation of $\mathcal{S}_2$. For the spin-1 Ising model the symmetry group is in general unchanged. The parameter $\hat{\rho}$ defined by (3.82), and hence ρ, is unaffected by the operator $\mathrm{Q}^{(m)}$. The density ρ is a one-dimensional symmetric representation of $\mathcal{S}_2$ and the Landau expansion is of the form

$$\phi_L(L_1,L_2,K_1,K_2,\rho) = -L_1 m - L_2\rho + B_0(K_1,K_2;\rho) + \sum_{i=1}^{\infty} B_{2i}(K_1,K_2;\rho)m^{2i}. \tag{3.84}$$

Since, in general, no new symmetry is associated with the parameter ρ there is no way, within the framework of Landau theory, of obtaining information about a possible transition with ρ as order parameter. There are two exceptions to this. One is the case of the 3-state Potts model, discussed in Sect. 3.5, and the other is when $J = \mathcal{H} = 0$. In the latter case site states are equally likely to be $+1$ and -1 and, in terms of the variable

$$\hat{n} = \hat{\rho} - \tfrac{1}{2}, \tag{3.85}$$

the Hamiltonian (3.83) becomes

$$\hat{H}(\mu,\varepsilon;\hat{n}) = \text{constant} - N\left[\tfrac{1}{2}z\varepsilon\hat{n}^2 + (\mu + \tfrac{1}{2}z\varepsilon)\hat{n}\right] . \tag{3.86}$$

This form is invariant under the operator $\mathrm{Q}^{(n)}$ which changes all site states $|\sigma| = 1$ to zero and sites states zero to $+1$ or -1 with equal probability. Apart from the trivial entropy factor associated with the two non-zero site states, the model has become the simple lattice fluid version of the spin-$\frac{1}{2}$ Ising model (see Chap. 4 and Volume 1, Sect. 5.3). The first-order transition line and critical point lie on the line $\mu = \mu_\mathrm{c} = -\frac{1}{2}z\varepsilon$ and the Landau expansion has the form given by (3.10) with n replacing m, $(\mu - \mu_\mathrm{c})/T$ replacing L and ε/T replacing K.

3.5 The 3-State Potts Model

The *ν-state Potts model* (Potts 1952, Wu 1982, Martin 1991) is a lattice model in which each microsystem, which occupies one lattice site, can exist in ν distinct states and for which the Hamiltonian is invariant under the permutation group $\mathcal{S}_\nu$, which permutes the ν microstates among themselves. According to the discussion of universality classes in Sect. 2.3 we should expect the critical behaviour of this model to be in a different universality class for each value of ν. The simplest, nearest-neighbour version, of the model is when a 'like' nearest-neighbour pair (i.e. both members in the same state) has interaction energy $-R$, with an 'unlike' nearest-neighbour pair has zero interaction energy. The Hamiltonian is then of the form

$$\widehat{H}(R;\sigma(\boldsymbol{r})) = -R \sum_{\{\boldsymbol{r},\boldsymbol{r}'\}}^{(\text{n.n.})} \delta^{\text{Kr}}(\sigma(\boldsymbol{r}) - \sigma(\boldsymbol{r}')) \,, \tag{3.87}$$

where $\sigma(\boldsymbol{r})$ can take some set of ν distinct integer values. The model is said to be *ferromagnetic* when $R > 0$, encouraging the members of a nearest-neighbour pair to be in the same state and *antiferromagnetic* when $R < 0$, encouraging the members of a nearest-neighbour pair to be in different states. For $\nu = 2$ the symmetry group is $\mathcal{S}_2$ and with $\sigma(\boldsymbol{r}) = \pm 1$

$$\delta^{\text{Kr}}(\sigma(\boldsymbol{r}) - \sigma(\boldsymbol{r}')) = \tfrac{1}{2}[1 + \sigma(\boldsymbol{r})\sigma(\boldsymbol{r}')] \,. \tag{3.88}$$

The model is, therefore, isomorphic to the spin-$\frac{1}{2}$ Ising model in zero field.

In this section we consider the 3-state Potts model using Landau theory. We first show that the model is a special case of the spin-1 Ising model. This can be done directly using the Hamiltonian (3.81) (Example 3.3) or in the context of Landau theory, as we shall demonstrate here. This means that, if the 3-state Potts model has a transition, this will occur in a subspace of the phase space of the spin-1 Ising model. We first need to reformulate the Hamiltonian (3.83) in terms of two variables which are zero at high temperatures. This is the case for $\hat{m}$, since the states $+1$ and -1 will be equally probable, but now we have three equally probable states so $\hat{\rho}$ needs to be replaced by $\hat{p}$ defined by

$$\hat{p} = \hat{\rho} - \tfrac{2}{3} \,, \tag{3.89}$$

when (3.83) becomes, with $\mathcal{H} = 0$,

$$\hat{H}(\mu, J, \varepsilon; \hat{m}, \hat{p}) = -\text{constant} - N\left[\tfrac{1}{2}zJ\hat{m}^2 + \tfrac{1}{2}z\varepsilon\hat{p}^2 + \left(\mu + \tfrac{2}{3}z\varepsilon\right)\hat{p}\right] . \tag{3.90}$$

The group $\mathcal{S}_3$ has six elements and two generators. One can be taken to be $\mathrm{Q}^{(m)}$, under which the Hamiltonian in (3.90) is clearly invariant. The second generator can be taken as $\mathrm{Q}^{(*)}$, defined by

$$\mathrm{Q}^{(*)} : -1 \to 0 \to +1 \to -1. \tag{3.91}$$

It is not difficult to show that

$$\begin{aligned} \mathrm{Q}^{(*)}\hat{m} &= -\tfrac{1}{2}\hat{m} + \tfrac{3}{2}\hat{p}\,, \\ \mathrm{Q}^{(*)}\hat{p} &= -\tfrac{1}{2}\hat{m} - \tfrac{1}{2}\hat{p}\,. \end{aligned} \tag{3.92}$$

The Hamiltonian of (3.90) is invariant under $\mathrm{Q}^{(*)}$ if $\mu = -2zJ$ and $\varepsilon = 3J$. The Landau potential in this case will, therefore, be of the form $\phi_{\mathrm{L}}(K; m, p)$, and the problem is to find polynomials in the two variables m and p which are invariant under the operations of the group $\mathcal{S}_3$. The order parameter is the vector (m, p) with the operations of the group generators given by

$$\mathbf{Q}^{(m)}\begin{pmatrix} m \\ p \end{pmatrix} = \begin{pmatrix} -1 & 0 \\ 0 & 1 \end{pmatrix}\begin{pmatrix} m \\ p \end{pmatrix}, \tag{3.93}$$

$$\mathbf{Q}^{(*)}\begin{pmatrix} m \\ \\ p \end{pmatrix} = \begin{pmatrix} -\frac{1}{2} & \frac{3}{2} \\ \\ -\frac{1}{2} & -\frac{1}{2} \end{pmatrix}\begin{pmatrix} m \\ \\ p \end{pmatrix}. \tag{3.94}$$

The two parameters m and p are intertwined by the operator $\mathrm{Q}^{(*)}$ and the matrices in (3.93) and (3.94), together with the unit matrix give a two-dimensional irreducible representation of $\mathcal{S}_3$. Clearly no linear expression $Am + Dp$ is invariant under $\mathcal{S}_3$, since (3.93) implies that $A = 0$, and, from (3.94), p on its own is not invariant. Consider now the polynomial $Am^2 + Dmp + Ep^2$. From (3.93) $D = 0$ and from (3.94) $3A = E$. We, therefore, have the second-degree invariant $m^2 + 3p^2$. Now examine the third-degree polynomial $Am^3 + Dm^2p + Emp^2 + Fp^3$. From (3.93), $A = E = 0$ and, from (3.94), $D = -F$, giving the third-degree invariant $p^3 - m^2p$. We already know that there is at least one fourth-degree invariant, namely $(m^2 + 3p^2)^2$. It is not difficult to show that there are no others. The Landau expansion for the 3-state Potts model to the fourth degree in the invariants is

$$\begin{aligned} \phi_{\mathrm{L}}(K; m, p) \simeq B_0(K) &+ B_2(K)(m^2 + 3p^2) + B_3(K)(p^3 - m^2 p) \\ &+ B_4(\tau)(m^2 + 3p^2)^2\,. \end{aligned} \tag{3.95}$$

Stable equilibrium states are given by minimizing this potential form with respect to m and p. The equilibrium conditions are

$$\begin{aligned} 0 &= 2B_2(K)m - 2B_3(K)mp + 4B_4(K)m(m^2 + 3p^2)\,, \\ 0 &= 6B_2(K)p + B_3(K)(3p^2 - m^2) + 12B_4(K)p(m^2 + 3p^2)\,, \end{aligned} \tag{3.96}$$

and the stability conditions are obtained by considering the determinant

$$\Phi(K; m, p) = \begin{vmatrix} \phi_{mm}(K; m, p) & \phi_{mp}(K; m, p) \\ \phi_{pm}(K; m, p) & \phi_{pp}(K; m, p) \end{vmatrix}, \tag{3.97}$$

where

$$\phi_{mm}(K;m,p) = 2B_2(K) - 2B_3(K)p + 12B_4(K)(m^2 + p^2)\,,$$

$$\phi_{mp}(K;m,p) = \phi_{pm}(K;m,p) = -2B_3(K)m + 24B_4(K)mp\,, \qquad (3.98)$$

$$\phi_{pp}(K;m,p) = 6B_2(K) + 6B_3(K)p + 12B_4(K)(m^2 + 9p^2)\,.$$

A solution to (3.96) is a maximum or a minimum of $\phi_L(K;m,p)$ if $\Phi(K;m,p) > 0$, otherwise it is a saddle point. If $\Phi(K;m,p) > 0$, the solution is a maximum or minimum according as $\phi_{mm}(K;m,p)$ is negative or positive. The high-temperature solution $m = p = 0$ is a maximum if $B_2(K) < 0$ and a minimum if $B_2(K) > 0$. Suppose that to leading order in $\Delta K = K - K_c$

$$B_2(K) \simeq -A_2\Delta K\,, \qquad B_3(K) \simeq A_3\,, \qquad B_4(K) \simeq A_4\,, \qquad (3.99)$$

where A_2, A_3 and A_4 are constants, A_2 being positive. Then the high-temperature solution is stable if $K < K_c$, $(T > T_c)$. In general, the derivation of non-zero solutions to (3.96) is quite complicated. We shall, however, be concerned only with solutions at the critical temperature, which are not in a neighbourhood of the high-temperature solution. In this case it is not difficult to show that, when $K = K_c$ $(B_2(K_c) = 0)$, there are three solutions

$$(m_1^{(\pm)}, p_1^{(\pm)}) = \frac{(\pm 3, 1)A_3}{24A_4}\,, \qquad (m_2, p_2) = \frac{(0, -2)A_3}{24A_4}\,. \qquad (3.100)$$

Given any solution to (3.96), other solutions can be obtained by the operations of the group $\mathcal{S}_3$. In general solutions will occur in sets of six, corresponding to the six elements of $\mathcal{S}_3$. However, in this case

$$\begin{aligned} &\mathrm{Q}^{(m)}(3,1) = (-3,1)\,, && \mathrm{Q}^{(*)}(3,1) = (0,-2)\,, \\ &\mathrm{Q}^{(m)}(0,-2) = (0,-2)\,, && \mathrm{Q}^{(*)}(0,-2) = (-3,1)\,, \\ &\mathrm{Q}^{(m)}(-3,1) = (3,1)\,, && \mathrm{Q}^{(*)}(-3,1) = (3,1)\,, \end{aligned} \qquad (3.101)$$

and, since the group is generated by $\mathrm{Q}^{(m)}$ and $\mathrm{Q}^{(*)}$, the solutions (3.100) form a closed set under the operations of the group. Since

$$\begin{aligned} &\Phi(K_c; m_1^{(\pm)}, p_1^{(\pm)}) = \frac{A_3^4}{16A_4^2}\,, \qquad \phi_{mm}(K_c; m_1^{(\pm)}, p_1^{(\pm)}) = \frac{A_3^2}{8A_4}\,, \\ &\phi_L(K_c; m_1^{(\pm)}, p_1^{(\pm)}) = B_0(K_c) - \frac{A_3^3}{6912A_4^3}\,, \end{aligned} \qquad (3.102)$$

$$\begin{aligned} &\Phi(K_c; m_2, p_2) = \frac{A_3^4}{16A_4^2}\,, \qquad \phi_{mm}(K_c; m_2, p_2) = \frac{A_3^2}{4A_4}\,, \\ &\phi_L(K_c; m_2, p_2) = B_0(K_c) - \frac{A_3^3}{6912A_4^3}\,, \end{aligned} \qquad (3.103)$$

it follows that, when $A_4 > 0$, the solutions (3.100) are lower minima of $\phi_L(K;m,p)$ than $m = p = 0$ at, and in a neighbourhood of, the critical temperature. This means that the system will have already passed, with equal probability, to one of these states at some temperature above the critical value. The phase transition will be first-order since it involves a discontinuous change in m and p.

The prediction by Landau theory of a first-order phase transition in the 3-state Potts model is an interesting example of the limitations of this method. The results of Landau theory for the order of phase transitions are the same as those obtained by any mean-field calculation and, for the general ν-state Potts model, these have been shown (Mittag and Stephen 1974) to give a first-order transition for $\nu > 2$ in any dimension. In contradiction to this it has been shown rigorously by Baxter (1973), for the square lattice (see Volume 1, Sect. 10.14) and by Baxter et al. (1978), for the triangular lattice, that the ferromagnetic ν-state Potts model has a first-order transition for $\nu > 4$ but a higher-order transition for $\nu \leq 4$. For the antiferromagnetic model the situation is in general more complicated (Wang et al. 1989) because of the presence of infinite ground-state degeneracy. There are two exceptions to this. On the square lattice the antiferromagnetic 3-state Potts model is, like the ferromagnetic case, equivalent to a staggered six-vertex model (Baxter 1982b, Volume 1, Sect. 10.14) with a higher-order transition although in this case it occurs at zero temperature. On the triangular lattice the ground state is six-fold degenerate (Schick and Griffiths 1977) like the triangular spin-$\frac{1}{2}$ Ising antiferromagnet in a field (see Sect. 6.11). Monte Carlo simulations (Grest 1981, Saito 1982, Adler et al. 1995), series methods (Enting and Wu 1982) and analytical investigations (Park 1994) indicate that the transition is first-order in agreement with the prediction of Landau theory.

3.6 Landau Theory for a Tricritical Point

In Sect. 2.12 we described a model for a metamagnet, which, for a certain parameter range, exhibits a critical line meeting a triple line at a tricritical point in the symmetry plane $L_{\mathrm{s}} = \mathcal{H}_{\mathrm{s}}/T = 0$. The model is a generalization of the spin-$\frac{1}{2}$ Ising model. Points in the symmetry plane are given in terms of the coupling $K = J_1/T$ and $L = \mathcal{H}/T$ and the Landau potential can be expressed as the generalization

$$\phi_{\mathrm{L}}(L_{\mathrm{s}}, L_3, L, K; m_{\mathrm{s}}) = -L_{\mathrm{s}} m_{\mathrm{s}} - L_3 m_{\mathrm{s}}^3 + B_0(L, K) + \sum_{i=1}^{\infty} B_{2i}(L, K) m_{\mathrm{s}}^{2i} \tag{3.104}$$

of (3.10), where m_{s} is the sublattice order parameter. For completeness we have included the cubic field $\mathcal{H}_3$, with $L_3 = \mathcal{H}_3/T$, but in a metamagnet, this field is not an independent quantity. It vanishes with $\mathcal{H}_{\mathrm{s}}$ usually in a linear way (Lawrie and Sarbach 1984). We now modify the analysis of Sect. 3.2 by including terms in this series up to the sixth degree in m_{s}. Thus

$$\phi_{\mathrm{L}}(L_{\mathrm{s}}, L_3, L, K; m_{\mathrm{s}}) \simeq -L_{\mathrm{s}} m_{\mathrm{s}} - L_3 m_{\mathrm{s}} + B_0(L, K) + B_2(L, K) m_{\mathrm{s}}^2 \\ + B_4(L, K) m_{\mathrm{s}}^4 + B_6(L, K) m_{\mathrm{s}}^6 \,, \tag{3.105}$$

and we shall consider only the case $B_6(L, K) > 0$. The conditions for minimizing ϕ_{L} as a function of m_{s} are

$$2B_2(L, K) m_{\mathrm{s}} + 4B_4(L, K) m_{\mathrm{s}}^3 + 6B_6(L, K) m_{\mathrm{s}}^5 \\ - L_{\mathrm{s}} - 3L_3 m_{\mathrm{s}}^2 = 0, \tag{3.106}$$

$$2B_2(L, K) + 12B_4(L, K) m_{\mathrm{s}}^2 + 30B_6(L, K) m_{\mathrm{s}}^4 \\ - 6L_3 m_{\mathrm{s}} > 0 \,. \tag{3.107}$$

In the symmetry plane when $L_{\mathrm{s}} = L_3 = 0$ the solutions of (3.106) are $m_{\mathrm{s}} = 0$, which is stable when $B_2(L, K) > 0$, and

$$m_{\mathrm{s}}^{(\pm)^2} = -\frac{B_4(L, K)}{3B_6(L, K)} \left[1 \pm \sqrt{1 - w(L, K)}\right] , \tag{3.108}$$

where

$$w(L, K) = \frac{3B_2(L, K) B_6(L, K)}{B_4^2(L, K)} \,. \tag{3.109}$$

From (3.106) and (3.107) the stability condition for the solutions (3.108) becomes

$$\mp B_4(L, K) \sqrt{1 - w(L, K)} > 0 \,. \tag{3.110}$$

It follows that the possible stable solutions from (3.108) are $\pm m_{\mathrm{s}}^{(-)}$ or $\pm m_{\mathrm{s}}^{(+)}$ according as $B_4(L, K) >, < 0$ respectively. In the limit $B_6(L, K) \to 0$ ($w(L, K) \to 0$), $m_{\mathrm{s}}^{(+)^2} \to \infty$, while $m_{\mathrm{s}}^{(-)^2}$ has the limit (3.14).

For the case $B_4(L, K) > 0$, $m_{\mathrm{s}}^{(-)}$ is real only when $w(L, K) < 0$. If

$$K = K^{(\mathrm{c})}(L) \tag{3.111}$$

is a solution of the equation

$$B_2(L, K) = 0, \tag{3.112}$$

giving a curve in the symmetry plane $L_{\mathrm{s}} = 0$. This curve represents a critical line $\mathcal{C}$ of continuous transitions with the ordered and disordered phases stable in the regions $B_2(L, K) < 0$ and $B_2(L, K) > 0$ respectively. On physical grounds we expect the ordered phase to occur on the side of $\mathcal{C}$ for which $K > K^{(\mathrm{c})}(L)$. This situation is very similar to that described in Sect. 3.2.

For the case $B_4(L, K) < 0$, the situation is quite different. Equation (3.108) gives a real value of $m_{\mathrm{s}}^{(+)}$ as long as $w(L, K) < 1$. From (3.105)

$$\phi_{\rm L}(0,L,K,m_{\rm s}^{(+)}) = B_0(L,K) + \tfrac{2}{3}B_4(L,K){m_{\rm s}^{(+)}}^4\left[\sqrt{1-w(L,K)}-\tfrac{1}{2}\right]. \tag{3.113}$$

The two minima $\pm m_{\rm s}^{(+)}$ are therefore metastable, with the stable solution given by $m_{\rm s}=0$, when $w(L,K)<\frac{3}{4}$. A first-order phase transition occurs at $w(L,K)=\frac{3}{4}$, or

$$B_2(L,K) = \frac{B_4^2(L,K)}{4B_6(L,K)}, \tag{3.114}$$

from the disordered state to the ordered state, with a discontinuous change of the order parameter to

$$m_{\rm s}^{(+)} = \pm\sqrt{\frac{-B_4(L,K)}{2B_6(L,K)}}, \tag{3.115}$$

the two values occurring with equal probability. The solution curve $\mathcal{T}$ to the equation (3.114) is a line of first-order transitions, which becomes a triple line is the full phase diagram in which the staggered coupling $L_{\rm s}$ is included. Suppose that $\mathcal{T}$ takes the form

$$K = K^{(\rm t)}(L). \tag{3.116}$$

Then the critical curve $\mathcal{C}$ given by (3.111) and the first-order curve $\mathcal{T}$ given by (3.116) meet at a tricritical point T with coordinates $(L_{\rm t},K_{\rm t})$ given as the solution of

$$B_2(L,K)=0, \qquad B_4(L,K)=0, \tag{3.117}$$

where

$$K_{\rm t} = K^{(\rm c)}(L_{\rm t}) = L^{(\rm t)}(L_{\rm t}). \tag{3.118}$$

It is straightforward to show that $\mathcal{C}$ and $\mathcal{T}$ together form a curve in the (L,K) plane with a continuous gradient at the T.

We now use the scaling procedure of Sect. 3.2 to obtain critical exponents along $\mathcal{C}$ and at T. In order to satisfy the critical curve condition (3.112), the tricritical condition (3.117) and with $B_4(L,K)$ and $B_6(L,K)$ having the appropriate signs we suppose that, to leading order,

$$\begin{aligned} &B_2(L,K) \simeq A_2[K^{(\rm c)}(L)-K], \\ &B_4(L,K) \simeq A_4 \mathrm{Sign}\{K-K_{\rm t}\}\sqrt{|K-K_{\rm t}|}, \\ &B_6(L,K) \simeq A_6, \end{aligned} \tag{3.119}$$

where A_2, A_4 and A_6 are positive constants (cf. equation (3.15)). The scaling analysis for a point on $\mathcal{C}$ proceeds exactly in the same way as in Sect. 3.2. It can be seen from equations (3.108) and (3.109) that, in a power-series expansion of the order parameter, $B_6(L,K)$ appears only with $B_2(L,K)$ in the combination given by $w(L, K)$. Thus A_6 constitutes only a higher-order correction to scaling and does not affect the exponents. The same analysis applied at the tricritical point yields, in place of the critical values of y_1, y_2 and x given by (3.18),

$$y_1 = \tfrac{5}{6}d\,, \qquad y_2 = \tfrac{2}{3}d\,, \qquad x = \tfrac{1}{6}d\,. \tag{3.120}$$

Substituting into the formulae (2.162) and (2.167) we obtain the tricritical exponents

$$\alpha_t = \tfrac{1}{2}\,, \qquad \beta_t = \tfrac{1}{4}\,, \qquad \gamma_t = 1\,, \qquad \delta_t = 5\,. \tag{3.121}$$

These exponent values do, of course, satisfy the Widom and Essam–Fisher scaling laws (2.169) and (2.170) (or their tricritical equivalents (2.244) and (2.245)).

An analysis of tricritical behaviour using the Gaussian approximation to Landau–Ginzberg theory can be made on the lines of Sects. 3.3 and 3.3.2. This yields $\nu = \frac{1}{2}$ and $\eta = 0$, the same values as for a critical point. The Fisher scaling law (2.203) is therefore satisfied. The Josephson hyper-scaling law (2.199) is satisfied only when $d = d_u = 3$, which is the upper borderline dimension for a tricritical point. It may, however, be shown (Sarbach and Fisher 1978) that the Gaussian critical exponents (3.70) and (3.71) apply to the tricritical point with the crossover from the classical values governed by a parameter with exponent $3 - d$. This parameter is thus relevant for $d < 3$, leading to the crossover, and marginal for $d = 3$. For $d > 3$ it is dangerously irrelevant, leading to an effective change of dimension in the system to $\bar{d} = 3$.

Examples

3.1 It is clear that equation (1.49) can be expressed in the form

$$\widehat{H}(\mathcal{H}, J; \sigma(\boldsymbol{r})) = -\frac{1}{2}J\sum_{\{\boldsymbol{r}\}}\sigma(\boldsymbol{r})\sum_{\{\boldsymbol{r}'\}}^{(\boldsymbol{r})}\sigma(\boldsymbol{r}') - \mathcal{H}\sum_{\{\boldsymbol{r}\}}\sigma(\boldsymbol{r})\,,$$

where the inner summation over $\boldsymbol{r}'$ is restricted to the z nearest-neighbour sites of $\boldsymbol{r}$. Show the form for the Hamiltonian (3.1) is obtained by using the mean-field assumption, where

$$\sum_{\{\boldsymbol{r}'\}}^{(\boldsymbol{r})}\sigma(\boldsymbol{r}') \qquad \text{is replaced by} \qquad \frac{z}{N}\sum_{\{\boldsymbol{r}\}}\sigma(\boldsymbol{r})\,,$$

together with (1.46). Use the same method to derive (3.83) from (3.81).

3.2 The response function

$$\varphi_T(L,K) = -\frac{\partial^2\phi}{\partial L^2} = \frac{\partial m}{\partial L}$$

is a dimensionless form of the susceptibility. Show using (3.12), (3.14) and (3.15) that

$$\lim_{K\to -K_c} \varphi_T(0,K) = [2A_2(K_c - K)]^{-1},$$

$$\lim_{K\to +K_c} \varphi_T(0,K) = [4A_2(K - K_c)]^{-1}.$$

3.3 The total contribution to the Hamiltonian (3.81) from a pair of nearest-neighbour sites $\boldsymbol{r}$ and $\boldsymbol{r}'$ is

$$-J\sigma(\boldsymbol{r})\sigma(\boldsymbol{r}') - \varepsilon\sigma^2(\boldsymbol{r})\sigma^2(\boldsymbol{r}')$$

$$-\frac{\mathcal{H}}{z}[\sigma(\boldsymbol{r}) + \sigma(\boldsymbol{r}')] - \frac{\mu}{z}[\sigma^2(\boldsymbol{r}') + \sigma^2(\boldsymbol{r}')],$$

Show that this quantity is invariant under the operations of the group $\mathcal{S}_3$ if $\mathcal{H} = 0$ and, for some constant R, $J = \frac{1}{2}R$, $\varepsilon = \frac{3}{2}R$ and $\mu = -Rz$. Hence show that the Hamiltonian (3.87) of the 3-state Potts model can be expressed in the form

$$\widehat{H}(R;\sigma(\boldsymbol{r})) = -R\sum_{\{\boldsymbol{r},\boldsymbol{r}'\}}^{(\text{n.n.})}\Big[1 + \tfrac{1}{2}\sigma(\boldsymbol{r})\sigma(\boldsymbol{r}')$$

$$+\tfrac{3}{2}\sigma^2(\boldsymbol{r})\sigma^2(\boldsymbol{r}') - \sigma^2(\boldsymbol{r}) - \sigma^2(\boldsymbol{r}')\Big].$$

4. Algebraic Methods in Statistical Mechanics

4.1 Introduction

There are rather few models in statistical mechanics which have been solved exactly and many of these can be formulated in vertex form (see Chap. 5 and Baxter 1982a). In the absence of such solutions, two alternative methods of attack have been to determine 'rigorous' conditions for the existence of phase transitions, or bounds on the regions of phase space within which phase transitions can exist (Ruelle 1969, Griffiths 1972, Sinai 1982). A common feature of many of these approaches is that information about the properties of an infinite system is obtained by investigating an equivalent finite system (or system of reduced dimensionality). This can be taken to be the unifying theme of the methods described in this chapter. In Sects. 4.3–4.8 it is convenient, although not essential, to described the ideas in terms of a lattice fluid model. The equivalence between the spin-$\frac{1}{2}$ Ising model with nearest-neighbour interactions and a simple lattice fluid with nearest-neighbour pair interactions between particles was first shown by Lee and Yang (1952). This relationship holds between any one-component lattice fluid of particles of chemical potential μ and a spin-$\frac{1}{2}$ system on the same lattice in a magnetic field $\mathcal{H}$. It follows that the methods and results described here can be 'translated' into a spin-$\frac{1}{2}$ formulation.

Consider a one-component lattice fluid on a finite lattice $\mathcal{N}$ of $|\mathcal{N}| = N$ sites. A state of the system can be described by specifying the subset $\mathcal{M} \subseteq \mathcal{N}$ of sites occupied by particles. The Hamiltonian for the model in the grand canonical distribution can be expressed in the form

$$\widehat{H}(\mathcal{M}) = E(\mathcal{M}) - \mu|\mathcal{M}|\,, \tag{4.1}$$

where $E(\mathcal{M})$ is the interaction energy of the system when the subset $\mathcal{M}$ is occupied by particles. The grand partition function is

$$\Xi(\mathcal{N}, T, \mu) = \sum_{\{\mathcal{M}\}} \exp[-\widehat{H}(\mathcal{M})/T]\,, \tag{4.2}$$

in terms of which the pressure and density of the fluid are given by

$$P(\mathcal{N}, T, \mu) = \frac{T}{N} \ln \Xi(\mathcal{N}, T, \mu) \tag{4.3}$$

and

$$\rho = \frac{\partial P}{\partial \mu}, \tag{4.4}$$

respectively. If the number $|\mathcal{M}| = M$ of particles on $\mathcal{N}$ were fixed, the distribution would be the canonical distribution with independent extensive variables N and M (see Volume 1, Sect. 2.1) and partition function

$$Z(\mathcal{N}, T, M) = \sum_{\{|\mathcal{M}|=M\}} \exp[-E(\mathcal{M})/T], \tag{4.5}$$

where the sum is over all subsets of $\mathcal{N}$ with M sites. The grand partition function can now be expressed in the form

$$\Xi(\mathcal{N}, T, \mathfrak{z}) = \sum_{\widehat{M}=0}^{N} Z(\mathcal{N}, T, \widehat{M}) \mathfrak{z}^{\widehat{M}}, \tag{4.6}$$

where $\mathfrak{z} = \exp(\mu/T)$ is the *fugacity* or *activity*.

In Sect. 1.4 we referred to the concept of the thermodynamic limit. In this case it corresponds to taking the limit as the lattice $\mathcal{N}$ becomes infinitely large. That this limit is important for the existence of phase transitions can be seen in the following way. A first-order phase transition in the system corresponds to a discontinuity in density. Higher-order phase transitions correspond to other sorts of non-smooth behaviour of P as a function of $\mathfrak{z}$. For finite $\mathcal{N}$ it is clear, from (4.3) and (4.6), that the only singular points of $P(\mathcal{N}, T, \mathfrak{z})$, as a function of $\mathfrak{z}$, are the N zeros of $\Xi(\mathcal{N}, T, \mathfrak{z})$, which is an N-th degree polynomial in $\mathfrak{z}$. Since $\Xi(\mathcal{N}, T, \mathfrak{z})$ has only positive coefficients none of its roots is real and positive. The lattice fluid on a finite lattice $\mathcal{N}$ can therefore have no phase transitions.

The *Peierls method* (Peierls 1936) provides one of the few procedures for proving the existence of a phase transition without explicit calculation of the transition point. The original work of Peierls contains a proof that the Ising model on the square lattice has a spontaneous magnetic moment at sufficiently low temperatures. The argument was given a rigorous formulation by Griffiths (1964) and Dobrushin (1965). It was extended to lattice fluids with nearest-neighbour repulsion or exclusion by Dobrushin (1968a, 1968b), for hypercubic lattices, and by Heilmann (1972), Abraham and Heilmann (1980) and Heilmann (1980) for triangular and face-centred cubic lattices. Bonded lattice fluid models for water, have been developed on the triangular lattice by Bell and Lavis (1970) and on the body-centred cubic by Bell (1972). The Peierls method has been used to established the existence of phase transitions at sufficiently low temperatures in these models by Heilmann and Huckaby (1979) and Lavis and Southern (1984) respectively.

An important insight into the theory of phase transitions, originating in the papers of Yang and Lee, (Yang and Lee 1952, Lee and Yang 1952) is based on the idea of taking the fugacity $\mathfrak{z}$ in the partition function $\Xi(\mathcal{N}, T, \mathfrak{z})$

to be a complex variable. They proved the well-known *circle theorem* that, for a spin-$\frac{1}{2}$ Ising ferromagnet with pairwise interactions of arbitrary range, the zeros of the grand partition function lie on the unit circle $|\mathfrak{Z}| = 1$ in the complex plane of the variable $\mathfrak{Z} = \exp(-2\mathcal{H}/T)$.[1] This result has extended to a larger class of models (Asano 1968, 1970, Suzuki and Fisher 1971, Newman 1975, Lieb and Sokal 1981) but this does not include the ν-state Potts model for $\nu > 2$ (Martin and Maillard 1986, O'Rourke et al. 1995). Lieb and Ruelle (1972) considered the distribution of zeros for the antiferromagnetic case and showed that, if the temperature is above the critical temperature, with an arc of the unit circle around $\mathfrak{Z} = 1$ free of zeros then there will also be no zeros in the circle containing this arc and orthogonal to the unit circle. Heilmann and Lieb (1972) were able to derive the circle theorem for both Ising and Heisenberg models from an investigation of the partition function zeros of a monomer–dimer system in the monomer fugacity plane. Nishimori and Griffiths (1983) have investigated the motion of the zeros in the complex plane as the temperature is changed. Ruelle (1971) used the contraction method of Asano (1970) to prove a result which provides, in principle, a method for obtaining regions containing the grand partition function zeros of one-component lattice fluids with finite-range interactions whether or not they lie on $|\mathfrak{Z}| = 1$.

The Yang–Lee circle theorem and Ruelle's theorem are both about the complex plane of the variable $\mathfrak{Z}$, whether it is understood as the Boltzmann factor of the magnetic field $\mathcal{H}$ or of the chemical potential μ. The important point about $\mathfrak{Z}$ is that it is the Boltzmann factor for a field associated with single site occupations. Any partition function of a lattice system can, however, be expressed as a sum of powers of a number of Boltzmann factors, some of which will arise from particle interactions. These quantities can also, in the spirit of Yang and Lee, be treated as complex variables. Thus, for example, for the spin-$\frac{1}{2}$ ferromagnetic Ising model with Hamiltonian (1.49) on the finite lattice $\mathcal{N}$ the grand partition function is a polynomial in $\mathfrak{X} = \exp(-2J/T)$ and $\mathfrak{Z} = \exp(-2\mathcal{H}/T)$. As indicated above the distribution of zeros in the complex plane of $\mathfrak{Z}$ satisfy the circle theorem. The distribution of zeros in the complex plane of $\mathfrak{X}$ has also been investigated (Fisher 1965, Jones 1966, Brascamp and Kunz 1974) and it has been shown that in the limit of an infinite system, when $\mathcal{H} = 0$ the asymptotic locus of zeros is given by the *Fisher circles*

$$\mathfrak{X} = \sqrt{2}\exp(\mathrm{i}\theta) \pm 1\,. \tag{4.7}$$

These circles are a result of the self-duality and spin-inversion symmetry of the Ising model (Volume 1, Chap. 8). The duality (but not spin inversion)

[1] There is, of course, an equivalent lattice fluid formulation of this result (see Sect. 4.7), but since it relies on the Ising model hole–particle symmetry condition $E(\mathcal{M}) = E(\mathcal{N} - \mathcal{M})$, where $\mathcal{N} - \mathcal{M}$ is the complement of $\mathcal{M}$ on the lattice $\mathcal{N}$, it has the rather artificial property that two empty sites interact with the same energy as two occupied sites.

generalizes to the ν-state Potts model (Wu 1982), which has a self-dual *Potts circle*

$$\mathfrak{X} = \sqrt{\nu}\exp(\mathrm{i}\theta) + 1\,. \tag{4.8}$$

For the square lattice Ising model on a semi-infinite lattice Wood (1985) showed that the Fisher circles could be computed as curves on which transfer matrix eigenvalues are degenerate in modulus and thus that these latter yield the asymptotic locus of partition function zeros. Following a description of the use of transfer matrices in Sect. 4.9 Wood's method is described in detail in Sect. 4.10.

4.2 The Thermodynamic Limit

We consider the thermodynamic limit of the one-component lattice fluid when $\mathcal{N}$ grows so that $|\mathcal{N}| = N \to \infty$ and $\mathcal{N}$ tends to an infinite lattice $\mathcal{L}$. To achieve a sensible outcome in this procedure it is necessary that

$$\lim_{\mathcal{N}\to\mathcal{L}} P(\mathcal{N}, T, \mathfrak{Z}) = P(\mathcal{L}, T, \mathfrak{Z})\,, \tag{4.9}$$

where $P(\mathcal{L}, T, \mathfrak{Z})$ is the finite pressure of the lattice fluid on the infinite lattice $\mathcal{L}$. Yang and Lee (1952) established the existence of the thermodynamic limit for the grand distribution of a one-component fluid of particles with hard cores and finite range interactions in a volume V. This work was extended by Witten (1954), who relaxed the condition of hard cores. The corresponding theorem for lattice systems is given by Ruelle (1969) (see also Griffiths 1972). Three ingredients are necessary for the existence of the thermodynamic limit:

(i) translational invariance of the interparticle interactions,

(ii) some restriction on the range of the interactions,

(iii) a limitation on the proportional number of boundary sites as the lattice grows.

These conditions can be stated more precisely in the following way:

Given any subset $\mathcal{X}$ of the lattice sites, let $\mathcal{E}(\mathcal{X})$ be the contribution to the energy of the system which would arise if every site of $\mathcal{X}$ were occupied, but which would be absent if one or more sites of $\mathcal{X}$ were empty. This means that $\mathcal{E}(\mathcal{X})$ cannot be decomposed in a sum of contributions from proper subsets of $\mathcal{X}$. If $|\mathcal{X}| = X > 1$ and $\mathcal{E}(\mathcal{X}) \neq 0$, then $\mathcal{X}$ is called an *interacting set*. The system is *translationally invariant* if, for all $\mathcal{X}$,

$\mathcal{E}(\mathcal{X})$ is dependent only on the geometric form of $\mathcal{X}$ and not on its location on the lattice. (4.10)

Now choose a site $\boldsymbol{r}_0$ belonging to every lattice $\mathcal{N}$ in the limit $\mathcal{N} \to \mathcal{L}$ and let $\{\mathcal{X}\}_0$ be all the subsets, of all possible $\mathcal{N}$, which contain $\boldsymbol{r}_0$. The necessary condition on the range of interactions is that

$$\sum_{\{\mathcal{X}\}_0} \frac{|\mathcal{E}(\mathcal{X})|}{|\mathcal{X}|} < \infty . \tag{4.11}$$

Finally suppose that $\mathcal{N}_h$ is the set of sites of $\mathcal{N}$ which are within some fixed distance h of the boundary. A limiting procedure $\mathcal{N} \to \mathcal{L}$ is said to be taken *in the sense of van Hove* (Ruelle 1969, p. 14) if

$$\lim_{\mathcal{N}\to\mathcal{L}} \frac{|\mathcal{N}_h|}{|\mathcal{N}|} = 0 , \qquad \text{for any fixed } h. \tag{4.12}$$

The following theorem can now be proved (Ruelle 1969, p. 22):

Theorem 4.2.1. *The limit (4.9) exists for* $\mathfrak{Z} > 0$ *with* $P(\mathcal{L}, T, \mathfrak{Z})$ *a continuous non-decreasing function of* $\mathfrak{Z}$ *if (4.10)–(4.12) are satisfied.*

Because of the equivalence between the one-component lattice fluid and spin-$\frac{1}{2}$ lattice models referred to above this result can be translated into an equivalent result for the latter, establishing the continuity and monotonicity of the free energy density as a function of the appropriate variable.

4.3 Lower Bounds for Phase Transitions: the Peierls Method

Phase transitions in statistical mechanical models correspond to the onset of some kind of order. Except in exceptional circumstances, of which the transition to the frozen ferroelectric state in the six-vertex model is an example (see Sect. 5.11 and Volume 1, Sect. 10.6), immediately below the phase transition this ordering is imperfect, with the perfectly ordered state occurring only at $T = 0$. An ordered phase can thus be understood by reference to its perfect realization at zero temperature and the types of ordered phases can in turn be understood by defining the concept of a *ground state*. In a rather general way the *order parameter* of a phase can be defined as the expectation value of a quantity which measures the extent to which the site occupations mirror those of the ground state of the phase. For the Ising ferromagnet on the square lattice with Hamiltonian (1.49) the magnetization density is

$$m(\mathcal{N}, L, K) = 2\left[\frac{\langle \widehat{N}(\uparrow)\rangle}{N} - \frac{1}{2}\right], \tag{4.13}$$

where $\widehat{N}(\uparrow)$ is the number from the N sites of $\mathcal{N}$ which are in the spin-up state. The order parameter for the all spin-up ferromagnetic phase in zero field on the infinite lattice $\mathcal{L}$ is

$$m(\mathcal{L},0,K) = \lim_{L\to 0+} \lim_{\mathcal{N}\to\mathcal{L}} m(\mathcal{N},L,K)\,. \tag{4.14}$$

Let

$$m_\uparrow(\mathcal{N},L,K) = 2\left[\frac{\langle \widehat{N}(\uparrow)\rangle_\uparrow}{N} - \frac{1}{2}\right], \tag{4.15}$$

where $\langle\cdot\rangle_\uparrow$ denotes the constrained expectation value with all the boundary sites containing up spins. There are two steps in the Peierls argument for the square lattice Ising ferromagnet:

(a) Show that, if there exists an $\alpha > 0$, independent of N, such that

$$m_\uparrow(\mathcal{N},0,K) \geq \alpha\,, \tag{4.16}$$

for all $K > K'$ and all $\mathcal{N}$, then

$$m(\mathcal{L},0,K) \geq \alpha\,, \tag{4.17}$$

for all $K > K'$.

(b) Establish the existence of α and K'.

There are many presentations of details of this argument (Thompson 1988). Rather than giving another we shall adopt the more general approach of Heilmann and Huckaby (1979) and concentrate on the derivation of a lower temperature bound for the existence of a phase transition. Although Heilmann and Huckaby (1979) present their development for some bonded lattice fluid models it is not in essence specific to those models and can be applied to any one-component lattice fluid satisfying conditions (4.10)–(4.12).

In the terminology of Sect. 1.4 suppose that the one-component lattice fluid has an energy $E(\mathcal{M})$, which, for all $\mathcal{M}$, is a linear function of the parameters $\varepsilon_1, \varepsilon_2, \ldots, \varepsilon_{n_e}$. Then the Hamiltonian is a linear function of this set and the chemical potential μ. A *ground state* G is defined to be a configuration of the system for which the set $\mathcal{G} \subseteq \mathcal{N}$ of sites is occupied by particles and for which:

(i) $\mathcal{G}$ is a periodic arrangement of particles on the lattice.

(ii) There is a set of values of $\varepsilon_1, \varepsilon_2, \ldots, \varepsilon_{n_e}, \mu$ for which

$$\widehat{H}(\mathcal{G}) \leq \widehat{H}(\mathcal{M})\,, \tag{4.18}$$

for all $\mathcal{M} \subseteq \mathcal{N}$, with equality holding, if at all, only when $\mathcal{M}$ is also periodic and thus itself a ground state.

G is said to be a *stable* ground state in the region of parameter space in which it satisfies (4.18). The number $\omega(\mathrm{G})$ of ground states which are stable in the region in which G is stable is called the *degeneracy* of G.

Consider now a ground state G_0. For any occupation $\mathcal{M}$, a site of the lattice is said to *belong to* G_0 if it is occupied in the way it would be if the system were in the ground state G_0. For any occupation of the lattice let $N_0(\mathcal{M})$ be the number of sites which belong to G_0 and define

$$\rho_0(\mathcal{N}, L, K_1, \ldots, K_{n_e}) = \frac{\langle N_0(\mathcal{M})\rangle_0}{N} - \frac{1}{\omega(\mathrm{G}_0)}, \tag{4.19}$$

where $K_\ell = \varepsilon_\ell/T$, $L = \mu/T$ and $\langle\cdot\rangle_0$ denotes the constrained expectation value taken over all $\mathcal{M}$ for which the boundary sites belong to G_0. When the parameters $\varepsilon_1, \varepsilon_2, \ldots, \varepsilon_{n_e}, \mu$ take values for which G_0 is a stable ground state the limit $\mathcal{N} \to \mathcal{L}$ of $\rho_0(\mathcal{N}, L, K_1, \ldots, K_{n_e})$ is the order parameter for G_0. The replacement for (4.16) is then

$$\rho_0(\mathcal{N}, L, K_1, \ldots, K_{n_e}) \geq \alpha > 0, \tag{4.20}$$

(Heilmann and Huckaby 1979, Dobrushin 1968a, 1968b). It is clear that (4.20) could never be satisfied if $\omega(\mathrm{G}_0) = 1$; non-degenerate ground states cannot lead to ordered phases. This would allow us, if we chose, to normalize $\rho_0(\mathcal{N}, L, K_1, \ldots, K_{n_e})$ at $T = 0$ by dividing by $1 - 1/\omega(\mathrm{G}_0)$. It is in fact this normalized quantity which is equivalent to $m_\uparrow(\mathcal{N}, 0, K)$ in the zero-field Ising ferromagnet. To obtain a useful lower bound on phase transitions we must now use (4.20) to derive an inequality involving the temperature.

Let the lattice $\mathcal{N}$ be divided into *units* $\mathcal{C}_\iota$, $\iota = 1, 2, \ldots, \eta$ which consist of the smallest subsets of the sites with the following properties:

(a) All the units have the same number of sites $C = |\mathcal{C}_\iota|$ and the same geometrical structure. That is, they can be transformed into each other by rotations and translations.

(b) For any ground state and for some labelling of the sites of the units each unit is identically occupied both with respect to the number of particles and the labelling.

(c) Every interacting set is contained within at least one unit.

It is not difficult to see that it is always possible to divide the lattice into units in the way described here as long as the boundaries of the lattice are constructed in a suitable way. For each unit $\mathcal{C}_\iota$, a *cell* $\mathfrak{C}_\iota$ is defined by the formula

$$\mathfrak{C}_\iota = \left\{ \boldsymbol{r} : \sum_{j=1}^{C} \lambda_j \boldsymbol{r}_j^{(\iota)} = \boldsymbol{r},\ \forall\ \lambda_j \geq 0 \text{ such that } \sum_{j=1}^{C} \lambda_j = 1 \right\}, \tag{4.21}$$

where $\boldsymbol{r}_1^{(\iota)}, \boldsymbol{r}_2^{(\iota)}, \ldots, \boldsymbol{r}_C^{(\iota)}$ are the sites of the unit $\mathcal{C}_\iota$, which are called the *vertices* of $\mathfrak{C}_\iota$. The cells of the lattice are closed convex sets of points in the space occupied by the lattice and spanned by their vertices.[2] In a d-dimensional lattice the dimension $d_{\mathfrak{C}}$ of the cells is less than or equal to d.

[2] If the points of the unit are linearly independent, as is the case for the models treated by Heilmann and Huckaby (1979), then (4.21) is the usual definition of a *simplex* (see, example, Munkres 1984).

The elements of the surfaces of the cells spanned by sets of $d_{\mathfrak{C}} - 1$ vertices are called *faces*.[3] We shall confine our attention to the case where the only overlaps between cells are the points on the faces and where each cell face lies in two cells which will be called *neighbours*. A simply-connected set of cells will be called a *cluster*. The *surface* of a cluster consists of those faces of the cells of the cluster which belong to only one cell of the cluster. The *boundary* of the cluster consists of those cells which have faces on the surface. An *interior site* of a cluster is one which does not lie on its surface. An *interior cell* of the cluster is one which is not part of the boundary. Suppose that all sites of the lattice not lying on the boundary belong to $q_0 \geq 2$ cells. Then a cluster with an interior site must contain at least q_0 cells.

Each cell of the lattice can be occupied in 2^C ways which we denote by $\mathcal{O}_k$, $k = 1, 2, \ldots, 2^C$. Among the particle arrangements when each cell is identically occupied will be, by definition, all the ground states of the system. However, it is not necessarily the case that all such configurations are ground states, since there may be some corresponding to particle occupations which do not satisfy the minimization condition (4.18) for any set of values of the energy parameters. Any particle occupation $\mathcal{M}$ of the lattice implies the occupation $\mathcal{C}_\iota \cap \mathcal{M}$ for the cell $\mathfrak{C}_\iota$. Each $\mathcal{C}_\iota \cap \mathcal{M}$ is one of the occupations $\mathcal{O}_k$ and if, for occupation $\mathcal{M}$, $\nu_k(\mathcal{M})$ of the cells have occupation $\mathcal{O}_k$ the Hamiltonian can be expressed in the form

$$\widehat{H}(\mathcal{M}) = \sum_{\iota=1}^{\eta} \widehat{H}_{\mathfrak{C}}(\mathcal{C}_\iota \cap \mathcal{M}) = \sum_{k=1}^{2^C} \nu_k(\mathcal{M}) \widehat{H}_{\mathfrak{C}}(\mathcal{O}_k) , \tag{4.22}$$

where $\widehat{H}_{\mathfrak{C}}(\cdot)$ is the contribution to the Hamiltonian from a single cell $\mathfrak{C}$. This contribution will contain all factors of the form $\mathcal{E}(\mathcal{X})$, for every fully occupied interacting set $\mathcal{X} \subseteq \mathcal{C}$. If $\mathcal{X}$ lies in more than one unit the contribution is divided equally between that subset of units. For any occupation $\mathcal{M}$ in which each of the cells is identically occupied each of the terms in the first sum in (4.22) will be equal and the second sum will have only one term. For the ground state G_0 in which each cell has occupation $\mathcal{C} \cap \mathcal{G}_0$

$$\widehat{H}(\mathcal{G}_0) = \eta \widehat{H}_{\mathfrak{C}}(\mathcal{C} \cap \mathcal{G}_0) . \tag{4.23}$$

A face of a cell is said to be G_0-*ordered* (or more briefly *ordered*, since we are concentrating on a fixed ground state) if all its sites belong to G_0. Otherwise it is said to be *disordered*. Given any occupation $\mathcal{M} \neq \mathcal{G}_0$, two cells of the lattice are said to be *linked* if they share a disordered face. Given the specific occupation $\mathcal{M}$ of the lattice a cell $\mathfrak{C}_\iota$ with some occupation $\mathcal{O}_k \neq \mathcal{C}_\iota \cap \mathcal{G}_0$ is called a *contour segment*. The boundary of a cluster is called a *contour* if:

(i) Each of its cells is a contour segment. (Some, but not necessarily all, of the interior cells of the cluster will also be contour segments.)

[3] For a simplex, all such linear combinations will give points on the surface, but this is not generally true for the definition of a cell given here.

(ii) The cluster contains at least one interior site not belonging to G_0. (A contour must, therefore, consist of at least q_0 cells.)

(iii) The cells of the contour are not linked to any cell not in the cluster. (The surface of the cluster must consist of ordered faces.)

(iv) The cluster cannot be decomposed into two disjoint clusters, each bounded by a contour.

We denote a particular contour by $[q;j]$, where q is the number of segments in the contour and $j = 1, 2, \ldots, m(q)$ will be the variable which differentiates between different contours which can be formed on the lattice with q segments. Let $n_j(q)$ be the number of interior sites of the cluster bounded by the contour $[q;j]$.

Suppose that the occupation $\mathcal{M}$ of the lattice has all the boundary sites belonging to G_0. We are interested in obtaining an upper bound on the number of sites of $\mathcal{M}$ which do not belong to G_0. This will clearly be given by the total number of interior sites contained within the contours present in $\mathcal{M}$. Thus

$$N - N_0(\mathcal{M}) \le \sum_{\{q\}} \sum_{j=1}^{m(q)} n_j(q)\delta[q;j] , \tag{4.24}$$

where the outer summation is over all possible contour sizes and

$$\delta[q;j] = \begin{cases} 1, & \text{if } [q;j] \text{ is present in } \mathcal{M}, \\ 0, & \text{otherwise.} \end{cases} \tag{4.25}$$

Let $\{\mathcal{M}\}_0$ be the set of lattice occupations for which the boundary sites of $\mathcal{N}$ belong to G_0 and let $\{\mathcal{M}\}_0^{[q;j]}$ be the set which satisfy the additional requirement that the contour $[q;j]$ is present. Then

$$\langle \delta[q;j] \rangle_0 = \frac{\displaystyle\sum_{\{\mathcal{M}\}_0^{[q;j]}} \exp[-\hat{H}(\mathcal{M})/T]}{\displaystyle\sum_{\{\mathcal{M}\}_0} \exp[-\hat{H}(\mathcal{M})/T]} . \tag{4.26}$$

For every $\mathcal{M} \in \{\mathcal{M}\}_0^{[q;j]}$ there is an $\mathcal{M}^* \in \{\mathcal{M}\}_0$, but $\notin \{\mathcal{M}\}_0^{[q;j]}$, which differs from $\mathcal{M}$ only in that all the interior points of $[q;j]$ not belonging to G_0 have had their occupations changed to put them into G_0. We denote this new set of occupations by $\{\mathcal{M}^*\}_0^{[q;j]}$. Thus

$$\langle\delta[q;j]\rangle_0 < \frac{\displaystyle\sum_{\{\mathcal{M}\}_0^{[q;j]}} \exp[-\widehat{H}(\mathcal{M})/T]}{\displaystyle\sum_{\{\mathcal{M}^*\}_0^{[q;j]}} \exp[-\widehat{H}(\mathcal{M}^*)/T]}$$

$$= \exp\left\{-\sum_{k=1}^{2^C} \theta_k[q;j][\widehat{H}_{\mathfrak{C}}(\mathcal{O}_k) - \widehat{H}_{\mathfrak{C}}(\mathcal{C}\cap\mathcal{G}_0)]/T\right\}, \tag{4.27}$$

where $\theta_k[q;j]$ is the number of cells with occupation $\mathcal{O}_k$ in contour $[q;j]$. To find an upper bound to the final term in (4.27) it is necessary to replace $\widehat{H}_{\mathfrak{C}}(\mathcal{O}_k)$ by its least possible value for each cell of the cluster bounded by $[q;j]$. For the interior cells of the cluster it is possible for $\widehat{H}_{\mathfrak{C}}(\mathcal{O}_k) = \widehat{H}_{\mathfrak{C}}(\mathcal{C}\cap\mathcal{G}_0)$, either because $\mathcal{O}_k = \mathcal{C}\cap\mathcal{G}_0$ or because $\mathcal{O}_k$ is the occupation appropriate to some other ground-state degenerate with G_0. The cells of the contour itself, however, are contour segments with occupations different from $\mathcal{C}\cap\mathcal{G}_0$. They must also be different from the cell occupation of some other ground-state degenerate with G_0. (If this were not the case it would be possible to 'mix' these two occupations to achieve a ground-state degeneracy of the order of N.) It follows that

$$\langle\delta[q;j]\rangle_0 < \exp[-\varepsilon_0 q/T], \tag{4.28}$$

where

$$\varepsilon_0 = \min_{\widehat{H}_{\mathfrak{C}}(\mathcal{O}_k)\neq\widehat{H}_{\mathfrak{C}}(\mathcal{C}\cap\mathcal{G}_0)} \{\widehat{H}_{\mathfrak{C}}(\mathcal{O}_k)\} - \widehat{H}_{\mathfrak{C}}(\mathcal{C}\cap\mathcal{G}_0). \tag{4.29}$$

From (4.19), (4.24) and (4.28)

$$1 - \sum_{\{q\}} Q(q)\exp(-\varepsilon_0 q/T) - \frac{1}{\omega(\mathrm{G}_0)} < \rho_0(K_1,\ldots,K_{n_e},L,\mathcal{N}), \tag{4.30}$$

where $Q(q)$ is some function of q such that

$$\frac{1}{N}\sum_{j=1}^{m(q)} n_j(q) \le Q(q). \tag{4.31}$$

If such a $Q(q)$ can be found such that the series in (4.30) is convergent for T in the range $[0, \varepsilon_0 t(\mathrm{G}_0)]$, where $t(\mathrm{G}_0)$ is the least positive solution of the equation

$$\sum_{\{q\}} Q(q)\exp\left[-\frac{q}{t(\mathrm{G}_0,\epsilon)}\right] = 1 - \frac{1}{\omega(\mathrm{G}_0)}, \tag{4.32}$$

then (4.20) is satisfied if

$$T < \varepsilon_0 t(\mathrm{G}_0), \tag{4.33}$$

and, in the thermodynamic limit, the system, with parameter values appropriate to the ground state G_0, will be in the corresponding ordered state.

Table 4.1. Cell occupations for nearest-neighbour pairs for the simple lattice fluid.

ι	Occupation	$\widehat{H}_{\mathfrak{C}}(\mathcal{C} \cap \mathcal{G}_\iota)$
1	●—●	$-\varepsilon - 2\mu/z$
2	●—○	$-\mu/z$
3	○—●	$-\mu/z$
4	○—○	0

4.4 Lower Bounds for the Simple Lattice Fluid

The Hamiltonian for the spin-$\frac{1}{2}$ Ising model is given by (1.49). By using the variable $\tau(\boldsymbol{r}) = (1 + \sigma(\boldsymbol{r}))/2$, where $\tau(\boldsymbol{r}) = 0, 1$, this Hamiltonian can be transformed (see Volume 1, Sect. 5.3) into the lattice fluid form

$$\hat{H}(\mu, \varepsilon; \tau(\boldsymbol{r})) = -\varepsilon \sum_{\{\boldsymbol{r},\boldsymbol{r}'\}}^{(\text{n.n.})} \tau(\boldsymbol{r})\tau(\boldsymbol{r}') - \mu \sum_{\{\boldsymbol{r}\}} \tau(\boldsymbol{r}) , \tag{4.34}$$

where, $\varepsilon = 4J$, the chemical potential $\mu = 2\mathcal{H} - 2Jz$, with z the coordination number of the lattice and we have omitted a constant term from (4.34). The cells of the lattice can be taken to be the lines joining nearest-neighbour pairs. There are four possible occupations of a cell. These are listed in Table 4.1. They are denoted by $\mathcal{C} \cap \mathcal{G}_\iota$, $\iota = 1, \ldots 4$, since for each there is a region of parameter space for which it corresponds to a stable ground state. These regions are shown in Fig. 4.1. In the region $2 \sim 3$ the two ground states 2 and 3 are degenerate so it is possible for an ordered phase to occur. Otherwise ordered states can occur only on the phase boundaries. In particular on the boundary $\mu = -\frac{1}{2}z\varepsilon$, $\varepsilon > 0$ the ground states 1 and 4 are degenerate. (This line corresponds to the zero-field axis $\mathcal{H} = 0$ in the spin representation and the

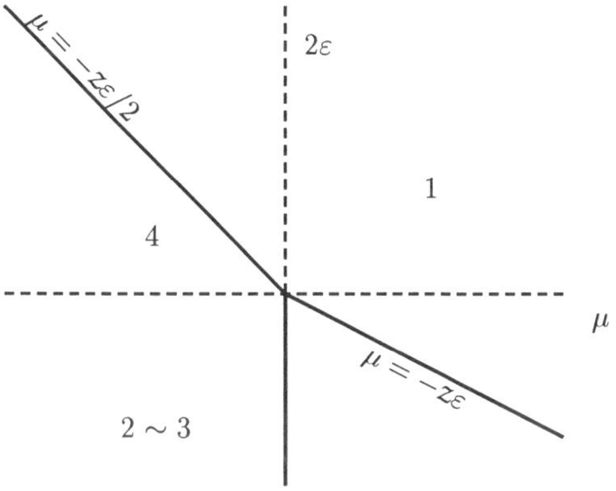

Fig. 4.1. The ground-state diagram for the simple lattice fluid. Regions of stability are labelled by ι, $\iota = 1, \ldots, 4$. The region $2 \sim 3$ is that in which the degenerate ground states 2 and 3 are both stable.

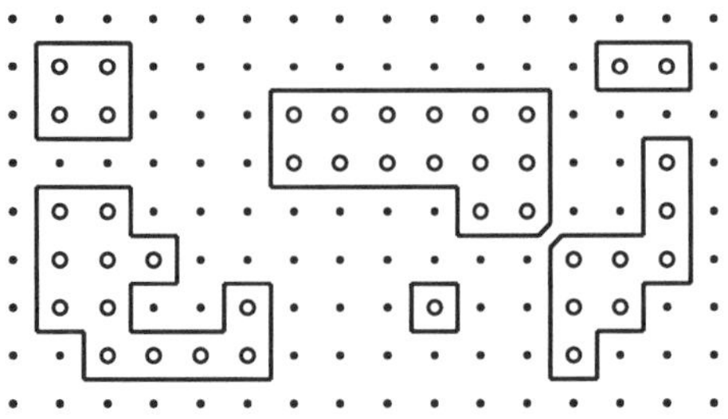

Fig. 4.2. The equivalence between contours and polygons on the square lattice. The maximum number of vacant sites corresponds to the case, as shown here, where all the sites enclosed by the polygons are empty.

model is equivalent to the zero-field Ising ferromagnet) We shall now apply the method of the previous section to the model along this phase boundary, to investigate the existence of the ordered phase corresponding to G_1. From (4.29) $\varepsilon_0 = -\mu/z = \varepsilon/2$ and, of course, $\omega(\mathrm{G}_1) = 2$, $q_0 = z$. It is now necessary to determine a quantity $Q(q)$ satisfying (4.31). This will depend on the lattice structure under consideration.

Although it is well-known that there is no phase transition for the linear lattice (Volume 1, Sect. 2.4), it is of interest to see how the theory of the previous section can be applied to this case. The procedure for doing this is outlined in Example 4.1. Here we consider the case of the square lattice. For a particular occupation, with all the boundary sites of the lattice occupied by particles, a contour can be represented by drawing a line through the centre of each boundary segment (each nearest-neighbour pair with occupation 2 or 3), between the centres of the two lattice squares of which the segment is a side. It is clear that an even number of lines will meet at the centre of each square and that the lines form closed polygons. Whenever four lines meet at the centre of a square the lines are cut to disconnected the polygon into two simply-connected parts. The effect of this procedure is to enclose all the interior sites of the clusters bounded by the contours within polygons (see Fig. 4.2). Now a contour of q segments is equivalent to a polygon consisting of q line elements drawn between the centres of lattice squares and the possible values of q are $4, 6, 8, \ldots$. A polygon encloses the maximum number of sites when it is a square. It follows that $n_j(q) \le (q/4)^2$. We now need an upper bound on the total number $m(q)$ of closed polygons of q line elements. Ignoring the boundaries of the lattice we chose the centre of one of the N squares and begin to draw a polygon. The first line element can be chosen in four ways and each subsequent line element in three ways. There are, therefore, $4N3^{q-1}$ ways of completing this construction. Among these will be all closed polygons, including those which cross themselves, and all open arcs. Each simply-connected closed polygon will have been counted q times and thus $m(q) < 4N3^{q-1}/q$. It follows that

$$\frac{1}{N}\sum_{j=1}^{m(q)} n_j(q) < \tfrac{1}{4}q3^{q-1} = Q(q)\,. \tag{4.35}$$

Substituting into equation (4.32) we obtain the equation

$$x^3 + x^2 - 6x + 3 = 0\,, \tag{4.36}$$

as long as $x = 9\exp[-2/t(\mathrm{G}_1)] < 1$. This equation has the single root $x = 0.5936$ in the range $0 < x < 1$ and from (4.33) we establish the existence of an ordered phase when $T < 0.3678\epsilon$. This can be compared with the exact result for the critical temperature $T_\mathrm{c} = \epsilon/[2\ln(1+\sqrt{2})] = 0.5673\epsilon$. The difference between these two values illustrates the point that the importance of the Peierls method consists not in is ability to yield good values for critical points, for which series methods are much better (Chap. 7), but rather in its ability to establish in a rigorous way the existence of an ordered phase.

4.5 Grand Partition Function Zeros and Phase Transitions

Suppose that $\mathbb{O}_\mathcal{N}(T) = \{\mathfrak{z}_1, \mathfrak{z}_2, \ldots, \mathfrak{z}_N\}$ is the set of zeros of the grand partition function (4.6) in the complex plane $\mathbb{C}_\mathfrak{z}$. The connection between the roots of the grand partition function and phase transitions in the system is established by the following theorem, due to Yang and Lee (1952), for which we shall provide a proof:

Theorem 4.5.1. *Let $\mathbb{W}(T)$ be a region in $\mathbb{C}_\mathfrak{z}$ which contains a segment of the positive real axis $\mathbb{R}^{(+)}$, but none of the members of $\mathbb{O}_\mathcal{N}(T)$, for any of the sequence of lattices involved in the limiting process described in Sect. 4.2. Then for all $\mathfrak{z} \in \mathbb{W}(T)$ the limit (4.9) is uniform and $P(\mathcal{L}, T, \mathfrak{z})$ is an analytic function[4] of $\mathfrak{z}$ in $\mathbb{W}(T)$.*

Proof: Let $\mathbb{V}(T) \subseteq \mathbb{W}(T)$ be a closed disc of finite radius with its centre on $\mathbb{R}^{(+)}$. We first prove the theorem for $\mathbb{V}(T)$ using the *Vitali convergence theorem.*[5] We consider the sequence of functions $P(\mathcal{N}, T, \mathfrak{z})$ of $\mathfrak{z}$ given by the set of lattices leading to the limit in Theorem 4.2.1. These are analytic functions in $\mathbb{V}(T)$ for the reasons given above. According to Theorem 4.2.1 they tend to a limit on $\mathbb{V}(T) \cap \mathbb{R}^{(+)}$. Now let $\mathfrak{z}^*$ be the right-hand extremity of the diameter of $\mathbb{V}(T)$. The sequence of finite real numbers $P(\mathcal{N}, T, \mathfrak{z}^*)$ tends to the limit $P(\mathcal{L}, T, \mathfrak{z}^*)$. It is therefore bounded above by some number B and since $|\mathfrak{z}| \le \mathfrak{z}^*$ on $\mathbb{V}(T)$ it follows that $P(\mathcal{N}, T, \mathfrak{z}) \le B$ for all lattices $\mathcal{N}$ in the sequence and all $\mathfrak{z} \in \mathbb{V}(T)$. This establishes the theorem for $\mathbb{V}(T)$.

[4] A complex valued function is said to be *analytic* in a region of the complex plane if it is defined and has a derivative at each point of that region.

[5] The statement of the theorem is a follows (for a proof see Titchmarsh 1939): Let $f_n(\mathfrak{z})$ be a sequence of functions which are analytic in a region $\mathbb{D} \subset \mathbb{C}_\mathfrak{z}$ and have the following properties: (i) $|f_n(\mathfrak{z})| \le B$ for all n and all $\mathfrak{z} \in \mathbb{D}$. (ii) $f_n(\mathfrak{z})$ tends to a limit as $n \to \infty$ at a set of points having a limit point inside $\mathbb{D}$. Then $f_n(\mathfrak{z})$ tends uniformly to a limit in any region bounded by a contour interior to $\mathbb{D}$, the limit being an analytic function of $\mathfrak{z}$.

Now take any disc $\mathbb{V}'(T) \subseteq \mathbb{W}(T)$ with centre in $\mathbb{V}(T)$. The conditions of the Vitali convergence theorem now hold in $\mathbb{V}'(T)$ with $\mathfrak{z}^*$ replaced by the real number equal to the maximum modulus of $\mathfrak{z}$ in $\mathbb{V}'(T)$. By repeating the process with a sequence of discs the whole of $\mathbb{W}(T)$ can be covered.

Now suppose that $\mathbb{Q}(T)$ is a closed region in $\mathbb{C}_\mathfrak{z}$ which contains the set $\mathbb{O}_\mathcal{N}(T)$ of zeros of $\Xi(\mathcal{N}, T, \mathfrak{z})$ for all of the sequence of finite lattices. It follows from Theorem 4.5.1 that $P(\mathcal{L}, T, \mathfrak{z})$ is an analytic function of $\mathfrak{z}$ in any $\mathbb{W}(T) \subseteq \mathbb{C}_\mathfrak{z}$ for which $\mathbb{Q}(T) \cap \mathbb{W}(T) = \emptyset$. Any phase transition of the system at temperature T must therefore occur at a chemical potential μ for which $\exp(\mu/T) \in \mathbb{Q}(T) \cap \mathbb{R}^{(+)}$.

We have now defined the main problem in this approach to the theory of phase transitions, namely, for any particular model leading to a set of partition functions $Z(\mathcal{N}, T, M)$, to determine the set $\mathbb{Q}(T)$. This set will not, of course, be unique and different choices may have different advantages according to circumstances. Under rather general conditions a lattice fluid will not have any phase transitions at sufficiently high temperatures. If a set $\mathbb{Q}(T)$ can be found which evolves on to $\mathbb{R}^{(+)}$ as late as possible as the temperature is decreased then a good upper bound on T at any fixed μ will be obtained. On the other hand, if a $\mathbb{Q}(T)$ can be found, for which the range of μ lying in $\mathbb{Q}(T) \cap \mathbb{R}^{(+)}$ is as narrow as possible then good bounds of μ at constant T will be obtained. These differing aims in the construction of $\mathbb{Q}(T)$ will not, as we shall see below, necessarily lead to the same set, although we can always form the intersection of all known sets of type $\mathbb{Q}(T)$ in order to improve the bounds.

There is one class of one-component lattice fluids for which it is not difficult to obtain some information about $\mathbb{Q}(T) \cap \mathbb{R}^{(+)}$ without any calculation. If

$$Z(\mathcal{N}, T, N - M) = Z(\mathcal{N}, T, M), \qquad M = 1, 2, \ldots, N - 1, \tag{4.37}$$

then, if $\mathfrak{z}_0 \in \mathbb{O}_\mathcal{N}(T)$, so does its conjugate complex $\bar{\mathfrak{z}}_0$ and $1/\mathfrak{z}_0$ and $1/\bar{\mathfrak{z}}_0$. This set of roots in invariant under inversion and conjugation and it follows that $\mathbb{O}_\mathcal{N}(T)$ must also have this property. Thus, if $\mathbb{O}_\mathcal{N}(T)$ is exterior to the disc $|\mathfrak{z}| < R < 1$ it is contained in the annulus $R \le |\mathfrak{z}| \le 1/R$. If condition (4.37) is replaced by

$$Z(\mathcal{N}, T, N - M) = Z(\mathcal{N}, T, M)[Z(\mathcal{N}, T, N)]^{1-\frac{2M}{N}}, \qquad M = 1, 2, \ldots, N - 1, \tag{4.38}$$

it can be shown that the grand partition function can be symmetrized by a change of variable. If the lattice fluid is expressed in terms of Ising spin variables then condition (4.38) is equivalent to the restriction that the only odd degree terms in the Hamiltonian are linear. This condition is satisfied for a lattice fluid with only pairwise interactions satisfying conditions (4.10) and (4.11). If condition (4.37) holds then, for there to be a phase transition at temperature T, the condition $1 \in \mathbb{Q}(T) \cap \mathbb{R}^{(+)}$ must be satisfied. In this case the region of phase transitions can be reduced to the largest closed interval $\mathbb{I}(T) \subset \mathbb{Q}(T) \cap \mathbb{R}^{(+)}$ which maps into itself under inversion. If (4.37) holds and the interactions are pairwise attractive the circle theorem of Lee and Yang (1952) proves that the set $\mathbb{O}_{\mathcal{N}}(T)$ lies on the unit circle $|\mathfrak{z}| = 1$. This result is a special case of the theorem of Ruelle (1971), which will now be discussed.

4.6 Ruelle's Theorem

Let a variable $\mathfrak{z}(\boldsymbol{r})$ be given at every site $\boldsymbol{r} \in \mathcal{N}$ and define the polynomial

$$\Lambda(\mathcal{N}, T, \{\mathfrak{z}(\boldsymbol{r})\}) = \sum_{\{\mathcal{M}\}} \left\{ \exp[-E(\mathcal{M})/T] \prod_{\{\boldsymbol{r}\in\mathcal{M}\}} \mathfrak{z}(\boldsymbol{r}) \right\}. \tag{4.39}$$

It is clear, from (4.5) and (4.6), that

$$\Lambda(\mathcal{N}, T, \{\mathfrak{z}(\boldsymbol{r}) = \mathfrak{z}\}) = \Xi(\mathcal{N}, T, \mathfrak{z}). \tag{4.40}$$

Suppose now that a set of closed subsets $\mathbb{S}_{\mathcal{N}}(T; \boldsymbol{r})$ of $\mathbb{C}_{\mathfrak{z}}$ can be found such that

$$\Lambda(\mathcal{N}, T, \{\mathfrak{z}(\boldsymbol{r})\}) \neq 0, \quad \text{if} \quad \mathfrak{z}(\boldsymbol{r}) \notin \mathbb{S}_{\mathcal{N}}(T; \boldsymbol{r}), \quad \forall \quad \boldsymbol{r} \in \mathcal{N}. \tag{4.41}$$

The zeros of $\Xi(\mathcal{N}, T, \mathfrak{z})$ must lie in the closed set

$$\mathbb{U}_{\mathcal{N}}(T) = \bigcup_{\{\boldsymbol{r}\in\mathcal{N}\}} \mathbb{S}_{\mathcal{N}}(T; \boldsymbol{r}) \tag{4.42}$$

and a possible set $\mathbb{Q}(T)$ would be one which satisfied the condition that

$$\mathbb{U}_{\mathcal{N}}(T) \subseteq \mathbb{Q}(T), \qquad \forall\, \mathcal{N} \text{ in the sequence } \mathcal{N} \to \mathcal{L}\,. \tag{4.43}$$

In the cases discussed in Sects. 4.7 and 4.8, below, the determination of $\mathbb{Q}(T)$ from $\mathbb{U}_{\mathcal{N}}(T)$ is trivial, since the application of Ruelle's theorem leads to a set $\mathbb{U}_{\mathcal{N}}(T)$, which is independent of $\mathcal{N}$.

We now define a *covering* $\mathfrak{Y}_{\mathcal{N}}$ of the lattice $\mathcal{N}$ to be the set $\mathcal{Y}_\kappa$, $\kappa = 1, 2, \ldots, \nu$ of subsets of $\mathcal{N}$ such that:

(i) Every site of $\mathcal{N}$ belongs to at least one member of $\mathfrak{Y}_{\mathcal{N}}$.

(ii) Every interacting set is contained within at least one member of $\mathfrak{Y}_{\mathcal{N}}$.

It is clear that the set of units, defined in Sect. 4.3, is one possible covering of the lattice, but for present purposes it is convenient to be less specific. For any occupation $\mathcal{M}$ of the lattice we wish to be able to write

$$E(\mathcal{M}) = \sum_{\kappa=1}^{\nu} E_\kappa(\mathcal{Y}_\kappa \cap \mathcal{M}), \tag{4.44}$$

where $E_\kappa(\mathcal{Y}_\kappa \cap \mathcal{M})$ is the contribution to the energy from $\mathcal{Y}_\kappa$. To do this we must avoid multiple counting of energies from interacting sets of particles which occur in more than one $\mathcal{Y}_\kappa$. One way to do this is to divide the energy of such a set equally between all $\mathcal{Y}_\kappa$ in which it is contained. Another way would be to define the covering so that every interacting set is a member of $\mathfrak{Y}_\mathcal{N}$ and, in the energy $E_\kappa(\mathcal{Y}_\kappa \cap \mathcal{M})$, to exclude contributions from proper subsets of $\mathcal{Y}_\kappa$. For each $\mathcal{Y}_\kappa \in \mathfrak{Y}_\mathcal{N}$ define the polynomial

$$\lambda(\mathcal{Y}_\kappa, T, \{\mathfrak{z}_\kappa(\boldsymbol{r})\}) = \sum_{\{\mathcal{X} \subseteq \mathcal{Y}_\kappa\}} \left\{ \exp[-E_\kappa(\mathcal{X})/T] \prod_{\{\boldsymbol{r} \in \mathcal{X}\}} \mathfrak{z}_\kappa(\boldsymbol{r}) \right\}, \tag{4.45}$$

where $\{\mathfrak{z}_\kappa(\boldsymbol{r})\}$ is a set of variables defined for $\boldsymbol{r} \in \mathcal{Y}_\kappa$. For later reference we note, from (4.44), that, when the members of the covering are disjoint sets,

$$\Lambda(\mathcal{N}, T, \{\mathfrak{z}(\boldsymbol{r})\}) = \prod_{\kappa=1}^{\nu} \lambda(\mathcal{Y}_\kappa, T, \{\mathfrak{z}_\kappa(\boldsymbol{r}) = \mathfrak{z}(\boldsymbol{r})\}). \tag{4.46}$$

Ruelle's theorem can now be expressed in the following way:

Theorem 4.6.1. *If,* $\forall\, \mathcal{Y}_\kappa \in \mathfrak{Y}_\mathcal{N}$ *and* $\forall\, \boldsymbol{r} \in \mathcal{Y}_\kappa$, $\exists$ *a closed set* $\mathbb{X}_\kappa(T; \boldsymbol{r}) \subset \mathbb{C}\mathfrak{Z}$ *such that* $0 \notin \mathbb{X}_\kappa(T; \boldsymbol{r})$ *and such that* $\lambda(\mathcal{Y}_\kappa, T, \{\mathfrak{z}_\kappa(\boldsymbol{r})\}) \neq 0$ *when* $\mathfrak{z}_\kappa(\boldsymbol{r}) \notin \mathbb{X}_\kappa(T; \boldsymbol{r})$ $\forall\, \boldsymbol{r} \in \mathcal{Y}_\kappa$, *then* $\Lambda(\mathcal{N}, T, \{\mathfrak{z}(\boldsymbol{r})\}) \neq 0$ *when*[6]

$$\mathfrak{z}(\boldsymbol{r}) \notin \mathbb{S}_\mathcal{N}(T; \boldsymbol{r}) = -\prod_{\{\kappa\}} [-\mathbb{X}_\kappa(T; \boldsymbol{r})], \tag{4.47}$$

$\forall\, \boldsymbol{r} \in \mathcal{N}$, *where the product in (4.47) is over those values of* κ *for which* $\boldsymbol{r} \in \mathcal{Y}_\kappa$.

Proof: An important step in the proof of this theorem is derivation of a formula for $\Lambda(\mathcal{N}, T, \{\mathfrak{z}(\boldsymbol{r})\})$ in terms of the members of the set $\lambda(\mathcal{Y}_\kappa, T, \{\mathfrak{z}_\kappa(\boldsymbol{r})\})$, $\kappa = 1, 2, \ldots, \nu$. When the covering $\mathfrak{Y}_\mathcal{N}$ consists of disjoint sets, this formula is (4.46) and the theorem is trivial. This is because each lattice site belongs to exactly one member of the covering, the product in (4.47) has one term with $\mathbb{S}_\mathcal{N}(T; \boldsymbol{r}) = \mathbb{X}_\kappa(T; \boldsymbol{r})$ and $\mathfrak{z}(\boldsymbol{r}) = \mathfrak{z}_\kappa(\boldsymbol{r})$. The result follows from the assumption $\lambda(\mathcal{Y}_\kappa, T, \{\mathfrak{z}(\boldsymbol{r})\}) \neq 0$ and (4.46).

When the covering contains overlapping sets (4.46) is no longer valid. Its replaced is given by the *contraction procedure* of Asano (1970):

Lemma 4.6.1(a) *Let*

$$\phi_\alpha = \phi_{\alpha-1} \circ \lambda(\mathcal{Y}_\alpha, T, \{\mathfrak{z}_\alpha(\boldsymbol{r})\}), \qquad \alpha = 2, 3, \ldots, \nu, \tag{4.48}$$

where

$$\phi_1 = \lambda(\mathcal{Y}_1, T, \{\mathfrak{z}_1(\boldsymbol{r})\}), \tag{4.49}$$

[6] Notation: For subsets $\mathbb{A}$ and $\mathbb{B}$ of $\mathbb{C}_\mathfrak{Z}$, $-\mathbb{A}$ is the set of all points $-\mathfrak{z}_A$, for $\mathfrak{z}_A \in \mathbb{A}$, and $\mathbb{A}\mathbb{B}$ is the set of all points $\mathfrak{z}_A\mathfrak{z}_B$, for $\mathfrak{z}_A \in \mathbb{A}$ and $\mathfrak{z}_B \in \mathbb{B}$.

and the contraction symbol $\circ$ *means that, when the product is taken, terms containing factors of the form* $\prod_{\{\kappa\}} \mathfrak{z}_\kappa(\boldsymbol{r})$ *are deleted unless the product contains all* $\mathfrak{z}_\kappa(\boldsymbol{r})$ *for which* $\boldsymbol{r} \in \mathcal{Y}_\kappa$ *for* $\kappa = 1, 2, \dots, \alpha$. *Then with the substitution*

$$\mathfrak{z}(\boldsymbol{r}) = \prod_{\{\kappa\}} \mathfrak{z}_\kappa(\boldsymbol{r}), \tag{4.50}$$

$$\Lambda(\mathcal{N}, T, \{\mathfrak{z}(\boldsymbol{r})\}) = \phi_\nu . \tag{4.51}$$

Proof of 4.6.1(a): From (4.45), each of the terms in ϕ_ν is of the form

$$\prod_{\kappa=1}^{\nu} \left\{ \exp[-E_\kappa(\mathcal{X}_\kappa)/T] \prod_{\{\boldsymbol{r} \in \mathcal{X}_\kappa\}} \mathfrak{z}_\kappa(\boldsymbol{r}) \right\}, \tag{4.52}$$

where $\mathcal{X}_\kappa \subseteq \mathcal{Y}_\kappa$. The sum is over certain subsets $\mathcal{X}_\kappa$ of the covering sets $\mathcal{Y}_\kappa$ with a constraint imposed by the contraction procedure of (4.48). This can be expressed by the condition that if, for some subsets $\mathcal{Y}_i, \mathcal{Y}_j, \mathcal{Y}_k, \dots$ of the covering $\mathfrak{Y}_\mathcal{N}$,

$$\mathcal{Y}_i \cap \mathcal{Y}_j \cap \mathcal{Y}_k \cap \cdots \neq \emptyset , \tag{4.53}$$

then

$$\mathcal{Y}_i \cap \mathcal{X}_i = \mathcal{Y}_j \cap \mathcal{X}_j = \mathcal{Y}_k \cap \mathcal{X}_k = \cdots . \tag{4.54}$$

With this condition we can define

$$\mathcal{M} = \bigcup_{\kappa=1}^{\nu} \mathcal{X}_\kappa , \tag{4.55}$$

with

$$\mathcal{X}_\kappa = \mathcal{M} \cap \mathcal{Y}_\kappa , \qquad \kappa = 1, 2, \dots, \nu , \tag{4.56}$$

and the lemma follows from (4.39) and (4.44).

The formula (4.47) for $\mathbb{S}_\mathcal{N}(T; \boldsymbol{r})$ can now be obtained by induction, using the contraction procedure of Lemma 4.6.1(a) and the following result, proved by Ruelle (1971):

Lemma 4.6.1(b): *Let* $\mathbb{A}$ *and* $\mathbb{B}$ *be closed subsets of* $\mathbb{C}_3$ *which do not contain* 0. *Suppose that the complex polynomial*

$$c + a\mathfrak{z}_A + b\mathfrak{z}_B + g\mathfrak{z}_A\mathfrak{z}_B$$

can vanish only when $\mathfrak{z}_A \in \mathbb{A}$ *and* $\mathfrak{z}_B \in \mathbb{B}$. *Then*

$$c + g\mathfrak{z}$$

can vanish only when $\mathfrak{z} \in -\mathbb{A}\mathbb{B}$.

To complete the theorem using induction on the contraction procedure we first define $\mathbb{S}_{\mathcal{N}}^{(\alpha)}(T;\boldsymbol{r})$ by a formula like (4.47), except that the product is restricted to values of κ in the range $[1,\alpha]$ for which $\boldsymbol{r} \in \mathcal{Y}_\kappa$. The statement of the theorem, applied to the lattice $\mathcal{N}_\alpha = \cup_{\kappa=1}^{\alpha}\mathcal{Y}_\kappa$, is that $\Lambda(\mathcal{N}_\alpha, T, \{\mathfrak{z}(\boldsymbol{r})\}) \neq 0$ when $\mathfrak{z}(\boldsymbol{r}) \notin \mathbb{S}_{\mathcal{N}}^{(\alpha)}(T;\boldsymbol{r})$, $\forall\ \boldsymbol{r} \in \mathcal{N}_\alpha$. This is true for $\alpha = 1$, when $\Lambda(\mathcal{N}_1, T, \{\mathfrak{z}(\boldsymbol{r})\}) = \lambda(\mathcal{Y}_1, T, \{\mathfrak{z}(\boldsymbol{r})\})$, $\mathbb{S}_{\mathcal{N}}^{(1)}(T;\boldsymbol{r}) = \mathbb{X}_1(T;\boldsymbol{r})$. Let it be assumed for $\alpha - 1$. It is now important to note that the substitution (4.50) can be made step by step in the contraction procedure with $\phi_{\alpha-1}$ having a single variable $\mathfrak{z}(\boldsymbol{r})$ defined for each site $\boldsymbol{r} \in \mathcal{N}_{\alpha-1}$. Consider now the formation of ϕ_α according to the formula (4.48). For a site $\boldsymbol{r}$ in $\mathcal{Y}_\alpha$ but not in $\mathcal{N}_{\alpha-1}$, $\mathbb{S}_{\mathcal{N}}^{(\alpha)}(T;\boldsymbol{r}) = \mathbb{X}_\alpha(T;\boldsymbol{r})$ and for a site $\boldsymbol{r}$ in $\mathcal{N}_{\alpha-1}$ but not in $\mathcal{Y}_\alpha$, $\mathbb{S}_{\mathcal{N}}^{(\alpha)}(T;\boldsymbol{r}) = \mathbb{S}_{\mathcal{N}}^{(\alpha-1)}(T;\boldsymbol{r})$. In both these cases no contraction is necessary for the terms in $\mathfrak{z}(\boldsymbol{r})$ and $\mathfrak{z}_\alpha(\boldsymbol{r})$. For a site which belongs to both $\mathcal{Y}_\alpha$ and $\mathcal{N}_{\alpha-1}$ the contraction procedure is applied. Linear terms in $\mathfrak{z}(\boldsymbol{r})$ and $\mathfrak{z}_\alpha(\boldsymbol{r})$ are deleted and $\mathfrak{z}(\boldsymbol{r})\mathfrak{z}_\alpha(\boldsymbol{r})$ is replaced by $\mathfrak{z}(\boldsymbol{r})$. According to Lemma 4.6.1(b) $\Lambda(\mathcal{N}_\alpha, T, \{\mathfrak{z}(\boldsymbol{r})\})$ vanishes as a function of $\mathfrak{z}(\boldsymbol{r})$ only when $\mathfrak{z}(\boldsymbol{r}) \in -\mathbb{S}_{\mathcal{N}}^{(\alpha-1)}(T;\boldsymbol{r})\mathbb{X}_\alpha(T;\boldsymbol{r}) = \mathbb{S}_{\mathcal{N}}^{(\alpha)}(T;\boldsymbol{r})$. This completes the proof of the theorem.

We now summarize the 'practical' procedure for obtaining the set $\mathbb{U}_{\mathcal{N}}(T)$:

(i) For the lattice $\mathcal{N}$ choose the covering $\mathfrak{Y}_{\mathcal{N}}$ of subsets $\mathcal{Y}_\kappa$, $\kappa = 1, 2, \ldots, \nu$.

(ii) For each $\mathcal{Y}_\kappa$ form the polynomial $\lambda(\mathcal{Y}_\kappa, T, \{\mathfrak{z}_\kappa(\boldsymbol{r})\})$.

(iii) For each $\boldsymbol{r} \in \mathcal{Y}_\kappa$ find a subset $\mathbb{X}_\kappa(T;\boldsymbol{r}) \subset \mathbb{C}_3$ and not containing 0 such that $\lambda(\mathcal{Y}_\kappa, T, \{\mathfrak{z}_\kappa(\boldsymbol{r})\}) \neq 0$ when $\mathfrak{z}_\kappa(\boldsymbol{r}) \notin \mathbb{X}_\kappa(T;\boldsymbol{r})$ for every $\boldsymbol{r} \in \mathcal{Y}_\kappa$.

(iv) For each $\boldsymbol{r} \in \mathcal{N}$ construct $\mathbb{S}_{\mathcal{N}}(T;\boldsymbol{r})$ according to the formula (4.47).

(v) Obtain $\mathbb{U}_{\mathcal{N}}(T)$ by taking the union, as in (4.42).

It is clear that the tractability of this procedure depends very much on the covering chosen. The number of geometrically different covering sets should be very small. In practice it is often convenient to use the lattice units, defined in Sect. 4.2, as the elements of the covering.

4.7 The Yang–Lee Circle Theorem

We consider the generalization

$$\widehat{H}(\mathcal{H}, J(\boldsymbol{r},\boldsymbol{r}'); \sigma(\boldsymbol{r})) = -\sum_{\{\boldsymbol{r},\boldsymbol{r}'\}} J(\boldsymbol{r},\boldsymbol{r}')\sigma(\boldsymbol{r})\sigma(\boldsymbol{r}') - \mathcal{H}\sum_{\{\boldsymbol{r}\}}\sigma(\boldsymbol{r}) \tag{4.57}$$

of the spin-$\frac{1}{2}$ Ising model Hamiltonian (1.49), where the first summation is over all pairs of sites, with $J(\boldsymbol{r},\boldsymbol{r}') = J(\boldsymbol{r}',\boldsymbol{r})$ and $J(\boldsymbol{r},\boldsymbol{r}) = 0$. Using the lattice fluid variables $\tau(\boldsymbol{r})$ of Sect. 4.4 and again omitting a constant term, (4.57) can be re-expressed in the form

$$\widehat{H}(\tilde{\mu},\tilde{\varepsilon}(\boldsymbol{r},\boldsymbol{r}');\tau(\boldsymbol{r})) = -\sum_{\{\boldsymbol{r},\boldsymbol{r}'\}} \tilde{\varepsilon}(\boldsymbol{r},\boldsymbol{r}')[2\tau(\boldsymbol{r})\tau(\boldsymbol{r}') - \tau(\boldsymbol{r}) - \tau(\boldsymbol{r}')] - \tilde{\mu}\sum_{\{\boldsymbol{r}\}}\tau(\boldsymbol{r}), \tag{4.58}$$

where $\tilde{\varepsilon}(\boldsymbol{r},\boldsymbol{r}') = 2J(\boldsymbol{r},\boldsymbol{r}')$ and $\tilde{\mu} = 2\mathcal{H}$. The terms in the first summation in (4.58) are the energies of a pair of sites. The linear terms are retained in this summation, rather than being absorbed in to the chemical potential, in order to maintain the hole-particle symmetry. This implies the rather artificial situation where the occupations ○—○ and ●—● both have zero interaction energy, with non-zero energies occurring for ○—● and ●—○. This can be regarded as a mathematical device for simplifying the application of Ruelle's theorem, with a transformation to more natural variables being applied at the end. The interacting sets for this system consist of all pairs of sites for which $\tilde{\varepsilon}(\boldsymbol{r},\boldsymbol{r}') \neq 0$ and thus, from (4.11), we must have

$$\tfrac{1}{2}\sum_{\{\boldsymbol{r}'\}} |\tilde{\varepsilon}(\boldsymbol{r},\boldsymbol{r}')| < \infty \qquad \forall \quad \boldsymbol{r} \in \mathcal{N}. \tag{4.59}$$

This is achieved by imposing the condition

$$\tilde{\varepsilon}(\boldsymbol{r},\boldsymbol{r}') = 0 \qquad \text{when } |\boldsymbol{r}-\boldsymbol{r}'| > \mathcal{R}, \tag{4.60}$$

where $\mathcal{R}$ is the *range of interaction.* We denote a particular pair of lattice sites by $[\boldsymbol{r},\boldsymbol{r}']$. The covering $\mathcal{Y}_{\mathcal{N}}$ can now be taken to be all the pairs $[\boldsymbol{r},\boldsymbol{r}']$ for which $\tilde{\varepsilon}(\boldsymbol{r},\boldsymbol{r}') \neq 0$ and the polynomials (4.45) take the form

$$\lambda([\boldsymbol{r},\boldsymbol{r}'],T,\mathfrak{Z}(\boldsymbol{r}';\boldsymbol{r}),\mathfrak{Z}(\boldsymbol{r};\boldsymbol{r}')) = 1 + \tilde{\mathfrak{X}}(T;\boldsymbol{r},\boldsymbol{r}')[\mathfrak{Z}(\boldsymbol{r}';\boldsymbol{r}) + \mathfrak{Z}(\boldsymbol{r};\boldsymbol{r}')] + \mathfrak{Z}(\boldsymbol{r}';\boldsymbol{r})\mathfrak{Z}(\boldsymbol{r};\boldsymbol{r}'), \tag{4.61}$$

where $\tilde{\mathfrak{X}}(T;\boldsymbol{r},\boldsymbol{r}') = \exp[-\tilde{\varepsilon}(\boldsymbol{r},\boldsymbol{r}')/T]$ and $\mathfrak{Z}(\boldsymbol{r}';\boldsymbol{r})$ is the variable for the site $\boldsymbol{r}$ and the covering set $[\boldsymbol{r},\boldsymbol{r}']$. The analysis of this problem now depends of whether the terms $\tilde{\mathfrak{X}}(T;\boldsymbol{r},\boldsymbol{r}')$ are greater or less than unity. The case to which the Yang–Lee circle theorem applies is when

$$\tilde{\mathfrak{X}}(T;\boldsymbol{r},\boldsymbol{r}') < 1, \qquad \forall\ [\boldsymbol{r},\boldsymbol{r}'], \tag{4.62}$$

which in spin formulation corresponds to all interactions being ferromagnetic. $\lambda([\boldsymbol{r},\boldsymbol{r}'],T,\mathfrak{Z}(\boldsymbol{r}';\boldsymbol{r}),\mathfrak{Z}(\boldsymbol{r};\boldsymbol{r}'))$ is of the form given by equation (A.152). The regions $\mathbb{X}_\kappa(T;\boldsymbol{r})$ are then given by (A.160) for all covering sets and all lattice sites for any choice of $x \neq 1$. As we shall see in our discussions in the next section, the difficult step in the application of Ruelle's theorem is (iv), the construction of $\mathbb{S}_{\mathcal{N}}(T)$ according to the formula (4.47). There is, however, one choice of x for which this process is straightforward, namely $x = 0$ for which we have

$$\mathbb{X}_\kappa(T;\boldsymbol{r}) = \mathbb{D}^{(+)} = \{\mathfrak{z} : |\mathfrak{z}| \geq 1\}. \tag{4.63}$$

To obtain $\mathbb{S}_\mathcal{N}(T;\boldsymbol{r})$ (4.47) must now be used. In general this can be difficult, particularly since boundary sites are members of fewer covering sets than interior sites and must be considered separately. However, in this case the form of $\mathbb{D}^{(+)}$ means that the product procedure has no effect and $\mathbb{S}_\mathcal{N}(T;\boldsymbol{r}) = \mathbb{D}^{(+)}$ for every site of the lattice. It then follows from (4.42) that $\mathbb{U}_\mathcal{N}(T) = \mathbb{D}^{(+)}$, which is independent of $\mathcal{N}$. Thus $\mathbb{D}^{(+)}$ is a possible choice of $\mathbb{Q}(T)$ for all T. However, in this case the symmetry condition (4.37) is satisfied. If some $\mathfrak{z}_0 \in \mathbb{O}_\mathcal{N}(T)$ with $|\mathfrak{z}_0| > 1$ then $1/\mathfrak{z}_0 \in \mathbb{O}_\mathcal{N}(T)$. This contradicts the result of Ruelle's theorem and it follows that

$$\mathbb{Q}(T) = \{\mathfrak{z} : |\mathfrak{z}| = 1\} \tag{4.64}$$

for all temperatures. This is the Yang–Lee circle theorem. For this formulation in lattice fluid terms, for which condition (4.62) is equivalent to an attractive potential between pair of particles (and pairs of holes), we have shown that, for an arbitrary but finite range of interaction on any lattice, phase transitions can occur only for $\tilde{\mu} = 0$. In the magnetic spin formulation this is equivalent to the result that the Ising ferromagnet with arbitrary range of interaction can have a phase transition only on the zero-field axis. Equation (4.58) can be re-expressed in the more usual lattice fluid form

$$\widehat{H}(\mu, \varepsilon(\boldsymbol{r},\boldsymbol{r}'); \tau(\boldsymbol{r})) = -\sum_{\{\boldsymbol{r},\boldsymbol{r}'\}} \varepsilon(\boldsymbol{r},\boldsymbol{r}')\tau(\boldsymbol{r})\tau(\boldsymbol{r}') - \mu \sum_{\{\boldsymbol{r}\}} \tau(\boldsymbol{r}), \tag{4.65}$$

where

$$\varepsilon(\boldsymbol{r},\boldsymbol{r}') = 2\tilde{\varepsilon}(\boldsymbol{r},\boldsymbol{r}') = 4J(\boldsymbol{r},\boldsymbol{r}'), \tag{4.66}$$

$$\mu = \tilde{\mu} - \sum_{\{\boldsymbol{r}\}} \epsilon(\boldsymbol{r},\boldsymbol{r}') = 2\mathcal{H} - 2\sum_{\{\boldsymbol{r}\}} J(\boldsymbol{r},\boldsymbol{r}'). \tag{4.67}$$

In this case it follows that, as long as the summations in (4.67) have the same value for each lattice site, a phase transition can occur only on the line

$$\mu = -\tfrac{1}{2}\sum_{\{\boldsymbol{r}\}} \varepsilon(\boldsymbol{r},\boldsymbol{r}'). \tag{4.68}$$

4.8 Systems with Pair Interactions

For the application of Ruelle's theorem, the choice of the sets $\mathbb{X}_\kappa(T;\boldsymbol{r})$ is, as we have seen above, not unique. For covering sets consisting of pairs of sites this is exemplified by the different allowed values of the parameters x and y of Appendix A.4. In particular cases one choice may have particular advantages, in either narrowing the intersection of $\mathbb{Q}(T)$ with $\mathbb{R}^{(+)}$ or obtaining a good estimate of the critical temperature by delaying as much as possible, as

T is decreased, the value of T at which $\mathbb{Q}(T)$ intersects $\mathbb{R}^{(+)}$. The circle theorem, obtained by the application of Ruelle's theorem and symmetry, narrows $\mathbb{Q}(T) \cap \mathbb{R}^{(+)}$ to the single point $\mathfrak{z} = 1$, This is an extreme case, where the value $\tilde{\mu} = 0$ for the phase transition (if it occurs) is predicted with complete accuracy. This result is, however, achieved at the expense of a total inability to obtain an upper bound for the critical temperature. In this section we consider other possible choices for $\mathbb{X}_\kappa(T; \boldsymbol{r})$ for systems with pair interactions. We first examine the situation where

$$\tilde{\varepsilon}(\boldsymbol{r}, \boldsymbol{r}') = \begin{cases} \tilde{\varepsilon} > 0\,, & \text{for every nearest-neighbour pair } [\boldsymbol{r}, \boldsymbol{r}'], \\ 0\,, & \text{otherwise,} \end{cases} \tag{4.69}$$

which gives

$$\tilde{\mathfrak{X}}(T; \boldsymbol{r}, \boldsymbol{r}') = \begin{cases} \tilde{\mathfrak{X}}(T) = \exp(-\tilde{\varepsilon}/T) < 1\,, & \text{for every nearest-neighbour pair } [\boldsymbol{r}, \boldsymbol{r}'], \\ 1\,, & \text{otherwise.} \end{cases} \tag{4.70}$$

This is a special case of the situation to which the circle theorem applies. We can however, from (A.160), choose

$$\mathbb{X}_\kappa(T; \boldsymbol{r}) = \mathbb{X}_x(T) = \left\{\mathfrak{z} : |2\tilde{\mathfrak{X}}(T)\mathfrak{z} + x| \le \sqrt{4\tilde{\mathfrak{X}}^2(T)(1-x) + x^2}\right\}, \tag{4.71}$$

for every lattice site, every covering set and any $x > 1$. The region $\mathbb{X}_x(T)$ is a disc not including $\mathfrak{z} = 0$ and with centre on $\mathbb{R}^{(-)}$. From (4.47) and (4.71),

$$\mathbb{S}_{\mathcal{N}}(T; \boldsymbol{r}) = -[-\mathbb{X}_x(T)]^{\mathrm{z}}, \tag{4.72}$$

for every interior lattice site, where z is the coordination number of the lattice. For boundary sites the formula is the same as (4.72) except that z is replaced by some $\mathrm{z}' < \mathrm{z}$, this being the number of nearest-neighbour sites of the particular boundary site. Since, however, $-1 \in \mathbb{X}_x(T)$ for all $x > 1$ it follows that

$$-[-\mathbb{X}_x(T)]^{\mathrm{z}'} \subset -[-\mathbb{X}_x(T)]^{\mathrm{z}} \tag{4.73}$$

and thus, from (4.42),

$$\mathbb{U}_{\mathcal{N}}(T) = -[-\mathbb{X}_x(T)]^{\mathrm{z}}. \tag{4.74}$$

Since this $\mathbb{U}_{\mathcal{N}}(T)$ does not involve $\mathcal{N}$ it can be taken to be $\mathbb{Q}(T)$. From (A.158) the points $\mathfrak{z}^{(\pm)} = -\tilde{\mathfrak{X}}(T) \pm \mathrm{i}\sqrt{1 - \tilde{\mathfrak{X}}^2(T)}$ lie on the boundary of $\mathbb{X}_x(T)$ for all $x > 1$. These points subtend an angle $\arccos[\tilde{\mathfrak{X}}(T)]$ with $\mathbb{R}^{(-)}$. It follows that the arc $\arg(\mathfrak{z}) \in [\mathrm{z}\{\pi - \arccos[\tilde{\mathfrak{X}}(T)]\}, \mathrm{z}\{\pi + \arccos[\tilde{\mathfrak{X}}(T)]\}]$ of the unit circle $|\mathfrak{z}| = 1$ is contained in $\mathbb{Q}(T)$. Since the zeros of the partition function lie on $|\mathfrak{z}| = 1$ an upper bound $T = T_{\mathrm{u}}$ for the occurrence of a phase transition is given by

$$\tilde{\mathfrak{X}}(T_{\mathrm{u}}) = \cos\left(\frac{\pi}{\mathrm{z}}\right), \qquad T_{\mathrm{u}} = \frac{\tilde{\varepsilon}}{-\ln(\pi/\mathrm{z})} = \frac{\varepsilon}{-2\ln(\pi/\mathrm{z})}, \tag{4.75}$$

where $\varepsilon = 2\tilde{\varepsilon}$ is the usually lattice fluid interaction energy given by the nearest-neighbour case of (4.66). For the square lattice ($z = 4$), $T_u = 1.4427\varepsilon$, which is consistent with the known critical temperature $T_c = 0.5673\varepsilon$.

We now consider the homogeneous nearest-neighbour lattice fluid with Hamiltonian (4.34). In this case

$$\lambda([\boldsymbol{r},\boldsymbol{r}'],T,\mathfrak{z},\mathfrak{z}') = 1 + \mathfrak{z} + \mathfrak{z}' + \mathfrak{X}(T)\mathfrak{z}\mathfrak{z}' , \tag{4.76}$$

where $\mathfrak{X}(T) = \exp(\varepsilon/T)$, for every covering set, consisting of a pair of nearest-neighbour sites. The expression (4.76) is of the form given by (A.163) and the sets $\mathbb{X}_\kappa(T;\boldsymbol{r})$ can be taken to be all the same and given by (A.165) or (A.166) according as $\mathfrak{X}(T) <,> 1$. These cases correspond respectively to repulsive and attractive nearest-neighbour interactions between particles.

Consider the case where the interparticle interaction is attractive. The possible covering sets $\mathbb{X}_\kappa(T,\boldsymbol{r})$ are given by (A.166). The useful choices are given by $x > 1$ and consist of discs not containing $\mathfrak{z} = 0$. The best upper bound for the critical temperature is given by (4.72) with $\mathbb{X}$ replaced by $\mathbb{X}'$ when x is chosen so that $\mathbb{S}_\mathcal{N}(T;\boldsymbol{r})$ reaches $\mathbb{R}^{(+)}$ as late as possible as the temperature is decreased. This involves the choice of the disc for which the tangent from the origin to the boundary subtends the smallest angle with $\mathbb{R}^{(-)}$. As indicated in Appendix A.4 this is given by choosing $x = 2$ with

$$\mathbb{X}_\kappa(T;\boldsymbol{r}) = \mathbb{X}'_2(T) = \left\{\mathfrak{z} : |\mathfrak{z}+1| \le \sqrt{1-\mathfrak{X}^{-1}(T)}\right\} , \tag{4.77}$$

for every lattice site and every covering set. The tangent from the origin to the boundary of this disc subtends angles $\pm\arccos[1/\sqrt{\mathfrak{X}(T)}]$ with $\mathbb{R}^{(-)}$. Again, since $-1 \in \mathbb{X}'_2(T)$, we can make the identification $\mathbb{Q}(T) = \mathbb{U}_\mathcal{N}(T) = \mathbb{S}_\mathcal{N}(T;\boldsymbol{r})$, where $\boldsymbol{r}$ is any interior site of the lattice. Using (4.72) the upper bound temperature T_u is then given by

$$T_u = -\frac{|\varepsilon|}{2\ln(\pi/z)} , \tag{4.78}$$

which, as is to be expected, is the same as the form given by (4.75).

In a similar way, when the interparticle interaction is repulsive, the possible sets $\mathbb{X}_\kappa(T;\boldsymbol{r})$ are given by (A.165) with the best choice being given by $y = 0$. Thus

$$\mathbb{X}_\kappa(T;\boldsymbol{r}) = \mathbb{Y}'_0(T) = \left\{\mathfrak{z} : |\mathfrak{X}(T)\mathfrak{z}+1| \le \sqrt{1-\mathfrak{X}(T)}\right\} , \tag{4.79}$$

for every lattice site and every covering set. The tangent from the origin to the boundary of the disc subtends angles $\pm\arccos[\sqrt{\mathfrak{X}(T)}]$ with $\mathbb{R}^{(-)}$ and the upper bound temperature is again given by (4.79), except that ε is now positive.

A particular case of this analysis, which was investigated by Runnels and Hubbard (1972), is that of the lattice fluid with nearest-neighbour exclusion. This corresponds to the limit $\mathfrak{X}(T) \to 0$, for all T, when equation (4.79) takes the form

$$\mathbb{X}_\kappa(T;\boldsymbol{r}) = \{\mathfrak{z} : \Re(\mathfrak{z}) \le -\tfrac{1}{2}\}\,. \tag{4.80}$$

The points on the boundary of this region are given by

$$\mathfrak{z} = -\frac{\exp(\mathrm{i}\theta)}{2\cos(\theta)}\,, \qquad -\pi \le \theta \le \pi\,. \tag{4.81}$$

The set $\mathbb{Q}$, containing the zeros of the grand partition function, will consist of the whole of the plane $\mathbb{C}_\mathfrak{z}$ except a region containing the origin. The boundary of this region will be obtained from products of z points given by (4.81). However, not all such products lie on the boundary. Suppose that $R(\psi)\exp(\mathrm{i}\psi)$ is a point on the boundary. It is clear, from (4.72), that

$$R(\psi) = -\frac{1}{2^\mathrm{z}}\min\prod_{j=1}^{\mathrm{z}}\sec(\theta_j)\,, \tag{4.82}$$

where the minimization is subject to the constraint

$$\sum_{j=1}^{\mathrm{z}}\theta_j = \psi\,. \tag{4.83}$$

Using a Lagrange undetermined multiplier it is not difficult to show that the required minimum is achieved when $\theta_j = \psi/\mathrm{z}$ for all j and the boundary curve is given by

$$\mathfrak{z} = -\frac{\exp(\mathrm{i}\psi)}{[2\cos(\psi/\mathrm{z})]^\mathrm{z}}\,, \qquad -\pi \le \psi \le \pi\,. \tag{4.84}$$

The point where the curve (4.84) meets $\mathbb{R}^{(+)}$ gives a lower bound $\mathfrak{z}_\mathrm{L}$ for the critical fugacity $\mathfrak{z}_\mathrm{c}$. The square lattice fluid model with nearest-neighbour exclusion is known as the *hard-square model.* From (4.84), setting $\psi = \pi$ and $\mathrm{z} = 4$, $\mathfrak{z}_\mathrm{L} = \frac{1}{4}$. The corresponding model on the triangular lattice is the *hard-hexagon model.* In this case $\mathrm{z} = 6$ and $\mathfrak{z}_\mathrm{L} = \frac{1}{27}$. The hard-square model has been shown to be equivalent to a sixteen-vertex model on the square lattice (Wood and Goldfinch 1980). Estimates of the critical fugacity were been obtained by transfer matrix (Runnels and Combs 1966) and series (Gaunt and Fisher 1965) methods. Baxter et al. (1980) used corner transfer matrices to obtain the high-density series of the order parameter to twenty-four terms giving an estimate 3.7962 ± 0.0001 for the critical fugacity. The hard-hexagon model is one of the few examples of a model with a known exact solution (Baxter 1980, 1981, 1982a, Joyce 1988a, 1988b), the critical fugacity being $\frac{1}{2}(5\sqrt{5}+11) = 11.09$. As Runnels and Hubbard (1972) observed for the hard-square model, the lower bounds derived by the procedure described here, although of course valid, differ substantially from the critical values. These authors obtained the improvement $\mathfrak{z}_\mathrm{L} = \frac{1}{2}$ for the hard-square model by choosing covering sets consisting of half the elementary squares of the lattice with each site belonging to two squares. It is of interest to adopt the same procedure for the hard-hexagon model (Example 4.2). In this case equation (4.84) is replaced by

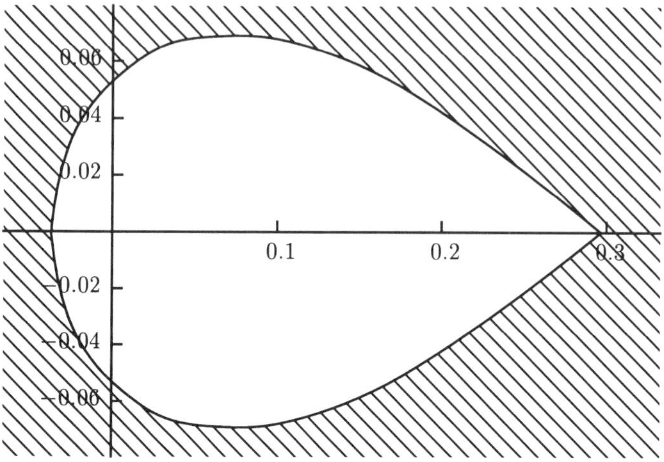

Fig. 4.3. The result given by equation (4.85) for the hard hexagon model. The zeros of the grand partition function lie in the shaded region.

$$\mathfrak{z} = -\frac{\exp(\mathrm{i}\psi)}{[3\cos(\psi/3)]^3}\,, \qquad -\pi \leq \psi \leq \pi\,, \tag{4.85}$$

which gives $\mathfrak{z}_{\mathrm{L}} = \frac{8}{27}$. The region free from zeros obtained from (4.85) is shown in Fig. 4.3. The analysis of this section has given no unexpected information, showing simply that for nearest-neighbour lattice fluids with either repulsive or attractive interactions there is no ordered phase at high temperatures and giving consistency with known critical fugacities for models with nearest-neighbour exclusion. In all but the last example we have chosen covering sets consisting only of pairs of sites. The last example gives some indication that improvements can be gained by increasing the number of sites within a covering.

4.9 Transfer Matrices

The standard solution of the one-dimensional Ising model (see, for example, Volume 1, Sects. 2.4) and many problems in two-dimensions including Onsager's solution of the two-dimensional Ising model (Onsager 1944, Kaufman 1949, Kaufman and Onsager 1949) and Baxter's solution of the eight-vertex model (Sect. 5.7) are based on the use of transfer matrices. In principle, however, the method can be applied to lattices of any dimension and in this section we shall describe the method as applied to a finite lattice $\mathcal{N}$ of dimension d.

4.9.1 The Partition Function

Suppose that this lattice can be decomposed into N_2 identical $(d-1)$-dimensional non-overlapping *slices* $\mathcal{N}_\mathrm{j}$, $\mathrm{j} = 0, 1, \ldots, N_2 - 1$. Each slice $\mathcal{N}_\mathrm{j}$ consists of N_1 sites. These will be labelled (i, j) for $\mathrm{i} = 0, 1, \ldots, N_1 - 1$ and $N_1 N_2 = |\mathcal{N}| = N$. We suppose that on each site of $\mathcal{N}$ there is a microsystem with ν states and the state of site (i, j) is denoted by $\tau_\mathrm{i}(\mathrm{j})$, with the N_1-dimensional vector $\boldsymbol{\tau}(\mathrm{j})$ denoting the state of slice $\mathcal{N}_\mathrm{j}$. The interactions between microsystems are restricted to sites which lie either in the same slice or in neighbouring slices, so that microsystems on the sites of $\mathcal{N}_\mathrm{j}$ interact with those on $\mathcal{N}_{\mathrm{j}-1}$ and $\mathcal{N}_{\mathrm{j}+1}$. The Hamiltonian can then be expressed in the form

$$\widehat{H}(\boldsymbol{\tau}) = \hat{b}(\boldsymbol{\tau}(N_2 - 1), \boldsymbol{\tau}(0)) + \sum_{\mathrm{j}=0}^{N_2-2} \hat{h}(\boldsymbol{\tau}(\mathrm{j}), \boldsymbol{\tau}(\mathrm{j}+1)), \tag{4.86}$$

where $\hat{h}(\boldsymbol{\tau}(\mathrm{j}), \boldsymbol{\tau}(\mathrm{j}+1))$ contains the contribution to the Hamiltonian from interactions between sites in $\mathcal{N}_\mathrm{j}$ and $\mathcal{N}_{\mathrm{j}+1}$ and half the contributions from interactions within these two slices, and $\hat{b}(\boldsymbol{\tau}(N_2-1), \boldsymbol{\tau}(0))$ contains any boundary contributions at $\mathcal{N}_0$ and $\mathcal{N}_{N_2-1}$. Let

$$V(\boldsymbol{\tau}(\mathrm{j}), \boldsymbol{\tau}(\mathrm{j}+1)) = \exp[-\hat{h}(\boldsymbol{\tau}(\mathrm{j}), \boldsymbol{\tau}(\mathrm{j}+1))/T]. \tag{4.87}$$

Any slice can be in any one of $\nu^{N_1} = M$ states which we denote by $\boldsymbol{\tau}_m$, $m = 1, 2, \ldots, M$. The particular ordering of these state will be discussed below. For the moment it is sufficient to observe that, for any ordering, $V(\boldsymbol{\tau}_m, \boldsymbol{\tau}_{m'})$ is an element of a $M \times M$ dimensional matrix. This is the *transfer matrix* for this problem and is denoted by $\boldsymbol{V}$.

Consider an operator Q which operates on the state $\boldsymbol{\tau}_m$ of any slice to produce another state $\boldsymbol{\tau}_{m'}$. The largest group $\mathcal{Q} = \{\mathrm{Q}^{(0)}, \mathrm{Q}^{(1)}, \mathrm{Q}^{(2)}, \ldots, \mathrm{Q}^{(\mathrm{n}-1)}\}$ of operators, such that

$$\hat{h}(\mathrm{Q}^{(\mathrm{s})}\boldsymbol{\tau}_m, \mathrm{Q}^{(\mathrm{s})}\boldsymbol{\tau}_{m'}) = \hat{h}(\boldsymbol{\tau}_m, \boldsymbol{\tau}_{m'}), \qquad \mathrm{s} = 0, 1, \ldots, \mathrm{n}-1, \tag{4.88}$$

is the *symmetry group of the Hamiltonian.* It also applies to the elements of the transfer matrix and we have

$$V(\mathrm{Q}^{(\mathrm{s})}\boldsymbol{\tau}_m, \mathrm{Q}^{(\mathrm{s})}\boldsymbol{\tau}_{m'}) = V(\boldsymbol{\tau}_m, \boldsymbol{\tau}_{m'}), \qquad \mathrm{s} = 0, 1, \ldots, \mathrm{n}-1. \tag{4.89}$$

For a two-dimensional lattice with periodic boundary conditions, $\mathcal{Q}$ will normally include the cyclic group $\mathcal{C}_{N_1}$ as a subgroup.[7] In a similar way, for a d-dimensional lattice with periodic boundary conditions, $\mathcal{Q}$ will contain the direct product of cyclic groups for the periodicity of the $d-1$ lattice directions in the slices $\mathcal{N}_\mathrm{j}$. Other subgroups of $\mathcal{Q}$ will correspond to any symmetry of the Hamiltonian arising from permutations of the states of

[7] Exceptions to this can occur in the case of sublattice ordering when the order of the cyclic group may be some factor of N_1 (see, for example, Lavis 1976).

the microsystems. For the ν-state Potts model (see Sect. 3.5) $\mathcal{Q}$ will contain $\mathcal{S}_\nu$, the permutation group of ν elements. Let the irreducible representations of $\mathcal{Q}$ be denoted by $\mathsf{E}(\mathsf{k})$, for $\mathsf{k} = 1, 2, \ldots, \kappa$, with $\mathsf{e}(\mathsf{k})$ denoting the dimension of $\mathsf{E}(\mathsf{k})$. As is described in Appendix A.3 the effect of the group $\mathcal{Q}$ on the states $\boldsymbol{\tau}_m$ is given, in terms of the permutation representation $\mathcal{R}(\mathcal{Q}) = \{\boldsymbol{I}, \boldsymbol{Q}^{(1)}, \boldsymbol{Q}^{(2)}, \ldots, \boldsymbol{Q}^{(\mathfrak{n}-1)}\}$ of M-dimensional matrices, by (A.78) and the symmetry property of the transfer matrix is then expressible in the form (A.79). For any statistical mechanical system of the type described above it is normally possible to order the states $\boldsymbol{\tau}_m$ into a set of *equivalence classes* $\mathfrak{E}(\alpha) = \{\boldsymbol{\tau}_1(\alpha), \boldsymbol{\tau}_2(\alpha), \ldots, \boldsymbol{\tau}_{\mathfrak{p}(\alpha)}(\alpha)\}$, $\alpha = 1, 2, \ldots, \gamma$ which are invariant and transitive under the operators of $\mathcal{Q}$. This means that the representation $\mathcal{R}(\mathcal{Q})$ divides into γ representations $\mathcal{R}(\alpha; \mathcal{Q})$, with $\mathcal{R}(\alpha; \mathcal{Q}) = \{\boldsymbol{I}(\alpha), \boldsymbol{Q}^{(1)}(\alpha), \boldsymbol{Q}^{(2)}(\alpha), \ldots, \boldsymbol{Q}^{(\mathfrak{n}-1)}(\alpha)\}$ a $\mathfrak{p}(\alpha)$-dimensional representation operating on $\mathfrak{E}(\alpha)$. The matrix $\boldsymbol{Q}^{(\mathsf{s})}$ has the block-diagonal form given in equation (A.87).

A procedure is described in Appendix A.3 for obtaining the matrix $\boldsymbol{W}(\alpha)$ which reduces the members of $\mathcal{R}(\alpha; \mathcal{Q})$ to block-diagonal sums each submatrix of which involves only one irreducible representation of $\mathcal{Q}$. Using the set of matrices $\boldsymbol{W}(\alpha)$, $\alpha = 1, 2, \ldots, \gamma$ and a permutation matrix $\boldsymbol{\Pi}$ the matrix $\boldsymbol{W}$ is constructed according to the formula (A.90). This has the effect of reducing $\boldsymbol{V}$ to block-diagonal form as shown in equation (A.82), where the matrix $\widetilde{\boldsymbol{V}}(\mathsf{k})$, which is associated with the irreducible representation $\mathsf{E}(\mathsf{k})$, is of dimension $\mathsf{w}(\mathsf{k}) = \mathsf{e}(\mathsf{k})\mathsf{m}(\mathsf{k})$. In fact, as is shown in Appendix A.3, for $\mathsf{e}(\mathsf{k}) > 1, \widetilde{\boldsymbol{V}}(\mathsf{k})$ can be further reduced to a direct sum of $\mathsf{e}(\mathsf{k})$ identical blocks of dimension $\mathsf{m}(\mathsf{k})$. This will, however, play only a minor role in our present discussion.

Equations (A.132)–(A.134) define the eigenvalues and left and right eigenvectors of $\boldsymbol{V}$. From (A.82) it follows that $\boldsymbol{V}$ and $\widetilde{\boldsymbol{V}}$ have the same eigenvalues and these are, therefore, associated with the irreducible representations $\mathsf{E}(\mathsf{k})$ and denoted by $\Lambda_\ell(\mathsf{k})$, $\ell = 1, 2, \ldots, \mathsf{w}(\mathsf{k})$. For the M dimensional vectors and $M \times M$ dimensional matrices used in this section there are three ways of labelling the elements of the rows and columns. This can be done either by a single index $r = 1, 2, \ldots, M$, by a pair of indices (k, ℓ), where $\mathsf{k} = 1, 2, \ldots, \kappa$, $\ell = 1, 2, \ldots, \mathsf{w}(\mathsf{k})$ refer to irreducible representations or by a pair (α, m), where $\alpha = 1, 2, \ldots, \gamma$, $m = 1, 2, \ldots, \mathfrak{p}(\alpha)$ refer to equivalence classes. The translation rules from the second and third forms of labelling to the first are respectively

$$r = \begin{cases} \ell, & \text{if } \mathsf{k} = 1, \\ \displaystyle\sum_{\mathsf{k}'=1}^{\mathsf{k}-1} \mathsf{w}(\mathsf{k}') + \ell & , \text{if } 1 < \mathsf{k} \leq \kappa, \end{cases} \tag{4.90}$$

$$r = \begin{cases} m\,, & \text{if } \alpha = 1, \\ \sum\limits_{\alpha'=1}^{\alpha-1} \mathsf{p}(\alpha') + m\,, & \text{if } 1 < \alpha \leq \gamma. \end{cases} \tag{4.91}$$

Let $\boldsymbol{\Lambda}$ be the diagonal matrix with the eigenvalues $\Lambda_\ell(\mathsf{k})$ along the diagonal. Thus[8]

$$\langle \mathsf{k}, \ell | \boldsymbol{\Lambda} | \mathsf{k}', \ell' \rangle = \delta^{\mathrm{Kr}}(\mathsf{k} - \mathsf{k}')\delta^{\mathrm{Kr}}(\ell - \ell')\Lambda_\ell(\mathsf{k})\,. \tag{4.92}$$

Let $\boldsymbol{X}$ and $\boldsymbol{Y}$ be the matrices with the left and right eigenvectors as rows and columns respectively. That is

$$\langle \mathsf{k}, \ell | \boldsymbol{X} = \boldsymbol{x}_\ell(\mathsf{k})\,, \qquad \boldsymbol{Y} | \mathsf{k}, \ell \rangle = [\boldsymbol{y}_\ell(\mathsf{k})]^{(\mathrm{T})}\,. \tag{4.93}$$

It follows, from (A.132)–(A.133), that

$$\boldsymbol{X}\boldsymbol{V} = \boldsymbol{\Lambda}\boldsymbol{X}\,, \tag{4.94}$$

$$\boldsymbol{V}\boldsymbol{Y} = \boldsymbol{Y}\boldsymbol{\Lambda}\,, \tag{4.95}$$

$$\boldsymbol{Y}^{-1} = \boldsymbol{X}\,, \tag{4.96}$$

from which

$$\boldsymbol{V} = \boldsymbol{Y}\boldsymbol{\Lambda}\boldsymbol{X}\,. \tag{4.97}$$

The partition function $Z(N_1, N_2, T)$ of the system is given by

$$\begin{aligned} Z(N_1, N_2, T) &= \sum_{\{\boldsymbol{\tau}\}} B(\boldsymbol{\tau}(N_2 - 1), \boldsymbol{\tau}(0)) \prod_{\mathrm{j}=0}^{N_2-2} V(\boldsymbol{\tau}(\mathrm{j}), \boldsymbol{\tau}(\mathrm{j}+1))\,, \\ &= \mathrm{Trace}\{\boldsymbol{V}^{N_2-1}\boldsymbol{B}\}\,, \end{aligned} \tag{4.98}$$

where $\boldsymbol{B}$ is the matrix with elements $B(\boldsymbol{\tau}_m, \boldsymbol{\tau}_{m'}) = \exp[-\hat{b}(\boldsymbol{\tau}_m, \boldsymbol{\tau}_{m'})/T]$. From (4.97),

$$Z(N_1, N_2, T) = \mathrm{Trace}\{\boldsymbol{X}\boldsymbol{B}\boldsymbol{Y}\boldsymbol{\Lambda}^{N_2-1}\} = \sum_{\{\mathsf{k},\ell\}} \beta_\ell(\mathsf{k})[\Lambda_\ell(\mathsf{k})]^{N_2-1}\,, \tag{4.99}$$

where for the sake of brevity '$\{\mathsf{k}, \ell\}$' will henceforth denote the double summation $\mathsf{k} = 1, 2, \ldots, \kappa$, $\ell = 1, 2, \ldots, \mathsf{w}(\mathsf{k})$ and the coefficients

$$\beta_\ell(\mathsf{k}) = \langle \mathsf{k}, \ell | \boldsymbol{X}\boldsymbol{B}\boldsymbol{Y} | \mathsf{k}, \ell \rangle = \boldsymbol{x}_\ell(\mathsf{k})\boldsymbol{B}[\boldsymbol{y}_\ell(\mathsf{k})]^{(\mathrm{T})}. \tag{4.100}$$

The partition function is dependent on N_2 only through the powers of the eigenvalues. Both the coefficients $\beta_\ell(\mathsf{k})$, the eigenvalues themselves and their number are, of course, dependent on N_1.

[8] Where convenient we shall, for any matrix $\boldsymbol{Q}$ denote the row and column vectors formed by its r-th row and j-th column by $\langle r|\boldsymbol{Q}$ and $\boldsymbol{Q}|j\rangle$ respectively and the (r, j)-th element by $\langle r|\boldsymbol{Q}|j\rangle$. With this notation we shall also use the pairs of suffices (k, ℓ) and (α, m).

4.9.2 Boundary Conditions

Formula (4.99) has a number of special cases of interest. When periodic toroidal boundary conditions are applied $\hat{b}(\boldsymbol{\tau}_m, \boldsymbol{\tau}_{m'}) = \hat{h}(\boldsymbol{\tau}_m, \boldsymbol{\tau}_{m'})$ giving $\boldsymbol{B} = \boldsymbol{V}$. From (4.96), (4.97) and (4.100) $\beta_\ell(\mathrm{k}) = \Lambda_\ell(\mathrm{k})$ and (4.99) becomes

$$Z(N_1, N_2, T) = \mathrm{Trace}\{\boldsymbol{\Lambda}^{N_2}\} = \sum_{\{\mathrm{k},\ell\}} [\Lambda_\ell(\mathrm{k})]^{N_2}. \tag{4.101}$$

Cases where the lattice has lines of boundary sites at both ends are given by

$$\hat{b}(\boldsymbol{\tau}_m, \boldsymbol{\tau}_{m'}) = \hat{b}^{(\mathrm{L})}(\boldsymbol{\tau}_{m'}) + \hat{b}^{(\mathrm{R})}(\boldsymbol{\tau}_m), \tag{4.102}$$

with the vectors $\boldsymbol{b}^{(\mathrm{L})}$ and $\boldsymbol{b}^{(\mathrm{R})}$ having elements $\exp[-\hat{b}^{(\mathrm{L})}(\boldsymbol{\tau}_m)/T]$ and $\exp[-\hat{b}^{(\mathrm{R})}(\boldsymbol{\tau}_m)/T]$ respectively

$$\boldsymbol{B} = [\boldsymbol{b}^{(\mathrm{R})}]^{(\mathrm{T})} \otimes \boldsymbol{b}^{(\mathrm{L})}. \tag{4.103}$$

From (4.100),

$$\beta_\ell(\mathrm{k}) = [\boldsymbol{x}_\ell(\mathrm{k}) \cdot \boldsymbol{b}^{(\mathrm{R})}][\boldsymbol{y}_\ell(\mathrm{k}) \cdot \boldsymbol{b}^{(\mathrm{L})}]. \tag{4.104}$$

We have shown in Appendix A.3 that any M-dimensional vector which is invariant under the operations of the matrices of the representation $\mathcal{R}(\mathcal{Q})$ will have zero scalar production with all the left and right eigenvectors of $\boldsymbol{V}$ except for those in the one-dimensional symmetric representation A ($\mathrm{k} = 1$). It follows that, if one or both of the boundary vectors $\boldsymbol{b}^{(\mathrm{R})}$ and $\boldsymbol{b}^{(\mathrm{L})}$ have this invariance property,

$$\beta_\ell(\mathrm{k}) = 0, \qquad \text{if } \mathrm{k} \neq 1, \tag{4.105}$$

with (4.99) reduced to

$$Z(N_1, N_2, T) = \sum_{\ell=1}^{w(1)} \beta_\ell(1)[\Lambda_\ell(1)]^{N_2-1}. \tag{4.106}$$

4.9.3 The Limit $N_2 \to \infty$

According to Perron's theorem (see Appendix A.3) the largest eigenvalue $\Lambda_1(1)$ is real, non-degenerate and greater in magnitude that any other eigenvalue and has left and right eigenvectors with real strictly positive elements. Since the elements of the boundary matrix $\boldsymbol{B}$ are also strictly positive it follows from (4.93) and (4.100) that

$$\beta_1(1) = \boldsymbol{x}_1(1)\boldsymbol{B}[\boldsymbol{y}_1(1)]^{(\mathrm{T})} > 0. \tag{4.107}$$

We define the *scaled eigenvalues*

$$\Omega_\ell(\mathrm{k}) = \Lambda_\ell(\mathrm{k})/\Lambda_1(1), \tag{4.108}$$

for which

$$\lim_{N_2\to\infty} |\Omega_\ell(\mathsf{k})|^{N_2} = 0\,, \qquad (\mathsf{k},\ell) \neq (1,1)\,. \tag{4.109}$$

From (4.99), in the limit $N_2 \to \infty$,

$$\begin{aligned} Z(N_1,N_2,T) &= [\Lambda_1(1)]^{N_2-1} \sum_{\{\mathsf{k},\ell\}} \beta_\ell(\mathsf{k})[\Omega_\ell(\mathsf{k})]^{N_2-1} \\ &\simeq \beta_1(1)[\Lambda_1(1)]^{N_2-1}. \end{aligned} \tag{4.110}$$

In the case of toroidal boundary conditions it follows from (4.101) that

$$Z(N_1,N_2,T) \simeq [\Lambda_1(1)]^{N_2}. \tag{4.111}$$

It can be seen that both (4.110) and (4.111) yield the result

$$\phi(N_1) = -\lim_{N_2\to\infty} \frac{\ln[Z(N_1,N_2,T)]}{N_1 N_2} = -\frac{1}{N_1}\ln[\Lambda_1(1)] \tag{4.112}$$

for the dimensionless free energy $\phi(N_1)$ per lattice site in the limit $N_2 \to \infty$, when the effects of the boundaries become negligible.

4.9.4 Correlation Functions

In this section we shall be ultimately concerned with the $N_2 \to \infty$ limit. It is therefore convenient to assume toroidal boundary conditions. The probability that the state $\boldsymbol{\tau}(\mathsf{j})$ of the j-th slice of the lattice is $\boldsymbol{\tau}_m(\alpha)$ is given by

$$\begin{aligned} \mathit{Pr}[\boldsymbol{\tau}(\mathsf{j}) = \boldsymbol{\tau}_m(\alpha)] &= \frac{\mathrm{Trace}\{\boldsymbol{V}^{\mathsf{j}-1}|\alpha,m\rangle\langle\alpha,m|\boldsymbol{V}^{N_2-\mathsf{j}+1}\}}{Z(N_1,N_2,T)} \\ &= \frac{\mathrm{Trace}\{\boldsymbol{Y}\boldsymbol{\Lambda}^{\mathsf{j}-1}\boldsymbol{X}|\alpha,m\rangle\langle\alpha,m|\boldsymbol{Y}\boldsymbol{\Lambda}^{N_2-\mathsf{j}+1}\boldsymbol{X}\}}{Z(N_1,N_2,T)} \\ &= \frac{\mathrm{Trace}\{\boldsymbol{\Theta}_\alpha^{(m)}\boldsymbol{\Lambda}^{N_2}\}}{Z(N_1,N_2,T)}\,, \end{aligned} \tag{4.113}$$

where

$$\boldsymbol{\Theta}_\alpha^{(m)} = \boldsymbol{X}|\alpha,m\rangle\langle\alpha,m|\boldsymbol{Y}\,. \tag{4.114}$$

The elements of the matrix $\boldsymbol{\Theta}_\alpha^{(m)}$, parameterized in terms of the irreducible group representations, are

$$\begin{aligned} \Theta_\alpha^{(m)}(\mathsf{k},\ell;\mathsf{k}',\ell') &= \langle\mathsf{k},\ell|\boldsymbol{X}|\alpha,m\rangle\langle\alpha,m|\boldsymbol{Y}|\mathsf{k}',\ell'\rangle \\ &= [\boldsymbol{x}_\ell(\mathsf{k})|\alpha,m\rangle][\boldsymbol{y}_{\ell'}(\mathsf{k}')|\alpha,m\rangle] \end{aligned} \tag{4.115}$$

and

$$\mathit{Pr}[\boldsymbol{\tau}(\mathsf{j}) = \boldsymbol{\tau}_m(\alpha)] = \frac{\sum_{\{\mathsf{k},\ell\}} \Theta_\alpha^{(m)}(\mathsf{k},\ell;\mathsf{k},\ell)[\Omega_\ell(\mathsf{k})]^{N_2}}{\sum_{\{\mathsf{k},\ell\}} [\Omega_\ell(\mathsf{k})]^{N_2}}\,. \tag{4.116}$$

As is to be expected this probability is independent of j. In a similar way it is not difficult to show that,

$$Pr\begin{bmatrix}\tau(\mathrm{j}') = \tau_{m'}(\alpha') \\ \tau(\mathrm{j}) = \tau_m(\alpha)\end{bmatrix} = \frac{\sum_{\{\mathrm{k},\ell\}} \Phi^{(m'm)}_{\alpha'\alpha}(\mathrm{j}-\mathrm{j}';\mathrm{k},\ell;\mathrm{k},\ell)[\Omega_\ell(\mathrm{k})]^{N_2+\mathrm{j}'-\mathrm{j}}}{\sum_{\{\mathrm{k},\ell\}} [\Omega_\ell(\mathrm{k})]^{N_2}},$$

$$\mathrm{j} \geq \mathrm{j}', \tag{4.117}$$

where

$$\begin{aligned}\Phi^{(m'm)}_{\alpha'\alpha}&(\mathrm{j};\mathrm{k},\ell;\mathrm{k}',\ell') \\ &= \langle \mathrm{k},\ell|\boldsymbol{X}|\alpha,m\rangle\langle\alpha,m|\boldsymbol{V}^{\mathrm{j}}|\alpha',m'\rangle\langle\alpha',m'|\boldsymbol{Y}|\mathrm{k}',\ell'\rangle/[\Lambda_1(1)]^{\mathrm{j}} \\ &= \sum_{\{\mathrm{k}'',\ell''\}} \Theta^{(m')}_{\alpha'}(\mathrm{k},\ell;\mathrm{k}'',\ell'')[\Omega_{\ell''}(\mathrm{k}'')]^{\mathrm{j}}\Theta^{(m)}_{\alpha}(\mathrm{k}'',\ell'';\mathrm{k}',\ell').\end{aligned} \tag{4.118}$$

From (4.115)

$$\Phi^{(m',m)}_{\alpha',\alpha}(0;\mathrm{k},\ell;\mathrm{k}',\ell') = \delta^{\mathrm{Kr}}(m-m')\delta^{\mathrm{Kr}}(\alpha-\alpha')\Theta^{(m)}_{\alpha}(\mathrm{k},\ell;\mathrm{k}',\ell'). \tag{4.119}$$

Similar results can be obtained for the joint probabilities for the states of other subsets of the slices. In particular we shall need

$$Pr\begin{bmatrix}\tau(\mathrm{j}'-1) & = \tau_{m_1'}(\alpha_1') \\ \tau(\mathrm{j}') & = \tau_{m'}(\alpha') \\ \tau(\mathrm{j}-1) & = \tau_{m_1}(\alpha_1) \\ \tau(\mathrm{j}) & = \tau_m(\alpha)\end{bmatrix}$$

$$= \frac{\sum_{\{\mathrm{k},\ell\}} \Psi^{(m_1'm'm_1m)}_{\alpha_1'\alpha'\alpha_1\alpha}(\mathrm{j}-\mathrm{j}';\mathrm{k},\ell)[\Omega_\ell(\mathrm{k})]^{N_2+\mathrm{j}'-\mathrm{j}-1}}{\sum_{\{\mathrm{k},\ell\}} [\Omega_\ell(\mathrm{k})]^{N_2}}, \qquad \mathrm{j} \geq \mathrm{j}', \tag{4.120}$$

where, for $\mathrm{j} > 0$,

$$\begin{aligned}\Psi^{(m_1'm'm_1m)}_{\alpha_1'\alpha'\alpha_1\alpha}(\mathrm{j};\mathrm{k},\ell) &= \langle \mathrm{k},\ell|\boldsymbol{X}|\alpha_1',m_1'\rangle\langle\alpha_1',m_1'|\boldsymbol{V}|\alpha',m'\rangle \\ &\quad \times \langle\alpha',m'|\boldsymbol{V}^{\mathrm{j}-1}|\alpha_1,m_1\rangle \\ &\quad \times \langle\alpha_1,m_1|\boldsymbol{V}|\alpha,m\rangle\langle\alpha,m|\boldsymbol{Y}|\mathrm{k},\ell\rangle/[\Lambda_1(1)]^{\mathrm{j}+1} \\ &= \sum_{\{\mathrm{k}',\ell'\}} \Phi^{(m_1'm')}_{\alpha_1'\alpha'}(1;\mathrm{k},\ell;\mathrm{k}',\ell')[\Omega_{\ell'}(\mathrm{k}')]^{\mathrm{j}-1} \\ &\quad \times \Phi^{(m_1m)}_{\alpha_1\alpha}(1;\mathrm{k}',\ell';\mathrm{k},\ell)\end{aligned} \tag{4.121}$$

and, for $\mathfrak{j} = 0$,

$$\Psi^{(m_1' m' m_1 m)}_{\alpha_1' \alpha' \alpha_1 \alpha}(0; \mathfrak{k}, \ell) = \delta^{\mathrm{Kr}}(m - m')\delta^{\mathrm{Kr}}(\alpha - \alpha')\delta^{\mathrm{Kr}}(m_1 - m_1') \times \delta^{\mathrm{Kr}}(\alpha_1 - \alpha_1')\Phi^{(m_1 m)}_{\alpha_1 \alpha}(1; \mathfrak{k}, \ell; \mathfrak{k}, \ell) . \quad (4.122)$$

The expectation value $\langle \mathfrak{c}_\mathfrak{j} \rangle$ for any function $\mathfrak{c}(\boldsymbol{\tau}(\mathfrak{j})) = \mathfrak{c}_\mathfrak{j}$ of the state of a single slice $\mathcal{N}_\mathfrak{j}$ is

$$\langle \mathfrak{c}_\mathfrak{j} \rangle = \sum_{\alpha=1}^{\gamma} \sum_{m=1}^{\mathfrak{p}(\alpha)} \mathit{Pr}[\boldsymbol{\tau}_m(\alpha)]\mathfrak{c}(\boldsymbol{\tau}_m(\alpha)) = \mathcal{C}_1(N_2) , \quad (4.123)$$

where

$$\mathcal{C}_1(N_2) = \frac{\sum_{\{\mathfrak{k},\ell\}} \theta(\mathfrak{k}, \ell; \mathfrak{k}, \ell)[\Omega_\ell(\mathfrak{k})]^{N_2}}{\sum_{\{\mathfrak{k},\ell\}} [\Omega_\ell(\mathfrak{k})]^{N_2}} , \quad (4.124)$$

$$\theta(\mathfrak{k}, \ell; \mathfrak{k}', \ell') = \sum_{\alpha=1}^{\gamma} \sum_{m=1}^{\mathfrak{p}(\alpha)} \Theta^{(m)}_\alpha(\mathfrak{k}, \ell; \mathfrak{k}', \ell')\mathfrak{c}(\boldsymbol{\tau}_m(\alpha)) . \quad (4.125)$$

It is also clear that

$$\langle \mathfrak{c}_{\mathfrak{j}'} \mathfrak{c}_\mathfrak{j} \rangle = \sum_{\alpha=1}^{\gamma} \sum_{\alpha'=1}^{\gamma} \sum_{m=1}^{\mathfrak{p}(\alpha)} \sum_{m'=1}^{\mathfrak{p}(\alpha')} \mathit{Pr}\begin{bmatrix} \boldsymbol{\tau}(\mathfrak{j}') = \boldsymbol{\tau}_{m'}(\alpha') \\ \boldsymbol{\tau}(\mathfrak{j}) = \boldsymbol{\tau}_m(\alpha) \end{bmatrix} \mathfrak{c}(\boldsymbol{\tau}_{m'}(\alpha'))\mathfrak{c}(\boldsymbol{\tau}_m(\alpha))$$

$$= \mathcal{C}_2(\mathfrak{j} - \mathfrak{j}'; N_2), \quad (4.126)$$

where

$$\mathcal{C}_2(\mathfrak{j}; N_2) = \frac{\sum_{\{\mathfrak{k},\ell\}} \sum_{\{\mathfrak{k}',\ell'\}} \theta(\mathfrak{k}, \ell; \mathfrak{k}', \ell')[\Omega_{\ell'}(\mathfrak{k}')]^\mathfrak{j}\theta(\mathfrak{k}', \ell'; \mathfrak{k}, \ell)[\Omega_\ell(\mathfrak{k})]^{N_2 - \mathfrak{j}}}{\sum_{\{\mathfrak{k},\ell\}} [\Omega_\ell(\mathfrak{k})]^{N_2}} . \quad (4.127)$$

The expectation value $\langle \mathfrak{c}_{\mathfrak{j}'} \mathfrak{c}_\mathfrak{j} \rangle$ is dependent on $\mathfrak{j}$ and $\mathfrak{j}'$ only through the difference $\mathfrak{j} - \mathfrak{j}'$. The pair correlation function between the values of $\mathfrak{c}(\boldsymbol{\tau})$ at slices $\mathcal{N}_{\mathfrak{j}'}$ and $\mathcal{N}_\mathfrak{j}$ is given (see (1.54)) by

$$\Gamma_2(\mathfrak{c}_\mathfrak{j}, \mathfrak{c}_{\mathfrak{j}'}) = \langle \mathfrak{c}_{\mathfrak{j}'} \mathfrak{c}_\mathfrak{j} \rangle - \langle \mathfrak{c}_{\mathfrak{j}'} \rangle \langle \mathfrak{c}_\mathfrak{j} \rangle$$

$$= \mathcal{C}_2(\mathfrak{j} - \mathfrak{j}'; N_2) - [\mathcal{C}_1(N_2)]^2 . \quad (4.128)$$

A similar analysis can be used to obtain the pair correlation function of quantities which are dependent on the states of a pair of neighbouring slices $\mathcal{N}_{\mathfrak{j}-1}$ and $\mathcal{N}_\mathfrak{j}$. Of particular interest is the case of functions of the form $\mathfrak{g}_\mathfrak{j} =$

$\mathfrak{g}(\boldsymbol{\tau}(\mathrm{j}-1),\boldsymbol{\tau}(\mathrm{j}))$ which are invariant under the operations of the symmetry group. Then we have

$$\Gamma_2(\mathfrak{g}_{\mathrm{j}},\mathfrak{g}_{\mathrm{j}'}) = \mathcal{G}_2(\mathrm{j}-\mathrm{j}';N_2) - [\mathcal{G}_1(N_2)]^2\,, \tag{4.129}$$

where

$$\mathcal{G}_1(N_2) = \frac{\sum\limits_{\{\mathrm{k},\ell\}} \phi_1(\mathrm{k};\ell,\ell)[\Omega_\ell(\mathrm{k})]^{N_2-1}}{\sum\limits_{\{\mathrm{k},\ell\}} [\Omega_\ell(\mathrm{k})]^{N_2}}\,, \tag{4.130}$$

$$\mathcal{G}_2(\mathrm{j};N_2) = \frac{\sum\limits_{\{\mathrm{k},\ell,\ell'\}} \phi_1(\mathrm{k};\ell,\ell')[\Omega_{\ell'}(\mathrm{k})]^{\mathrm{j}-1}\phi_1(\mathrm{k};\ell',\ell)[\Omega_\ell(\mathrm{k})]^{N_2-\mathrm{j}-1}}{\sum\limits_{\{\mathrm{k},\ell\}} [\Omega_\ell(\mathrm{k})]^{N_2}}\,, \tag{4.131}$$

for $\mathrm{j}>0$,

$$\mathcal{G}_2(0;N_2) = \frac{\sum\limits_{\{\mathrm{k},\ell\}} \phi_2(\mathrm{k};\ell,\ell)[\Omega_\ell(\mathrm{k})]^{N_2-1}}{\sum\limits_{\{\mathrm{k},\ell\}} [\Omega_\ell(\mathrm{k})]^{N_2}}\,, \tag{4.132}$$

and

$$\phi_q(\mathrm{k};\ell,\ell') = \sum_{\alpha=1}^{\gamma}\sum_{\alpha'=1}^{\gamma}\sum_{m=1}^{\mathfrak{p}(\alpha)}\sum_{m'=1}^{\mathfrak{p}(\alpha')} \Phi_{\alpha,\alpha'}^{(m,m')}(1;\mathrm{k},\ell;\mathrm{k},\ell')[\mathfrak{g}(\boldsymbol{\tau}_m(\alpha),\boldsymbol{\tau}_{m'}(\alpha'))]^q\,. \tag{4.133}$$

The fact that (4.133) involves only one irreducible representation follows from Theorem A.3.13 and the supposed symmetry property of $\mathfrak{g}(\boldsymbol{\tau}_m(\alpha),\boldsymbol{\tau}_{m'}(\alpha'))$. With $\mathfrak{g}(\boldsymbol{\tau}_m(\alpha),\boldsymbol{\tau}_{m'}(\alpha')) = \hat{h}(\boldsymbol{\tau}_m(\alpha),\boldsymbol{\tau}_{m'}(\alpha'))$ (4.129) gives the energy–energy correlation function for pairs of neighbouring slices of the lattice.

Consider now the limit as $N_2 \to \infty$. From (4.109), (4.111), (4.124), (4.127), (4.130) and (4.131),

$$\mathcal{C}_1(N_2) \simeq \theta(1,1;1,1)\,, \tag{4.134}$$

$$\mathcal{C}_2(\mathrm{j};N_2) \simeq \sum_{\{\mathrm{k},\ell\}} \theta(1,1;\mathrm{k},\ell)[\Omega_\ell(\mathrm{k})]^{\mathrm{j}}\theta(\mathrm{k},\ell;1,1)\,, \tag{4.135}$$

$$\mathcal{G}_1(N_2) \simeq \phi_1(1;1,1)\,, \tag{4.136}$$

$$\mathcal{G}_2(\mathrm{j};N_2) \simeq \begin{cases} \sum\limits_{\ell=1}^{w(1)} \phi_1(1;1,\ell)\phi_1(1;\ell,1)[\Omega_\ell(1)]^{\mathrm{j}-1}\,, & \mathrm{j}>0,\\ \phi_2(1;1,1)\,, & \mathrm{j}=0. \end{cases} \tag{4.137}$$

Thus, from (4.128) and (4.129),

$$\Gamma_2(\mathfrak{c}_{\rm j},\mathfrak{c}_{\rm j'}) \simeq \sum_{\{{\rm k},\ell\}}^{[1,1]} \theta(1,1;{\rm k},\ell)\theta({\rm k},\ell;1,1)[\Omega_\ell({\rm k})]^{{\rm j}-{\rm j}'}, \tag{4.138}$$

$$\Gamma_2(\mathfrak{g}_{\rm j},\mathfrak{g}_{\rm j'}) = \begin{cases} \displaystyle\sum_{\ell=2}^{w(1)} \phi_1(1;1,\ell)\phi_1(1;\ell,1)[\Omega_\ell(1)]^{{\rm j}-{\rm j}'-1}, & {\rm j} > {\rm j}', \\ \phi_2(1;1,1) - [\phi_1(1;1,1)]^2, & {\rm j} = {\rm j}', \end{cases} \tag{4.139}$$

where '[1,1]' indicates that the pair of indices $\mathrm{k} = 1$, $\ell = 1$ is excluded from the summation. It should be noted that, whereas correlation functions like (4.138) depends on all the scaled eigenvalues, the energy-energy correlation which is an example of the type given by (4.139) depends only on the scaled eigenvalues in the one-dimensional symmetric representation. This result is important for Wood's method described in Sect. 4.10.

4.9.5 Correlation Lengths and Phase Transitions

Given that the elements of the transfer matrix can be expressed in terms of some suitable Boltzmann factor $\mathfrak{X}$, it follows that the partition function $Z(N_1, N_2, T)$ for finite N_1 and N_2 is a polynomial in $\mathfrak{X}$ with positive coefficients. By the argument used in Sect. 4.1 for the simple lattice fluid it follows that the system can have no phase transitions. A modified version of Theorem 4.2 can then be used to show that no phase transitions occur in the one-dimensionally infinite (N_1 finite, $N_2 \to \infty$) system. Since, however, phase transitions do occur in systems more than one-dimensionally infinite this must be possible in the double limit when both N_1 and N_2 become large. Ree and Chesnut (1966) have argued that a necessary condition for a phase transition at $T = T_{\rm c}$ is that, in the double limit when both N_2 and N_1 become large,

$$|\Omega_\ell(\mathrm{k})|^{N_1} \to L(T), \qquad \text{for some } (\mathrm{k},\ell) \neq (1,1), \tag{4.140}$$

where $L(T_{\rm c}) \neq 0$. To achieve a condition of the type (4.140) it is of course necessary to specify the precise way in which the double limit is taken. In any event the condition

$$\lim_{N_1\to\infty} |\Omega_\ell(\mathrm{k})| = 1, \qquad \text{for some } (\mathrm{k},\ell) \neq (1,1) \tag{4.141}$$

is implied by (4.140) and is in turn a necessary condition for a phase transition. We now consider one-dimensionally infinite systems in which the limit $N_2 \to \infty$ has already been taken and the partition function and correlation function are given by (4.111) and (4.138) respectively. Using the latter we shall obtain an asymptotic formula for the correlation lengths $\mathfrak{r}(\mathfrak{c})$ and $\mathfrak{r}(\mathfrak{g})$. From (2.6),

$$\mathfrak{r}^2(\mathfrak{c}) = \frac{a^2}{2} \frac{\sum_{j=0}^{\infty} j^2 \Gamma_2(\mathfrak{c}_j, \mathfrak{c}_0)}{\sum_{j=0}^{\infty} \Gamma_2(\mathfrak{c}_j, \mathfrak{c}_0)}, \tag{4.142}$$

where a is a length parameter measuring the distance between $\mathcal{N}_j$ and $\mathcal{N}_{j+1}$ for any j and we have used the result $c(\mathfrak{d}) = c(1) = \frac{1}{2}$ obtained in Sect. 3.3.2. The correlation length for $\mathfrak{g}$ is given by a similar formula and, from (4.138) and (4.139),

$$\mathfrak{r}^2(\mathfrak{c}) = \frac{a^2}{2} \frac{\displaystyle\sum_{\{k,\ell\}}^{[1,1]} \frac{\theta(1,1;k,\ell)\theta(k,\ell;1,1)\Omega_\ell(k)[1+\Omega_\ell(k)]}{[1-\Omega_\ell(k)]^3}}{\displaystyle\sum_{\{k,\ell\}}^{[1,1]} \frac{\theta(1,1;k,\ell)\theta(k,\ell;1,1)}{1-\Omega_\ell(k)}}, \tag{4.143}$$

$$\mathfrak{r}^2(\mathfrak{g}) = \frac{a^2}{2} \frac{\displaystyle\sum_{\ell=2}^{w(1)} \frac{\phi(1;1,\ell)\phi(1;\ell,1)[1+\Omega_\ell(k)]}{[1-\Omega_\ell(k)]^3}}{\Gamma_2(\mathfrak{g}_0, \mathfrak{g}_0) + \displaystyle\sum_{\ell=2}^{w(1)} \frac{\phi(1;1,\ell)\phi(1;\ell,1)}{1-\Omega_\ell(k)}}. \tag{4.144}$$

The scaled eigenvalues $\Omega_\ell(k)$ are, in general, complex numbers lying within a disc, centre the origin and radius unity in a complex plane $\mathbb{C}_\Omega$. The number of scaled eigenvalues $M = \nu^{N_1}$ increases with increasing N_1 and they will move in the disc as the temperature, or any other thermodynamic variable, is varied. It is clear from (4.143) that the correlation length becomes large when a scaled eigenvalue or a number of scaled eigenvalues approach the point $1 \in \mathbb{C}_\Omega$. For such a scaled eigenvalue

$$\Omega_\ell(k) = \exp\{\ln[\Omega_\ell(k)]\} \simeq 1 + \ln[\Omega_\ell(k)]. \tag{4.145}$$

The summations in (4.143) are dominated by the terms to which (4.145) applies and thus

$$\mathfrak{r}^2(\mathfrak{c}) \simeq a^2 \frac{\displaystyle\sum_{\{k,\ell\}}^{[1,1]} \theta(1,1;k,\ell)\theta(k,\ell;1,1)\{\ln[\Omega_\ell(k)]\}^{-3}}{\displaystyle\sum_{\{k,\ell\}}^{[1,1]} \theta(1,1;k,\ell)\theta(k,\ell;1,1)\{\ln[\Omega_\ell(k)]\}^{-1}}. \tag{4.146}$$

This formula gives a real result since complex eigenvalues will approach $1 \in \mathbb{C}_\Omega$ in conjugate pairs. Since the coefficients $\theta(1,1;k,\ell)\theta(k,\ell;1,1)$ are constructed from the eigenvectors $\boldsymbol{x}_\ell(k)$ and $\boldsymbol{y}_\ell(k)$ of the real matrix $\boldsymbol{V}$ it follows that they are conjugate complex if the corresponding eigenvalues are conjugate complex. If the most singular terms in the summations in (4.146) correspond to a set of degenerate real eigenvalues then the formula reduces to

$$\mathfrak{r}(\mathfrak{c}) \simeq -a\{\ln[\Omega_\ell(\mathrm{k})]\}^{-1}, \tag{4.147}$$

for any $\mathfrak{c}(\tau)$, which is a function of the state τ of a single slice. We now understand that a phase transition to an ordered phase in a system with interactions only between neighbouring slices $\mathcal{N}_\mathrm{j}$ can occur only in the limit of an infinite system, where both N_1 and N_2 become large. We have also seen, from (4.143), (4.146) and (4.147), that, for a semi-infinite system ($N_2 \to \infty$), the correlation length becomes large when one or more of the scaled eigenvalues approaches $1 \in \mathbb{C}_\Omega$. For N_1 finite this can be regarded as an *incipient phase transition*, which will become a true phase transition with a singularity in the correlation length if

$$\lim_{N_1\to\infty} \Omega_\ell(\mathrm{k}) = 1\,, \qquad \text{for some } (\mathrm{k},\ell) \neq (1,1). \tag{4.148}$$

It is now interesting to compare conditions (4.141) and (4.148). The former, which is a necessary condition, relates the occurrence of a phase transition to a scaled eigenvalue approaching any point on the unit circle in $\mathbb{C}_\Omega$, whereas the latter is concerned only with the single point where the unit circle cuts the real axis. The limit (4.148) leads to a singularity of the correlation length (4.142). This is constructed using the formula (4.128) for the pair correlation function and measures the degree of correlation between slices at a separation $\mathrm{j}-\mathrm{j}'$ which tends to align them *in the same state*. However, not all ordered phases are related to this type of homogeneous ground state. Suppose that the phase transition were to an ordered phase with a periodic ground state of period η. In this case we need a correlation length based on the correlation between sites at a lattice distance which is an integer multiple of η. The expressions (4.142), (4.143), (4.145) and (4.146) are replaced by

$$\mathfrak{r}^2_\eta(\mathfrak{c}) = \frac{a^2}{2}\frac{\sum_{\mathrm{j}=0}^{\infty}(\mathrm{j}\eta)^2\Gamma_2(\mathfrak{c}_{\mathrm{j}\eta},\mathfrak{c}_0)}{\sum_{\mathrm{j}=0}^{\infty}\Gamma_2(\mathfrak{c}_{\mathrm{j}\eta},\mathfrak{c}_0)}, \tag{4.149}$$

$$\mathfrak{r}^2_\eta(\mathfrak{c}) = \frac{(\eta a)^2}{2}\frac{\displaystyle\sum_{\{\mathrm{k},\ell\}}^{[1,1]}\frac{\theta(1,1;\mathrm{k},\ell)\theta(\mathrm{k},\ell;1,1)[\Omega_\ell(\mathrm{k})]^\eta\{1+[\Omega_\ell(\mathrm{k})]^\eta\}}{\{1-[\Omega_\ell(\mathrm{k})]^\eta\}^3}}{\displaystyle\sum_{\{\mathrm{k},\ell\}}^{[1,1]}\frac{\theta(1,1;\mathrm{k},\ell)\theta(\mathrm{k},\ell;1,1)}{1-[\Omega_\ell(\mathrm{k})]^\eta}}, \tag{4.150}$$

$$[\Omega_\ell(\mathrm{k})]^\eta = \exp[\ln\{[\Omega_\ell(\mathrm{k})]^\eta\}] \simeq 1+\ln\{[\Omega_\ell(\mathrm{k})]^\eta\}\,, \tag{4.151}$$

$$\mathfrak{r}^2(\mathfrak{c}) \simeq (\eta a)^2 \frac{\sum\limits_{\{\mathrm{k},\ell\}}^{[1,1]} \theta(1,1;\mathrm{k},\ell)\theta(\mathrm{k},\ell;1,1)[\ln\{[\Omega_\ell(\mathrm{k})]^\eta\}]^{-3}}{\sum\limits_{\{\mathrm{k},\ell\}}^{[1,1]} \theta(1,1;\mathrm{k},\ell)\theta(\mathrm{k},\ell;1,1)[\ln\{[\Omega_\ell(\mathrm{k})]^\eta\}]^{-1}} . \tag{4.152}$$

The correlation length $\mathfrak{r}_\eta(\mathfrak{c})$ will become large if, for some (ℓ, k), $\Omega_\ell(\mathrm{k})$ is close to $\exp(2\pi \mathrm{i} q/\eta)$ for some $q = 0, 1, \ldots, \eta - 1$ and a phase transition to an ordered state with a periodic ground state of period η will occur if

$$\lim_{N_1 \to \infty} \Omega_\ell(\mathrm{k}) = \exp(2\pi \mathrm{i} q/\eta) , \qquad \text{for some } (\mathrm{k}, \ell) \neq (1,1). \tag{4.153}$$

This approach to the unit circle in $\mathbb{C}_\Omega$ will, of course, occur for a conjugate complex pair of eigenvalues if $\Omega_\ell(\mathrm{k})$ is not real. Suppose that the set of reduced eigenvalues is dominated in magnitude by some subset all of whose members have magnitude Ω_{Max}. For these eigenvalues

$$[\Omega_\ell(\mathrm{k})]^\eta = \exp\left[\mathrm{i}\mathrm{Arg}\{[\Omega_\ell(\mathrm{k})]^\eta\} + \eta \ln(\Omega_{\mathrm{Max}})\right] , \tag{4.154}$$

and, if

$$|\mathrm{Arg}\{[\Omega_\ell(\mathrm{k})]^\eta\}| \ll \eta |\ln(\Omega_{\mathrm{Max}})| , \tag{4.155}$$

(4.147) generalizes to

$$\mathfrak{r}_\eta(\mathfrak{c}) \simeq -a[\ln(\Omega_{\mathrm{Max}})]^{-1} . \tag{4.156}$$

This is the form for the correlation length used in many investigations which employ transfer matrices (e.g. the phenomenological renormalization method of Sect. 6.13). It is, however, predicated on (4.155) being true for cases where the second largest eigenvalue is not real.

4.10 The Wood Method

In this section we describe the method, developed by Wood and his coworkers (Wood 1987, Wood et al. 1987, Wood and Turnbull 1988, Wood et al. 1989, Wood and Ball 1990), for understanding the critical properties of a system by considering the limit as $N_1 \to \infty$ of the locus of the complex zeros of the partition function of the finite $N_1 \times N_2$ system with toroidal boundary conditions.

In Sect. 1.4 we presented a formulation of thermodynamics in terms of a set of internal couplings K_j, $j = 1, 2, \ldots, n_{\mathrm{e}}$ and external couplings L_i, $i = 1, 2, \ldots n_{\mathrm{f}}$. Given that there are no independent densities ($r = n_{\mathrm{f}} - 1$) and that all the couplings are expressed in terms of one coupling K by the parameters a_j and b_i with $K_j = a_j K$ and $L_i = b_i K$ the temperature dependence of the system, for fixed values of a_j and b_i, arises from dependence on the variable

$\mathfrak{X} = \exp(\kappa)$. The characteristic equation for the transfer matrix now takes the form

$$\mathrm{Det}\{\boldsymbol{V} - \Lambda \boldsymbol{I}_M\} = \mathcal{F}(N_1, \mathfrak{X}; \Lambda) = 0\,, \tag{4.157}$$

where $M = \nu^{N_1}$ and

$$\mathcal{F}(N_1, \mathfrak{X}; \Lambda) = \mathsf{f}_M(N_1, \mathfrak{X})\Lambda^M + \mathsf{f}_{M-1}(N_1, \mathfrak{X})\Lambda^{M-1} + \cdots + \mathsf{f}_0(N_1, \mathfrak{X})\,. \tag{4.158}$$

In general the coefficients $\mathsf{f}_j(N_1, \mathfrak{X})$, $j = 0, 1, \ldots, M$, can be sums of monomials in $\mathfrak{X}$ with negative and positive exponents. We can however, without loss of generality, suppose that a constant has been added to all the interaction energies so that the coefficients are analytic polynomial functions of $\mathfrak{X}$. (For an example of this see the discussion of the square Ising model below.)

As in the methods of Yang and Lee (Yang and Lee 1952, Lee and Yang 1952) and Ruelle (1971), described in Sects. 4.5–4.8, the present analysis investigates the system using a complex plane, which in this case is $\mathbb{C}_{\mathfrak{X}}$, the complex plane of $\mathfrak{X}$. For the moment we shall denote the eigenvalues of the transfer matrix $\boldsymbol{V}$ by $\Lambda_r(N_1, \mathfrak{X})$, $r = 1, 2, \ldots, M$, without distinguishing their association with the irreducible representations of the symmetry group of the system. Then, from (4.101), the partition function is given by

$$Z(N_1, N_2, \mathfrak{X}) = \sum_{r=1}^{M} [\Lambda_r(N_1, \mathfrak{X})]^{N_2}. \tag{4.159}$$

Equation (4.159) is exact and, for finite N_1 and N_2, can be expanded to give a polynomial in $\mathfrak{X}$ with positive real coefficients. Since $\mathsf{f}_M(N_1, \mathfrak{X}) = (-1)^M$ the eigenvalues will be finite in $\mathbb{C}_{\mathfrak{X}}$, except possibly at the point a infinity. They may however, have a complicated branch-point structure. This cancels from the summation in (4.159) to produce an analytic function with a finite set of zeros. We denote this set by $\mathbb{O}(N_1, N_2)$.

4.10.1 Evolution of Partition Function Zeros

Given any two eigenvalues $\Lambda_r(N_1, \mathfrak{X})$ and $\Lambda_{r'}(N_1, \mathfrak{X})$ it is clear that they will have the same magnitude on the line for which $\mathfrak{X}$ satisfies the equation

$$\Lambda_r(N_1, \mathfrak{X}) = \varpi \Lambda_{r'}(N_1, \mathfrak{X})\,, \qquad \varpi = \exp(\mathrm{i}\vartheta)\,, \quad 0 \le \vartheta < 2\pi\,. \tag{4.160}$$

In general the order of the relative magnitudes of the eigenvalues will interchange across this line. The set of all curves defined by (4.160) for all r and all $r' \neq r$ is denoted by $\mathcal{D}_{N_1}$. This forms a complicated network in the plane $\mathbb{C}_{\mathfrak{X}}$. Let $\mathcal{M}_{N_1}$ be the subset of $\mathcal{D}_{N_1}$ for which the eigenvalues satisfying (4.160) are of maximum magnitude. It follows that, for any $\mathfrak{X} \notin \mathcal{M}_{N_1}$, there is a unique eigenvalue of maximum magnitude. This we denote by $\Lambda_{\mathrm{Max}}(N_1, \mathfrak{X})$.[9] We

[9] Bearing in mind that the definition is ambiguous on $\mathcal{M}_{N_1}$ and implies a change of analytic form across a curve of $\mathcal{M}_{N_1}$.

shall now show that in the limit $N_2 \to \infty$ the zeros $\mathbb{O}(N_1, N_2)$ converge onto the curves $\mathcal{M}_{N_1}$ To make this idea precise we prove the following theorem:

Theorem 4.10.1. *For every $\mathfrak{X} \notin \mathcal{M}_{N_1}$ and any ϵ satisfying the condition $|\Lambda_{\text{Max}}(N_1, \mathfrak{X})| > \epsilon > 0$, there exists an $N_0 > 0$ such that, for all $N_2 > N_0$, $|Z(N_1, N_2, \mathfrak{X})|^{1/N_2} > \epsilon$.*

Proof: From (4.159),

$$|Z(N_1, N_2, \mathfrak{X})|^{1/N_2} = |\Lambda_{\text{Max}}(N_1, \mathfrak{X})| \left| \sum_{r=1}^{M} [\Omega_r(N_1, \mathfrak{X})]^{N_2} \right|^{1/N_2}, \tag{4.161}$$

where now the definition (4.108), for the scaled eigenvalues, is generalized to

$$\Omega_r(N_1, \mathfrak{X}) = \Lambda_r(N_1, \mathfrak{X}) / \Lambda_{\text{Max}}(N_1, \mathfrak{X}) . \tag{4.162}$$

Since the transfer matrix $\boldsymbol{V}$ is non-singular for any $\mathfrak{X}$, not all of its eigenvalues are zero. In particular it is possible, for the given ϵ, to find a $B > 0$ such that $|\Lambda_{\text{Max}}(N_1, \mathfrak{X})| > B > \epsilon$. Since $\mathfrak{X} \notin \mathcal{M}_{N_1}$ only one scaled eigenvalue has modulus of unity. Choosing

$$\eta = \frac{B - \epsilon}{B(M-1)} > 0 , \tag{4.163}$$

it is, therefore, possible to find an N_0 such that

$$|\Omega_r(N_1, \mathfrak{X})|^{N_2} < \eta , \tag{4.164}$$

for all $N_2 > N_0$ and every scaled eigenvalue with modulus less than unity. It follows, from equations (4.161)–(4.164), that

$$\begin{aligned} |Z(N_1, N_2, \mathfrak{X})|^{1/N_2} &> B[1 - (M-1)\eta]^{1/N_2} \\ &> B[1 - (M-1)\eta] = \epsilon . \end{aligned} \tag{4.165}$$

This theorem establishes the result that no point $\mathfrak{X} \notin \mathcal{M}_{N_1}$ can be a member of $\mathbb{O}(N_1, N_2)$ for arbitrarily large N_2. Thus that the set

$$\lim_{N_2 \to \infty} \mathbb{O}(N_1, N_2) = \mathbb{O}(N_1, \infty) \in \mathcal{M}_{N_1} . \tag{4.166}$$

Further insight into this situation can be gained if we consider a curve defined by (4.160), for particular values of r and r'. If this curve belongs to $\mathcal{M}_{N_1}$, then in the limit $N_2 \to \infty$

$$Z(N_1, N_2, \mathfrak{X}) \simeq [\Lambda_r(N_1, \mathfrak{X})]^{N_2} [1 + \exp(\mathrm{i}\vartheta N_2)] . \tag{4.167}$$

This asymptotic expression is zero at the values $\vartheta = (2r+1)\pi/N_2$, $r = 0, 1, \ldots, N_2 - 1$, which correspond to a set of points on the curve with density which increases with N_2. For $M = 2$, as would be the case for the one-dimensional Ising model, this argument establishes the fact that the zeros of the partition function lie on the single curve given by (4.160) *for all* N_2. For

$M > 2$, it adds understanding rather than rigour to the result established by Theorem. 4.10.1.

On $\mathbb{R}^{(+)}$, the positive real axis of $\mathbb{C}_{\mathfrak{X}}$, Perron's theorem applies and $\Lambda_{\text{Max}}(N_1, \mathfrak{X})$ denotes (unambiguously) a real positive non-degenerate eigenvalue. Let the eigenvalues be ordered so that $\Lambda_1(N_1, \mathfrak{X})$ is identified by the condition that $\Lambda_1(N_1, \mathfrak{X}) = \Lambda_{\text{Max}}(N_1, \mathfrak{X})$, for $\mathfrak{X} \in \mathbb{R}^{(+)}$. Although curves belonging to $\mathcal{D}_{N_1}$ can cross or touch $\mathbb{R}^{(+)}$, none that do can belong to $\mathcal{M}_{N_1}$. We now define $\mathcal{E}_{N_1}$ as the subset of $\mathcal{M}_{N_1}$ consisting of points which can be reached by a continuous path from some point of $\mathbb{R}^{(+)}$ without first crossing some other member of $\mathcal{M}_{N_1}$. It must be the case that, along the curves of $\mathcal{E}_{N_1}$, $\Lambda_1(N_1, \mathfrak{X})$ is one of the eigenvalues which satisfies (4.160). The dimensionless free energy per lattice site of the system for finite N_1 and N_2 is

$$\phi(N_1, N_2, \mathfrak{X}) = -\frac{1}{N_1 N_2} \ln Z(N_1, N_2, \mathfrak{X}) . \tag{4.168}$$

Suppose that $\mathbb{W}$ is a region in $\mathbb{C}_{\mathfrak{X}}$ containing a segment of $\mathbb{R}^{(+)}$, but none of the members of $\mathbb{O}(N_1, N_2)$ for all N_1 and all N_2 greater than some N_0. Then, by a slight modification of Theorem 4.5.1, it follows that the free energy per lattice site $\phi(\mathcal{L}, \mathfrak{X})$, defined by

$$\phi(\mathcal{L}, \mathfrak{X}) = \lim_{N_1 \to \infty} \lim_{N_2 \to \infty} \phi(N_1, N_2, \mathfrak{X}) , \tag{4.169}$$

for the infinite system $\mathcal{L}$ is a regular function of $\mathfrak{X}$ in $\mathbb{W}$ and no singularity of $\phi(\mathcal{L}, \mathfrak{X})$ can occur at a real value of $\mathfrak{X} \in \mathbb{W} \cap \mathbb{R}^{(+)}$. For any fixed value of N_1 it follows from Theorem 4.10.1 that N_0 can be chosen so that, if $\mathbb{W} \cap \mathcal{E}_{N_1} = \emptyset$, then $\mathbb{O}(N_1, N_2) \cap \mathbb{W} = \emptyset$ for $N_2 > N_0$. The final conclusion of this argument is that singularities of the infinite system $\mathcal{L}$ are possible only at real values of $\mathfrak{X}$ to which some points of $\mathcal{E}_{N_1}$ converge as $N_1 \to \infty$.

4.10.2 Connection Curves and Cross-Block Curves

Suppose, as in Sect. 4.9, that the irreducible representations of the symmetry group $\mathcal{Q}$ are $\mathsf{E}(\mathsf{k})$, $\mathsf{k} = 1, 2, \ldots, \mathsf{\kappa}$. The transfer matrix can be block diagonalized into blocks corresponding to these representations.It is also shown in Appendix A.3 that, for a representation $\mathsf{E}(\mathsf{k})$ of dimension $\mathsf{e}(\mathsf{k}) > 1$, a further reduction is possible into $\mathsf{e}(\mathsf{k})$ identical blocks. The characteristic equation take the form

$$\prod_{\mathsf{k}=1}^{\mathsf{\kappa}} [\text{Det}\{\boldsymbol{V}'(\mathsf{k}) - \Lambda \boldsymbol{I}_{\mathsf{m}(\mathsf{k})}\}]^{\mathsf{e}(\mathsf{k})} = \mathcal{F}(N_1, \mathfrak{X}; \Lambda) = 0 , \tag{4.170}$$

where $\boldsymbol{V}'(\mathsf{k})$ is the block of the transfer matrix which appears in the direct sum in (A.127). The factors in the product in equation (4.170) are polynomials of degree $\mathsf{m}(\mathsf{k})$ with

$$\text{Det}\{\boldsymbol{V}'(\mathsf{k}) - \Lambda \boldsymbol{I}_{\mathsf{m}(\mathsf{k})}\} = \mathcal{G}(\mathsf{k}; N_1, \mathfrak{X}; \Lambda) , \tag{4.171}$$

where

$$\mathcal{G}(\mathrm{k};N_1,\mathfrak{X};\Lambda)=\mathrm{g}_{\mathrm{m(k)}}(\mathrm{k};N_1,\mathfrak{X})\Lambda^{\mathrm{m(k)}}+\mathrm{g}_{\mathrm{m(k)}-1}(\mathrm{k};N_1,\mathfrak{X})\Lambda^{\mathrm{m(k)}-1}+\cdots+\mathrm{g}_0(\mathrm{k};N_1,\mathfrak{X})\,. \tag{4.172}$$

Thus we have

$$\mathcal{F}(N_1,\mathfrak{X};\Lambda)=\prod_{\mathrm{k}=1}^{\kappa}[\mathcal{G}(\mathrm{k};N_1,\mathfrak{X};\Lambda)]^{\mathrm{e(k)}}. \tag{4.173}$$

The coefficients in (4.172) are polynomials in $\mathfrak{X}$ and, if the factorization (4.173) were complete in the sense that no further reduction into factors with polynomial coefficients were possible, then the eigenvalue solutions $\Lambda_\ell(\mathrm{k};N_1,\mathfrak{X})$, $\ell=1,2,\ldots,\mathrm{m(k)}$, of

$$\mathcal{G}(\mathrm{k};N_1,\mathfrak{X};\Lambda)=0 \tag{4.174}$$

would be the branches of an *(irreducible) algebraic function.* There are two reasons why this might not be the case. Firstly, it is sometimes quite difficult to determine the complete symmetry group of the system and any further 'hidden' symmetry would lead to a factorization of $\mathcal{G}(\mathrm{k};N_1,\mathfrak{X};\Lambda)$. Secondly, we have no guarantee that the full symmetry group will lead to a complete factorization of the characteristic equation. A example of this is provided by the square lattice Ising model which is discussed below. For the moment we shall suppose that $\mathcal{G}(\mathrm{k};N_1,\mathfrak{X};\Lambda)$ *is* irreducible. The only modification needed if this were not the case, would be an additional index to distinguish the separate factors. We have now reverted to the notation of Sect. 4.9, denoting the eigenvalues by the pair of indices k and ℓ and (4.159) is re-expressed as

$$Z(N_1,N_2,\mathfrak{X})=\sum_{\mathrm{k}=1}^{\kappa}\sum_{\ell=1}^{\mathrm{m(k)}}[\Lambda_\ell(\mathrm{k};N_1,\mathfrak{X})]^{N_2}. \tag{4.175}$$

The curves in $\mathcal{D}_{N_1}$ are now distinguished by their representations. The set for which $\mathfrak{X}$ satisfies the equation

$$\Lambda_\ell(\mathrm{k};N_1,\mathfrak{X})=\varpi\Lambda_{\ell'}(\mathrm{k}';N_1,\mathfrak{X})\,,\qquad \varpi=\exp(\mathrm{i}\vartheta)\,,\quad 0\le\vartheta<2\pi\,, \tag{4.176}$$

for all $\ell=1,2,\ldots,\mathrm{m(k)}$, $\ell'=1,2,\ldots,\mathrm{m(k')}$ is denoted by $\mathcal{C}_{N_1}(\mathrm{k},\mathrm{k}')$. For reasons made clear in Appendix A.5, $\mathcal{C}_{N_1}(\mathrm{k},\mathrm{k})$ are called *connection curves* and $\mathcal{C}_{N_1}(\mathrm{k},\mathrm{k}')$, with $\mathrm{k}\neq\mathrm{k}'$ are called *cross-block curves.* The total set

$$\bigcup_{\{\mathrm{k},\mathrm{k}'\}}\mathcal{C}_{N_1}(\mathrm{k},\mathrm{k}')=\mathcal{D}_{N_1}\,. \tag{4.177}$$

On $\mathbb{R}^{(+)}$, $\Lambda_{\mathrm{Max}}(N_1,\mathfrak{X})=\Lambda_1(1;N_1,\mathfrak{X})$. So $\mathcal{E}_{N_1}$ consists of curves in the set $\mathcal{C}_{N_1}(\mathrm{k},1)$ for some k.

In principle an investigation of the evolution of the set $\mathcal{E}_{N_1}$ for increasing N_1 should provide some indication of the critical point of the system in the limit $N_1\to\infty$. However, the dimension $M=\nu^{N_1}$ of the transfer matrix will grow very rapidly with N_1, with all the resulting computational difficulties. These can be lessened by using the block diagonalization of the transfer matrix, particularly if $\mathcal{E}_{N_1}$ consisted of connection curves in $\mathcal{C}_{N_1}(1,1)$ rather than cross-block curves in $\mathcal{C}_{N_1}(\mathsf{k},1)$ with $\mathsf{k} \neq 1$. Then it would be necessary to calculate only the one-dimensional symmetric block of the transfer matrix, which is straightforward using formula (A.122). Unfortunately this is not necessarily the case, as we shall see below for the square lattice nearest-neighbour spin-$\frac{1}{2}$ Ising model and as was also shown for the square Ising model with equal nearest- and second-neighbour interactions by Wood (1987),. The interesting and important feature of Wood's method is the argument that, as $N_1 \to \infty$, $\mathcal{E}_{N_1}$ converges on to that part $\mathcal{C}^*_{N_1}(1,1)$ of $\mathcal{C}_{N_1}(1,1)$ which involves $\Lambda_1(1; N_1, \mathfrak{X})$. This means that, for finite N_1, the curves of $\mathcal{C}^*_{N_1}(1,1)$ *rather than* $\mathcal{E}_{N_1}$ can be used to obtain an approximation to the critical properties of the infinite system. Two arguments can be advanced for the truth of Wood's contention. The first of a general nature and the second based on calculations for particular systems. In the latter category we shall present below a discussion of the square lattice spin-$\frac{1}{2}$ Ising model.

First, in a general context, consider the two different finite systems, one with toroidal boundary conditions and the partition function (4.175) and the other with the boundary matrix $\boldsymbol{B}$ given by

$$\langle \alpha, m|\boldsymbol{B}|\alpha', m'\rangle = \delta^{\mathrm{Kr}}(\alpha - \alpha')/\mathfrak{p}(\alpha). \tag{4.178}$$

It then follows from (4.99), (4.100) and (A.145) that

$$\beta_\ell(\mathsf{k}) = \delta^{\mathrm{Kr}}(\mathsf{k} - 1), \tag{4.179}$$

and

$$Z(N_1, N_2, \mathfrak{X}) = \sum_{\ell=1}^{\mathfrak{m}(1)} [\Lambda_\ell(1; N_1, \mathfrak{X})]^{N_2 - 1}. \tag{4.180}$$

Both this partition function and (4.175) give the same limit (4.112) for the free energy per lattice site. This means that the critical properties of the two systems in the limit $N_1 \to \infty$ are identical and are also the same as that of a *reduced* system with toroidal boundary conditions, transfer matrix consisting of just the one-dimensional symmetric block of that for the full system and $N_2 - 1$ rather than N_2 slices. A sequence of finite cases of each of these three systems can be understood to give a succession of approximations to the same infinite system. Support for this argument is given by considering the pair correlation functions (4.138) and (4.139). Summation of these over j will yield according to the fluctuation-response function relation (1.62) response functions, which have a maximum on $\mathbb{R}^{(+)}$ when a scaled eigenvalue approaches unity. For the full system this corresponds to the point where the magnitude of the second largest scaled eigenvalue attains a maximum. For the reduced system the response function is the heat capacity which attains a maximum where the magnitude of the second largest scaled eigenvalue *in the one-dimensional symmetric representation* attains a maximum. Since both these points develop into the same singularity at the critical point as $N_1 \to \infty$, they must converge with increasing N_1.

4.10.3 The Spin-$\frac{1}{2}$ Square Lattice Ising Model

In this case the symmetry group is $\mathcal{G} = \mathcal{C}_{N_1\mathrm{v}} \times \mathcal{S}_2$, where $\mathcal{C}_{N_1\mathrm{v}}$ is the group of rotations of the lattice and reflections through planes containing the lattice axis (see, for example, Leech and Newman 1969 or McWeeny 1963 for further details) and $\mathcal{S}_2 = \{\mathrm{I}, \mathrm{S}\}$ corresponds to spin inversion. To ensure that the coefficients in the characteristic equation (4.157) are analytic polynomials in $\mathfrak{X}$ we add the constant $-J$ to each nearest-neighbour pair interaction, giving $-2J$ and zero for a pair of parallel and anti-parallel spins, respectively. The eigenvalues are given in two sets, corresponding to direct products of the representations of $\mathcal{C}_{N_1\mathrm{v}}$ with the one-dimensional anti-symmetric and symmetric representations of $\mathcal{S}_2$. In our notation, these are respectively

$$\Lambda^{(\mathrm{e})}(\pm,\ldots,\pm,N_1;\mathfrak{X}) = \exp\left\{\tfrac{1}{2}[\pm\gamma_0(N_1;\mathfrak{X})\cdots\pm\gamma_{2N_1-2}(N_1;\mathfrak{X})]\right\} \times [\mathfrak{X}(\mathfrak{X}^2-1)]^{\frac{1}{2}N_1}, \tag{4.181}$$

$$\Lambda^{(\mathrm{o})}(\pm,\ldots,\pm,N_1;\mathfrak{X}) = \exp\left\{\tfrac{1}{2}[\pm\gamma_1(N_1;\mathfrak{X})\cdots\pm\gamma_{2N_1-1}(N_1;\mathfrak{X})]\right\} \times [\mathfrak{X}(\mathfrak{X}^2-1)]^{\frac{1}{2}N_1}, \tag{4.182}$$

with the choice of an even number of negative signs (Onsager 1944, Kaufman 1949, Domb 1960), where the $\gamma_r(N_1;\mathfrak{X})$, $r = 0, 1, \ldots, 2N_1 - 1$ are given by

$$\cosh[\gamma_r(N_1;\mathfrak{X})] = \frac{(\mathfrak{X}^2+1)^2}{2\mathfrak{X}(\mathfrak{X}^2-1)} - \cos\left(\frac{r\pi}{N_1}\right). \tag{4.183}$$

The sign ambiguity of the $\gamma_r(N_1;\mathfrak{X})$ in (4.183) is resolved by imposing the condition that, on $\mathbb{R}^{(+)}$, they are all smooth functions of $\mathfrak{X}$ which are positive at low temperatures ($\mathfrak{X}$ large). Using this condition the appropriate root for $\exp[\gamma_r(N_1;\mathfrak{X})]$ in (4.183) is

$$\exp[\gamma_r(N_1;\mathfrak{X})] = \frac{f_r(N_1;\mathfrak{X}) + g_r(N_1;\mathfrak{X}) + 2\sqrt{f_r(N_1;\mathfrak{X})g_r(N_1;\mathfrak{X})}}{4\mathfrak{X}(\mathfrak{X}^2-1)}, \tag{4.184}$$

where

$$f_r(N_1;\mathfrak{X}) = (\mathfrak{X}^2+1)^2 - 4\cos^2\left(\frac{r\pi}{2N_1}\right)\mathfrak{X}(\mathfrak{X}^2-1),$$
$$g_r(N_1;\mathfrak{X}) = (\mathfrak{X}^2+1)^2 + 4\sin^2\left(\frac{r\pi}{2N_1}\right)\mathfrak{X}(\mathfrak{X}^2-1). \tag{4.185}$$

These functions are perfect square only for

$$\begin{aligned} f_0(N_1;\mathfrak{X}) &= (\mathfrak{X}^2-2\mathfrak{X}-1)^2\,, & g_0(N_1;\mathfrak{X}) &= (\mathfrak{X}^2+1)^2\,, \\ g_{N_1}(N_1;\mathfrak{X}) &= (\mathfrak{X}^2+2\mathfrak{X}-1)^2\,, & f_{N_1}(N_1;\mathfrak{X}) &= (\mathfrak{X}^2+1)^2\,. \end{aligned} \tag{4.186}$$

In all other cases the zeros of $f_r(N_1;\mathfrak{X})$ and $g_r(N_1;\mathfrak{X})$ give eight branch-points for $\exp[\gamma_r(N_1;\mathfrak{X})]$. These branch-points will appear in the eigenvalues given by (4.181) or (4.182), according to the parity of r, unless $\gamma_{2N_1-r}(N_1;\mathfrak{X})$ and $\gamma_r(N_1;\mathfrak{X})$ have opposite signs. In that case, since $\gamma_{2N_1-r}(N_1;\mathfrak{X}) = \gamma_r(N_1;\mathfrak{X})$, the branch-points are eliminated. With φ defined in terms of $\mathfrak{X}$ by

$$\mathfrak{X} = \exp(\mathrm{i}\varphi) \pm \sqrt{\exp(2\mathrm{i}\varphi)+1}\,, \tag{4.187}$$

the zeros of $f_r(N_1;\mathfrak{X})$ and $g_r(N_1;\mathfrak{X})$ are given by

$$\cos\varphi = \begin{cases} \cos^2\left(\dfrac{r\pi}{2N_1}\right), \\ -\sin^2\left(\dfrac{r\pi}{2N_1}\right). \end{cases} \tag{4.188}$$

If the characteristic polynomial of any model is expressed, in the form (4.173), as a product of irreducible polynomials, then we know that every branch of an algebraic function must be an eigenvalue in the same irreducible representation. However, as we shall now see the converse is not necessarily the case. Not all eigenvalues in the same irreducible representation are branches of the same algebraic function. For the present model the eigenvalues which are the branches of a single algebraic function can be identified in the sets (4.181) and (4.182) from their branch-points. Given any eigenvalue, another branch of the same algebraic function can be obtained by changing the signs of both members of a pair $\gamma_{2N_1-r}(N_1;\mathfrak{X})$ and $\gamma_r(N_1;\mathfrak{X})$. On $\mathbb{R}^{(+)}$ the largest eigenvalue is $\Lambda^{(\mathrm{o})}(+,\ldots,+,N_1;\mathfrak{X})$ and this is, therefore, a branch of an algebraic function of degree

$$\mathrm{d}(N_1) = \begin{cases} 2^{\frac{1}{2}N_1}, & N_1 \text{ even}, \\ 2^{\frac{1}{2}(N_1-1)}, & N_1 \text{ odd}. \end{cases} \tag{4.189}$$

The multiplicity $\mathfrak{m}(1;N_1)$ of the one-dimensional symmetric representation is given by the formulae in Example 4.3. From these we see that $\mathfrak{m}(1;N_1) = \mathrm{d}(N_1)$ for $N_1 = 2,3,\ldots,6$, but $\mathfrak{m}(1;7) = 9 > \mathrm{d}(7) = 8$ and $\mathfrak{m}(1;8) = 18 > \mathrm{d}(8) = 16$. For $N_1 > 6$ the one-dimensional symmetric representation contains eigenvalues, with a different branch-point structure from $\Lambda^{(\mathrm{o})}(+,\ldots,+,N_1;\mathfrak{X})$. We do, however, know that the eigenvalues of any representation will be either entirely in the set (4.181) or the set (4.182) according to whether the representation in question is formed from the one-dimensional anti-symmetric or one-dimensional symmetric representation of $\mathcal{S}_2$. It follows that all the eigenvalues of the one-dimensional symmetric representation are members of (4.182).

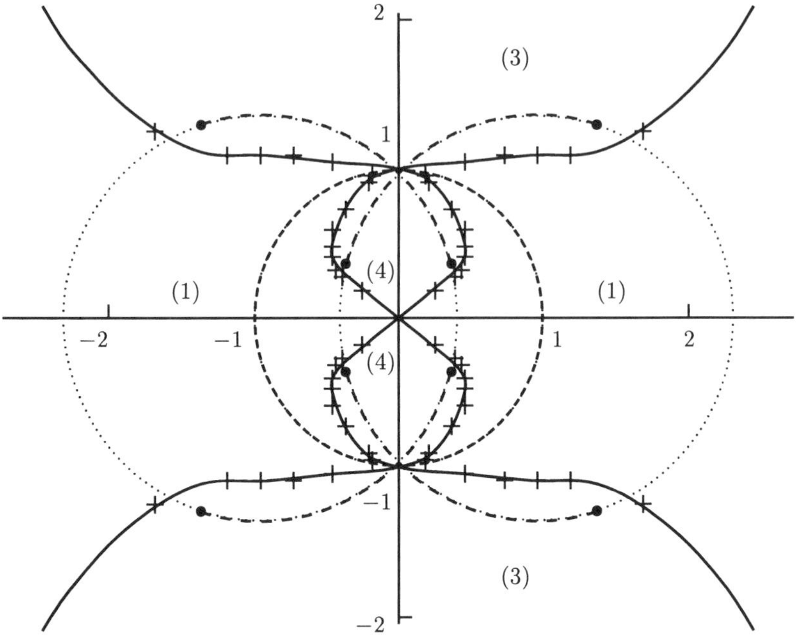

Fig. 4.4. The connection and cross-block curves, in the plane $\mathbb{C}_{\mathfrak{X}}$, for the Ising model with $N_1 = 2$. The zeros of $Z(2, 16; \mathfrak{X})$ are marked by + and the region in which $\Lambda_1(\mathrm{k}; 2, \mathfrak{X}$ is the eigenvalue of largest magnitude is labelled (k).

We now return to (4.187) and (4.188) which define the branch-points of the eigenvalues. If φ_0 is a solution of the first of the equations (4.188) in the interval $[0, \frac{1}{2}\pi]$, then φ_1 in the interval $[\frac{1}{2}\pi, \pi]$ satisfies the second equation when $\cos\varphi_0 - \cos\varphi_1 = 1$. The connection curve in $\mathbb{C}_{\mathfrak{X}}$ described by (4.187) with the positive sign and $\varphi \in [\varphi_0, \varphi_1]$ has a cusp at $\varphi = \frac{1}{2}\pi$, $\mathfrak{X} = \mathrm{i}$. Another connection curve is given by taking $\varphi \in [-\varphi_1, -\varphi_0]$. The two connection curves corresponding to the negative sign in (4.187) can be obtained by applying the mapping $\mathfrak{X} \to -1/\mathfrak{X}$. It is not difficult to show that these four curves form pairs of crossing curves lying on the Fisher circles (4.7). These connection curves for $r = 1$, $N_1 = 2$, when $\varphi_0 = \frac{1}{3}\pi$, $\varphi_1 = \frac{2}{3}\pi$ are shown in Fig. 4.4 as chain lines terminating at the eight branch-points which are denoted by •. The Fisher circles (4.7) are completed with dotted lines.

The second largest eigenvalue is $\Lambda^{(\mathrm{e})}(+, \ldots, +, N_1; \mathfrak{X})$ (Kaufman 1949) which cannot belong to the one-dimensional symmetric representation. $\mathcal{E}_{N_1}$ therefore consists of cross-block curves. Although it is possible that the zeros of $Z(N_1, N_2, \mathfrak{X})$ all lie on the part of $\mathcal{M}_{N_1}$ which is not $\mathcal{E}_{N_1}$ it is more plausible to expect, (as in the case $N_1 = 2$ investigated below), that some of the zeros $\mathbb{O}(N_1, N_2)$ for any N_2 will lie on $\mathcal{E}_{N_1}$. It can, however, be shown (Fisher 1965, Wood 1985) that the limiting locus of partition function zeros for this model is the pair of Fisher circles (4.7). It must, therefore, be concluded that, in the double limit $N_1 \to \infty$, $N_2 \to \infty$, the curve $\mathcal{E}_{N_1}$ closes in on the circles (4.7), which according to Theorem 4.10.1 contain the zeros of the partition function $Z^{(1)}(N_1, N_2, \mathfrak{X})$, defined by (4.180). These zeros lie on the connection curve $\mathcal{C}^*_{N_1}(1,1)$, defined by

$$\Lambda^{(\mathrm{o})}(+,+,\ldots,+,+,N_1;\mathfrak{X}) = \varpi\Lambda^{(\mathrm{o})}(-,+,\ldots,+,-,N_1;\mathfrak{X}), \tag{4.190}$$

which has branch-points on the circle $\mathfrak{X} = \sqrt{2}\exp(\mathrm{i}\theta) + 1$ at the pair of points

$$\mathfrak{X} = \exp\left\{\pm\mathrm{i}\arccos\left[\cos^2\left(\frac{\pi}{2N_1}\right)\right]\right\} + \sqrt{\exp\left\{\pm 2\mathrm{i}\ \arccos\left[\cos^2\left(\frac{\pi}{2N_1}\right)\right]\right\} + 1}\,. \tag{4.191}$$

As $N_1 \to \infty$ $\mathfrak{X} \to 1 + \sqrt{2}$, which is equivalent to $\sinh(2J/T) = 1$, the formula for the critical temperature of the square lattice Ising ferromagnet (Sect. 5.10 and Volume 1, Chap. 8). All the connection curves form a set of overlapping arcs on the circles (4.7) and in this limit branch-points of all the connection curves approach the critical point which becomes for them a point of accumulation. The critical point can also be seen as a result of analytically continuing $\mathcal{C}^*_{N_1}(1,1)$ through the branch-points to $\mathbb{R}^{(+)}$.

The remarkable feature of this result is that, although the critical point is for the system in the thermodynamic limit of infinite N_2 and N_1, it applies to the eigenvalues for any finite N_1. This can be seen by a more detailed examination of the case $N_1 = 2$. The block diagonalization of the transfer matrix for this case is the problem posed in Example 4.4. The characteristic equation for $\boldsymbol{V}'(1)$ is

$$\Lambda^2 - (\mathfrak{X}^2 + 1)^2\Lambda + \mathfrak{X}^2(\mathfrak{X}^2 - 1)^2 = 0\,, \tag{4.192}$$

which gives

$$\left.\begin{array}{l}\Lambda_1(1;2,\mathfrak{X})\\[1ex] \Lambda_2(1;2,\mathfrak{X})\end{array}\right\} = \frac{1}{2}\left[(\mathfrak{X}^2+1)^2 \pm \sqrt{(\mathfrak{X}^2+1)^4 - 4\mathfrak{X}^2(\mathfrak{X}^2-1)^2}\right]. \tag{4.193}$$

From $\boldsymbol{V}'(3)$ and $\boldsymbol{V}'(4)$ in Example 4.4

$$\begin{aligned}\Lambda_1(3;2,\mathfrak{X}) &= \mathfrak{X}^2(\mathfrak{X}^2 - 1),\\ \Lambda_1(4;2,\mathfrak{X}) &= \mathfrak{X}^2 - 1.\end{aligned} \tag{4.194}$$

The eigenvalues $\Lambda_1(1;2,\mathfrak{X})$ and $\Lambda_2(1;2,\mathfrak{X})$ can be obtain from (4.182) according to the prescription

$$\begin{aligned}\Lambda_1(1;2,\mathfrak{X}) &= \Lambda^{(\mathrm{o})}(+,+,2;\mathfrak{X}),\\ \Lambda_2(1;2,\mathfrak{X}) &= \Lambda^{(\mathrm{o})}(-,-,2;\mathfrak{X}).\end{aligned} \tag{4.195}$$

The question of the derivation of $\Lambda_1(3;2,\mathfrak{X})$ and $\Lambda_1(4;2,\mathfrak{X})$ from (4.181) is discussed below. The connection curves $\mathcal{C}_2(1,1)$ are given, either using (4.160) and (4.193) or the resultant formula (A.179) with the quadratic in (4.192). They have the structure described above for the general N_1 case and are shown in Fig. 4.4 by chain lines. From (4.194) the cross-block curve $\mathcal{C}_2(3,4)$ is the circle $|\mathfrak{X}| = 1$ (represented by a broken line in Fig. 4.4). From (4.193)–(4.194) the cross-block curves $\mathcal{C}_2(1,3)$ and $\mathcal{C}_2(1,4)$ are both given by the solutions of

$$\mathfrak{X}^2(\mathfrak{X}^2 - 1)\exp(2\mathrm{i}\vartheta) - (\mathfrak{X}^2 + 1)^2\exp(\mathrm{i}\vartheta) + \mathfrak{X}^2 - 1 = 0\,. \tag{4.196}$$

These curves are represented by continuous lines in Fig. 4.4. On that part of the curves for which $|\mathfrak{X}| > 1$, $|\Lambda_1(1;2,\mathfrak{X})| = |\Lambda_1(3;2,\mathfrak{X})|$ and when $|\mathfrak{X}| < 1$, $|\Lambda_1(1;2,\mathfrak{X})| = |\Lambda_1(4;2,\mathfrak{X})|$. The exterior and interior of the circle $|\mathfrak{X}| = 1$ correspond respectively to the ferromagnetic ($J > 0$) and antiferromagnetic ($J < 0$) models. The inversion symmetry of Fig. 4.4 with respect to this circle corresponds to the invariance of the free energy of the square lattice Ising model with respect to interchange of the sign of J. It follows from the topological structure of Fig. 4.4 that, for $J > 0$, the curve $|\Lambda_1(1;2,\mathfrak{X})| = |\Lambda_1(3;2,\mathfrak{X})|$ forms the subset $\mathcal{M}_2$ along which there are two eigenvalues of maximum magnitude. In a similar way $|\Lambda_1(1;2,\mathfrak{X})| = |\Lambda_1(4;2,\mathfrak{X})|$ gives the subset $\mathcal{M}_2$ when $J < 0$. The eigenvalues of maximum magnitude in each of the regions formed by the curves $\mathcal{M}_2$ are shown in Fig. 4.4. The curves of $\mathcal{E}_2$ consist of that part of $\mathcal{M}_2$ to the right of the imaginary axis. The conclusion for general N_1 that $\mathcal{E}_{N_1}$ consists of cross-block curves is, therefore, explicitly confirmed in this case. According to Theorem 4.10.1 the zeros of $Z(2, N_2, \mathfrak{X})$ converge on to $\mathcal{M}_2$ as $N_2 \to \infty$. The zeros of $Z(2,16,\mathfrak{X})$ are indicated by crosses in Fig. 4.4. Even with this comparatively small value of N_2 the conclusion of the theorem is supported.

We need at this point to distinguish those features of the behaviour of the square lattice Ising model which can be expected to appear in a wide class of lattice models and those which arise because of its special properties. The characteristic of general validity is the convergence with increasing N_1 of $\mathcal{E}_{N_1}$ on to $\mathcal{C}^*_{N_1}(1,1)$, with the analytic continuation of $\mathcal{C}^*_{N_1}(1,1)$ to $\mathbb{R}^{(+)}$ yielding an approximation to the critical point. A special feature of the Ising model is that $\mathcal{C}^*_{N_1}(1,1)$ lies on the *same* curves (4.7) for all values on N_1. This means that the *exact* critical temperature can be obtain using any finite value of N_1. The reason for this can be found in the self-duality of this model. With

$$\mathfrak{S} = \tfrac{1}{2}(\mathfrak{X} - \mathfrak{X}^{-1})\,, \tag{4.197}$$

the duality transformation for real $\mathfrak{X} > 1$ (and $\mathfrak{S} > 0$) is given by

$$\mathfrak{S} \leftrightarrow \mathfrak{S}^* = \frac{1}{\mathfrak{S}}\,, \qquad \text{or} \qquad \mathfrak{X} \leftrightarrow \mathfrak{X}^* = \frac{\mathfrak{X}+1}{\mathfrak{X}-1}\,, \tag{4.198}$$

(see, for example, Volume 1, Sect. 8.3). The curves given by

$$\tfrac{1}{2}(\mathfrak{X} - \mathfrak{X}^{-1}) = \exp(\mathrm{i}\varphi)\,, \qquad -\pi \le \varphi < \pi\,, \tag{4.199}$$

are invariant under the duality transformation, which takes $\varphi \to -\varphi$. Equations (4.187) give the solutions of (4.199) and it follows that all the branch-points and all the connection curves lie on the self-dual curves (4.199) for all N_1. The underlying reason for this, as can be seen from equations (4.184)–(4.185), is that, for $r = 1, 2, \ldots, 2N_1 - 1$, the functions $\exp[\gamma_r(N_1; \mathfrak{X})]$ are invariant under the duality transformation. It then follows that all the eigenvalues defined by (4.182), which include all those of the one-dimensional symmetric representation, satisfy the formula

$$\frac{\Lambda(N_1; \mathfrak{X})}{[\mathfrak{X}^2 \mathfrak{S}]^{\frac{1}{2}N_1}} = \frac{\Lambda(N_1; \mathfrak{X}^*)}{[\mathfrak{X}^{*2} \mathfrak{S}^*]^{\frac{1}{2}N_1}} . \tag{4.200}$$

The eigenvalues of (4.181) can also be made to satisfy (4.200) if, when defining $\exp[\gamma_0(N_1; \mathfrak{X})]$, care is taken in the choice of the sign of the square root in (4.184). From (4.184) and (4.186),

$$\exp[\gamma_0(N_1; \mathfrak{X})] = \begin{cases} \dfrac{\mathfrak{X}(\mathfrak{X} - 1)}{\mathfrak{X} + 1}, & \mathfrak{S} > 1, \\[2ex] \dfrac{\mathfrak{X} + 1}{\mathfrak{X}(\mathfrak{X} - 1)}, & \mathfrak{S} < 1. \end{cases} \tag{4.201}$$

Using this definition in (4.181) will, from (4.175), gives the usual duality formula for the partition function (see Volume 1, Sect. 8.3)[10] for all values of N_1 and N_2. This is, however, at the expense of a change of analytic form at the self-dual point $\mathfrak{X} = 1 + \sqrt{2}$. For $N_1 = 2$ this means that the formulae (4.194) are restricted to $\mathfrak{S} > 1$ with

$$\begin{aligned} \Lambda_1(3; 2, \mathfrak{X}) &= \mathfrak{X}(\mathfrak{X} + 1)^2, \\ \Lambda_1(4; 2, \mathfrak{X}) &= \mathfrak{X}(\mathfrak{X} - 1)^2, \end{aligned} \tag{4.202}$$

for $\mathfrak{S} < 1$. If $Z(2, N_2, \mathfrak{X})$ is calculated for some finite N_2 from first principles, simply by counting configurations on the lattice, then the result achieved is that obtained using the eigenvalues (4.193)–(4.194), which alone does not satisfy duality (see Example 4.5).

Duality also leads to the appearance of a curve which contains $\mathcal{C}^*_{N_1}(1,1)$ for all N_1 in the case of the triangular Ising model and the square lattice ν-state Potts model (Wood et al. 1987). The interesting question of using the method to uncover approximate, or hitherto unsuspected symmetries, leading to critical points given in terms of algebraic numbers has been discussed by Wood and Ball (1990).

4.10.4 Critical Points and Exponents

As has been indicated above, the procedure, in the absence of any special symmetry properties, is to analytically continue the connection curve $\mathcal{C}^*_{N_1}(1,1)$ to

[10] The extra factor $\mathfrak{X}^{N_1}$ arises from the choice of energies; $-2J$ and zero for parallel and antiparallel pairs, respectively.

$\mathbb{R}^{(+)}$ to give an approximation to the critical point. Two different approaches can be taken to this task.

In the first it is supposed that the appropriate branch of the connection curve has been identified by

$$\Lambda_1(1; N_1, \mathfrak{X}) = \varpi \Lambda_2(1; N_1, \mathfrak{X}), \tag{4.203}$$

with known expressions in terms of $\mathfrak{X}$ for the eigenvalues. With

$$f(\mathfrak{X}) = \frac{\Lambda_1(1; N_1, \mathfrak{X})}{\Lambda_2(1; N_1, \mathfrak{X})}, \tag{4.204}$$

the connection curve is given by $|f(\mathfrak{X})| = 1$, with $f(\mathfrak{X}) = 1$ at the branch-points. The analytic continuation of the connection curve through the branch-points, called the *extension curve*, is defined by $\text{Arg}\{f(\mathfrak{X})\} = 0$. Now let

$$\begin{aligned} \mathfrak{X} &= \mathrm{r} \exp(\mathrm{i}\theta), \\ f(\mathfrak{X}) &= \mathfrak{F}(\mathrm{r}, \theta) \exp[\mathrm{i}\Theta(\mathrm{r}, \theta)]. \end{aligned} \tag{4.205}$$

The extension curve is given by

$$\Theta(\mathrm{r}, \theta) = 0, \tag{4.206}$$

with normal in the direction

$$\hat{n}(\mathrm{r}, \theta) = \left(\frac{\partial \Theta}{\partial \mathrm{r}}, \frac{1}{\mathrm{r}} \frac{\partial \Theta}{\partial \theta} \right). \tag{4.207}$$

Except at the branch-points the Cauchy–Riemann conditions

$$\frac{\partial \mathfrak{F}}{\partial \mathrm{r}} = \frac{\mathfrak{F}}{\mathrm{r}} \frac{\partial \Theta}{\partial \theta}, \qquad \frac{1}{\mathrm{r}} \frac{\partial \mathfrak{F}}{\partial \theta} = -\mathfrak{F} \frac{\partial \Theta}{\partial \mathrm{r}}, \tag{4.208}$$

are satisfied. The approximation to the critical point is given by the intersection of the extension curve and $\mathbb{R}^{(+)}$. If this intersection is orthogonal then, from (4.207) and (4.208)

$$\frac{\partial \Theta}{\partial \theta} = 0, \qquad \frac{\partial \mathfrak{F}}{\partial \mathrm{r}} = 0. \tag{4.209}$$

The second of these conditions, identifies $\mathfrak{X}_\mathrm{c}$ as a turning point of $|f(\mathfrak{X}|$ along $\mathbb{R}^{(+)}$.

The second approach, which maps out the whole of $\mathcal{C}_{N_1}(1,1)$ and its extension curves, is to begin with the defining polynomial

$$\begin{aligned} \mathcal{G}(1; N_1, \mathfrak{X}; \Lambda) = \mathrm{g}_{\mathrm{m}(1)}(1; N_1, \mathfrak{X}) \Lambda^{\mathrm{m}(1)} + \mathrm{g}_{\mathrm{m}(1)-1}(1; N_1, \mathfrak{X}) \Lambda^{\mathrm{m}(1)-1} \\ + \cdots + \mathrm{g}_0(1; N_1, \mathfrak{X}). \end{aligned} \tag{4.210}$$

for the eigenvalues in the one-dimensional symmetric representation. According to the method described in Appendix A.5 we now form the polynomial

$$\mathcal{H}(1; N_1, \mathfrak{X}; \Lambda) = \frac{\mathcal{G}(1; N_1, \varpi\mathfrak{X}; \Lambda) - \mathcal{G}(1; N_1, \mathfrak{X}; \Lambda)}{\Lambda(\varpi - 1)}. \tag{4.211}$$

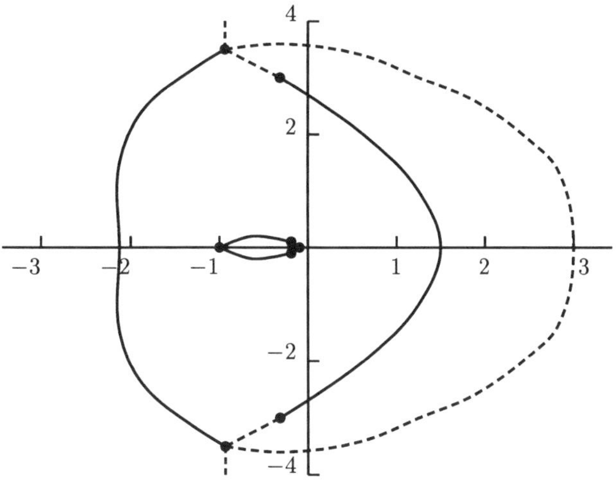

Fig. 4.5. The connection curves $\mathcal{C}_4(1,1)$ and their extensions for the hard square lattice fluid. The connection curves are shown by continuous lines and their extensions by broken lines.

The connection curves and their extensions are given by solving the equation

$$\mathrm{Res}\{\mathcal{G}, \mathcal{H}, \mathfrak{X}\} = 0\,. \tag{4.212}$$

As a simple example of the second procedure, we consider the hard-square lattice fluid of particles of fugacity $\mathfrak{Z} = \exp(\mu/T)$ on the square lattice (see Sect. 4.8). With $N_1 = 4$ the symmetry group is $\mathcal{C}_{4\mathrm{v}}$ and it is not difficult to show that the eigenvalues of the one-dimensional symmetric representation satisfy the cubic equation

$$\Lambda^3 - \Lambda^2(\mathfrak{Z}^2 + 3\mathfrak{Z} + 1) + \Lambda\mathfrak{Z}(\mathfrak{Z}^2 - \mathfrak{Z} - 1) + \mathfrak{Z}^3 = 0\,. \tag{4.213}$$

In this case the resultant equation (4.212) gives

$$\begin{aligned}
0 = {}& \mathfrak{Z}^8 + 4\mathfrak{Z}^7 + 2(9 - 2\varsigma)(1 + \varsigma)\mathfrak{Z}^6 + 8(5 + 5\varsigma - \varsigma^2)\mathfrak{Z}^5 \\
& + 2(41 + 52\varsigma - 4\varsigma^3)\mathfrak{Z}^4 + 4(24 + 25\varsigma + 4\varsigma^2)\mathfrak{Z}^3 \\
& + 4(13 + 8\varsigma + \varsigma^2)\mathfrak{Z}^2 + 4(3 + \varsigma)\mathfrak{Z} + 1\,,
\end{aligned} \tag{4.214}$$

where

$$\varsigma = \tfrac{1}{2}(\varpi + \varpi^{-1}) = \cos\vartheta\,. \tag{4.215}$$

The connection curves $\mathcal{C}_4(1,1)$ are the roots of (4.214) with $|\varsigma| \le 1$. The extension curves are given by continuing the curves into the region $|\varsigma| > 1$, where ϑ becomes imaginary. These curves are shown in Fig. 4.5. The extensions of the connection curves through the branch-points at $-0.881 \pm \mathrm{i}3.473$ meet $\mathbb{R}^{(+)}$ at $\mathfrak{Z}_{\mathrm{c}} = 3.016$, with $\varsigma_{\mathrm{c}} = 5.186$. These values can be computed from the condition that $\mathfrak{Z}_{\mathrm{c}}$ is the meeting point of two extension curves approaching from above and below. It, therefore, corresponds to a value of ς for which (4.214) has a repeated root. This can be found by the using the resultant with respect to $\mathfrak{Z}$ of (4.214) and its derivative. The estimate $\mathfrak{Z}_c = 3.016$ for the critical fugacity can be compared (rather unfavourably) with the high-density series result $\mathfrak{Z}_c = 3.7962 \pm 0.0001$ quoted in Sect. 4.8. However, it has been shown by Wood et al. (1987) that for $N_1 = 6$ the comparison is much improved with $\mathfrak{Z}_c = 3.730$.

In the picture of the development of critical behaviour originating in the work of Yang and Lee the critical point of a one parameter lattice system corresponds, in the thermodynamic limit, to a point of accumulation of the zeros of the partition function. We have now shown that the critical point can also be identified as the limit of some of the branch-points of $\mathcal{C}^*_{N_1}(1,1)$ as $N_1 \to \infty$. Suppose that there are s of these branch-points each with cycle number two. They will all be the termination of connection curves involving $\Lambda_1(1; N_1, \mathfrak{X})$ which coalesce at $\mathfrak{X}_c$ as $N_1 \to \infty$. Wood (1987) has argued that the result is to form a branch-point of $\Lambda_1(1; \infty, \mathfrak{X})$ with cycle number $p \le s$. It follows, from the discussion of algebraic functions in Appendix A.5 and of scaling at a critical point in Sect. 2.9, that the scaling exponent y_2 in a two-dimensional system is given by

$$\frac{2}{y_2} = \frac{q}{p} \,. \tag{4.216}$$

Wood (1987) has suggested that p is likely to take its maximum value s and that q should be some multiple t of $2(s-1)$, giving the result

$$y_2 = \frac{s}{t(s-1)} \,. \tag{4.217}$$

This then gives, from (2.167) and (2.198),

$$\alpha = \frac{2[s - t(s-1)]}{s} \,, \qquad \nu = \frac{t(s-1)}{s} \,. \tag{4.218}$$

The first of these formulae can give rise to a singularity in heat capacity only when $t = 1$. For this case the hard-hexagon and 3-state Potts models correspond to $s = 6$, $\alpha = \frac{1}{3}$ (see Volume 1, Sect. 10.14) and the K-type ($\alpha = \frac{1}{2}$) transition exhibited by the dimer model described in Sect. 8.2.2 corresponds to $s = 4$. In the case of the three-spin Ising model on the triangular lattice the polynomial of which $\Lambda_1(1; \infty, \mathfrak{X})$ is a branch is known (Baxter and Wu 1973, 1974, Baxter 1974, Joyce 1975a, 1975b). This leads directly to the result (Wood 1987) $2/y_2 = \frac{4}{3}$, $\alpha = \frac{2}{3}$ corresponding to $t = 1$, $s = 3$ in formulae (4.217) and (4.218). We have already seen that, for the spin-$\frac{1}{2}$ Ising model on a square lattice, the critical point corresponds, in the limit $N_1 \to \infty$, to a point of accumulation of an infinite number of branch-points. Thus we have $s = \infty$, which together with $t = 1$ yields $\alpha = 0$, which is consistent with the logarithmic singularity in the heat capacity.

A theory of finite-size scaling of partition function zeros was developed by Itzykson et al. (1983). For a two-dimensional $N_1 \times N_1$ lattice system they proposed the scaling form

$$|\mathfrak{X}_0 - \mathfrak{X}_c| \sim N_1^{-y}, \tag{4.219}$$

for the zero $\mathfrak{X}_0$ closest to the critical value $\mathfrak{X}_c$ in the complex plane $\mathbb{C}_{\mathfrak{X}}$ of some Boltzmann variable $\mathfrak{X}$ for which the scaling exponent is y. Given that boundary conditions ensure that $\mathcal{E}_{N_1}$ involves only eigenvalues in the one-dimensional symmetric representation, the points of this set closest to $\mathbb{R}^{(+)}$ will be branch-points of $\Lambda_1(1; N_1, \mathfrak{X})$, occurring in conjugate pairs. They correspond to bounds on the asymptotic distribution of partition function zeros. Using this insight Williams and Lavis (1996) proposed a modified version of the formula of Itzykson et al. (1983) for a $N_1 \times N_2$ lattice in the limit $N_2 \to \infty$. With the conjugate pair of branch-points closest to $\mathfrak{X}_c$ in $\mathbb{C}_{\mathfrak{X}}$ denoted by $\mathfrak{X}_*$ and $\overline{\mathfrak{X}}_*$, the proposed form was

$$|\mathfrak{X}_* - \mathfrak{X}_c| \sim N_1^{-y}, \tag{4.220}$$

for the calculation of y. Support for this method, when $\mathfrak{X}$ is the thermal Boltzmann factor, is given by the behaviour of the rounding exponent w defined in (2.313). The quantity N_1^{-w} is of the order of the size of the neighbourhood in which the correlation length exceeds the lattice width (Fisher and Barber 1972). This should be of the same order as the radius of convergence of a power-series expansion of the free energy about the critical point, which will be given by the location of the nearest branch-point of $\Lambda_1(1; N_1, \mathfrak{X})$. The equality (2.314), which in the present notation is equivalent to $w = y$, is thought to be satisfied in a wide class of systems, particularly in the case of a cylinder with free boundary conditions at the ends (Fisher and Ferdinand 1967).

Examples

4.1 Consider a linear lattice of N sites with the two boundary sites occupied by particles. The only possible contours are ones for which $q = 2$. Such contours can bound clusters with n interior sites for $n = 1, 2, \ldots, N-2$ and a cluster with n interior sites can be placed on the lattice in $N - n - 1$ ways. Using this information and equation (4.31) show that it is possible to take $Q(2) = \frac{1}{6}N^2$. By substituting into (4.32) show that $t(\mathrm{G}_1) = 1/\ln(N/\sqrt{3})$. Since $t(\mathrm{G}_1) \to 0$ as $N \to \infty$ the method, as is to be expected, fails to establish the existence of an ordered phase.

4.2 Apply the method of Sect. 4.8 to the hard-hexagon model. Show that, by choosing half the elementary triangles of the lattice, with each site belonging to three triangles,

$$\lambda([\boldsymbol{r}, \boldsymbol{r}', \boldsymbol{r}''], T, 3, 3', 3'') = 1 + 3 + 3' + 3''.$$

Hence, show that,

$$\mathbb{X}_\kappa(T;\boldsymbol{r}) = \{\mathfrak{z} : \Re(\mathfrak{z}) \le -\tfrac{1}{3}\}$$

is a possible choice, leading to

$$\mathfrak{z} = -\frac{\exp(\mathrm{i}\psi)}{[3\cos(\psi/3)]^3}, \qquad -\pi \le \psi \le \pi .$$

4.3 Consider the spin-$\frac{1}{2}$ Ising model on an $N_1 \times N_2$ square lattice wrapped on a torus. By counting the number of lattice states left invariant by the operations of the group $\mathcal{Q} = \mathcal{C}_{N_1\mathrm{v}} \times \mathcal{S}_2$, or otherwise, show that the multiplicity $\mathfrak{m}(1;N_1)$ of the one-dimensional symmetric representation, for $N_1 \ge 2$, is

$$\mathfrak{m}(1;N_1) = \begin{cases} 2^{\frac{1}{2}(N_1-3)} + \dfrac{1}{4N_1}\displaystyle\sum_{s=0}^{N_1-1} 2^{[s,N_1]}, & N_1 \text{ odd}, \\[2ex] 2^{\frac{1}{2}N_1-1} + \dfrac{1}{4N_1}\displaystyle\sum_{s=0}^{N_1-1} 2^{[s,N_1]}\Delta(N_1,s), & N_1 \text{ even}, \end{cases}$$

where

$$\Delta(N_1,s) = 1 + \delta\left(\frac{N_1}{[s,N_1]}, \text{even}\right),$$

and $[s,N_1]$ is the greatest common divisor of s and N_1 and $\delta(x,\text{even})$ is one or zero according to whether x is even or odd.

4.4 Consider the special case $N_1 = 2$ of the model described in question 3. The coefficients in the characteristic equation (4.157) are made analytic polynomial functions of $\mathfrak{X}$ by adding the constant $-J$ to every nearest-neighbour pair interaction, giving interactions $-2J$ and zero for a pair of parallel or anti-parallel spins, respectively. With the states of pairs ordered in the equivalence classes $\{(\uparrow,\uparrow),(\downarrow,\downarrow)\},\{(\uparrow,\downarrow),(\downarrow,\uparrow)\}$ and with $\kappa = 2J/T$, $\mathfrak{X} = \exp(2J/T)$ show that the transfer matrix takes the form

$$\boldsymbol{V} = \begin{pmatrix} \mathfrak{X}^4 & \mathfrak{X}^2 & \mathfrak{X}^2 & \mathfrak{X}^2 \\ \mathfrak{X}^2 & \mathfrak{X}^4 & \mathfrak{X}^2 & \mathfrak{X}^2 \\ \mathfrak{X}^2 & \mathfrak{X}^2 & \mathfrak{X}^2 & 1 \\ \mathfrak{X}^2 & \mathfrak{X}^2 & 1 & \mathfrak{X}^2 \end{pmatrix}.$$

The groups $\mathcal{C}_{2\mathrm{v}} = \mathcal{C}_2$ and $\mathcal{S}_2$ are isomorphic with a symmetric irreducible representation $\mathsf{E}(1)$ and an antisymmetric irreducible representation $\mathsf{E}(2)$. The irreducible representations of $\mathcal{Q}$ are $\mathsf{E}(1) = \mathsf{E}_{\mathcal{C}}(1) \otimes \mathsf{E}_{\mathcal{S}}(1)$, $\mathsf{E}(2) = \mathsf{E}_{\mathcal{C}}(2) \otimes \mathsf{E}_{\mathcal{S}}(1)$, $\mathsf{E}(3) = \mathsf{E}_{\mathcal{C}}(1) \otimes \mathsf{E}_{\mathcal{S}}(2)$ and $\mathsf{E}(4) = \mathsf{E}_{\mathcal{C}}(2) \otimes \mathsf{E}_{\mathcal{S}}(2)$. Construct the character table of $\mathcal{Q}$ and show that the multiplicities of the irreducible representations in the four-dimensional representation of states are $\mathfrak{m}(1) = 2$, $\mathfrak{m}(2) = 0$, $\mathfrak{m}(3) = 1$ and $\mathfrak{m}(4) = 1$ in the order $\mathsf{E}(1)$, $\mathsf{E}(3)$, $\mathsf{E}(1)$, $\mathsf{E}(4)$. Show, from (A.116), that the matrices $\boldsymbol{W}(1)$ and $\boldsymbol{W}(2)$ used to reduce the transfer matrix to block-diagonal form are

$$\boldsymbol{W}(1) = \boldsymbol{W}(2) = \begin{pmatrix} \frac{1}{\sqrt{2}} & \frac{1}{\sqrt{2}} \\ \frac{1}{\sqrt{2}} & -\frac{1}{\sqrt{2}} \end{pmatrix} .$$

Hence show that

$$\boldsymbol{V}'(1) = \begin{pmatrix} \mathfrak{X}^4 + \mathfrak{X}^2 & 2\mathfrak{X}^2 \\ 2\mathfrak{X}^2 & \mathfrak{X}^2 + 1 \end{pmatrix} ,$$

$$\boldsymbol{V}'(3) = (\mathfrak{X}^4 - \mathfrak{X}^2) ,$$

$$\boldsymbol{V}'(4) = (\mathfrak{X}^2 - 1) .$$

4.5 Using the eigenvalues (4.193)–(4.194) show that the partition functions (4.175) and (4.180) yield respectively, show that

$$Z(2,3;\mathfrak{X}) = 2\mathfrak{X}^{12} + 18\mathfrak{X}^8 + 26\mathfrak{X}^6 + 12\mathfrak{X}^4 + 6\mathfrak{X}^2 ,$$

$$Z(2,3;\mathfrak{X}) = \mathfrak{X}^{12} + 3\mathfrak{X}^{10} + 15\mathfrak{X}^8 + 26\mathfrak{X}^6 + 15\mathfrak{X}^4 + 3\mathfrak{X}^2 + 1 .$$

Verify the former by counting the configurations on the six sites of the lattice and show that the latter satisfies the duality relationship

$$\frac{Z(2,3;\mathfrak{X})}{[\mathfrak{X}^2\mathfrak{S}]^3} = \frac{Z(2,3;\mathfrak{X}^*)}{[\mathfrak{X}^{*2}\mathfrak{S}^*]^3} .$$

Use (4.202) in place of (4.195) to obtain the partition function

$$Z'(2,3;\mathfrak{X}) = \mathfrak{X}^{12} + 3\mathfrak{X}^{10} + 2\mathfrak{X}^9 + 15\mathfrak{X}^8 + 30\mathfrak{X}^7 + 26\mathfrak{X}^6 + 30\mathfrak{X}^5 + 15\mathfrak{X}^4 + 2\mathfrak{X}^3 + 3\mathfrak{X}^2 + 1 .$$

Show that

$$\frac{Z'(2,3;\mathfrak{X})}{[\mathfrak{X}^2\mathfrak{S}]^3} = \frac{Z'(2,3;\mathfrak{X}^*)}{[\mathfrak{X}^{*2}\mathfrak{S}^*]^3} .$$

5. The Eight-Vertex Model

5.1 Introduction

The name *vertex model* is used to denote a lattice model in which the microstates are represented by putting an arrow on each *edge* (line connecting a pair of nearest-neighbour sites) of the lattice. Such models can be constructed on any lattice, but those for the square lattice have received the most attention. It is clear that the most general model of this type on the square lattice is the *sixteen-vertex model*, where the different *vertex types* correspond to all 2^4 possible directions of the arrows on the four edges meeting at a vertex. This model, which can be shown to be equivalent to an Ising model with two, three and four site interactions and with an external field (Suzuki and Fisher 1971), is unsolved.[1] By applying the *ice rule* which restricts the vertex types to those with the same number of arrows in as out, the model becomes the *six-vertex model* which was solved by Lieb (1967a, 1967b, 1967c). It is discussed in Volume 1, Chap. 10. In the present chapter we consider the model where the ice rule is replaced by the rule restricting the vertex type to those with an even number of arrows pointing in and out. This allows the eight vertex types shown in the first line of Fig. 5.1. The first six of these vertex types correspond to those of the six-vertex model. In the eight-vertex model, vertices with four inward or four outward arrows are also permitted, The new vertices are labelled 7 and 8. If the model is regarded as a two-dimensional analogue of a hydrogen bonded network with the arrows representing the sense of polarization, then vertex types 7 and 8 are ionized but unpolarized and represent 'defects' in the hydrogen-bonded network.

Consider a square lattice of N sites wound on a torus and define N_1 and N_2 to be the number of sites in each row and column respectively, $(N_1 N_2 = N)$. Suppose that $\mathfrak{n}_\mathfrak{p}$ of the sites are vertices of type $\mathfrak{p}$, $(\mathfrak{p} = 1, \ldots, 8)$. Now vertices of types 5 and 7 change the flow of arrows along any row of edges from left to right and types 6 and 8 from right to left. From the application of periodic boundary conditions it follows that $\mathfrak{n}_5 + \mathfrak{n}_7 = \mathfrak{n}_6 + \mathfrak{n}_8$. In a similar way, by considering the flow of arrows along any column $\mathfrak{n}_5 + \mathfrak{n}_8 = \mathfrak{n}_6 + \mathfrak{n}_7$. Combining these two results, it follows that

[1] As may be expected, since the Ising model with just a two-site interaction and an external field is unsolved.

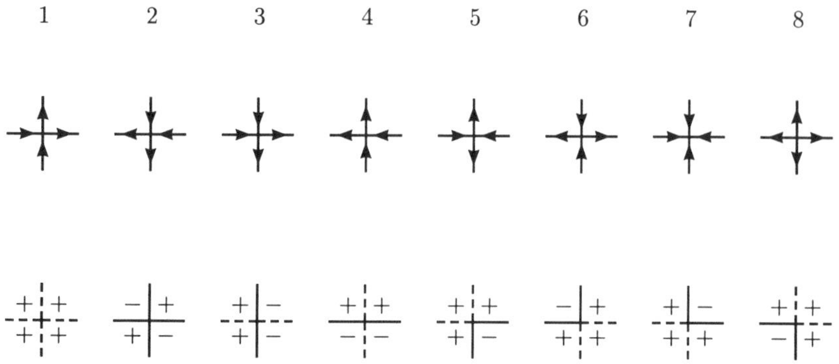

Fig. 5.1. The eight vertex configurations.

$$\mathfrak{n}_5 = \mathfrak{n}_6 , \qquad \mathfrak{n}_7 = \mathfrak{n}_8 . \tag{5.1}$$

As in the six-vertex model (Volume 1, Chap. 10) the configurational energy is taken to be the sum of vertex energies $e_\mathfrak{p}$ with the corresponding Boltzmann factors

$$\mathfrak{z}(\mathfrak{p}) = \exp(-e_\mathfrak{p}/T) , \qquad \mathfrak{p} = 1, \ldots, 8 . \tag{5.2}$$

We shall assume that[2]

$$e_1 = e_2 , \qquad e_3 = e_4 , \qquad e_5 = e_6 , \qquad e_7 = e_8 . \tag{5.3}$$

Since similar equalities then hold for the corresponding Boltzmann factors, it is convenient to define

$$\begin{aligned} \mathfrak{z}(1) &= \mathfrak{z}(2) = \mathfrak{a} , \qquad & \mathfrak{z}(3) &= \mathfrak{z}(4) = \mathfrak{b} , \\ \mathfrak{z}(5) &= \mathfrak{z}(6) = \mathfrak{c} , \qquad & \mathfrak{z}(7) &= \mathfrak{z}(8) = \mathfrak{d} . \end{aligned} \tag{5.4}$$

The six-vertex model can then be recovered by setting $e_7 = e_8 = \infty$ ($\mathfrak{d} = 0$). From (5.1) the last two relations of (5.3) involve no loss of generality, but the first two imply that the energy is invariant under the reversal of all arrows and hence the absence of any electric field.

For the six-vertex model certain exact results can be obtained, by perturbation methods, in the presence of an electric field (Volume 1, Sects. 10.12 and 10.13). In the eight-vertex model exact results are available for zero fields, but not for the general model with a field. For $T > 0$, it might be expected that the presence of two extra vertex types in the eight-vertex model, as compared to the six-vertex model, would have the effect of 'loosening up' the available configurations. This is confirmed by the configuration graphs which, as for the six-vertex model, are constructed by placing a line on any edge where the

[2] Any one of the vertex energies e_1, e_3, e_5 and e_7 can be set to zero without loss of generality. However, this is not advantageous, except in special cases, because of the symmetry of the eight-vertex free energy transformations.

arrow points to the left or downwards. Vertex patterns 7 and 8 are shown in the second line of Fig. 5.1 and supply the top right and bottom left 'corners' missing in six-vertex model patterns. Thus any graph is permissible in which the number of lines incident at each vertex is zero or even. The configuration graphs are thus *polygon graphs*.[3] It follows that small perturbations on the completely polarized ground states (with all vertices of type 1, 2, 3, or 4) are possible. For instance, with the ground state with all vertices of type 1, there can be a perturbation represented by a single elementary square of four lines. The six-vertex phenomenon of the 'frozen in' ground state (Volume 1, Sect. 10.6) will thus not occur and the behaviour of the eight-vertex model is likely to be closer to that of the Ising model than that of the six-vertex model. In fact, equivalence with a modified Ising model will be shown in the next section.

5.2 Spin Representations

Suppose that an Ising spin is placed at the centre of each face of the eight-vertex model square lattice. Then the spin sites form another square lattice which is the dual of the original lattice. For any vertex configuration on the original lattice the configuration graph divides the dual lattice into polygon regions. If each region is assigned a homogeneous spin state different from its neighbours then it follows that each polygon graph structure of the eight-vertex model corresponds to two Ising spin states on the dual lattice. Each line of the polygon graph separates the spins of an unlike pair. The plus and minus signs in the second line of Fig. 5.1 show one of the pair of spin configurations at the corners of a spin lattice face which corresponds to each vertex type at the centre.[4]

A Hamiltonian will now be constructed for the spin model, which makes it equivalent to the eight-vertex model, and enables us to construct a relationship of the form

$$Z_s = 2Z \tag{5.5}$$

between the partition function Z_s of the spin model and Z of the eight-vertex model. For this equivalence to exist the Hamiltonian must be invariant under the reversal of all spins and thus can contain no odd-degree spin terms. The operation of reversing the spins on one of the interpenetrating sublattices of the square lattice, while leaving those on the other unchanged, interchanges like and unlike nearest-neighbour spin pairs and corresponds to reversing all arrows in the eight-vertex model. The Hamiltonian must be invariant under

[3] These are the high-temperature zero-field graphs of Sect. 7.4.

[4] The correspondence between spin configurations and polygon graphs on dual lattices was used in Volume 1, Sects. 8.2 and 8.3, in deriving the Ising model dual transformation.

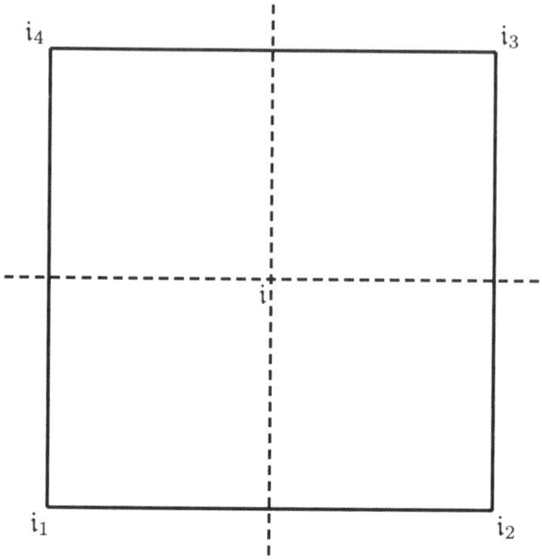

Fig. 5.2. A face of the dual (spin) lattice with sites $\mathfrak{i}_1, \ldots, \mathfrak{i}_4$ and the original lattice site $\mathfrak{i}$ at the centre.

this operation and can, therefore, contain no nearest-neighbour terms. Given that the eight-vertex model has four independent energies e_1, e_3, e_5 and e_7 we must, for consistency, have the same number of independent energy parameters in the Hamiltonian. The form

$$\widehat{H}(J_0, J_{21}, J_{22}, J_4; \sigma) = -\sum_{\{\mathfrak{i}\}}(J_0 + J_{21}\sigma_{\mathfrak{i}_1}\sigma_{\mathfrak{i}_3} + J_{22}\sigma_{\mathfrak{i}_2}\sigma_{\mathfrak{i}_4} + J_4\sigma_{\mathfrak{i}_1}\sigma_{\mathfrak{i}_2}\sigma_{\mathfrak{i}_3}\sigma_{\mathfrak{i}_4}), \tag{5.6}$$

where the sum is over all the faces $\mathfrak{i} = 1, \ldots, N$ of the spin lattice corresponding to the sites $\mathfrak{i} = 1, \ldots, N$ of the original lattice, was suggested independently by Wu (1971) and Kadanoff and Wegner (1971). The corners labelled $\mathfrak{i}_1, \ldots, \mathfrak{i}_4$ are as shown in Fig. 5.2 with spins $\sigma_{\mathfrak{i}_1}$ and $\sigma_{\mathfrak{i}_3}$ belonging to one sublattice and $\sigma_{\mathfrak{i}_2}$ and $\sigma_{\mathfrak{i}_4}$ to the other. By equating the spin energy for a single face of the spin lattice to that of the vertex type at the centre it follows that

$$\begin{aligned} e_1 &= -J_0 - J_{21} - J_{22} - J_4\,, & e_3 &= -J_0 + J_{21} + J_{22} - J_4\,, \\ e_5 &= -J_0 - J_{21} + J_{22} + J_4\,, & e_7 &= -J_0 + J_{21} - J_{22} + J_4\,, \end{aligned} \tag{5.7}$$

which can be inverted to give

$$\begin{aligned} J_0 &= -\tfrac{1}{4}(e_1 + e_3 + e_5 + e_7)\,, & J_{21} &= \tfrac{1}{4}(-e_1 + e_3 - e_5 + e_7)\,, \\ J_{22} &= -\tfrac{1}{4}(-e_1 + e_3 + e_5 - e_7)\,, & J_4 &= \tfrac{1}{4}(-e_1 - e_3 + e_5 + e_7)\,. \end{aligned} \tag{5.8}$$

These relationships define the constant coupling $K_0 = J_0/T$, two-spin couplings $K_{21} = J_{21}/T$ and $K_{22} = J_{22}/T$ and four-spin coupling $K_4 = J_4/T$ of a modified Ising model whose partition function $Z_s(K_0, K_{21}, K_{22}, K_4)$ satisfies (5.5). If we impose the condition $K_4 = 0$ then, from (5.8), (5.2) and (5.4),

$$e_1 + e_3 = e_5 + e_7\,, \qquad \mathfrak{a}\mathfrak{b} = \mathfrak{c}\mathfrak{d}\,. \tag{5.9}$$

The two sublattices are decoupled and each carries an Ising model with unequal two-spin couplings K_{21} and K_{22} in the two lattice directions. Equation (5.5) becomes

$$2Z(\mathfrak{a},\mathfrak{b},\mathfrak{c},\mathfrak{d}) = Z_{\mathrm{s}}(K_0, K_{21}, K_{22}, 0)$$

$$= \exp(NK_0)\left[Z^{(\mathrm{I.M.})}\left(\tfrac{1}{2}N, K_{21}, K_{22}\right)\right]^2 . \tag{5.10}$$

Each Ising model becomes isotropic when $K_{21} = K_{22} = K = J/T$. This condition and (5.9) are satisfied when

$$e_1 = -J_0 - 2J\,, \qquad e_3 = -J_0 + 2J\,, \qquad e_5 = e_7 = -J_0\,. \tag{5.11}$$

To introduce nearest-neighbour interactions on the spin lattice the first two conditions of (5.3) must be dropped (Lieb and Wu 1972 and Example 5.1). Except in certain special cases this more general model becomes equivalent to a triangular lattice Ising model with unequal interactions when *free-fermion conditions*[5] are satisfied (Example 5.2).

An alternative Ising spin representation of the eight-vertex model was developed by Jüngling and Obermair (1974) and Jüngling (1975). In this formulation spins σ_{i_0} are situated at the centres of the faces of the dual lattice (at the same positions as the vertices) and the Hamiltonian is taken to be

$$\begin{aligned}&\widehat{H}(J_0, J_{21}, J_{22}, W_1, W_2, W_3, W_4; \sigma)\\ &\quad = -\sum_{\{\mathrm{i}\}}[J_0 + J_{21}\sigma_{\mathrm{i}_1}\sigma_{\mathrm{i}_3} + J_{22}\sigma_{\mathrm{i}_2}\sigma_{\mathrm{i}_4}\\ &\qquad + \sigma_{\mathrm{i}_0}(W_1\sigma_{\mathrm{i}_1} + W_2\sigma_{\mathrm{i}_2} + W_3\sigma\mathrm{i}_3 + W_4\sigma\mathrm{i}_4)]\,.\end{aligned} \tag{5.12}$$

The spins at the centres of the faces interact only with the corners of faces on which they are placed. They are therefore *decoration* sites (Volume 1, Chap. 9) and this type of decoration is known as the *Union Jack lattice*. A generalization of the star–triangle called the *star–square transformation* can now be applied in which the partition function is summed over the states of the decoration sites. In terms of the couplings $L_i = W_i/T$, $i = 1, \ldots, 4$, this yields the formulae

$$\begin{aligned} e_{1,2} &= -J_0 - J_{21} - J_{22} - T\ln\left[2\cosh\left(L_1 \pm L_2 + L_3 \pm L_4\right)\right]\,,\\ e_{3,4} &= -J_0 + J_{21} + J_{22} - T\ln\left[2\cosh\left(L_1 \pm L_2 - L_3 \mp L_4\right)\right]\,,\\ e_{5,6} &= -J_0 - J_{21} + J_{22} - T\ln\left[2\cosh\left(L_1 \pm L_2 + L_3 \mp L_4\right)\right]\,,\\ e_{7,8} &= -J_0 + J_{21} - J_{22} - T\ln\left[2\cosh\left(L_1 \pm L_2 - L_3 \pm L_4\right)\right]\end{aligned} \tag{5.13}$$

for the vertex energies, where for each pair denoted by $e_{\mathrm{p},\mathrm{q}}$ the first and second indices correspond to the upper and lower signs on the right-hand

[5] This terminology is used because of the mathematical equivalence to a system of non-interacting fermions (Lieb and Wu 1972).

side. The isotropic case $J_{21} = J_{22}$, $L_1 = L_2 = L_3 = L_4$ was considered by Jüngling and Obermair (1974). This does not satisfy the first two relations of (5.3) and the model is equivalent to a Ising spin model with four-spin, second-neighbour and nearest-neighbour interactions (Example 5.3). It is easy to see that, in order for the first two of equations (5.3) to be satisfied, either $L_1 = L_3$ and $L_2 = -L_4$ or $L_1 = -L_3$ and $L_2 = L_4$. The special case considered by Jüngling (1975) is when $L_1 = -L$, $L_2 = L_3 = L_4 = L$. From (5.13) this gives $e_1 = e_2$, $e_3 = e_4$, $e_5 = e_6$, but $e_7 \neq e_8$. However, in view of the first of conditions (5.1) this is not a problem and e_7 and e_8 can be replaced by their arithmetic mean. Once this is done the vertex energies can be expressed in the form (5.7), where (distinguishing these new values by a tilde)

$$\tilde{J}_0 = J_0 + \tfrac{1}{8}T \ln\left[256 \cosh^4(2L)\cosh(4L)\right] , \tag{5.14}$$

$$\tilde{J}_{21} = J_{21} + \tfrac{1}{8}T \ln\left[\cosh(4L)\right] , \tag{5.15}$$

$$\tilde{J}_{22} = J_{22} - \tfrac{1}{8}T \ln\left[\cosh(4L)\right] , \tag{5.16}$$

$$\tilde{J}_4 = -\tfrac{1}{8}T \ln\left[1 - \tanh^4(2L)\right] . \tag{5.17}$$

Again the model consists of two interpenetrating square lattice Ising models linked by a four-spin coupling, but all the couplings are now temperature dependent.

5.3 Parameter Space and Ground States

The $(\mathfrak{a}, \mathfrak{b}, \mathfrak{c}, \mathfrak{d})$ parameter space can be divided into four regions defined by

(i) $\{e_1 < e_3,\ e_1 < e_5,\ e_1 < e_7\} \Rightarrow \{\mathfrak{a} > \mathfrak{b},\ \mathfrak{a} > \mathfrak{c},\ \mathfrak{a} > \mathfrak{d}\}$,

(ii) $\{e_3 < e_1,\ e_3 < e_5,\ e_3 < e_7\} \Rightarrow \{\mathfrak{b} > \mathfrak{a},\ \mathfrak{b} > \mathfrak{c},\ \mathfrak{b} > \mathfrak{d}\}$,

(iii) $\{e_5 < e_1,\ e_5 < e_3,\ e_5 < e_7\} \Rightarrow \{\mathfrak{c} > \mathfrak{a},\ \mathfrak{c} > \mathfrak{b},\ \mathfrak{c} > \mathfrak{d}\}$,

(iv) $\{e_7 < e_1,\ e_7 < e_3,\ e_7 < e_5\} \Rightarrow \{\mathfrak{d} > \mathfrak{a},\ \mathfrak{d} > \mathfrak{b},\ \mathfrak{d} > \mathfrak{c}\}$.

When $\mathfrak{d} = 0$ ($e_7 = \infty$), cases (i), (ii) and (iii) reduce to six-vertex models. The ground states are the same as for the corresponding six-vertex models, so that (i) and (ii) are ferroelectric and (iii) is antiferroelectric (see Volume 1, Sect. 10.3). Case (iv) is also antiferroelectric with vertices of types 7 and 8 respectively occupying the two interpenetrating sublattices of the square lattice (see Volume 1, Appendix A.1) in the ground state. Case (iv) is therefore sublattice ordered and can be obtained from the case (iii) ground state by replacing vertices 5 and 6 by 7 and 8 respectively. The antiferroelectric ground state configuration graphs are shown in Fig. 5.3. We now consider the corresponding spin ground states, noting that, from the form of the Hamiltonian in (5.6), the energy is invariant under the reversal of all spins on either or both of the sublattices. For (i) the ground state vertices are either all of type

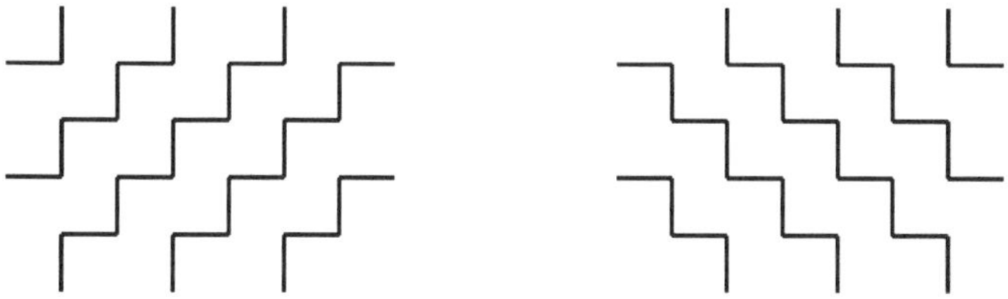

(a) (b)

Fig. 5.3. Configuration graphs for **(a)** the case (iii) ground state (vertices of types 5 and 6) and **(b)** the case (iv) ground state (vertices of types 7 and 8).

1 or all of type 2. For type 1 all the spins have the same sign, but for type 2 the spins on the two sublattices have opposite signs (see Fig. 5.1). In either case there is complete ferromagnetic order on each sublattice. This is the situation when (5.11) applies with $J > 0$. For (ii), the ground state vertices are of type 3 or type 4 and the corresponding spin lattice ground states consist either of alternate columns or alternate rows of + and − spins. The reader can verify, with the aid of a sketch, that this gives complete antiferromagnetic order on each sublattice. This is the situation when (5.11) applies with $J < 0$. For (iii), the spin ground state is shown in Fig. 5.4(a). The corners of each face of the spin lattice are occupied by three spins of one sign and one of the other. Fig. 5.4(b) is the same as Fig. 5.4(a) except that the sites of one sublattice are circled; it can be seen that there are alternate rows of + and − spins in each sublattice. From (5.7) and the inequalities (iii), this ground state is obtained for $J_{21} > 0$, $J_{22} < 0$ $J_4 < J_{21}$ and $J_4 < |J_{22}|$. The ground state spin configuration for case (iv) can be obtained from Fig. 5.4(a) by a rotation of 180° about a row or column of sites.

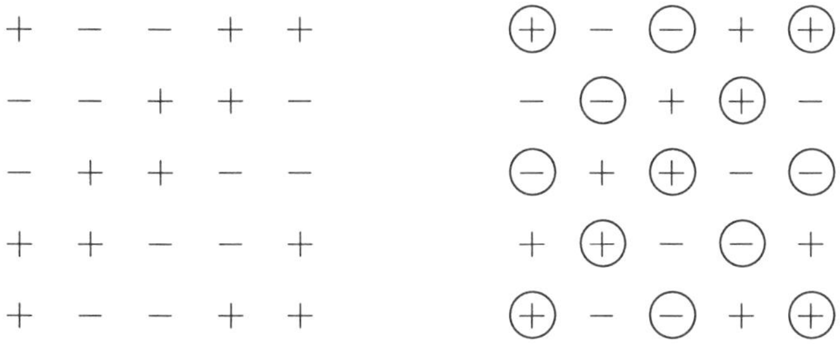

(a) (b)

Fig. 5.4. (a) Equivalent spin state for case (iii) ground state. **(b)** As for (a) with the sites of one sublattice circled.

It is easy to derive the entropy of eight-vertex ice. Putting $e_{\mathfrak{p}} = 0$, $\mathfrak{p} = 1, \ldots, 8$ gives $J_0 = J_{21} = J_{22} = J_4 = 0$ from (5.8) and the spins in the equivalent Ising model are uncoupled. It then follows, from (5.10), that

$$Z(\mathfrak{a}, \mathfrak{b}, \mathfrak{c}, \mathfrak{d}) = Z(1,1,1,1) = \tfrac{1}{2} Z_{\mathrm{s}}(0,0,0,0) = 2^{N-1}. \tag{5.18}$$

For large N, in the absence of interactions, the reduced entropy per site s, defined by equations (1.2) and (1.19) (with $Q_{n_{\mathrm{f}}} = N$), is thus given by

$$s = N^{-1} \ln Z(1,1,1,1) = \ln(2) = N^{-1} \ln(\Omega), \tag{5.19}$$

where Ω is the number of possible configurations with the eight vertex types shown in Fig. 5.1. From (5.19), $\Omega = 2^N = 8^N/4^N$. The denominator of this expression 8^N would be the value of Ω if the eight vertex types could be freely distributed over the N lattice sites. The restriction that any nearest-neighbour pair of vertices must consist of types with compatible arrows along the connecting bond therefore has the effect of reducing the number of vertex configurations by a factor 4^N. The Pauling method for computing Ω described in Volume 1, Sect. 10.1 gives the correct result $\Omega = 2^{2N}(8/16)^N = 2^N$ for the eight-vertex model.

5.4 Some Transformations

Analysis of the eight-vertex model is considerably simplified by the use of transformations which map between the four regions of parameter space defined in Sect. 5.3. In this section the transformations will be described. They are not applicable to the six-vertex model since they interchange other vertex types with types 7 and 8 (Fan and Wu 1970).

Suppose that, in some configuration of the eight-vertex model, all horizontal arrows are reversed. This transformation $\mathcal{T}_1$ corresponds, according to Fig. 5.1, to the interchange of vertex types $(1,2,3,4,5,6,7,8) \to (4,3,2,1,8,7,6,5)$. The transformation can be characterized as

$$\mathcal{T}_1: \qquad (\mathfrak{a}, \mathfrak{b}, \mathfrak{c}, \mathfrak{d}) \to (\mathfrak{b}, \mathfrak{a}, \mathfrak{d}, \mathfrak{c}). \tag{5.20}$$

When the horizontal arrow operation is applied to all eight-vertex model configurations it reproduces them in a different order and hence, from (5.20)

$$Z(\mathfrak{a}, \mathfrak{b}, \mathfrak{c}, \mathfrak{d}) = Z(\mathfrak{b}, \mathfrak{a}, \mathfrak{d}, \mathfrak{c}). \tag{5.21}$$

The partition function is thus invariant under $\mathcal{T}_1$. It can also be seen that the transformation $\mathcal{T}_1$ can be achieved by reversing all vertical arrows. The transformation $J \to -J$ ($J > 0$) in (5.11) converts the ferromagnet on each of the decoupled sublattices to an antiferromagnet. It also, from (5.4), interchanges $\mathfrak{a}$ and $\mathfrak{b}$. Since in this case $\mathfrak{c} = \mathfrak{d}$ the zero-field ferromagnet-antiferromagnet equivalence in the Ising model is a special case of (5.21). The mapping $\mathcal{T}_1$ maps between regions with (i) $\leftrightarrow$ (ii) and (iii) $\leftrightarrow$ (iv). It therefore maps ferroelectric to ferroelectric and antiferroelectric to antiferroelectric. We now consider transformations which relate ferroelectric models to antiferroelectric models.

Suppose that, for some configuration of the eight-vertex model, the arrows on the edges corresponding to the full lines in the case (iii) diagram of Fig. 5.3 are reversed. Since two lines are incident at each vertex the 'even in, even out' rule is preserved. This transformation $\mathcal{T}_2$ has the effect $(1,2,3,4,5,6,7,8) \to (5,6,7,8,1,2,3,4)$ on one sublattice and $(1,2,3,4,5,6,7,8) \to (6,5,8,7,2,1,4,3)$ on the other. The transformation can thus be characterized by

$$\mathcal{T}_2 : \qquad (\mathfrak{a},\mathfrak{b},\mathfrak{c},\mathfrak{d}) \to (\mathfrak{c},\mathfrak{d},\mathfrak{a},\mathfrak{b})\,, \tag{5.22}$$

under which the partition function is invariant giving

$$Z(\mathfrak{a},\mathfrak{b},\mathfrak{c},\mathfrak{d}) = Z(\mathfrak{c},\mathfrak{d},\mathfrak{a},\mathfrak{b})\,. \tag{5.23}$$

It will be noted that $\mathcal{T}_1^{-1} = \mathcal{T}_1$ and $\mathcal{T}_2^{-1} = \mathcal{T}_2$ and, if we define the transformation

$$\mathcal{T}_3 : \qquad (\mathfrak{a},\mathfrak{b},\mathfrak{c},\mathfrak{d}) \to (\mathfrak{d},\mathfrak{c},\mathfrak{b},\mathfrak{a})\,, \tag{5.24}$$

which corresponds to the reversal of the arrows on the lines in the case (iv) diagram of Fig. 5.3, then $\mathcal{T}_3^{-1} = \mathcal{T}_3$ and $\mathcal{T}_1\mathcal{T}_2 = \mathcal{T}_3$.[6] It follows that

$$Z(\mathfrak{a},\mathfrak{b},\mathfrak{c},\mathfrak{d}) = Z(\mathfrak{d},\mathfrak{c},\mathfrak{b},\mathfrak{a})\,, \tag{5.25}$$

and the set $\{\mathcal{I},\mathcal{T}_1,\mathcal{T}_2,\mathcal{T}_3\}$, where $\mathcal{I}$ is the identity, forms a commutative subgroup of the permutation group of four objects. It will be seen that $\mathcal{T}_2$ maps between regions with (i) $\leftrightarrow$ (iii) and (ii) $\leftrightarrow$ (iv). It follows that the effect of $\mathcal{T}_3$ is (i) $\leftrightarrow$ (iv) and (ii) $\leftrightarrow$ (iii). Thus an expression for the partition function in one region immediately yields an expression for the partition function in the whole parameter space. Equations (5.23) and (5.25) give an equivalence, at any temperature T, between free energies of eight-vertex zero-field ferroelectrics and antiferroelectrics. They imply that, for any ferroelectric, there is an antiferroelectric with the same critical properties and vice-versa. This is in contrast to the properties of six-vertex models.

In Volume 1, Sect. 10.2 a transformation was introduced in which all vertex configurations were rotated 90° in the anticlockwise direction. In the form

$$\mathcal{T} : \qquad (\mathfrak{a},\mathfrak{b},\mathfrak{c},\mathfrak{d}) \to (\mathfrak{b},\mathfrak{a},\mathfrak{c},\mathfrak{d})\,, \tag{5.26}$$

this can also be applied to the eight-vertex model and we have

$$Z(\mathfrak{a},\mathfrak{b},\mathfrak{c},\mathfrak{d}) = Z(\mathfrak{b},\mathfrak{a},\mathfrak{c},\mathfrak{d})\,. \tag{5.27}$$

Equation (5.27) reduces to (10.4) of Volume 1 when $\mathfrak{c} = 1$, $\mathfrak{d} = 0$ ($e_5 = 0$, $e_7 = \infty$). The transformation $\mathcal{T}\mathcal{T}_1$ gives $Z(\mathfrak{a},\mathfrak{b},\mathfrak{c},\mathfrak{d}) = Z(\mathfrak{a},\mathfrak{b},\mathfrak{d},\mathfrak{c})$ and setting $\mathfrak{d} = 0$ gives

[6] In all products of transformations we adopt the convention that the order of application is from right to left.

$$Z(\mathfrak{a},\mathfrak{b},\mathfrak{c},0) = Z(\mathfrak{a},\mathfrak{b},0,\mathfrak{c})\,. \tag{5.28}$$

The free energy of the six-vertex model is thus equal to that of an alternative model in which vertex types 7 and 8 take the place of 5 and 6.

From (5.1) only even powers of $\mathfrak{c}$ and $\mathfrak{d}$ occur in any term of $Z(\mathfrak{a},\mathfrak{b},\mathfrak{c},\mathfrak{d})$. From (5.23) and (5.25) the same must be true for $\mathfrak{a}$ and $\mathfrak{b}$. Thus the mathematical form of $Z(\mathfrak{a},\mathfrak{b},\mathfrak{c},\mathfrak{d})$ is invariant under any combination of the non-physical transformations

$$\begin{aligned} &\mathfrak{T}_\mathfrak{a}: \quad \mathfrak{a}\to-\mathfrak{a}\,, && \mathfrak{T}_\mathfrak{b}: \quad \mathfrak{b}\to-\mathfrak{b}\,,\\ &\mathfrak{T}_\mathfrak{c}: \quad \mathfrak{c}\to-\mathfrak{c}\,, && \mathfrak{T}_\mathfrak{d}: \quad \mathfrak{d}\to-\mathfrak{d}\,, \end{aligned} \tag{5.29}$$

and we have

$$Z(\mathfrak{a},\mathfrak{b},\mathfrak{c},\mathfrak{d}) = Z(\pm\mathfrak{a},\pm\mathfrak{b},\pm\mathfrak{c},\pm\mathfrak{d})\,. \tag{5.30}$$

5.5 The Weak-Graph Transformation

The transformations of the previous section are equivalent to permutations of the vertex energies $e_\mathfrak{p}$, applicable at any value of T. The weak-graph transformation (Wu 1969) is more akin to the Ising model dual transformation (Volume 1, Sect. 8.3) and indeed it can be shown that the dual transformation is a special case (Example 5.5).

Vertex weights $\mathfrak{Z}(\mathfrak{p})$ are defined in (A.239) for the sixteen-vertex model. The special case of the eight-vertex model has vertex weights given by (5.4), for $\mathfrak{p}=1,\ldots,8$ and

$$\mathfrak{Z}(\mathfrak{p}) = 0\,, \qquad \mathfrak{p}=9,\ldots 16 \tag{5.31}$$

(cf. Figs. 5.1 and A.3). According to the discussion in Sect. 5.1 the configuration graphs for the eight-vertex model are polygon graphs and the partition function (A.239) in this case becomes

$$Z(\mathfrak{a},\mathfrak{b},\mathfrak{c},\mathfrak{d}) = \sum_{\{\mathfrak{g}\}}\prod_{\mathfrak{i}=1}^{N}\mathfrak{Z}(\mathfrak{p}_\mathfrak{i})\,, \tag{5.32}$$

where $\{\mathfrak{g}\}$ is the set of polygon subgraphs of the lattice graph $\mathcal{N}$. In Appendix A.8 it is shown that the partition function is invariant under a weak-graph transformation from the set of vertex weights $\{\mathfrak{Z}(\mathfrak{p})\}$ to a new set $\{\mathfrak{N}(\mathfrak{p})\}$. This linear transformation is given by (A.247) and (A.248). In the case of the eight-vertex model with vertex weights given by (5.4) and (5.31) it is straightforward to confirm that the new set of vertex weights also satisfy the eight-vertex conditions. With

$$\begin{aligned} &\mathfrak{N}(1) = \mathfrak{N}(2) = \mathfrak{a}'\,, && \mathfrak{N}(3) = \mathfrak{N}(4) = \mathfrak{b}'\,, \\ &\mathfrak{N}(5) = \mathfrak{N}(6) = \mathfrak{c}'\,, && \mathfrak{N}(7) = \mathfrak{N}(8) = \mathfrak{d}'\,, \end{aligned} \tag{5.33}$$

$$\mathfrak{N}(\mathfrak{p}) = 0\,, \qquad \mathfrak{p} = 9, \ldots, 16\,, \tag{5.34}$$

the transformation reduces to

$$\begin{aligned} &\mathfrak{a}' = \tfrac{1}{2}(\mathfrak{a} + \mathfrak{b} + \mathfrak{c} + \mathfrak{d})\,, && \mathfrak{b}' = \tfrac{1}{2}(\mathfrak{a} + \mathfrak{b} - \mathfrak{c} - \mathfrak{d})\,, \\ &\mathfrak{c}' = \tfrac{1}{2}(\mathfrak{a} - \mathfrak{b} + \mathfrak{c} - \mathfrak{d})\,, && \mathfrak{d}' = \tfrac{1}{2}(\mathfrak{a} - \mathfrak{b} - \mathfrak{c} + \mathfrak{d}) \end{aligned} \tag{5.35}$$

and, from (A.249),

$$Z(\mathfrak{a}', \mathfrak{b}', \mathfrak{c}', \mathfrak{d}') = Z(\mathfrak{a}, \mathfrak{b}, \mathfrak{c}, \mathfrak{d})\,. \tag{5.36}$$

The partition function is, therefore, invariant under the transformation

$$\mathcal{T}' : \qquad (\mathfrak{a}, \mathfrak{b}, \mathfrak{c}, \mathfrak{d}) \to (\mathfrak{a}', \mathfrak{b}', \mathfrak{c}', \mathfrak{d}')\,, \tag{5.37}$$

defined by (5.35). The inverse transformation ${\mathcal{T}'}^{-1} = \mathcal{T}'$ since, from (5.35),

$$\begin{aligned} &\mathfrak{a} = \tfrac{1}{2}(\mathfrak{a}' + \mathfrak{b}' + \mathfrak{c}' + \mathfrak{d}')\,, && \mathfrak{b} = \tfrac{1}{2}(\mathfrak{a}' + \mathfrak{b}' - \mathfrak{c}' - \mathfrak{d}')\,, \\ &\mathfrak{c} = \tfrac{1}{2}(\mathfrak{a}' - \mathfrak{b}' + \mathfrak{c}' - \mathfrak{d}')\,, && \mathfrak{d} = \tfrac{1}{2}(\mathfrak{a}' - \mathfrak{b}' - \mathfrak{c}' + \mathfrak{d}')\,. \end{aligned} \tag{5.38}$$

For regions (ii), (iii) and (iv) the transformation $\mathcal{T}'$ is replaced by $\mathcal{T}_1\mathcal{T}'\mathcal{T}_1$, $\mathcal{T}_2\mathcal{T}'\mathcal{T}_2$ and $\mathcal{T}_3\mathcal{T}'\mathcal{T}_3$, respectively. These transformations have the form (Example 5.5)

$$\mathcal{T}_1\mathcal{T}'\mathcal{T}_1 : \qquad (\mathfrak{a}, \mathfrak{b}, \mathfrak{c}, \mathfrak{d}) \quad \to \quad (\mathfrak{b}', \mathfrak{a}', -\mathfrak{d}', -\mathfrak{c}')\,, \tag{5.39}$$

$$\mathcal{T}_2\mathcal{T}'\mathcal{T}_2 : \qquad (\mathfrak{a}, \mathfrak{b}, \mathfrak{c}, \mathfrak{d}) \quad \to \quad (\mathfrak{c}', -\mathfrak{d}', \mathfrak{a}', -\mathfrak{b}')\,, \tag{5.40}$$

$$\mathcal{T}_3\mathcal{T}'\mathcal{T}_3 : \qquad (\mathfrak{a}, \mathfrak{b}, \mathfrak{c}, \mathfrak{d}) \quad \to \quad (\mathfrak{d}', -\mathfrak{c}', -\mathfrak{b}', \mathfrak{a}')\,, \tag{5.41}$$

where $\mathfrak{a}'$, $\mathfrak{b}'$, $\mathfrak{c}'$ and $\mathfrak{d}'$ are again given by (5.35). The partition function is, of course, invariant under these transformations.

5.6 Transition Surfaces

From (5.35) it follows that $\mathfrak{a}' = \mathfrak{a}$, $\mathfrak{b}' = \mathfrak{b}$, $\mathfrak{c}' = \mathfrak{c}$ and $\mathfrak{d}' = \mathfrak{d}$ if, and only if,

$$\mathfrak{a} = \mathfrak{b} + \mathfrak{c} + \mathfrak{d}\,. \tag{5.42}$$

Equation (5.42) defines a surface $\mathcal{C}_1$ of fixed points of the transformation $\mathcal{T}'$, defined by (5.37). In Fig. 5.5 region (i) corresponds to the interior of the unit cube OBCDB′C′D′A. This cube is divided into two regions by the shaded planar triangle BCD, which corresponds to the surface $\mathcal{C}_1$. These regions are

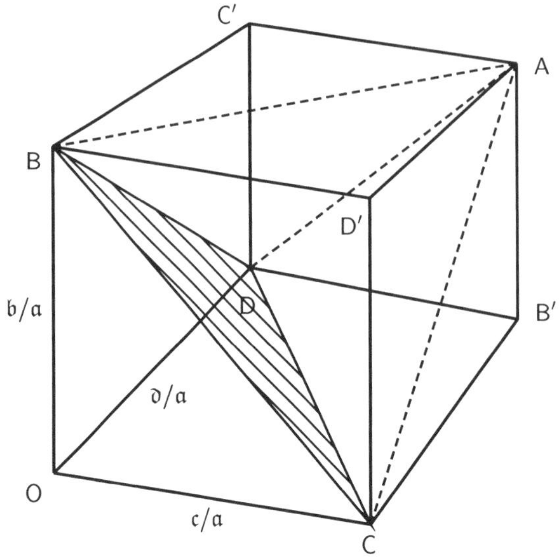

Fig. 5.5. The unit cube corresponding to region (i) in the space $(\mathfrak{b}/\mathfrak{a}, \mathfrak{c}/\mathfrak{a}, \mathfrak{d}/\mathfrak{a})$. The surface $\mathcal{C}_1$ defined by $\mathfrak{a} = \mathfrak{b}+\mathfrak{c}+\mathfrak{d}$ is hatched.

$$\mathcal{R}_{\rm L}: \qquad \mathfrak{a} > \mathfrak{b} + \mathfrak{c} + \mathfrak{d}\,, \tag{5.43}$$

$$\mathcal{R}_{\rm H}: \qquad \mathfrak{a} < \mathfrak{b} + \mathfrak{c} + \mathfrak{d}\,, \tag{5.44}$$

corresponding respectively to the tetrahedron OBCD and the remaining part of the cube.

Now consider the trajectory traced out by the point $(\mathfrak{a}, \mathfrak{b}, \mathfrak{c}, \mathfrak{d})$, representing a model with values of the $e_{\rm p}$ satisfying conditions (i) of Sect. 5.3. This trajectory remains in the unit cube and, as the temperature is decreased, it passes from A, where $T = \infty$ and $\mathfrak{a} = \mathfrak{b} = \mathfrak{c} = \mathfrak{d} = 1$ to O, where $T = 0$ and $\mathfrak{b}/\mathfrak{a} = \mathfrak{c}/\mathfrak{a} = \mathfrak{d}/\mathfrak{a} = 0$. Thus, as T decreases from infinity to zero, the trajectory for any type (i) model passes from the high-T zone $\mathcal{R}_{\rm H}$ of region (i) to the low-T zone $\mathcal{R}_{\rm L}$ of region (i). It must, therefore, intersect the surface $\mathcal{C}_1$.

Take a point in $\mathcal{R}_{\rm L}$ and perform the weak-graph transformation $\mathcal{T}'$. From (5.35) and (5.43) $\mathfrak{a}' > \mathfrak{b}' > 0$, $\mathfrak{a}' > \mathfrak{c}' > 0$ and $\mathfrak{a}' > \mathfrak{d}' > 0$ with

$$(\mathfrak{b}' + \mathfrak{c}' + \mathfrak{d}') - \mathfrak{a}' = \mathfrak{a} - (\mathfrak{b} + \mathfrak{c} + \mathfrak{d})\,. \tag{5.45}$$

It follows that $\mathcal{T}'$ maps a point in $\mathcal{R}_{\rm L}$ into a point in $\mathcal{R}_{\rm H}$. In general the points are not on the same trajectory (i.e. they do not correspond to the same values of the $e_{\rm p}$). An exception to this is the case where the vertex energies are given by (5.11) and the weak-graph transformation $\mathcal{T}'$ becomes the duality transformation (Example 5.4). The image of $\mathcal{R}_{\rm L}$ under $\mathcal{T}'$ is not the whole of $\mathcal{R}_{\rm H}$. Since $\mathfrak{a}$, $\mathfrak{b}$, $\mathfrak{c}$ and $\mathfrak{d}$ are positive

$$\mathfrak{a}' + \mathfrak{b}' > \mathfrak{c}' + \mathfrak{d}'\,, \qquad \mathfrak{a}' + \mathfrak{c}' > \mathfrak{b}' + \mathfrak{d}'\,, \qquad \mathfrak{a}' + \mathfrak{d}' > \mathfrak{b}' + \mathfrak{c}'\,. \tag{5.46}$$

The inequalities (5.46) define the tetrahedron ABCD of Fig. 5.5, which will be denoted by $\mathcal{R}_{\mathfrak{o}}$. From (5.38), an application of $\mathcal{T}'$ to a point $(\mathfrak{a}', \mathfrak{b}', \mathfrak{c}', \mathfrak{d}')$ in $\mathcal{R}_{\mathfrak{o}}$ takes it back to a point $(\mathfrak{a}, \mathfrak{b}, \mathfrak{c}, \mathfrak{d})$ in $\mathcal{R}_{\mathrm{L}}$. However, this is the case only for points in $\mathcal{R}_{\mathfrak{o}}$. Since $\mathfrak{a}' > \mathfrak{b}'$, $\mathfrak{a}' > \mathfrak{c}'$ and $\mathfrak{a}' > \mathfrak{d}'$, not more than one of the three inequalities (5.46) can be violated at any one point. The zone $\mathcal{R}_{\mathrm{H}}$ can thus be divided into four subzones. Referring to Fig. 5.5 these are the tetrahedra:

ABCD, denoted by $\mathcal{R}_{\mathfrak{o}}$, where (5.46) applies,

AB′CD, denoted by $\mathcal{R}_{\mathfrak{b}}$, where $\mathfrak{a}' + \mathfrak{b}' < \mathfrak{c}' + \mathfrak{d}'$,

ABC′D, denoted by $\mathcal{R}_{\mathfrak{c}}$, where $\mathfrak{a}' + \mathfrak{c}' < \mathfrak{b}' + \mathfrak{d}'$,

ABCD′, denoted by $\mathcal{R}_{\mathfrak{d}}$, where $\mathfrak{a}' + \mathfrak{d}' < \mathfrak{b}' + \mathfrak{c}'$.

It can easily be verified that the entire high-T part of the trajectory of the isotropic Ising model, whose vertex energies are given by (5.11), lies in $\mathcal{R}_{\mathfrak{o}}$. As an example of a trajectory which passes outside $\mathcal{R}_{\mathfrak{o}}$ consider a type (i) model where $e_1 + e_3 > e_5 + e_7$. By expanding $\mathfrak{a}$, $\mathfrak{b}$, $\mathfrak{c}$ and $\mathfrak{d}$ in inverse powers of T, it can be shown that, after leaving the $T = \infty$ point A, it follows a trajectory which traverses $\mathcal{R}_{\mathfrak{b}}$ before entering $\mathcal{R}_{\mathfrak{o}}$ and crossing the surface $\mathcal{C}_1$. From (5.38), a point in $\mathcal{R}_{\mathfrak{b}}$ is mapped by $\mathcal{T}'$ into a 'non-physical' point $(\mathfrak{a}, \mathfrak{b}^*, \mathfrak{c}, \mathfrak{d})$, with $\mathfrak{a}$, $\mathfrak{c}$ and $\mathfrak{d}$ positive, but $\mathfrak{b}^*$ negative. The transformation $\mathcal{T}_{\mathfrak{b}}$ defined in (5.29) maps this point into $(\mathfrak{a}, \mathfrak{b}, \mathfrak{c}, \mathfrak{d})$, with $\mathfrak{b} = -\mathfrak{b}^*$, so that all the coordinates are positive. Using (5.38), (5.29) and the conditions $\mathfrak{a}' > \mathfrak{b}'$, $\mathfrak{a}' > \mathfrak{c}'$, $\mathfrak{a}' > \mathfrak{d}'$ it can be shown that the transformation $\mathcal{T}_{\mathfrak{b}}\mathcal{T}'$ takes a point in the subzone $\mathcal{R}_{\mathfrak{b}}$ of $\mathcal{R}_{\mathrm{H}}$ into a point in $\mathcal{R}_{\mathrm{L}}$ where $\mathfrak{b} < \mathfrak{c}$ and $\mathfrak{b} < \mathfrak{d}$. The inverse process is effected by the transformation $\mathcal{T}'\mathcal{T}_{\mathfrak{b}}$ and the high-T subzones $\mathcal{R}_{\mathfrak{c}}$ and $\mathcal{R}_{\mathfrak{d}}$ are similarly connected to $\mathcal{R}_{\mathrm{L}}$ by the transformations $\mathcal{T}_{\mathfrak{c}}\mathcal{T}'$ and $\mathcal{T}_{\mathfrak{d}}\mathcal{T}'$ respectively. It follows, from (5.30) and (5.36), that the partition function $Z(\mathfrak{a}, \mathfrak{b}, \mathfrak{c}, \mathfrak{d})$ can be deduced for the entire region (i) if it is known for the low-T zone.

Suppose that, on some trajectory, a transition occurs at a point in $\mathcal{R}_{\mathrm{L}}$. Making reasonable continuity assumptions, variation of the trajectory (produced by changing the e_{p}) generates a transition surface in $\mathcal{R}_{\mathrm{L}}$. Since this is a surface of singularities in the free energy, it follows, from the discussion of the transformation $\mathcal{T}'$ in the earlier part of this section, that there is a corresponding surface of singularities in $\mathcal{R}_{\mathfrak{o}}$. However, the boundary surface $\mathcal{C}_1$ between $\mathcal{R}_{\mathrm{L}}$ and $\mathcal{R}_{\mathrm{H}}$, given by (5.42), is transformed into itself by $\mathcal{T}'$. That is to say, $\mathcal{C}_1$ is the invariant surface of the transformation $\mathcal{T}'$. So, if it is assumed that there is a unique transition surface in region (i) (Sutherland 1970) then this surface may coincide with $\mathcal{C}_1$. However, we know that the transformations $\mathcal{T}'\mathcal{T}_{\mathfrak{b}}$, $\mathcal{T}'\mathcal{T}_{\mathfrak{c}}$ and $\mathcal{T}'\mathcal{T}_{\mathfrak{d}}$ also take points from $\mathcal{R}_{\mathrm{L}}$ to $\mathcal{R}_{\mathrm{H}}$ and it is necessary to examine whether the corresponding invariant surfaces are possible surfaces of transitions. The invariant points for $\mathcal{T}'\mathcal{T}_{\mathfrak{b}}$ are obtained by replacing $\mathfrak{b}$ by

$-\mathfrak{b}$ on the right-hand sides of (5.35) and $\mathfrak{a}'$, $\mathfrak{b}'$, $\mathfrak{c}'$ and $\mathfrak{d}'$ by $\mathfrak{a}$, $\mathfrak{b}$, $\mathfrak{c}$ and $\mathfrak{d}$ on the left-hand sides. This yields $\mathfrak{a} = \mathfrak{c} + \mathfrak{d}$, $\mathfrak{b} = 0$, which specifies the line CD on Fig. 5.5, which is the intersection of the invariant surface $\mathcal{C}_1$ of $\mathcal{T}'$, given by (5.42), and the invariant surface $\mathfrak{b}/\mathfrak{a} = 0$ of $\mathcal{T}_\mathfrak{b}$. Now a trajectory cannot cross the plane $\mathfrak{b} = 0$, since this would involve entering a region with $\mathfrak{b} < 0$. In fact, for $T > 0$, no trajectory can touch this plane except in the limiting case $e_3 = \infty$, when it lies entirely in the plane. It can similarly be shown that the transformations $\mathcal{T}'\mathcal{T}_\mathfrak{c}$ and $\mathcal{T}'\mathcal{T}_\mathfrak{d}$ do not generate any possible transition surfaces. Hence, if there is only one transition surface in region (i) it must be the plane surface $\mathcal{C}_1$, which implies that there is exactly one transition point on any trajectory in region (i), occurring at its intersection with $\mathcal{C}_1$. It also implies the existence of ferroelectric type (i) order throughout $\mathcal{R}_\mathrm{L}$, while $\mathcal{R}_\mathrm{H}$ is a zone of disorder. The assumption that there is only one transition surface in region (i) is confirmed from the exact expression for the zero-field free energy of the eight-vertex model derived in Sect. 5.8. It is also in agreement with two special cases. For the Ising model, substituting from (5.11) into (5.42) gives $\sinh(2J/T) = 1$, which is the formula for the critical temperature (see Volume 1, Sect. 8.3 and Sect. 5.10 below). For the six-vertex model ($e_7 = e_8 = \infty$) $\mathfrak{d} = 0$ and (5.42) reduces to $\mathfrak{a} = \mathfrak{b} + \mathfrak{c}$, which is equivalent to the $\mathfrak{b} = \mathfrak{a} - 1$ of Volume 1, Sect. 10.6.

From the invariance of the free energy under the transformations $\mathcal{T}_1$, $\mathcal{T}_2$ and $\mathcal{T}_3$ it follows that the hypothesis that $\mathcal{C}_1$, given by $\mathfrak{a} = \mathfrak{b} + \mathfrak{c} + \mathfrak{d}$, is the unique transition surface in region (i) implies that

$$\mathcal{C}_2 : \qquad \mathfrak{b} = \mathfrak{a} + \mathfrak{c} + \mathfrak{d}\,, \tag{5.47}$$

$$\mathcal{C}_3 : \qquad \mathfrak{c} = \mathfrak{a} + \mathfrak{b} + \mathfrak{d}\,, \tag{5.48}$$

$$\mathcal{C}_4 : \qquad \mathfrak{d} = \mathfrak{a} + \mathfrak{b} + \mathfrak{c} \tag{5.49}$$

are respectively the unique transition surfaces for regions (ii), (iii) and (iv). In region (iii), for example, there is type (iii) antiferroelectric sublattice order in the low-T zone, where $\mathfrak{c} > \mathfrak{a} + \mathfrak{b} + \mathfrak{d}$, and disorder in the high-T zone, where $\mathfrak{c} < \mathfrak{a} + \mathfrak{b} + \mathfrak{d}$. When the least two values in the set (e_1, e_2, e_3, e_4) are degenerate the trajectory lies on the boundary between two regions. For example, suppose that $e_1 = e_3$. Then $\mathfrak{a} = \mathfrak{b}$, $\mathfrak{c} < \mathfrak{a}$ and $\mathfrak{d} < \mathfrak{a}$ at all points on the trajectory, which lies in the boundary between regions (i) and (ii). In such cases none of the transition conditions can be satisfied and hence no transition occurs for $T > 0$. This behaviour is similar to that of the IKDP and IF models, (see Volume 1, Table 10.3 and Example 10.3).

In the high-T (disordered) zone $\mathcal{R}_\mathrm{H}$ it is of interest to look at the properties of the model on the boundaries between subzone $\mathcal{R}_\mathrm{o}$, where all the conditions of (5.46) apply, and the other subzones. For the subzone $\mathcal{R}_\mathfrak{d}$ the boundary in question is the triangle ABC, where

$$\mathfrak{a}' + \mathfrak{d}' = \mathfrak{b}' + \mathfrak{c}' \tag{5.50}$$

and, from (5.38), the transformation $\mathcal{T}'$ maps a point on ABC into a point $(\mathfrak{a}, \mathfrak{b}, \mathfrak{c}, \mathfrak{d})$ of the triangle OBC, for which $\mathfrak{d} = 0$ and $\mathfrak{a} > \mathfrak{b} + \mathfrak{c}$. This represents a subcritical temperature state in a six-vertex ferroelectric of type (i). Such a state is perfectly ordered with all vertices identical. In the frozen low-temperature ferroelectric state the dimensionless free energy per site of the six-vertex model is

$$\phi(\mathfrak{a}, \mathfrak{b}, \mathfrak{c}, \mathfrak{d}) = -\ln(\mathfrak{a}) \tag{5.51}$$

(see Volume 1, Sect. 10.6). Thus, from (5.38) and (5.51) the free energy per site for a point on ABC is

$$\phi(\mathfrak{a}', \mathfrak{b}', \mathfrak{c}', \mathfrak{d}') = -\ln\left[\tfrac{1}{2}(\mathfrak{a}' + \mathfrak{b}' + \mathfrak{c}' + \mathfrak{d}')\right] . \tag{5.52}$$

It can be shown, using (5.28), that (5.52) also applies on the triangle ABD where

$$\mathfrak{a}' + \mathfrak{c}' = \mathfrak{b}' + \mathfrak{d}'. \tag{5.53}$$

For large N, (5.52) is equivalent to

$$Z(\mathfrak{a}', \mathfrak{b}', \mathfrak{c}', \mathfrak{d}') = \frac{1}{4^N}(2\mathfrak{a}' + 2\mathfrak{b}' + 2\mathfrak{c}' + 2\mathfrak{d}')^N . \tag{5.54}$$

Comparison with the eight-vertex ice results given in Sect. 5.3 shows that (5.54) corresponds to an equilibrium state in which the vertex types are distributed as randomly as possible without breaking the compatibility rules for nearest-neighbour vertices. Hence (5.50) and (5.53) appear to be surfaces of *disorder points*. A disorder point in a system is defined to be one, for finite temperature, at which all correlation functions between the states on pairs of sites are zero, as they would be at infinite temperature (see Volume 1, Sect. 9.10). That this is indeed the case in the circumstances described here was confirmed by correlation function calculations of Baxter (1982a). Thus trajectories starting in $\mathcal{R}_\mathfrak{c}$ or $\mathcal{R}_\mathfrak{d}$ pass through a point of disorder in the range between $T = \infty$ and the critical temperature. For the Ising model the energy relations (5.7) with $J_4 = 0$ are compatible with (5.50) only when $J_{22} = 0$ and the entire trajectory then lies in the surface (5.50). Since $J_{21} \neq 0$ there are correlations between spins on any line of sites in the J_{21}-direction, but with $J_{22} = 0$ the spins on different lines of this kind are uncorrelated. The system is thus essentially one-dimensional and substitution of (5.7) with $J_0 = J_{22} = J_4 = 0$ into (5.4) and (5.52) gives

$$\phi(K_{21}) = -\ln[2\cosh(K_{21})] , \tag{5.55}$$

which is equivalent to the formula for the free energy of the one-dimensional zero-field Ising model (Volume 1, Sect. 2.4). There is no transition for $T > 0$, which is to be expected since (5.50) has no points in common with (5.42). The situation on the surface (5.53) is similar, with $J_{21} = 0$ in the Ising model case. Points on the triangle ACD, where $\mathfrak{a}' + \mathfrak{b}' = \mathfrak{c}' + \mathfrak{d}'$, have different properties from those on ABC and ABD since the transformation $\mathcal{T}'$ maps them into points $(\mathfrak{a}, \mathfrak{b}, \mathfrak{c}, \mathfrak{d})$ where $\mathfrak{b} = 0$ and $\mathfrak{a} > \mathfrak{c} + \mathfrak{d}$. From (5.22) the free energy is equal to that of a six-vertex antiferroelectric below its critical temperature. However the six-vertex antiferroelectric state is not one of perfect order and no equation like (5.52) can be deduced.

5.7 The Transfer Matrix

An exact expression for the free energy of the zero-field eight-vertex model was derived by Baxter (1972) from the largest eigenvalue of the appropriate transfer matrix. For the six-vertex model the number $\mathfrak{n}$ of downward pointing arrows in each row of vertical edges is conserved (Volume 1, Sect. 10.3). This is no longer the case for the eight-vertex model, although the odd or even parity is conserved. This is one of the factors which makes the solution of the eight-vertex model rather complicated. In this and the following section we shall present only a basic outline of the analysis referring the reader to Baxter (1972) or Baxter (1982a) for further details.

In Sect. 5.1 N_1 and N_2 were defined as the number of sites in each row and column respectively, it being assumed that the lattice is wound on a torus. Starting at some arbitrary site, the sites of the lattice can be labelled by the cartesian coordinates $(\mathfrak{i}, \mathfrak{j})$, $\mathfrak{i} = 1, \ldots, N_1$, $\mathfrak{j} = 1, \ldots, N_2$. For convenience we adopt the convention of labelling the vertical edge *below* the site $(\mathfrak{i}, \mathfrak{j})$ as $(\mathfrak{i}, \mathfrak{j})$ and the horizontal edge to the *left of* the site $(\mathfrak{i}, \mathfrak{j})$ as $(\mathfrak{i}, \mathfrak{j})$. Then the states of the vertical edges are given by

$$\alpha_{\mathfrak{i}}(\mathfrak{j}) = \begin{cases} +1, & \text{if the arrow on } (\mathfrak{i},\mathfrak{j}) \text{ points upwards,} \\ -1, & \text{if the arrow on } (\mathfrak{i},\mathfrak{j}) \text{ points downwards.} \end{cases} \tag{5.56}$$

In a similar way the states of the horizontal edges are given by

$$\lambda_{\mathfrak{j}}(\mathfrak{i}) = \begin{cases} +1, & \text{if the arrow on } (\mathfrak{i},\mathfrak{j}) \text{ points to the right,} \\ -1, & \text{if the arrow on } (\mathfrak{i},\mathfrak{j}) \text{ points to the left.} \end{cases} \tag{5.57}$$

The state of the $\mathfrak{j}$-th row of vertical edges is specified by the vector

$$\boldsymbol{\alpha}(\mathfrak{j}) = (\alpha_1(\mathfrak{j}), \alpha_2(\mathfrak{j}), \ldots, \alpha_{N_1}(\mathfrak{j})) \tag{5.58}$$

and let

$$V(\boldsymbol{\alpha}(\mathfrak{j}), \boldsymbol{\alpha}(\mathfrak{j}+1)) = \sum \exp(-E_{\mathfrak{j}}/T) \,, \tag{5.59}$$

where $E_{\mathfrak{j}}$ is the sum of the vertex energies of row $\mathfrak{j}$, the summation being over all the states of these vertices compatible with $\boldsymbol{\alpha}(\mathfrak{j})$ and $\boldsymbol{\alpha}(\mathfrak{j}+1)$. $V(\boldsymbol{\alpha}, \boldsymbol{\alpha}')$ is an element of the $2^{N_1} \times 2^{N_1}$ dimensional transfer matrix $\boldsymbol{V}$ and as in the case of the one-dimensional Ising model (see Volume 1, Sect. 2.4) and for the more general analysis of Sect. 4.9, since toroidal boundary conditions are applied, the partition function can be expressed in the form

$$Z(\mathfrak{a}, \mathfrak{b}, \mathfrak{c}, \mathfrak{d}) = \mathrm{Trace}\{\boldsymbol{V}^{N_2}\} \,. \tag{5.60}$$

The type of vertex at site $(\mathfrak{i},\mathfrak{j})$ is specified by the four numbers $\alpha_\mathfrak{i}(\mathfrak{j})$, $\alpha_\mathfrak{i}(\mathfrak{j}+1)$, $\lambda_\mathfrak{j}(\mathfrak{i})$ and $\lambda_\mathfrak{j}(\mathfrak{i}+1)$. Let

$$R(\alpha_\mathfrak{i}(\mathfrak{j}),\alpha_\mathfrak{i}(\mathfrak{j}+1);\lambda_\mathfrak{j}(\mathfrak{i}),\lambda_\mathfrak{j}(\mathfrak{i}+1)) = \mathfrak{z}_{\mathfrak{i}\mathfrak{j}}\,, \tag{5.61}$$

where $\mathfrak{z}_{\mathfrak{i}\mathfrak{j}}$ is the Boltzmann factor for site $(\mathfrak{i},\mathfrak{j})$. For the eight-vertex model these are given by (5.4), but we shall include the possibility that the values of the αs and λs give a configuration incompatible with the eight-vertex restriction, when the Boltzmann factor is zero according to equation (5.31). For fixed $\alpha_\mathfrak{i}(\mathfrak{j})$ and $\alpha_\mathfrak{i}(\mathfrak{j}+1)$ (5.61) defines four 2×2 matrices with $\lambda_\mathfrak{j}(\mathfrak{i})$ and $\lambda_\mathfrak{j}(\mathfrak{i}+1)$ as the row and column indices respectively. From (5.4) and (5.31),

$$\begin{aligned}
\boldsymbol{R}(1,1) &= \begin{pmatrix}\mathfrak{a} & 0\\ 0 & \mathfrak{b}\end{pmatrix}, &\qquad \boldsymbol{R}(1,-1) &= \begin{pmatrix}0 & \mathfrak{d}\\ \mathfrak{c} & 0\end{pmatrix},\\
\boldsymbol{R}(-1,1) &= \begin{pmatrix}0 & \mathfrak{c}\\ \mathfrak{d} & 0\end{pmatrix}, &\qquad \boldsymbol{R}(-1,-1) &= \begin{pmatrix}\mathfrak{b} & 0\\ 0 & \mathfrak{a}\end{pmatrix}.
\end{aligned} \tag{5.62}$$

From (5.59)

$$V(\boldsymbol{\alpha},\boldsymbol{\alpha}') = \mathrm{Trace}\{\boldsymbol{R}(\alpha_1,\alpha_1')\boldsymbol{R}(\alpha_2,\alpha_2')\cdots\boldsymbol{R}(\alpha_{N_1},\alpha_{N_1}')\}\,. \tag{5.63}$$

The first step of Baxter's analysis is to obtain conditions under which transfer matrices with different values of $\mathfrak{a}$, $\mathfrak{b}$, $\mathfrak{c}$ and $\mathfrak{d}$ commute. This procedure is most easily described in terms of a set of new variables w_1, w_2, w_3 and w_4, which satisfy the conditions

$$w_1 \geq w_2 \geq w_3 \geq |w_4|\,. \tag{5.64}$$

These new variables are given in terms of $\mathfrak{a}$, $\mathfrak{b}$, $\mathfrak{c}$ and $\mathfrak{d}$, but their definitions differ according to the region of parameter space under consideration and according to whether we are concerned with the part of the region above or below the critical surface.

For the low-T zone of region (i) in which $\mathfrak{a} > \mathfrak{b}+\mathfrak{c}+\mathfrak{d}$ we define

$$\begin{aligned}
w_1 &= \tfrac{1}{2}(\mathfrak{a}+\mathfrak{b})\,, &\qquad w_2 &= \tfrac{1}{2}(\mathfrak{a}-\mathfrak{b})\,,\\
w_3 &= \tfrac{1}{2}(\mathfrak{c}+\mathfrak{d})\,, &\qquad w_4 &= \tfrac{1}{2}(\mathfrak{c}-\mathfrak{d})
\end{aligned} \tag{5.65}$$

and conditions (5.64) are satisfied.

To obtain the appropriate definitions for the low-T zones of the remaining regions we use the transformations $\mathcal{T}_1$, $\mathcal{T}_2$ and $\mathcal{T}_3$ of Sect. 5.4. Thus
for region (ii)

$$\begin{aligned}
w_1 &= \tfrac{1}{2}(\mathfrak{b}+\mathfrak{a})\,, &\qquad w_2 &= \tfrac{1}{2}(\mathfrak{b}-\mathfrak{a})\,,\\
w_3 &= \tfrac{1}{2}(\mathfrak{d}+\mathfrak{c})\,, &\qquad w_4 &= \tfrac{1}{2}(\mathfrak{d}-\mathfrak{c})\,,
\end{aligned} \tag{5.66}$$

for region (iii)

$$w_1 = \frac{1}{2}(\mathfrak{c}+\mathfrak{d}), \qquad w_2 = \frac{1}{2}(\mathfrak{c}-\mathfrak{d}), \qquad w_3 = \frac{1}{2}(\mathfrak{a}+\mathfrak{b}), \qquad w_4 = \frac{1}{2}(\mathfrak{a}-\mathfrak{b}), \tag{5.67}$$

for region (iv)

$$w_1 = \frac{1}{2}(\mathfrak{d}+\mathfrak{c}), \qquad w_2 = \frac{1}{2}(\mathfrak{d}-\mathfrak{c}), \qquad w_3 = \frac{1}{2}(\mathfrak{b}+\mathfrak{a}), \qquad w_4 = \frac{1}{2}(\mathfrak{b}-\mathfrak{a}). \tag{5.68}$$

For the high-T zones of the regions the appropriate forms for w_1, w_2, w_3 and w_4 are obtained by transforming equations (5.65)–(5.68). The basic mappings are given by the weak-graph transformation (5.37) for (5.65) and the transformations (5.39)–(5.41) for (5.66)–(5.68) respectively. However, an initial use of one of the transformations (5.29) may be necessary to ensure that conditions (5.64) are satisfied. For future reference two special cases should be considered.

The spin representation to be discussed in Sect. 5.10, lies in region (i) and the inequalities (5.46) are satisfied. The transformation $\mathcal{T}'$ can therefore be applied to (5.65) to give

$$w_1 = \frac{1}{2}(\mathfrak{a}+\mathfrak{b}), \qquad w_2 = \frac{1}{2}(\mathfrak{c}+\mathfrak{d}), \qquad w_3 = \frac{1}{2}(\mathfrak{a}-\mathfrak{b}), \qquad w_4 = \frac{1}{2}(\mathfrak{c}-\mathfrak{d}) \tag{5.69}$$

for the Ising model at high temperatures. Comparison of (5.65) and (5.69) shows that the definitions of w_2 and w_3 'cross over' at $w_2 = w_3$, which is the definition (5.42) of the critical surface $\mathcal{C}_1$.

The six-vertex model to be discussed in Sect. 5.11 is the limiting case of the eight-vertex model as $e_7 \to \infty$. It can lie in any of the regions (i)–(iii). In region (i) at high temperatures it violates the last inequality of (5.46). It is, therefore, necessary to apply the transformation $\mathcal{T}'\mathcal{T}_{\mathfrak{d}}$ to (5.65) giving

$$w_1 = \frac{1}{2}(\mathfrak{a}+\mathfrak{b}), \qquad w_2 = \frac{1}{2}(\mathfrak{c}+\mathfrak{d}), \qquad w_3 = \frac{1}{2}(\mathfrak{c}-\mathfrak{d}), \qquad w_4 = \frac{1}{2}(\mathfrak{a}-\mathfrak{b}). \tag{5.70}$$

In a similar way, for regions (ii) and (iii), the necessary transformations are $\mathcal{T}_1\mathcal{T}'\mathcal{T}_1\mathcal{T}_{\mathfrak{d}}$ and $\mathcal{T}_2\mathcal{T}'\mathcal{T}_2\mathcal{T}_{\mathfrak{d}}$. The resulting formulae are identical to (5.70).

By expressing $\boldsymbol{R}(\alpha, \alpha')$ in terms of *Pauli matrices* it can now be shown (Baxter 1982a) that two transfer matrices, corresponding to two different sets of w_i values, commute when the ratios $(w_j^2 - w_k^2)/(w_l^2 - w_m^2)$, where (j, k, l, m) is any permutation of $(1, 2, 3, 4)$, are the same for both sets of w_i. In fact it is not difficult to see that only two ratios of this form can be chosen independently. We define

$$\frac{w_1^2 - w_4^2}{w_3^2 - w_2^2} = \frac{1 + \Delta}{1 - \Delta}, \tag{5.71}$$

$$\frac{w_1^2 - w_2^2}{w_3^2 - w_4^2} = \frac{1 + \Gamma}{1 - \Gamma}, \tag{5.72}$$

and from (5.64), (5.71) and (5.72)

$$\Delta \geq 1, \qquad -1 \leq \Gamma \leq 1. \tag{5.73}$$

Transfer matrices with the same values of Δ and Γ commute and, for fixed values of Δ and Γ, there remains one degree of freedom in the ratios of the values of $w_1, \ldots, w_4$. Let

$$w_1 : w_2 : w_3 : w_4 = \frac{\mathrm{cn}(\mathsf{U}|\ell)}{\mathrm{cn}(\zeta|\ell)} : \frac{\mathrm{dn}(\mathsf{U}|\ell)}{\mathrm{dn}(\zeta|\ell)} : 1 : \frac{\mathrm{sn}(\mathsf{U}|\ell)}{\mathrm{sn}(\zeta|\ell)}, \tag{5.74}$$

using the Jacobian elliptic functions defined in Appendix A.6 (see also Volume 1, Appendix A.2) with ζ, U and ℓ real numbers and $0 \leq \ell \leq 1$. It can now be shown, using (A.196), that all the ratios $(w_j^2 - w_k^2)/(w_l^2 - w_m^2)$ are functions of ζ and ℓ, but not of U and that

$$\Delta = \frac{1}{\mathrm{dn}(2\zeta|\ell)}, \qquad \Gamma = -\frac{\mathrm{cn}(2\zeta|\ell)}{\mathrm{dn}(2\zeta|\ell)} \tag{5.75}$$

with the inverse relations

$$\mathrm{sn}^2(\zeta|\ell) = \frac{\Delta + \Gamma}{\Delta + 1}, \qquad \ell^2 = \frac{\Delta^2 - 1}{\Delta^2 - \Gamma^2}. \tag{5.76}$$

So when ζ and ℓ are fixed, transfer matrices with different values of U commute. It follows, from (A.197), that the formulae (5.75) for Δ and Γ satisfy (5.73) and

$$\begin{aligned} &\tfrac{1}{2}\mathcal{K}(\ell) \leq \zeta \leq \mathcal{K}(\ell), \qquad && \Gamma \geq 0, \\ &0 \leq \zeta \leq \tfrac{1}{2}\mathcal{K}(\ell), && \Gamma \leq 0, \end{aligned} \tag{5.77}$$

where $\mathcal{K}(\ell)$ is the complete elliptic integral defined in Appendix A.6. It remains necessary to examine the conditions under which (5.74) satisfies the inequalities (5.64). For this a more detailed use of the properties of the elliptic functions is necessary (see Gradshteyn and Ryzhik 1980 or Abramowitz and Segun 1965). It may be shown that consistency is achieved if

$$\begin{aligned} &0 \leq \mathsf{U} \leq \zeta, \qquad && w_4 \geq 0, \\ &-\zeta \leq \mathsf{U} \leq 0, && w_4 \leq 0. \end{aligned} \tag{5.78}$$

5.8 The Free Energy and Magnetization

We introduce a new modulus ℓ_1 given by

$$\ell_1 = \frac{2\sqrt{\ell}}{1+\ell}, \qquad \ell_1' = \frac{1-\ell}{1+\ell}, \tag{5.79}$$

which are equivalent to (A.213), and the new variables

$$\mathrm{u} = \frac{\mathrm{i}U}{1+\ell_1'}, \qquad \xi = \frac{\mathrm{i}\zeta}{1+\ell_1'}. \tag{5.80}$$

For fixed values of ℓ_1 and ξ transfer matrices $\boldsymbol{V}(\mathrm{u})$ and $\boldsymbol{V}(\mathrm{u}')$, for $\mathrm{u} \neq \mathrm{u}'$ commute. Let the eigenvalues of $\boldsymbol{V}(\mathrm{u})$ be $\Lambda_j(\mathrm{u})$, $j = 1, \ldots, 2^{N_1}$. Then, according to the analysis of Sect. 4.9,

$$\mathrm{Trace}\,\{\boldsymbol{V}^{N_2}\} = \sum_{j=1}^{2^{N_1}} [\Lambda_j(\mathrm{u})]^{N_2}. \tag{5.81}$$

Just as in the case of the six-vertex model (Volume 1, Sect. 10.4) the largest eigenvalue of $\boldsymbol{V}$ is real, positive and of magnitude larger than any other eigenvalue. It follows, from (5.60) and (5.81) that, in the limit of large N_2, the dimensionless free energy per site is given by

$$\phi(w_1, w_2, w_3, w_4) = -\lim_{N_1\to\infty} \frac{\ln\,[\Lambda_j(\mathrm{u})]_{\max}}{N_1}. \tag{5.82}$$

The problem is to find expressions for the eigenvalues $\Lambda_j(\mathrm{u})$ and then to determine the maximum member of the set. Baxter's strategy (suggested by a similar approach to the six-vertex case) is to find a matrix $\boldsymbol{Q}(\mathrm{u})$ and a scalar function $\psi(\mathrm{u})$ such that

$$\begin{aligned} &\boldsymbol{V}(\mathrm{u})\boldsymbol{Q}(\mathrm{u}) = \psi(\mathrm{u}-\xi)\boldsymbol{Q}(\mathrm{u}+2\xi) + \psi(\mathrm{u}+\xi)\boldsymbol{Q}(\mathrm{u}-2\xi)\,, \\ &\boldsymbol{V}(\mathrm{u})\boldsymbol{Q}(\mathrm{u}) = \boldsymbol{Q}(\mathrm{u})\boldsymbol{V}(\mathrm{u})\,, \\ &\boldsymbol{Q}(\mathrm{u})\boldsymbol{Q}(\mathrm{u}') = \boldsymbol{Q}(\mathrm{u}')\boldsymbol{Q}(\mathrm{u})\,. \end{aligned} \tag{5.83}$$

It follows from (5.83) that $\boldsymbol{V}(\mathrm{u})$, $\boldsymbol{Q}(\mathrm{u})$ and $\boldsymbol{Q}(\mathrm{u}')$ are simultaneously diagonalizable and that, if $\Omega_j(\mathrm{u})$, $j = 1, \ldots, 2^{N_1}$ are the eigenvalues of $\boldsymbol{Q}(\mathrm{u})$, then

$$\Lambda_j(\mathrm{u}) = \frac{\psi(\mathrm{u}-\xi)\Omega_j(\mathrm{u}+2\xi) + \psi(\mathrm{u}+\xi)\Omega_j(\mathrm{u}-2\xi)}{\Omega_j(\mathrm{u})}. \tag{5.84}$$

By a considerable amount of analysis (Baxter 1982a) the forms of $\psi(\mathrm{u})$ and $\Omega_j(\mathrm{u})$ can be found and an expression for $\phi(w_1, w_2, w_3, w_4)$ deduced. This is

$$\begin{aligned} \phi(w_1, w_2, w_3, w_4) = &- \ln(w_1 + w_2) \\ &- \sum_{n=1}^{\infty} \frac{(x^{2n} - q^n)^2 (x^n + x^{-n} - z^n - z^{-n})}{n x^n (1 - q^{2n})(1 + x^{2n})}, \end{aligned} \tag{5.85}$$

where

$$x = \exp[-\mathrm{i}\pi\xi/\mathcal{K}(\ell_1')] = \exp[-\pi\zeta/\mathcal{K}(\ell')],$$

$$q = \exp[-\pi\mathcal{K}(\ell_1)/\mathcal{K}(\ell_1')] = \exp[-2\pi\mathcal{K}(\ell)/\mathcal{K}(\ell')], \tag{5.86}$$

$$z = \exp[-\mathrm{i}u/\mathcal{K}(\ell_1')] = \exp[-\pi\mathsf{U}/\mathcal{K}(\ell')].$$

From (A.205) it will be observed that q is the nome of ℓ'. The spontaneous relative magnetization m_s in the spin interpretation of the eight-vertex model can be derived using 'corner' transfer matrices (Baxter 1982a) instead of 'row' transfer matrices like those defined in Sect. 5.7. It is found that

$$m_\mathrm{s} = \prod_{n=1}^{\infty} \frac{1 - x^{4n-2}}{1 + x^{4n-2}}. \tag{5.87}$$

If k_m is defined by the relation

$$\frac{\mathcal{K}(k_m')}{\mathcal{K}(k_m)} = \frac{2\zeta}{\mathcal{K}(\ell')}, \tag{5.88}$$

then it follows from (5.86) and (A.205) that x^2 is the nome corresponding to k_m and, from (A.209), that

$$m_\mathrm{s} = (k_m')^{\frac{1}{4}} = (1 - k_m^2)^{\frac{1}{8}}. \tag{5.89}$$

5.9 Critical Behaviour

From (5.85) the dimensionless free energy density is, through its dependence on $\mathsf{w}_1, \ldots, \mathsf{w}_4$, an entire function of T on any trajectory in the domain defined by (5.64).[7] Thus the singularities associated with transition points can occur only on the boundary of the domain, which corresponds to two successive w_i in the sequence of inequalities in (5.64) becoming equal. From (5.65)–(5.68) it can be seen that the cases $\mathsf{w}_1 = \mathsf{w}_2$ and $\mathsf{w}_3 = \pm\mathsf{w}_4$ correspond to special models with one or other of the energies e_1, e_3, e_5 or e_7 being infinite. On the other hand

$$\mathsf{w}_2 = \mathsf{w}_3 \tag{5.90}$$

corresponds in each case to the transition surface $\mathcal{C}_i$, $i = 1, \ldots, 4$, given respectively by (5.42) and (5.47)–(5.49). Since there are no singularities of the free energy inside the domain (5.64) the arguments of Sect. 5.6 indicate that the surface (5.90) is the only possible surface of singularities. This does not, however, prove that singularities exist. Let

$$\upmu = \frac{\pi\zeta}{\mathcal{K}(\ell)}, \qquad \upvarphi = \frac{\pi\mathsf{U}}{\mathcal{K}(\ell)}, \qquad \varpi = \frac{\mathcal{K}(\ell)}{\mathcal{K}(\ell')}. \tag{5.91}$$

---

[7] In contravention of the notation for the coupling-density representation introduced in Sect. 1.4 we shall here denote the dependence of ϕ on T by using $\phi(T)$.

Then (5.85) can be re-expressed in the form

$$\phi(T) = -\ln(w_1 + w_2) - \varpi \sum_{n=1}^{\infty} F(n\varpi), \tag{5.92}$$

where

$$F(y) = \frac{2\sinh^2[y(\pi - \mu)][\cosh(\mu y) - \cosh(\varphi y)]}{y \sinh(2\pi y)\cosh(\mu y)}. \tag{5.93}$$

Using the Poisson summation formula (A.15), and noting that $F(y)$ is an even function with $F(0) = 0$, we have

$$\phi(T) = -\ln(w_1 + w_2) - \tfrac{1}{2}G(0) - \sum_{n=1}^{\infty} G(2\pi n/\varpi), \tag{5.94}$$

where

$$G(s) = \int_{-\infty}^{\infty} \exp(\mathrm{i}ys)F(y)\mathrm{d}y \tag{5.95}$$

is also an even function. $F(y)$ has simple poles at $y = \mathrm{i}m/2$ and $y = \mathrm{i}\pi(2m-1)/(2\mu)$ for $m = 1, 2, \ldots$ in the upper half of the complex y plane. The integral (5.95) can, therefore, be evaluated using the residue method to give

$$\begin{aligned} G(s) = {} & 2\sum_{m=1}^{\infty} \frac{\exp(-ms)[\cos(m\mu) - (-1)^m]\left[\cos\left(\frac{m\mu}{2}\right) - \cos\left(\frac{m\varphi}{2}\right)\right]}{m\cos\left(\frac{m\mu}{2}\right)} \\ & + 2\sum_{m=1}^{\infty} \frac{(-1)^m \exp\left[-\frac{(2m-1)\pi s}{2\mu}\right]\left\{\cos\left[\frac{(2m-1)\pi^2}{\mu}\right] + 1\right\}\cos\left[\frac{(2m-1)\pi\varphi}{2\mu}\right]}{\left(m - \frac{1}{2}\right)\sin\left[\frac{(2m-1)\pi^2}{\mu}\right]}. \end{aligned} \tag{5.96}$$

On the critical surface (5.90) it follows from (5.74) and (A.199) that $\ell = 0$. Then, from (5.91) and (A.190), $\mu = 2\zeta$ and $\varphi = 2\mathrm{U}$, giving, from (5.74) and (A.199),

$$w_1 : w_2 : w_3 : w_4 = \frac{\cos\left(\frac{1}{2}\varphi\right)}{\cos\left(\frac{1}{2}\mu\right)} : 1 : 1 : \frac{\sin\left(\frac{1}{2}\varphi\right)}{\sin\left(\frac{1}{2}\mu\right)}. \tag{5.97}$$

From (A.190) and (A.191), $\varpi \to 0$ as $\ell \to 0$ and $G(2\pi n/\varpi) \to 0$. On the critical surface, therefore, from (5.93)–(5.95),

$$\begin{aligned} \phi(T) &= [\phi(T)]_{\mathrm{c}} \\ &= -\ln(w_1 + w_2) - \tfrac{1}{2}G(0) \\ &= -\ln(w_1 + w_2) \\ &\quad - \int_{-\infty}^{\infty} \frac{\sinh^2[y(\pi - \mu)][\cosh(\mu y) - \cosh(\varphi y)]}{y\sinh(2\pi y)\cosh(\mu y)}\mathrm{d}y, \end{aligned} \tag{5.98}$$

with μ and φ taking the critical values μ_{c} and φ_{c}, obtained by solving (5.97) for some fixed values of e_1, e_3, e_5 and e_7. For such a set of fixed values, as the temperature is increased, the system describes a trajectory in parameter space which crosses the critical surface as some critical temperature T_c. From (5.71) and (5.73) $\Delta \geq 1$, with $\Delta = 1$ on the critical surface. It also follows, from (5.72) and (5.97), that on the critical surface

$$\Gamma_c = -\cos(\mu_c)\,. \tag{5.99}$$

At a temperature $T = T_c - \delta T$ we have, therefore,

$$\Delta \simeq 1 + \Delta_0|\delta T|\,, \qquad \Gamma \simeq -\cos(\mu_c) + \Gamma_0 \delta T\,, \tag{5.100}$$

where Δ_0 and Γ_0 are some functions of e_1, e_3, e_5 and e_7 with $\Delta_0 \geq 0$. From these equations and (5.76),

$$\ell^2 \simeq \frac{2\Delta_0|\delta T|}{\sin^2(\mu_c)} \tag{5.101}$$

and, from (5.91) and (A.191),

$$\exp\left(-\frac{\pi}{\varpi}\right) \simeq \exp[-2\mathcal{K}(\ell')] \simeq \left(\frac{\ell}{4}\right)^2 . \tag{5.102}$$

Combining these results with (5.94) and (5.96) we see that, at the temperature $T = T_c - \delta T$, near to the critical surface, $\phi(T)$ can, as in (2.8), be divided into a sum of smooth and singular parts with

$$\phi_{\text{smth}}(T) \simeq [\phi(T)]_c - 4\left[\frac{\Delta_0|\delta T|}{8\sin^2(\mu_c)}\right]^2 \cos\left(\frac{\mu_c}{2}\right)\left[\cos\left(\frac{\mu_c}{2}\right) + \cos\left(\frac{\varphi_c}{2}\right)\right], \tag{5.103}$$

$$\phi_{\text{sing}}(T) \simeq 4\left[\frac{\Delta_0|\delta T|}{8\sin^2(\mu_c)}\right]^{\pi/\mu_c} \cot\left(\frac{\pi^2}{2\mu_c}\right)\cos\left(\frac{\pi\varphi_c}{2\mu_c}\right). \tag{5.104}$$

Unless π/μ_c is an integer

$$\phi_{\text{sing}}(T) \sim |\delta T|^{\pi/\mu_c}. \tag{5.105}$$

Exceptional cases occur if $\mu_c = \pi/p$, where p is an integer. If p is an even integer the factor $\cot(\pi^2/2\mu_c)$ in (5.104) is infinite. This is due to the coincidence of two poles of $F(y)$. If the residue of the resulting double pole is calculated, (5.105) is replaced by

$$\phi_{\text{sing}}(T) \sim |\delta T|^{\pi/\mu_c} \ln|\delta T|\,. \tag{5.106}$$

If p is an odd integer $\cot(\pi^2/2\mu_c) = 0$ and it is necessary to consider the dependence of μ on T.

When π/μ_c is non-integer (5.105) implies that the heat capacity exponents α and α' defined in (2.117) are given by

$$\alpha = \alpha' = 2 - \frac{\pi}{\mu_c}\,. \tag{5.107}$$

The critical value μ_c of μ is finite and non-zero and, from (5.91), $\zeta_c = \mu_c/2$. It follows from (5.88) that, as $\ell \to 0$, $k_m \to 1$ and, using (A.191),

$$k'_m \simeq 4\left(\frac{\ell}{4}\right)^{\pi/2\mu_c} . \tag{5.108}$$

From (5.89), (5.101) and (5.108) the magnetic exponent β, defined in (2.118) (with $\rho = m_s$, $\rho_c = 0$) is given by

$$\beta = \frac{\pi}{16\mu_c} . \tag{5.109}$$

If the Widom and Essam–Fisher scaling laws, given by (2.169) and (2.170) respectively, are assumed then it follows from (5.108) and (5.109) that

$$\gamma = \frac{7\pi}{8\mu_c} , \qquad \delta = 15 . \tag{5.110}$$

To complete the picture of the critical properties of the eight-vertex model it should be noted that the correlation length $\mathfrak{r}$ can be shown to take the form

$$\mathfrak{r} = \frac{2}{\ln(1/k_m)} , \tag{5.111}$$

(Baxter 1982a). From (5.101) and (5.108) it then follows that the critical exponents ν and ν', defined in (2.197), are given by

$$\nu = \nu' = \frac{\pi}{2\mu_c} . \tag{5.112}$$

From (5.107) and (5.112) the Josephson hyper-scaling law (2.199), with $d = 2$ is satisfied. If the Fisher scaling law (2.203) is assumed then, from (5.110) and (5.112), the exponent η, defined in (2.200), is given by

$$\eta = \tfrac{1}{4} . \tag{5.113}$$

Since μ_c is given by (5.99) and (5.72) it varies continuously as a function of e_1, e_3, e_5 and e_7. We, therefore, have a situation where the critical exponents, except δ and η, vary as continuous functions of the energy parameters of the model. The implication of this for the concept of universality is discussed in Sect. 5.12.

5.10 The Spin Representation and the Ising Model Limit

We consider the spin representation of the eight-vertex model given by (5.7). Let $K_{21} = K + K'$ and $K_{22} = K - K'$. From (5.2), (5.4) and (5.7),

$$\begin{aligned} \mathfrak{a} &= \exp(K_0 + 2K + K_4) , & \mathfrak{b} &= \exp(K_0 - 2K + K_4) , \\ \mathfrak{c} &= \exp(K_0 + 2K' - K_4) , & \mathfrak{d} &= \exp(K_0 - 2K' - K_4) . \end{aligned} \tag{5.114}$$

When $K_4 = 0$ the model reduces to two identical non-interacting Ising models, each with coupling $K + K'$ along one axis direction and $K - K'$ along the other axis direction. With $K_4 \neq 0$ the two Ising models are linked by a four-spin coupling K_4. When $K > K'$ and $K_4 \geq 0$ the model is ferromagnetic and in region (i) of the eight-vertex model and the inequalities (5.46) are satisfied. For $T < T_c$, $w_1, \ldots, w_4$ are defined by (5.65) and for $T > T_c$ by (5.69). It follows from (5.71) and (5.72) that

$$\Delta = \begin{cases} \Theta(K, K', K_4)\,, & T < T_c\,, \\ 1/\Theta(K, K', K_4)\,, & T > T_c\,, \end{cases} \tag{5.115}$$

$$\Gamma = \begin{cases} \tanh(2K_4)\,, & T < T_c\,, \\ \tanh(2K_4)/\Theta(K, K', K_4)\,, & T > T_c\,, \end{cases} \tag{5.116}$$

where

$$\Theta(K, K', K_4) = \tanh(2K_4)\cosh^2(2K) + \sinh^2(2K) + \sinh^2(2K')[\tanh(2K_4) - 1]\,. \tag{5.117}$$

In the three-dimensional phase space of the independent couplings the critical surface $\mathcal{C}$ is given by $\Delta = 1$ which is equivalent to

$$\Theta(K, K', K_4) = 1\,. \tag{5.118}$$

On this surface the critical exponents vary as functions of μ_c, which from, (5.99) and (5.116), is given by

$$\mu_c = \pi - \arccos[\tanh(2K_4)]\,. \tag{5.119}$$

It has been shown by Kadanoff and Wegner (1971) that the four-spin interaction in the eight-vertex model scales as $1/r^2$. This means, in terms of our discussion in Sect. 2.6.2, that K_4 is a marginal coupling, and the variation of the critical exponents α, β and γ, given by (5.107), (5.109) and (5.110) respectively, as functions of K_4 does not conflict with scaling theory. We now consider three special cases of this analysis.

5.10.1 The Isotropic Ising Model With a Four-Spin Coupling

When $K' = 0$ the system is isotropic. The critical surface $\mathcal{C}$ cuts the plane $K' = 0$ in a curve $\mathcal{C}_0$ given by $\Theta(K, 0, K_4) = 1$. This curve is shown in Fig. 5.6. It meets the $K_4 = 0$ axis at $K = K_c$ given by

$$\sinh(2K_c) = 1, \tag{5.120}$$

which is the standard formula for the critical temperature of the square lattice Ising model (Volume 1, Sect. 8.3). We also note that, as $K \to 0$ on $\mathcal{C}_0$, $K_4 \to \infty$, which shows that there is no phase transition in a spin-$\frac{1}{2}$ model on a square lattice with a purely four-spin interaction.

5.10.2 The Jüngling Spin Representation

The Jüngling spin representation is given by (5.14)–(5.17) where $J_{21} = J_{22} = J$, and K' and K_4 are expressed in terms of the coupling L by

$$\tanh(4K') = \tanh^2(2L)\,, \qquad \tanh(4K_4) = \frac{\tanh^4(2L)}{2-\tanh^4(2L)}\,. \tag{5.121}$$

Substituting into (5.118) gives the formula

$$\tanh^4(2L) = 1 - [2\sinh^2(2K) - 1]^2\,, \tag{5.122}$$

for the critical surface. The critical exponents are functions of L through (5.119) and the second of equations (5.121).

5.10.3 The Isotropic Ising Model Without a Four-Spin Coupling

Now $K' = K_4 = 0$ and, from (5.114), (5.65), (5.69) and (5.74), $\mathsf{U} = 0$ for all temperatures. From (5.116), $\Gamma = 0$ and thus, from (5.75) and (A.198),

$$\zeta = \tfrac{1}{2}\mathcal{K}(\ell)\,, \tag{5.123}$$

or, from (5.91),

$$\mu = \tfrac{1}{2}\pi\,. \tag{5.124}$$

From (5.76), (5.115) and (5.117),

$$\ell' = \begin{cases} \sinh^2(2K)\,, & T < T_c\,, \\ \sinh^{-2}(2K)\,, & T > T_c\,. \end{cases} \tag{5.125}$$

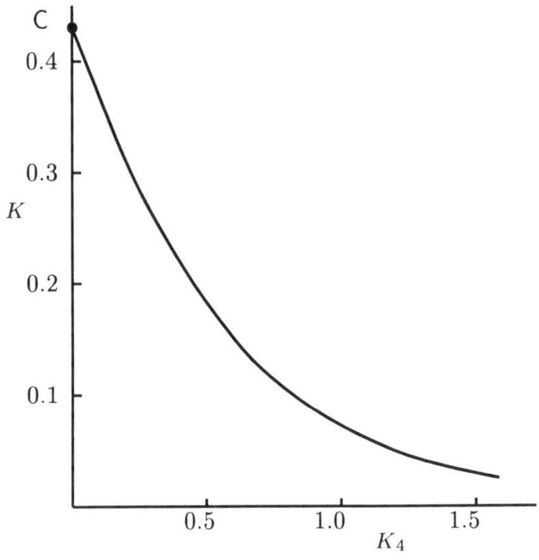

Fig. 5.6. The critical curve $\mathcal{C}_0$ for the eight-vertex model, when $J_{21} = J_{22} = J$, in the space of the couplings $K = J/T$ and $K_4 = J_4/T$. The critical point C of the isotropic Ising model lies on the $K_4 = 0$ axis at $K = 0.4407$.

At the critical temperature $\ell = \ell_c = 0$, which agrees with the result of Sect. 5.9, and yields the formula (5.120) for the critical temperature. From (5.124), (5.107), (5.109), (5.110), (5.112) and (5.113)

$$\begin{aligned} &\alpha = 0\,, \qquad &&\beta = \tfrac{1}{8}\,, \qquad &&\gamma = \tfrac{7}{4}\,, \\ &\delta = 15\,, \qquad &&\nu = 1\,, \qquad &&\eta = \tfrac{1}{4}\,, \end{aligned} \tag{5.126}$$

which are the well-known results for the two-dimensional Ising model. (Volume 1, Sect. 8.9). The fact that $\pi/\mu_c = 2$ shows, from the discussion of Sect. 5.9, that the heat capacity has a logarithmic singularity at the critical point. From (5.88) and (5.123),

$$\frac{\mathcal{K}(k'_m)}{\mathcal{K}(k_m)} = \frac{\mathcal{K}(\ell)}{\mathcal{K}(\ell')}\,, \tag{5.127}$$

from which it follows that $k_m = \ell'$. Then, from (5.89) and (5.125), the spontaneous magnetization of the $d = 2$ Ising model

$$m_s = [1 - \sinh^{-4}(2K)]^{\frac{1}{8}}\,, \qquad T < T_c \tag{5.128}$$

is recovered.

The series expansion (5.85) will now be used to obtain an expression for the internal energy of the Ising model following the procedure of Lavis (1996). Since $\mathsf{U} = 0$ and ζ is given by (5.123), equations (5.86) reduce to

$$z = 1\,, \qquad x = q^{\frac{1}{4}} = \exp[-\pi\mathcal{K}(\ell)/2\mathcal{K}(\ell')]\,. \tag{5.129}$$

Using the transformation

$$k_1 = \frac{2\sqrt{\ell'}}{1+\ell'}\,, \qquad k'_1 = \frac{1-\ell'}{1+\ell'}\,, \tag{5.130}$$

it follows, from (A.213) and (A.214) (with $m = \ell$, $m_1 = k'_1$), that

$$\mathcal{K}(k_1) = (1+\ell')\mathcal{K}(\ell')\,, \qquad \mathcal{K}(k'_1) = \tfrac{1}{2}(1+\ell')\mathcal{K}(\ell)\,. \tag{5.131}$$

From (5.129) and (A.205),

$$x = \exp[-\pi\mathcal{K}(k'_1)/\mathcal{K}(k_1)] \tag{5.132}$$

is the nome of k_1. From (5.70), (5.85), (5.114), (5.129) and (5.132),

$$\phi(K) = \begin{cases} -2K - \displaystyle\sum_{n=1}^{\infty} h_n(x)\,, & T < T_c\,, \\[2ex] -\ln[\cosh(2K)+1] - \displaystyle\sum_{n=1}^{\infty} h_n(x)\,, & T > T_c\,, \end{cases} \tag{5.133}$$

where

$$h_n(x) = \frac{x^{2n}(1-x^{2n})^2(1-x^n)^2}{n(1-x^{8n})(1+x^{2n})}\,. \tag{5.134}$$

Since

$$\frac{u}{J} = \frac{\partial \phi}{\partial K}, \tag{5.135}$$

it follows, from (5.125), (5.133) and (A.206), that

$$\frac{u}{J} = \begin{cases} -2 + \dfrac{B(K)}{\mathcal{K}^2(k_1)} \displaystyle\sum_{n=1}^{\infty} x \frac{\mathrm{d}h_n(x)}{\mathrm{d}x}, & T < T_\mathrm{c}, \\ -\dfrac{2\sinh(2K)}{1+\cosh(2K)} + \dfrac{B(K)}{\mathcal{K}^2(k_1)} \displaystyle\sum_{n=1}^{\infty} x \frac{\mathrm{d}h_n(x)}{\mathrm{d}x}, & T > T_\mathrm{c}, \end{cases} \tag{5.136}$$

where

$$B(K) = \frac{\pi^2 \cosh^3(2K)}{\sinh(2K)[\sinh^2(2K) - 1]}. \tag{5.137}$$

From (5.134)

$$x\frac{\mathrm{d}h_n(x)}{\mathrm{d}x} = \frac{x^n}{1+x^{2n}} + \frac{2x^{2n}}{(1+x^{2n})^2} - \frac{8x^{3n}}{(1+x^{2n})^3} - \frac{x^n(1+x^{2n})}{1+x^{4n}} - \frac{4x^{4n}}{(1+x^{4n})^2} + \frac{4x^{3n}(1+x^{2n})}{(1+x^{2n})^3}, \tag{5.138}$$

and hence, from equations (A.210)–(A.212), (A.215)–(A.217),

$$\sum_{n=1}^{\infty} x\frac{\mathrm{d}h_n(x)}{\mathrm{d}x} = \frac{k_1'\mathcal{K}^2(k_1)}{\pi^2}\left[\sqrt{2(1+k_1')} - 1\right] - \frac{2k_1'^2\mathcal{K}^3(k_1)}{\pi^3}. \tag{5.139}$$

Since, from (5.125) and (5.130),

$$k_1 = 2\sinh(2K)\mathrm{sech}^2(2K), \tag{5.140}$$

$$k_1' = \pm\left[\frac{1-\sinh^2(2K)}{\cosh^2(2K)}\right], \qquad T \lessgtr T_\mathrm{c}, \tag{5.141}$$

by substituting from (5.140) and (5.141) into (5.139) and then into (5.136) we obtain

$$u = -J\coth(2K)\left\{1 + \frac{2}{\pi}\mathcal{K}(k_1)[2\tanh^2(2K) - 1]\right\}, \tag{5.142}$$

which is the formula for the internal energy of the Ising model on the square lattice at all temperatures (Volume 1, Sect. 8.10).

5.11 The Six-Vertex Model as a Special Case

The six-vertex model corresponds to the limiting case $e_7 \to \infty$ ($\mathfrak{d} \to 0$) of the eight-vertex model. It therefore falls on the boundaries of regions (i), (ii) or (iii), but, not of (iv). At low temperatures in these regions the parameters $w_1, \ldots, w_4$ are given by (5.65)–(5.67) respectively. At high temperatures, in the part of parameter space corresponding to the six-vertex model, $w_1, \ldots, w_4$ are given by (5.70) for all regions. Let

$$\Delta^* = \frac{\mathfrak{a}^2 + \mathfrak{b}^2 - \mathfrak{c}^2 - \mathfrak{d}^2}{2(\mathfrak{a}\mathfrak{b} + \mathfrak{c}\mathfrak{d})}, \qquad \Gamma^* = \frac{\mathfrak{a}\mathfrak{b} - \mathfrak{c}\mathfrak{d}}{\mathfrak{a}\mathfrak{b} + \mathfrak{c}\mathfrak{d}}. \tag{5.143}$$

From (5.65)–(5.67), (5.70), (5.71) and (5.72)

$$\Delta = \begin{cases} \Delta^*, & \text{low-}T\text{, regions (i) and (ii),} \\ -\Delta^*, & \text{low-}T\text{, region (iii),} \\ 1/\Gamma^*, & \text{high-}T, \end{cases} \tag{5.144}$$

$$\Gamma = \begin{cases} \Gamma^*, & \text{low-}T\text{, regions (i) and (ii),} \\ -\Gamma^*, & \text{low-}T\text{, region (iii),} \\ \Delta^*/\Gamma^*, & \text{high-}T. \end{cases} \tag{5.145}$$

The variables Δ and Γ satisfy the conditions (5.73) and since, for the six-vertex model, when $\mathfrak{d} = 0$, $\Gamma^* = 1$, we have three cases to consider:[8]

5.11.1 Low-Temperature Regions (i) and (ii)

In these regions

$$\Gamma = 1, \qquad \Delta^* \geq 1, \qquad (\Delta = \Delta^*), \tag{5.146}$$

and, from (5.77) and (5.146), $\mathcal{K}(\ell) \leq 2\zeta \leq 2\mathcal{K}(\ell)$. Defining

$$\lambda = 2\mathcal{K}(\ell) - 2\zeta, \qquad \text{with} \qquad 0 \leq \lambda \leq \mathcal{K}(\ell), \tag{5.147}$$

it follows, from (5.75), (A.198), (A.200) and (A.201), that

$$\Delta = \frac{1}{\mathrm{dn}(\lambda|\ell)}, \qquad \Gamma = \frac{\mathrm{cn}(\lambda|\ell)}{\mathrm{dn}(\lambda|\ell)}. \tag{5.148}$$

The result $\Gamma = 1$ is achieved in the limit $\ell \to 1$, when, from (A.199), (5.146) and (5.148),

$$\Delta^* = \cosh(\lambda). \tag{5.149}$$

[8] The Δ defined in Volume 1, Chap. 10, is the present Δ^* with $\mathfrak{d} = 0$.

From (5.65) and (5.66), for small $\mathfrak{d}$, w_4 is positive in region (i) and negative in region (ii). So, from (5.78), $\mathfrak{U}$ is positive in region (i) and negative in region (ii) with $|\mathfrak{U}| \le \zeta$. Let

$$\vartheta = 2\mathcal{K}(\ell) - 2|\mathfrak{U}|, \qquad \text{with} \qquad 0 \le \lambda \le \vartheta \le \tfrac{1}{2}\mathcal{K}(\ell)\,. \tag{5.150}$$

In the limit $\ell \to 1$, from (5.65), (5.66), (5.74) and (A.199),

$$\mathfrak{a}+\mathfrak{b} : |\mathfrak{a}-\mathfrak{b}| : \mathfrak{c}+\mathfrak{d} : \mathfrak{c}-\mathfrak{d} = \frac{\sinh\left(\frac{1}{2}\vartheta\right)}{\sinh\left(\frac{1}{2}\lambda\right)} : \frac{\cosh\left(\frac{1}{2}\vartheta\right)}{\cosh\left(\frac{1}{2}\lambda\right)} : 1 : 1, \tag{5.151}$$

giving $\mathfrak{d} = 0$ and

$$\max(\mathfrak{a},\mathfrak{b}) = \frac{\mathfrak{c}\sinh\left[\frac{1}{2}(\vartheta+\lambda)\right]}{\sinh\left(\frac{1}{2}\lambda\right)}, \qquad \min(\mathfrak{a},\mathfrak{b}) = \frac{\mathfrak{c}\sinh\left[\frac{1}{2}(\vartheta-\lambda)\right]}{\sinh\left(\frac{1}{2}\lambda\right)}. \tag{5.152}$$

From (5.86), (5.147) and (5.150) in the limit $\ell \to 1$,

$$\begin{aligned} x &= \exp\left\{\pi\left[\mathcal{K}(\ell) - \tfrac{1}{2}\lambda\right]/\mathcal{K}(\ell')\right\} \to 0\,,\\ q &= \exp\{-2\pi\mathcal{K}(\ell)/\mathcal{K}(\ell')\} \to x^2 \to 0\,,\\ z &= \begin{cases} \exp\left\{-\pi\left[\mathcal{K}(\ell) - \tfrac{1}{2}\vartheta\right]/\mathcal{K}(\ell')\right\} \to x^2 \to 0\,, & \text{region (i)}\,,\\ \exp\left\{\pi\left[\mathcal{K}(\ell) - \tfrac{1}{2}\vartheta\right]/\mathcal{K}(\ell')\right\} \to x^{-2}\,, \to \infty & \text{region (ii)}\,. \end{cases} \end{aligned} \tag{5.153}$$

From (5.85) we have, therefore,

$$\begin{aligned} \phi(\mathfrak{a},\mathfrak{b},\mathfrak{c}) &= -\ln(w_1 + w_2)\\ &= -\ln\left[\tfrac{1}{2}(\mathfrak{a}+\mathfrak{b}) + \tfrac{1}{2}|\mathfrak{a}-\mathfrak{b}|\right], \end{aligned} \tag{5.154}$$

which is equivalent to Volume 1, (10.49), in region (i) ($\mathfrak{a} > \mathfrak{b}$).

From (5.99) and (5.146) $\mu_c = \pi$ in this case and, from (5.107), $\alpha = \alpha' = 1$. It thus follows that the scaling exponent y_2 of (2.167) is equal to $d = 2$, indicative of a first-order transition. This is of course as it should be since the transition is to the frozen completely ordered ferroelectric state.

5.11.2 Low-Temperature Region (iii)

In this region

$$\Gamma = -1\,, \qquad \Delta^* \le -1\,, \qquad (\Delta = \Delta^*) \tag{5.155}$$

and from (5.77) and (5.155), $0 \le 2\zeta \le \mathcal{K}(\ell)$. In this case we set

$$\lambda = 2\zeta\,, \qquad \vartheta = 2\mathfrak{U}\,, \qquad \text{with} \qquad 0 \le \vartheta \le \lambda \le \mathcal{K}(\ell)\,. \tag{5.156}$$

From (5.75) the result $\Gamma = -1$ is achieved in the limit $\ell \to 1$, when, from (A.199),

$$\Delta^* = -\cosh(\lambda)\,. \tag{5.157}$$

From (5.67) and (5.74), in the limit $\ell \to 1$,

$$\frac{\mathfrak{a}}{\mathfrak{c}} = \frac{\sinh\left[\frac{1}{2}(\lambda + \vartheta)\right]}{\sinh(\lambda)}\,, \qquad \frac{\mathfrak{b}}{\mathfrak{c}} = \frac{\sinh\left[\frac{1}{2}(\lambda - \vartheta)\right]}{\sinh(\lambda)} \tag{5.158}$$

and in the same limit, from (5.86),

$$x = \exp(-\lambda)\,, \qquad q = 0\,, \qquad z = \exp(\vartheta)\,. \tag{5.159}$$

Substituting into (5.85) gives

$$\phi(\mathfrak{a},\mathfrak{b},\mathfrak{c}) = -\ln(\mathfrak{c}) - \sum_{n=1}^{\infty} \frac{\exp(-2n\lambda)[\cosh(n\lambda) - \cosh(n\vartheta)]}{n\cosh(n\lambda)}\,. \tag{5.160}$$

Substituting for $\mathfrak{c}$ from the first of equations (5.158) and expanding the resulting logarithmic series gives the formula obtained from (10.120) and (10.123) of Volume 1.

5.11.3 High-Temperature Regions (i), (ii) and (iii)

In these regions

$$\Delta = 1\,, \qquad -1 \le \Delta^* \le 1\,, \qquad (\Gamma = \Delta^*) \tag{5.161}$$

and in this case we use the variables μ and φ defined in (5.91). The parameters $w_1, \ldots, w_4$ are given by (5.70) and the limit $\mathfrak{d} \to 0$ is equivalent to $\ell \to 0$. This then gives, from (5.74), (5.91) and (5.143),

$$\Delta^* = -\cos(\mu)\,, \tag{5.162}$$

$$\frac{\mathfrak{a}}{\mathfrak{c}} = \frac{\sin\left[\frac{1}{2}(\mu + \varphi)\right]}{\sin(\mu)}\,, \qquad \frac{\mathfrak{b}}{\mathfrak{c}} = \frac{\sin\left[\frac{1}{2}(\mu - \varphi)\right]}{\sin(\mu)}\,. \tag{5.163}$$

Equations (5.70) with $\mathfrak{d} = 0$ give $w_2 = w_3$, which is the expression (5.90) for the critical surface. So the form for the free energy is that given by (5.98) for the critical surface of the eight-vertex model. With $\mathfrak{d} = 0$ this becomes

$$\phi(\mathfrak{a},\mathfrak{b},\mathfrak{c}) = -\ln\left[\tfrac{1}{2}(\mathfrak{a}+\mathfrak{b}+\mathfrak{c})\right]$$
$$- \int_{-\infty}^{\infty} \frac{\sinh^2[y(\pi - \mu)][\cosh(\mu y) - \cosh(\varphi y)]}{y\sinh(2\pi y)\cosh(\mu y)}\,dy\,. \tag{5.164}$$

The fact that points on the critical surface are related by the transformations $\mathcal{T}'\mathcal{T}_{\mathfrak{b}}$, $\mathcal{T}'\mathcal{T}_{\mathfrak{c}}$ and $\mathcal{T}'\mathcal{T}_{\mathfrak{d}}$ to points on the surfaces $\mathfrak{b} = 0$, $\mathfrak{c} = 0$ and $\mathfrak{d} = 0$

respectively has been noted in Sect. 5.6. Equation (5.164) is the form for the free energy given by Baxter (1972). To show the equivalence between it and the forms given by Lieb and Wu (1972) and Volume 1, equations (10.76) and (10.80), it is necessary to establish that

$$\ln\left[\tfrac{1}{2}(\mathfrak{a}+\mathfrak{b}+\mathfrak{c})\right] = \ln(\mathfrak{a}) - \int_{-\infty}^{\infty} \frac{\sinh[y(\pi-\mu)]\{\sinh[y(\pi-\mu)]-\sinh[y(\pi-\varphi)]\}}{y\sinh(2\pi y)}\mathrm{d}y\,. \tag{5.165}$$

This is achieved by substituting on the left from (5.163), expanding the logarithmic series and evaluating the integral as a series using the residue theorem.

5.12 The Eight-Vertex Model and Universality

The concept of universality classes was discussed in Sect. 2.3, where we listed Kadanoff's three criteria for the division of phases transitions into classes. It was also pointed out that any particular model may well exhibit phase transitions in different universality classes within different parts of its phase space. We saw in Sect. 3.4 that the symmetry group of the zero-field spin-1 Ising model is in general $\mathcal{S}_2$, the same as that of the standard zero-field spin-$\frac{1}{2}$ Ising model and thus, for any particular value of the dimension d, it will exhibit a phase transition in the universality class of the latter. However, when the interaction energy parameters attain certain special values there is a change of symmetry group to $\mathcal{S}_3$, that of the 3-state Potts model (Sect. 3.5), for which the transition lies in a different universality class. Examples can also be given when a change in the energy parameters will lead to a change in universality class by means of an effective change in the dimension of the system rather than the symmetry group. We could, for example, consider an anisotropic lattice model on a d-dimensional hypercubic lattice. If the energy parameters connecting microsystems in the direction of one axis tend to zero there will, from the point of view of cooperative phenomena, be an effective change of dimension of the system from d to $d-1$, with a resulting change of universality class for the phase transition.

At this point it is useful to distinguish between the *universality hypothesis*, in whatever form it is represented, and the *scaling hypothesis*, introduced in Sect. 2.4. It is a consequence of the latter and the corresponding hypothesis for correlations of Sect. 2.11 that the critical exponents satisfy the four scaling laws (2.169), (2.170), (2.199) and (2.203). It is sometimes asserted that the eight-vertex model contradicts the universality hypothesis because its critical exponents vary as a function of a parameter. We have shown in Sect. 2.6.2 that the scaling hypothesis encompasses the possibility of exponents varying as functions of a marginal coupling and in Sect. 5.10 it was shown that the four-spin coupling in the spin representation of the eight-vertex model is of this type. The interesting question in relation to universality is whether the variation of the four-spin coupling corresponds to a change of symmetry and thus a change of universal class according to Kadanoff's criterion (b), or whether the symmetry group remains the same. As was pointed out by Kadanoff and Wegner (1971) and Baxter (1982a) there is certainly a change of symmetry when the conditions $e_1 = e_2$ and $e_3 = e_4$ are applied. These conditions, are equivalent to setting $L_1 = -L_3$, $L_2 = L_4$ or $L_1 = L_3$, $L_2 = -L_4$ in equations (5.13), or $J_h = J_v = 0$ in Example 5.1. Without them the symmetry group is simply the spin-reversal symmetry group $\mathcal{S}_2$ of the Ising model. With them the group becomes the direct product $\mathcal{S}_2 \times \mathcal{S}_2$, where the reversal symmetry applies to the two interpenetrating sublattices independently. Although the general eight-vertex model, without the conditions $e_1 = e_2$ and $e_3 = e_4$, has not been solved, the solution of the special free-fermion case by Fan and Wu (1970) give a strong indication that the critical exponents are those of the Ising model. According to Kadanoff's criteria (a) and (b) for universality classes this is to be expected, since the dimensions and symmetry groups of the models are the same. By the same token we expect the exponents of the eight-vertex model with conditions (5.3) to be different. That they vary is understandable in terms of scaling. To set them in the context of Kadanoff's criteria of universality classes two approaches are possible:

(i) The model with varying exponents can be taken as one universality class.

(ii) Criterion (c) can be invoked with the marginal coupling taken as another criterion differentiating between classes.

The latter has some advantages since the concept of constant exponents *within* classes is preserved. Also since, in renormalization group theory (see Chap. 6) a marginal exponent is associated with a line of fixed points and a fixed point is usually associated with a universality class, to associate each value of K_4 with a different class has some value.

Finally, the proposal of *weak universality* by Suzuki (1974) should be mentioned. He observed that the exponents in the eight-vertex model which vary are α, β, γ and ν (see equations (5.107), (5.109), (5.110) and (5.112) respectively). These are precisely those which represent asymptotic behaviour in terms of $|T - T_c|$ (see equations (2.117)–(2.119) and (2.197)). If, however, asymptotic behaviour is measured in terms of the inverse correlation length $\mathfrak{r}^{-1}$ then this leads to the use of the exponents

$$\hat{\varphi} = (2 - \alpha)/\nu , \qquad \hat{\beta} = \beta/\nu , \qquad \hat{\gamma} = \gamma/\nu . \tag{5.166}$$

The four scaling laws (2.169), (2.170), (2.199) and (2.203) are replaced by

$$\begin{aligned} \hat{\gamma} &= \hat{\beta}(\delta - 1) , & 2\hat{\beta} + \hat{\gamma} &= \hat{\varphi} , \\ d &= \hat{\varphi} , & 2 - \eta &= \hat{\gamma} . \end{aligned} \tag{5.167}$$

For the eight-vertex model the complete set of exponents is now

$$\hat{\varphi} = 2\,, \qquad \hat{\beta} = \tfrac{1}{8}\,, \qquad \hat{\gamma} = \tfrac{7}{4}\,, \qquad \delta = 15\,, \qquad \eta = \tfrac{1}{4}\,. \tag{5.168}$$

These are, of course, the same for the two-dimensional Ising model which unfortunately obscures the important symmetry difference between the two models.

Examples

5.1 If the first two relations of (5.3) are dropped, a term

$$-\tfrac{1}{2}J_{\mathrm{h}}(\sigma_{\mathrm{i}_1}\sigma_{\mathrm{i}_2} + \sigma_{\mathrm{i}_3}\sigma_{\mathrm{i}_4}) - \tfrac{1}{2}J_{\mathrm{v}}(\sigma_{\mathrm{i}_1}\sigma_{\mathrm{i}_4} + \sigma_{\mathrm{i}_2}\sigma_{\mathrm{i}_3})$$

can be added to each term over faces in (5.6), J_{h} and J_{v} being nearest-neighbour horizontal and vertical energy interaction parameters. The factor $\frac{1}{2}$ arises because each edge is on the boundary of two faces.) Using Fig. 5.2 show that

$$e_1 = -J_0 - J_{\mathrm{h}} - J_{\mathrm{v}} - J_{21} - J_{22} - J_4\,,$$

$$e_2 = -J_0 + J_{\mathrm{h}} + J_{\mathrm{v}} - J_{21} - J_{22} - J_4\,,$$

$$e_3 = -J_0 + J_{\mathrm{h}} - J_{\mathrm{v}} + J_{21} + J_{22} - J_4\,,$$

$$e_4 = -J_0 - J_{\mathrm{h}} + J_{\mathrm{v}} + J_{21} + J_{22} - J_4\,,$$

$$e_5 = e_6 = -J_0 - J_{21} + J_{22} + J_4\,,$$

$$e_7 = e_8 = -J_0 + J_{21} - J_{22} + J_4\,.$$

Invert these relations and verify that $e_1 = e_2$ and $e_3 = e_4$ imply that $J_{\mathrm{h}} = J_{\mathrm{v}} = 0$.

5.2 For the model of question (1) show that, when the free-fermion conditions

$$e_1 + e_2 = 2e_5\,, \qquad e_3 + e_4 = 2e_7$$

apply then $J_{22} = J_4 = 0$. Evaluate J_0, J_{h}, J_{v} and J_{21} in terms of the e_{p}. Illustrate by a sketch that, when $e_5 \neq e_7$ and $e_3 - e_4 \neq \pm(e_1 - e_2)$, the model is equivalent to the triangular lattice Ising model with unequal interactions in three lattice directions. Show that

$$e_1 + e_2 = 2e_7\,, \qquad e_3 + e_4 = 2e_5$$

are also free-fermion conditions producing equivalence to a triangular lattice Ising model.

5.3 Show that spin representation of Jüngling and Obermair (1974), which corresponds to equations (5.13) with $L_1 = L_2 = L_3 = L_4 = L$ is a model of the type given in question (1) with

$$\tilde{J}_0 = J_0 + \tfrac{1}{8}T \ln\left[256\cosh^4(2L)\cosh(4L)\right] ,$$

$$\tilde{J}_\mathrm{h} = \tilde{J}_\mathrm{v} = \tfrac{1}{4}T \ln\left[\cosh(4L)\right] ,$$

$$\tilde{J}_{21} = J_{21} ,$$

$$\tilde{J}_{22} = J_{22} ,$$

$$\tilde{J}_4 = -\tfrac{1}{8}T \ln\left[1 - \tanh^4(2L)\right] ,$$

where, as in (5.14)–(5.17), a tilde is used to distinguish the new set of values.

5.4 Take $\mathfrak{a}$, $\mathfrak{b}$, $\mathfrak{c}$ and $\mathfrak{d}$ as given by (5.2), (5.4) and (5.7) with $J_4 = K_4 = 0$ and apply the transformation (5.35) to show that

$$\mathfrak{a}' = B(K_0, K_{21}, K_{22}) \exp(K'_{21} + K'_{22}) ,$$

$$\mathfrak{b}' = B(K_0, K_{21}, K_{22}) \exp(-K'_{21} - K'_{22}) ,$$

$$\mathfrak{c}' = B(K_0, K_{21}, K_{22}) \exp(K'_{21} - K'_{22}) ,$$

$$\mathfrak{d}' = B(K_0, K_{21}, K_{22}) \exp(-K'_{21} + K'_{22}) ,$$

where

$$B(K_0, K_{21}, K_{22}) = \exp(K_0)\sqrt{\sinh(2K_{21})\sinh(2K_{22})} ,$$

$$\sinh(2K_{21})\sinh(2K'_{22}) = 1 ,$$

$$\sinh(2K'_{21})\sinh(2K_{22}) = 1 .$$

Show that $\mathfrak{a} = \mathfrak{a}'$, $\mathfrak{b} = \mathfrak{b}'$, $\mathfrak{c} = \mathfrak{c}'$ and $\mathfrak{d} = \mathfrak{d}'$ when

$$\sinh(2K_{21})\sinh(2K_{22}) = 1$$

and that this last equation can also be obtained by substituting the expressions for $\mathfrak{a}$, $\mathfrak{b}$, $\mathfrak{c}$ and $\mathfrak{d}$ in terms of K_{21} and K_{22} into the formula $\mathfrak{a} = \mathfrak{b} + \mathfrak{c} + \mathfrak{d}$. Obtain the dual transformation formulae

$$\sinh(2K)\sinh(2K') = 1 ,$$

$$\frac{Z(K)}{[\sinh(2K)]^{\frac{1}{2}N}} = \frac{Z(K')}{[\sinh(2K')]^{\frac{1}{2}N}}$$

for the case $K_{21} = K_{22} = K$.

5.5 Express the transformations $\mathcal{T}$, $\mathcal{T}'$, $\mathcal{T}_1$, $\mathcal{T}_2$ and $\mathcal{T}$ as permutation matrices operating on the vector $(\mathfrak{a}, \mathfrak{b}, \mathfrak{c}, \mathfrak{d})$. Show that $\{\mathcal{I}, \mathcal{T}_1, \mathcal{T}_2, \mathcal{T}_2\}$ form a commutative group. Obtain the matrix $\mathcal{T}\mathcal{T}_1$ and show that it has the effect $(\mathfrak{a}, \mathfrak{b}, \mathfrak{c}, \mathfrak{d}) \to (\mathfrak{a}, \mathfrak{b}, \mathfrak{d}, \mathfrak{c})$. Obtain the matrices $\mathcal{T}_i\mathcal{T}'\mathcal{T}_i$, $i = 1, 2, 3$ and show that they yield the transformations (5.39)–(5.41).

5.6 The modified F model (Wu 1969) is a zero-field eight-vertex model with $e_1 = e_3 = \varepsilon > 0$, $e_5 = 0$ and $e_7 = 2\varepsilon$. Show that this is equivalent to an Ising model ferromagnet with $J = \frac{1}{2}\varepsilon$ and that the critical temperature in $T_c = \varepsilon/[\ln(\sqrt{2}+1)]$.

5.7 Show that just inside the low-temperature zone of region (iii), when $T = T_c - \delta T$, $\delta T > 0$ and T_c satisfies (5.48),

$$\mathfrak{c} - (\mathfrak{a} + \mathfrak{b} + \mathfrak{d}) \sim \delta T\,, \qquad \text{as} \qquad \delta T \to 0\,.$$

6. Real-Space Renormalization Group Theory

6.1 Introduction

The approach to the renormalization group in this chapter is usually referred to as the use of the *real-space renormalization group*. This is in contrast to renormalization group methods for continuous spin distributions, using, for example, the Landau–Ginzburg local free energy density of Sect. 3.3. These latter methods, in which renormalization is performed in wave-vector space, were initiated by Wilson (1975) and owe much to the influence of quantum field theory. Real-space methods, which are conceptually rather simpler, are based, in many cases, on the block-spin method of Kadanoff (1966) described in Sect. 2.3.[1]

The essential feature of all renormalization group methods is the realization that critical phenomena are associated with long wave length (small wave vector) fluctuations in wave-vector space or long-range correlations in real space. This means that an integration out of large wave vectors or a block averaging of short-range interactions in real space will leave the essential critical features of the system unchanged. It was this perception which lead to a remarkable development of critical phenomena in the last two decades.

As in Sect. 2.3 we consider a d-dimensional lattice $\mathcal{N}$, although in this case we do not assume that the lattice type is necessarily hypercubic. With periodic boundary conditions applied, a lattice site is given, in terms of a set of independent unit vectors $\hat{\boldsymbol{r}}_1, \hat{\boldsymbol{r}}_2, \ldots, \hat{\boldsymbol{r}}_d$, by

$$\boldsymbol{r} = a_1 m_1 \hat{\boldsymbol{r}}_1 + a_2 m_2 \hat{\boldsymbol{r}}_2 + \ldots + a_d m_d \hat{\boldsymbol{r}}_d \,, \tag{6.1}$$

where m_ℓ is an integer modulo N_ℓ and

$$N = \prod_{\ell=1}^{d} N_\ell \tag{6.2}$$

---

[1] A comprehensive collection of articles on both types of renormalization group methods is contained in the volume edited by Domb and Green (1976) and on real-space methods in the volume edited by Burkhardt and van Leeuwen (1982). For a description of renormalization methods in wave-vector space the reader is referred to Ma (1976a) and Amit (1978). Accounts of both approaches are given by Binney et al. (1993) and Cardy (1996).

is the total number of sites in the lattice. The constant a_ℓ is the lattice spacing in the direction of the unit vector $\hat{\boldsymbol{r}}_\ell$. At each lattice site $\boldsymbol{r}$ there is a microsystem with state specified by the variable $\sigma(\boldsymbol{r})$. For ease of presentation we shall suppose that the state variable at each lattice site is the same and can take ν discrete values, although this can be modified in particular applications. As in Sect. 2.4 we assume that the system has n independent (non-trivial) couplings. Here we need not distinguish between internal and external couplings and they will be denoted by $\zeta_1, \zeta_2, \ldots, \zeta_n$, with the so-called 'trivial coupling' denoted by ζ_0. The dimensionless Hamiltonian $\widehat{H} = \widehat{H}/T$ is given by

$$\widehat{H}(N, \zeta_i; \sigma(\boldsymbol{r})) = -N\zeta_0 - \sum_{i=1}^{n} \widehat{Q}_i(\sigma(\boldsymbol{r}))\zeta_i \,, \tag{6.3}$$

where each $\widehat{Q}_i$ is some function of the microstates $\sigma(\boldsymbol{r})$. The extensive variable corresponding to the trivial coupling is N. Although this coupling can be neglected in the usual formulation of statistical mechanical probabilities in equation (1.12), it plays an important role in renormalization group methods. The partition function $Z(N, \zeta_i)$ is now given (see equation (1.13)) by

$$Z(N, \zeta_i) = \sum_{\{\sigma(\boldsymbol{r})\}} \exp[-\widehat{H}(N, \zeta_i; \sigma(\boldsymbol{r}))] \,. \tag{6.4}$$

The trivial coupling ζ_0 effectively sets the zero-point of the Hamiltonian and consequently of the dimensionless free energy $\Phi = F/T$. We define Φ for the Hamiltonian with ζ_0 set equal to zero. This then gives

$$\Phi(N, \zeta_i) = -\ln[\exp(-N\zeta_0) Z(N, \zeta_i)] \,, \tag{6.5}$$

or equivalently,

$$\Phi(N, \zeta_i) = N\zeta_0 - \ln[Z(N, \zeta_i)] \,. \tag{6.6}$$

An important point about renormalization group theory is that it is an implementation of the scaling procedure described in Chap. 2. However, while scaling theory is able to predict only relations between critical exponents, renormalization group methods are able to give values for these exponents. This procedure can be carried out in a variety of ways. Even within the context of real-space methods it would be very difficult to present a formulation which applies to all possible schemes for all possible systems. We have not attempted this. Instead a block-spin procedure is described, which includes most simple applications of the theory. Although it will need modification in some cases, these changes will not radically affect the development which is presented.

6.2 The Basic Elements of the Renormalization Group

The steps in the development of a real-space renormalization group procedure are as follows:

(i) The lattice $\mathcal{N}$ is divided into blocks of the form of d-dimensional parallelepipeds whose edges are given by the vectors $\lambda a_\ell \hat{\boldsymbol{r}}_\ell$ for some integer $\lambda > 1$ and such that in each block there are λ^d sites. A lattice $\widetilde{\mathcal{N}}$ is formed of the same structure as $\mathcal{N}$ by associating a lattice site $\tilde{\boldsymbol{r}} \in \widetilde{\mathcal{N}}$ with every block of $\mathcal{N}$. We denote the block of sites in $\mathcal{N}$ associated with $\tilde{\boldsymbol{r}}$ by $\mathcal{B}(\tilde{\boldsymbol{r}})$. The number $\widetilde{N}$ of sites in $\widetilde{\mathcal{N}}$ is given by

$$\widetilde{N} = \lambda^{-d} N \,, \tag{6.7}$$

and the lattice spacings are

$$\tilde{a}_\ell = \lambda a_\ell \,, \qquad \ell = 1, 2, \ldots, d \,. \tag{6.8}$$

(ii) The size of the lattice $\widetilde{\mathcal{N}}$ is reduced by a length scaling

$$|\tilde{\boldsymbol{r}}| = \lambda^{-1} |\boldsymbol{r}| \,. \tag{6.9}$$

This means that $\widetilde{\mathcal{N}}$ now differs from $\mathcal{N}$ only in number of lattice sites.

(iii) The conditional probability

$$p(\sigma(\tilde{\boldsymbol{r}}) | \sigma(\boldsymbol{r}), \boldsymbol{r} \in \mathcal{B}(\tilde{\boldsymbol{r}})) \geq 0 \,, \tag{6.10}$$

that the microstate at $\tilde{\boldsymbol{r}}$ is $\sigma(\tilde{\boldsymbol{r}})$, given a particular configuration of microstates in the block $\mathcal{B}(\tilde{\boldsymbol{r}})$, is called the *weight function* and it satisfies the usual formula

$$\sum_{\{\sigma(\tilde{\boldsymbol{r}})\}} p(\sigma(\tilde{\boldsymbol{r}}) | \sigma(\boldsymbol{r}), \boldsymbol{r} \in \mathcal{B}(\tilde{\boldsymbol{r}})) = 1 \,, \tag{6.11}$$

where the summation is over all the microstates of $\tilde{\boldsymbol{r}}$. The probability $p(\zeta_i; \sigma(\boldsymbol{r}))$ that there is a particular configuration of microstates on $\mathcal{N}$ is, from (1.12),

$$p(\zeta_i; \sigma(\boldsymbol{r})) = \frac{\exp[-\widehat{H}(N, \zeta_i; \sigma(\boldsymbol{r}))]}{Z(N, \zeta_i)} \,. \tag{6.12}$$

Since the mapping is between two lattices which differ only in the number of lattice sites, the Hamiltonian function $\widehat{H}$ and partition function Z will be the same in each case and the probability $p(\tilde{\zeta}_i; \sigma(\tilde{\boldsymbol{r}}))$ that there is a particular configuration of microstates on $\widetilde{\mathcal{N}}$, with coupling constants $\tilde{\zeta}_i$, is

$$p(\tilde{\zeta}_i; \sigma(\tilde{\boldsymbol{r}})) = \frac{\exp[-\widehat{H}(\widetilde{N}, \tilde{\zeta}_i; \sigma(\tilde{\boldsymbol{r}}))]}{Z(\widetilde{N}, \tilde{\zeta}_i)} \,. \tag{6.13}$$

(iv) The probabilities given by (6.12) and (6.13) can be related by a conditional probability of the form (6.10) using the standard formula of probability theory. In general this means that the conditional probability will be a function of both sets of couplings ζ_i and $\tilde{\zeta}_i$ with no specified relationship between the two. In renormalization group theory, however, the probability formula is used to *impose* the relationship

$$p(\tilde{\zeta}_i;\sigma(\tilde{\boldsymbol{r}})) = \sum_{\{\sigma(\boldsymbol{r})\}} \Big[\prod_{\{\tilde{\boldsymbol{r}}\}} p(\sigma(\tilde{\boldsymbol{r}})|\sigma(\boldsymbol{r}), \boldsymbol{r} \in \mathcal{B}(\tilde{\boldsymbol{r}}))\Big] p(\zeta_i;\sigma(\boldsymbol{r}))\,, \tag{6.14}$$

between the two sets of couplings, where the conditional probability is independent of the values of the members of both sets.

(v) Although the probabilities $p(\zeta_i;\sigma(\boldsymbol{r}))$ and $p(\tilde{\zeta}_i;\sigma(\tilde{\boldsymbol{r}}))$, as they are defined in (6.12) and (6.13), involve the trivial couplings ζ_0 and $\tilde{\zeta}_0$ respectively, this dependence is purely formal. The factor $\exp(-N\zeta_0)$ can be cancelled from the numerator and denominator of (6.12) and similarly for $\exp(-\widetilde{N}\tilde{\zeta}_0)$ in (6.13). We, therefore, have two arbitrary parameters in (6.14). Imposing that condition

$$Z(\widetilde{N},\tilde{\zeta}_i) = Z(N,\zeta_i) \tag{6.15}$$

fixes the difference $\widetilde{N}\tilde{\zeta}_0 - N\zeta_0$ and, from (6.12)–(6.15),

$$\exp[-\widehat{H}(\widetilde{N},\tilde{\zeta}_i;\sigma(\tilde{\boldsymbol{r}}))] = \sum_{\{\sigma(\boldsymbol{r})\}} \Big[\prod_{\{\tilde{\boldsymbol{r}}\}} p(\sigma(\tilde{\boldsymbol{r}})|\sigma(\boldsymbol{r}), \boldsymbol{r} \in \mathcal{B}(\tilde{\boldsymbol{r}}))\Big] \times \exp[-\widehat{H}(N,\zeta_i;\sigma(\boldsymbol{r}))]\,. \tag{6.16}$$

6.3 Renormalization Transformations and Weight Functions

Equation (6.16) is the starting point for the development of a renormalization transformation. Ideally, by allowing the microstates of $\widetilde{\mathcal{N}}$ to range over all their possible values and performing the summation on the right-hand side, we hope to obtain formulae of the form

$$\tilde{\zeta}_0 = \lambda^d\zeta_0 + \mathcal{K}_0(\zeta_j)\,, \tag{6.17}$$

$$\tilde{\zeta}_i = \mathcal{K}_i(\zeta_j)\,, \qquad i = 1,2,\ldots,n\,. \tag{6.18}$$

These are known as *recurrence relationships.* Equation (6.17) can be neglected from present consideration as it serves only to obtain thermodynamic properties (see Sect. 6.5). If the relationships (6.18) can be obtained they are used to perform a sequence of iterations indexed by $s = 0,1,2,\ldots$ in the space $\mathcal{S}$ of non-trivial couplings. This produces a trajectory $\mathsf{P}_0 \to \mathsf{P}_1 \to \mathsf{P}_2 \to \cdots$ in $\mathcal{S}$. It satisfies the property

$$\zeta_i^{(s+s')} = \mathcal{K}_i^{s'}(\zeta_j^{(s)}) \tag{6.19}$$

and is therefore a *semi-group*.[2] As in the case of general dynamic systems (see, for example, Devaney 1989) the behaviour of renormalization group trajectories in $\mathcal{S}$ could be very complicated, exhibiting fixed points, periodic points and even chaos. However, in most simple examples, it might be anticipated that the behaviour is governed by a relatively small set of fixed points. A point $\mathsf{P}^* \in \mathcal{S}$ with coupling values $\zeta_1^*, \zeta_2^*, \ldots,$ is a *fixed point* of the transformation if it satisfies

$$\zeta_i^* = \mathcal{K}_i(\zeta_j^*)\,, \qquad i = 1, 2, \ldots, n\,. \tag{6.20}$$

If it exists, the set $\mathfrak{B}(\mathsf{P}^*)$ of all points P_0 which on iteration converge to P^* is called the *basin of attraction* of P^*. If an n-dimensional neighbourhood of P^* contains only points of $\mathfrak{B}(\mathsf{P}^*)$ then P^* is called a *sink*. In general fixed points can have neighbourhoods containing points for which they are *attractive*, points for which they are *repulsive* and points for which trajectories iterated neither towards nor away from the fixed point. This last type of fixed point is called *marginal* and this characteristic is often associated with the existence of a region (line or surface) of fixed points.

In Sect. 6.2 we referred to the lattice $\mathcal{N}$ having N sites with the number of sites of $\widetilde{\mathcal{N}}$ being $\widetilde{N} = \lambda^{-d}N < N$. However, it is clear that development of trajectories in $\mathcal{S}$, leading to fixed points, implicitly assumes that a large number (potentially infinite) number of iterations can be applied to the system. This could be the case only if the thermodynamic limit $N \to \infty$ is assumed. The method described in Sect. 6.11, whereby the form of the transformation derived for finite N is applied an indefinite number of times, is in these terms by its very structure an approximation method, even though the calculation of (6.16) for finite N may be exact. It is clear that the form of an transformation will be in part determined by the choice of weight function. Three commonly used forms are:

(a) The decimation weight function

Choose a particular site $\mathring{\boldsymbol{r}} \in \mathcal{B}(\tilde{\boldsymbol{r}})$ so that the set $\mathring{\mathcal{N}}$ of all sites $\mathring{\boldsymbol{r}}$ for all the blocks of $\mathcal{N}$ forms a lattice of the same type as $\mathcal{N}$. Then define

$$p(\sigma(\tilde{\boldsymbol{r}})|\sigma(\boldsymbol{r}), \boldsymbol{r} \in \mathcal{B}(\tilde{\boldsymbol{r}})) = \delta^{\mathrm{Kr}}(\sigma(\tilde{\boldsymbol{r}}) - \sigma(\mathring{\boldsymbol{r}}))\,. \tag{6.21}$$

It is clear that the effect of choosing this weight function is that the summation on the right-hand side of (6.16) is a partial sum over all the sites of the lattice $\mathcal{N}$ except those of $\mathring{\mathcal{N}}$. The effect is to 'thin out' the sites of $\mathcal{N}$ leaving $\mathring{\mathcal{N}}$ which is now identified as $\widetilde{\mathcal{N}}$.[3] For the spin-$\frac{1}{2}$ model ($\nu = 2$, $\sigma(\boldsymbol{r}) = \pm 1$) equation (6.21) takes the simple form

$$p(\sigma(\tilde{\boldsymbol{r}})|\sigma(\boldsymbol{r}), \boldsymbol{r} \in \mathcal{B}(\tilde{\boldsymbol{r}})) = \tfrac{1}{2}[1 + \sigma(\tilde{\boldsymbol{r}})\sigma(\mathring{\boldsymbol{r}})]\,. \tag{6.22}$$

---

[2] Not a full group since no inverse transformation is defined.

[3] Strictly speaking the term 'decimation' is more properly used when one tenth of the sites is removed but here it is used more loosely to denote any proportion. In the case where half the sites are removed the more accurate term 'deimation' is sometimes used.

(b) The majority-rule weight function

This weight function was introduced by Niemeijer and van Leeuwen (1973, 1974). The first step in assigning $\sigma(\tilde{\boldsymbol{r}})$ for the block $\mathcal{B}(\tilde{\boldsymbol{r}})$ can be described in terms of the 'winner takes all' voting procedure used in some democracies. Given that among the sites of $\mathcal{B}(\tilde{\boldsymbol{r}})$ one of the ν microstates occurs more that any other then $\sigma(\tilde{\boldsymbol{r}})$ is assigned to this value. Unless $\nu = 2$ and the number of sites λ^d in a block is odd it is clear that this is not sufficient to determine $\sigma(\tilde{\boldsymbol{r}})$ for every configuration of the block. A 'tie' can occur in the voting procedure and a strategy must be adopted to deal with such cases. One possibility is to assign to $\sigma(\tilde{\boldsymbol{r}})$ one of these predominating values on the basis of equal probabilities. In the special case of a spin-$\frac{1}{2}$ model this weight function would take the form

$$p(\sigma(\tilde{\boldsymbol{r}})|\sigma(\boldsymbol{r}), \boldsymbol{r} \in \mathcal{B}(\tilde{\boldsymbol{r}})) = \frac{1}{2}\left[1 + \sigma(\tilde{\boldsymbol{r}}) S\Big(\sum_{\{\boldsymbol{r}\in\mathcal{B}(\tilde{\boldsymbol{r}})\}} \sigma(\boldsymbol{r})\Big)\right], \tag{6.23}$$

where

$$S(m) = \begin{cases} -1, & \text{if } m < 0\ , \\ \ \ 0, & \text{if } m = 0\ , \\ \ \ 1, & \text{if } m > 0\ . \end{cases} \tag{6.24}$$

In some cases this may not, however, be the most appropriate choice. In their work on the Ising model using square nearest-neighbour blocks Nauenberg and Nienhuis (1974a, 1974b) divided the configurations with equal numbers of up and down spins between block spins up and down with probability one. The rule (one of four) which they chose ensured that the reversal of all the spins in the block reversed the block spin.

(c) The Kadanoff–Houghton weight function

For the spin-$\frac{1}{2}$ model the weight function

$$p(\sigma(\tilde{\boldsymbol{r}})|\sigma(\boldsymbol{r}), \boldsymbol{r} \in \mathcal{B}(\tilde{\boldsymbol{r}})) = \frac{\exp\left[\tau\, \sigma(\tilde{\boldsymbol{r}}) \sum_{\{\boldsymbol{r}\in\mathcal{X}(\tilde{\boldsymbol{r}})]} \sigma(\boldsymbol{r})\right]}{2\cosh\left[\tau \sum_{\{\boldsymbol{r}\in\mathcal{X}(\tilde{\boldsymbol{r}})]} \sigma(\boldsymbol{r})\right]} \tag{6.25}$$

was introduced by Kadanoff and Houghton (1975). In this formula τ is a real number and $\mathcal{X}(\tilde{\boldsymbol{r}})$ is some subset of $\mathcal{B}(\tilde{\boldsymbol{r}})$. It can be shown (Example 6.1) that, in the limit $\tau \to \infty$, we recover the majority-rule weight function (6.23), if $\mathcal{X}(\tilde{\boldsymbol{r}}) = \mathcal{B}(\tilde{\boldsymbol{r}})$, and the decimation weight function (6.22), if $\mathcal{X}(\tilde{\boldsymbol{r}})$ is the single site $\mathring{\boldsymbol{r}} \in \mathcal{B}(\tilde{\boldsymbol{r}})$.

In itself the weight function chosen does not lead to any approximation as long as the summation on the left-hand side of (6.16) can be evaluated exactly. This means that the calculation of thermodynamic functions (see Sect. 6.5) would be exact. What can be lost are fixed points associated with particular parts of phase space. This leads to the inability to map the part of the phase space associated with these fixed points. The usual rule employed is to choose a weight function, which, when applied to the ground state in $\mathcal{N}$ associated with a phase of interest, leads to the same ground state in the lattice $\widetilde{\mathcal{N}}$. This idea will be illustrated by reference to the one-dimensional Ising model in Sect. 6.6.

Beginning in the early 1970s the general pattern of development described here was used to construct approximate renormalization group transformations (see Niemeijer and van Leeuwen 1976) and to gain a new perspective on the known solutions of some exactly solvable models (see Nelson and Fisher 1975). One problem associated with this development was recognized at an early stage. This was the difficulty of obtaining a consistent set of equations of the form of (6.17) and (6.18) from (6.16). The number of equations generated by choosing all possible states of the microsystems on $\widetilde{\mathcal{N}}$ is $\nu^{\tilde{N}}$. Because of the symmetries of the system many of these equations will be identical. However, within this set, there must be exactly $n+1$ independent equations which express the new set of couplings as functions of old set. It is not guaranteed that this can be achieved. In fact, as we shall see in Sect. 6.7, for the two-dimensional Ising model on a square lattice with nearest-neighbour couplings alone using a decimation weight function, this is impossible. Consistent equations can be obtained by adding second-neighbour and four-site interactions in $\widetilde{\mathcal{N}}$. These, however, would in their turn demand the inclusion of further couplings at the next iteration. This *proliferation* of couplings led to the development of a number of approximation methods (Sects. 6.7, 6.9, 6.10) designed to 'cut-off' the proliferation in a more or less *ad hoc* manner.

Equations (6.17) and (6.18) must not only be consistent, but the functions $\mathcal{K}_i$, $i = 0, 1, \ldots, n$ must be smooth functions of the couplings ζ_j, $j = 1, 2, \ldots, n$. For explicit approximation methods for finite systems this is not normally a problem. However, as was first recognized by Griffiths and Pearce (1978, 1979) (see also Griffiths 1981), this condition is not obviously true in the thermodynamic limit. The fundamental point is that, whereas we need, for the existence of phase transitions, non-smooth behaviour in the partition function (6.4), we require smooth behaviour from the summation on the left-hand side of (6.16), which differs from $Z(N, \zeta_i)$ only by the presence of the weight function. It is by no means assured that the weight function is able to suppress all non-smooth behaviour. As an example Griffiths and Pearce (1979) considered the case, for a spin-$\frac{1}{2}$ model, of the Kadanoff–Houghton weight function where $\mathcal{X}(\tilde{\boldsymbol{r}}) = \mathring{\boldsymbol{r}} \in \mathcal{B}(\tilde{\boldsymbol{r}})$ yielding

$$p(\sigma(\tilde{\boldsymbol{r}})|\sigma(\boldsymbol{r}), \boldsymbol{r} \in \mathcal{B}(\tilde{\boldsymbol{r}})) = \frac{\exp\left[\tau\,\sigma(\tilde{\boldsymbol{r}})\sigma(\mathring{\boldsymbol{r}})\right]}{2\cosh(\tau)}\,. \tag{6.26}$$

This gives, apart from a multiplicative constant, for the right-hand side of (6.16) the partition function of the model with the addition of an external field-type coupling $\tau\,\sigma(\tilde{\boldsymbol{r}})$ on the subset of sites $\mathring{\mathcal{N}}$. Given that the model exhibits phase transitions it is not necessarily the case that the extra external field is sufficient to lead to their suppression. In particular if the decimation limit $\tau \to \infty$ is chosen it is possible by a judicious choice of the signs of the states on $\mathring{\mathcal{N}}$ (Griffiths 1981) to cancel out the additional field leading to zero-field models know to have phase transitions. Transformations which give unwelcome phase transitions in the modified partition function on the right-hand side of (6.16) are said by Griffiths and Pearce (1979) to exhibit *peculiarities.* Examples are given by these authors of schemes with and without such behaviour. One simple observation with respect to the decimation weight function can be made. This is that if, as in Sects. 6.6–6.7, the sites of $\mathcal{N} - \mathring{\mathcal{N}}$ form disconnected clusters then no peculiarities will occur. This is clear from the fact that each cluster constitutes an effective finite lattice system, which cannot have a phase transition.

A number of authors (Wilson 1975, Kadanoff and Houghton 1975, van Leeuwen 1975) have drawn attention to another problem with the decimation weight function. Consider any subset $\{\boldsymbol{r}_1, \boldsymbol{r}_2, \ldots\}$ of $\mathring{\mathcal{N}} \subset \mathcal{N}$. These sites will remain as sites of $\widetilde{\mathcal{N}}$. It is clear that the expectation value $\langle\sigma(\boldsymbol{r}_1)\sigma(\boldsymbol{r}_2)\cdots\rangle$ will have the same value in both $\mathcal{N}$ and $\widetilde{\mathcal{N}}$. In particular the pair correlation function, defined as in (2.5), for the states at $\boldsymbol{r}_1$ and $\boldsymbol{r}_2$ satisfies the scaling form

$$\Gamma_2(\theta_1, \ldots, \theta_n; \sigma(\boldsymbol{r}_2 - \boldsymbol{r}_1)) = \Gamma_2(\lambda^{y_1}\theta_1, \ldots, \lambda^{y_n}\theta_n; \sigma(\lambda^{-1}[\boldsymbol{r}_2 - \boldsymbol{r}_1])) . \tag{6.27}$$

It follows from (2.194) that the scaling exponent associated with the coupling conjugate to the spin variable is equal to d, the physical dimension of the system. This means that everywhere on a critical surface the pair correlation function is equal to its value at the fixed point controlling the surface and, from (2.162) and (2.201), $\beta = 0$, $\delta = \infty$ and $\eta = 2-d$. While these results are true for the one-dimensional Ising model, where the phase transition occurs at $T = 0$ with a discontinuity in field (see Sect. 6.6), they are certainly untrue for the two-dimensional Ising model (see (5.126)) and conflict with series estimates for the three-dimensional Ising model according to which $\beta \simeq 0.33$, $\delta \simeq 4.8$ and $\eta \simeq 0.04$. Although it may be argued that this failure of an *exact* decimation transformation is hardly relevant to real situations, where in most cases decimation is used in conjunction with further approximation, it is of some interest to understand why it occurs. The basis of scaling and of the use of the renormalization group is, as has already been remarked, the perception that critical behaviour arises from long-range correlations. The application of the renormalization group in most cases averages over the short-range structure of the system leaving the long-range behaviour and thus the critical properties unaffected. As has been observed by Sneddon and Barber (1977), this is not the case with the decimation weight function which sums over some short and some long-range effects. This may be understand as a loss of information about critical behaviour leading to the absence of asymptotic decay of correlations.

6.4 Fixed Points and the Linear Renormalization Group

To analyze the nature of the fixed point given by (6.20) we linearize the recurrence relationships (6.18) about P^*. With

$$\Delta\boldsymbol{\zeta}^{(s)} = (\zeta_1^{(s)} - \zeta_1^*, \zeta_2^{(s)} - \zeta_2^*, \ldots, \zeta_n^{(s)} - \zeta_n^*), \tag{6.28}$$

we have

$$[\Delta\boldsymbol{\zeta}^{(s+1)}]^{(\mathrm{T})} \simeq \boldsymbol{L}^*[\Delta\boldsymbol{\zeta}^{(s)}]^{(\mathrm{T})}, \tag{6.29}$$

where '(T)' denotes the column vector transpose of a row vector and $\boldsymbol{L}^*$ is the fixed point value of

$$\boldsymbol{L} = \begin{pmatrix} \dfrac{\partial \mathcal{K}_1}{\partial \zeta_1} & \dfrac{\partial \mathcal{K}_1}{\partial \zeta_2} & \cdots & \dfrac{\partial \mathcal{K}_1}{\partial \zeta_n} \\ \dfrac{\partial \mathcal{K}_2}{\partial \zeta_1} & \dfrac{\partial \mathcal{K}_2}{\partial \zeta_2} & \cdots & \dfrac{\partial \mathcal{K}_2}{\partial \zeta_n} \\ \vdots & \vdots & \ddots & \vdots \\ \dfrac{\partial \mathcal{K}_n}{\partial \zeta_1} & \dfrac{\partial \mathcal{K}_n}{\partial \zeta_2} & \cdots & \dfrac{\partial \mathcal{K}_n}{\partial \zeta_n} \end{pmatrix}. \tag{6.30}$$

In general $\boldsymbol{L}^*$ is not symmetric and thus we have the left and right eigenvalue equations,

$$\boldsymbol{w}_j \boldsymbol{L}^* = \Lambda_j \boldsymbol{w}_j , \tag{6.31}$$

$$\boldsymbol{L}^* \boldsymbol{x}_j^{(\mathrm{T})} = \boldsymbol{x}_j^{(\mathrm{T})} \Lambda_j , \tag{6.32}$$

respectively, where $\boldsymbol{w}_j$ and $\boldsymbol{x}_j$ denote the left and right eigenvectors, for the eigenvalue Λ_j. It is simple to show that $\boldsymbol{w}_j$ and $\boldsymbol{x}_{j'}$ are orthogonal when $\Lambda_j \neq \Lambda_{j'}$. We shall assume that, within subspaces of eigenvectors for degenerate eigenvalues, a choice of basis vectors can be made for which this condition also holds, and thus we have

$$\boldsymbol{w}_j \cdot \boldsymbol{x}_{j'} = \delta^{\mathrm{Kr}}(j - j') . \tag{6.33}$$

From (6.29) and (6.31),

$$\boldsymbol{w}_j \cdot \Delta\boldsymbol{\zeta}^{(s+1)} \simeq \Lambda_j \boldsymbol{w}_j \cdot \Delta\boldsymbol{\zeta}^{(s)} . \tag{6.34}$$

We now suppose, as we did for scaling in Sect. 2.4, that there exist a set of scaling fields

$$\theta_j = \theta_j(\Delta\zeta_i) , \tag{6.35}$$

which are smooth functions of the couplings (and thus of $\triangle\zeta_j$, $j = 1, 2, \dots, n$) and such that

$$\theta_j(0) = 0\,, \tag{6.36}$$

$$\theta_j^{(s+1)} = \Lambda_j \theta_j^{(s)}\,. \tag{6.37}$$

From (6.33), (6.34) and (6.37) it follows that

$$\theta_j \simeq \boldsymbol{w}_j \,.\, \triangle\boldsymbol{\zeta}\,, \tag{6.38}$$

$$\triangle\boldsymbol{\zeta} \simeq \sum_{j=1}^{n} \boldsymbol{x}_j \theta_j\,. \tag{6.39}$$

Iterating (6.37)

$$\theta_j^{(s)} = \Lambda_j^s \theta_j^{(0)} \tag{6.40}$$

and thus, from (6.39),

$$\triangle\boldsymbol{\zeta}^{(s)} \simeq \sum_{j=1}^{n} \Lambda_j^s \boldsymbol{x}_j \theta_j^{(0)}\,. \tag{6.41}$$

Given that the scaling fields exist, they are a set of curvilinear coordinates in a region surrounding P^*. They have the following properties:

(i) $\theta_j = 0$ is an $(n-1)$-dimensional subspace in $\mathcal{S}$, which, from (6.37), is invariant under the transformation. From (6.38) the left eigenvector $\boldsymbol{w}_j$ is normal to this subspace at P^*.

(ii) $\{\theta_i = 0, \forall\, i \neq j\}$ is an invariant line in $\mathcal{S}$, which, from (6.39), is tangential to $\boldsymbol{x}_j$ at P^*.

(iii) Suppose that all the eigenvalues Λ_j, $j = 1, 2, \dots, n$ are real and positive. Then

(a) If $\Lambda_j > 1$ and $\theta_j^{(0)} \neq 0$ then $\theta_j^{(s)}$ increases as s increases. The trajectory iterates away from P^* with a component in the direction of $\boldsymbol{x}_j$. The eigenvalue is called *relevant.*

(b) If $\Lambda_j < 1$ and $\theta_j^{(0)} \neq 0$ then $\theta_j^{(s)}$ decreases as s increases. The trajectory approaches the subspace $\theta_j = 0$. It will, however, approach P^* only if there is no relevant scaling θ_i with $\theta_i^{(0)} \neq 0$. The eigenvalue Λ_j is called *irrelevant.*

(c) If $\Lambda_j = 1$ it is called *marginal* and θ_j remains unchanged under iteration.

The exponents y_j for P^* are defined by the formulae

$$\Lambda_j = \lambda^{y_j}\,, \qquad j = 1, 2, \dots, n\,. \tag{6.42}$$

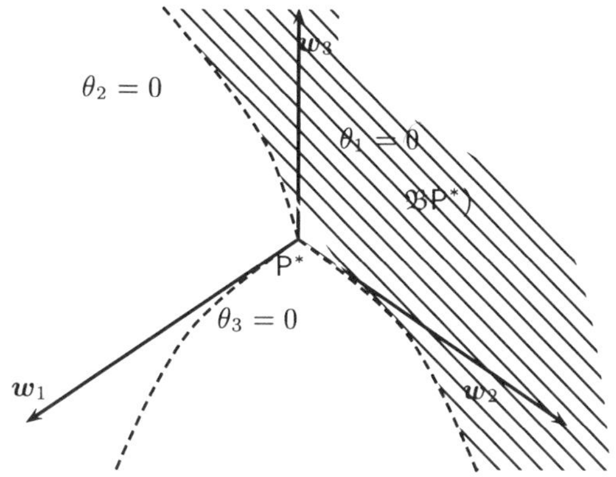

Fig. 6.1. The region around a fixed point P^* for $n = 3$, with one relevant and two irrelevant eigenvalues.

Exponents are named in the same way as their respective eigenvalues. Since $\lambda > 1$, an exponent is relevant if it is positive, zero if it is marginal and negative if it is irrelevant (see Sect. 2.4). The region around P^* with tangent space spanned by the set of right eigenvectors corresponding to irrelevant eigenvalues is the basin of attraction $\mathfrak{B}(P^*)$. A sink has only irrelevant eigenvalues. A fixed point which has a neighbourhood containing no other fixed point is called *isolated*. As indicated above this usually means that it has no marginal eigenvalues.

We shall now illustrate this analysis by considering two examples for the case $n = 3$.

(1) $\Lambda_1 > 1$, $\Lambda_2 < 1$, $\Lambda_3 < 1$. The region around P^* for this case is shown in Fig. 6.1. The local part of $\mathfrak{B}(P^*)$ is the surface $\theta_1 = 0$. Trajectories on this surface will converge on P^*. Any trajectory near to the fixed point, but not on $\theta_1 = 0$ will iterate away from the fixed point, tending towards the curve $\theta_2 = \theta_3 = 0$.

(2) $\Lambda_2 > \Lambda_1 > 1$, $\Lambda_3 < 1$. Λ_2 and Λ_1 are called respectively the *strong* and *weak* relevant eigenvalues. This is the case shown in Fig. 6.2. The local part of $\mathfrak{B}(P^*)$ is the curve $\theta_1 = \theta_2 = 0$. Any trajectory beginning on this curve will converge on P^*. Any other trajectory will be repelled by P^*. If it begins at a point where $\theta_2 \neq 0$ then it will tend towards the strong direction $\theta_1 = \theta_3 = 0$. Otherwise it will tend towards the weak direction $\theta_2 = \theta_3 = 0$.

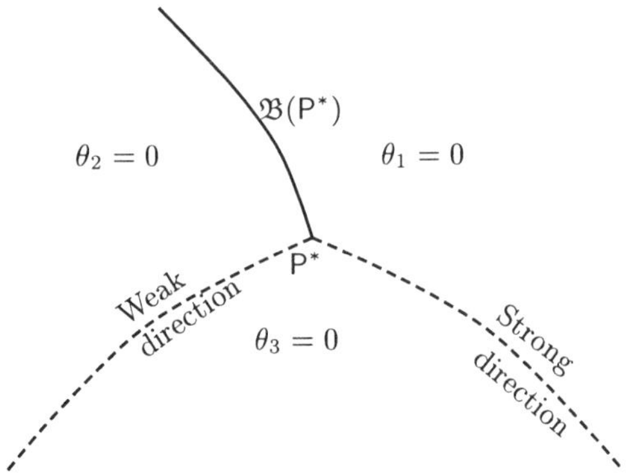

Fig. 6.2. The region around a fixed point P^* for $n = 3$, with one irrelevant and two relevant eigenvalues.

6.5 Free Energy and Densities

From (6.6) and (6.15)

$$\Phi(\widetilde{N}, \tilde{\zeta}_i) - \widetilde{N}\tilde{\zeta}_0 = \Phi(N, \zeta_i) - N\zeta_0 \,. \tag{6.43}$$

If we now define the free energy per lattice site

$$\phi(\zeta_i) = \Phi(N, \zeta_i)/N \,, \tag{6.44}$$

assuming this to be, at least for large N, independent of N (see Sect. 1.4), then

$$\phi(\tilde{\zeta}_i) - \tilde{\zeta}_0 = \lambda^d \phi(\zeta_i) - \lambda^d \zeta_0 \,, \tag{6.45}$$

which, from (6.17), becomes

$$\phi(\tilde{\zeta}_i) = \lambda^d \phi(\zeta_i) + \mathcal{K}_0(\zeta_i) \,. \tag{6.46}$$

Given that the couplings in a neighbourhood of a fixed point can, as we have described them in Sect. 6.4, be given in terms of smooth scaling fields, equation (6.46) can be written in the form

$$\phi(\lambda^{y_1}\theta_1, \ldots, \lambda^{y_n}\theta_n) = \lambda^d \phi(\theta_1, \ldots, \theta_n) + \mathcal{K}_0(\theta_1, \ldots, \theta_n) \,. \tag{6.47}$$

This equation is of the form of the Nightingale–'T Hooft scaling hypothesis (2.48). The only difference is that here $\lambda > 1$ is an integer, whereas in scaling theory it is assumed to have any real value in this range. This difference can be resolved if it is assumed that (6.47) can be analytically continued into non-integer values of λ. We have assumed above that $\mathcal{K}_0$ is a smooth function and the Kadanoff scaling hypothesis (2.17) now follows if, as in (2.8), we divide ϕ into smooth and singular parts.

Iterating equation (6.46) along a trajectory from a point with couplings $\zeta_i^{(0)}$, $i = 1, 2, \ldots, n$ gives

$$\phi(\zeta_i^{(0)}) = \frac{1}{\lambda^{kd}}\phi(\zeta_i^{(k)}) - \frac{1}{\lambda^d}\sum_{s=0}^{k-1}\frac{1}{\lambda^{sd}}\mathcal{K}_0(\zeta_i^{(s)})\,. \tag{6.48}$$

If we now assume that

$$\lim_{k\to\infty}\frac{1}{\lambda^{kd}}\phi(\zeta_i^{(k)}) = 0\,, \tag{6.49}$$

then it follows from (6.48) that the free energy per lattice site, at the initial point $\zeta_i^{(0)}$, is given by

$$\phi(\zeta_i^{(0)}) = -\frac{1}{\lambda^d}\sum_{s=0}^{\infty}\frac{1}{\lambda^{sd}}\mathcal{K}_0(\zeta_i^{(s)})\,. \tag{6.50}$$

In practice, it is usually found that this series converges after a very few iterations (see, for example, Southern and Lavis 1980). From (2.90) the densities at the initial point on the trajectory are given by

$$\rho_i^{(0)} = -\left(\frac{\partial\phi}{\partial\zeta_i}\right)^{(0)}\,. \tag{6.51}$$

To obtain these functions a chain differentiation must be performed along the trajectory. Let

$$\kappa_{\ell m} = \frac{\partial\mathcal{K}_m}{\partial\zeta_\ell}\,, \tag{6.52}$$

$$\tau_{im}^{(0)} = \kappa_{im}(\zeta_j^{(0)})\,, \tag{6.53}$$

$$\tau_{im}^{(k)} = \sum_{\ell=1}^{n}\tau_{i\ell}^{(k-1)}\kappa_{\ell m}(\zeta_j^{(k)})\,, \qquad k = 1, 2, \ldots, s\,. \tag{6.54}$$

Then

$$\rho_i^{(0)} = -\frac{1}{\lambda^d}\sum_{s=0}^{\infty}\frac{1}{\lambda^{sd}}\tau_{i0}^{(s)}\,. \tag{6.55}$$

The response functions, which correspond to the second-order partial derivatives of the free energy density with respect to the couplings, can be obtained in a similar way.

6.6 Decimation of the One-Dimensional Ising Model

The one-dimensional Ising model can be solved quite simply using the transfer matrix method described in Sect. 4.9 (see Volume 1, Sect. 2.4). We shall now present an real-space renormalization group treatment of the same problem using a decimation weight function. Suppose that $\mathcal{N}$ is a ring of N sites labelled $\mathrm{j} = 0, 1, \ldots, N-1$, with the first and last sites as nearest neighbours. Let the spin state at site j be $\sigma_\mathrm{j} = \pm 1$. The partition function is

$$Z(N, C, L, K) = \sum_{\{\sigma_\mathrm{j}\}} \sum_{\mathrm{j}=0}^{N-1} \exp\left[C + \tfrac{1}{2}L(\sigma_\mathrm{j} + \sigma_{\mathrm{j}+1}) + K\sigma_\mathrm{j}\sigma_{\mathrm{j}+1}\right], \tag{6.56}$$

where C is the trivial coupling, $L = \mathcal{H}/T$ and $K = J/T$. The lattice $\mathcal{N}$ is divided into blocks of λ neighbouring sites. The application of the decimation weight function corresponds to fixing the spin state on one site of each block (the first, say) and performing a partial sum over the remaining spin states. The cases $\lambda = 2$ and $\lambda = 3$ have been considered in detail by Nelson and Fisher (1975). The general case can be most conveniently formulated in terms of the transfer matrix.

It has been shown in Sect. 4.9 that the partition function (6.56) can be expressed in the form

$$Z(N, C, L, K) = \mathrm{Trace}\{\boldsymbol{V}^N\}, \tag{6.57}$$

in terms of the transfer matrix

$$\boldsymbol{V} = \mathfrak{z}_0 \begin{pmatrix} \mathfrak{z}_1\mathfrak{z}_2 & \mathfrak{z}_2^{-1} \\ \mathfrak{z}_2^{-1} & \mathfrak{z}_1^{-1}\mathfrak{z}_2 \end{pmatrix}, \tag{6.58}$$

where

$$\mathfrak{z}_0 = \exp(C), \qquad \mathfrak{z}_1 = \exp(L), \qquad \mathfrak{z}_2 = \exp(K). \tag{6.59}$$

The partition function for $\widetilde{\mathcal{N}}$ is now given by

$$Z(\widetilde{N}, \widetilde{C}, \widetilde{L}, \widetilde{K}) = \mathrm{Trace}\{\widetilde{\boldsymbol{V}}^{\widetilde{N}}\}, \tag{6.60}$$

where

$$\widetilde{\boldsymbol{V}} = \boldsymbol{V}^\lambda. \tag{6.61}$$

Computing the elements of this formula for any λ yields the recurrence relationships.

We first investigate in detail the case $\lambda = 2$. It is not difficult to see that (6.61) yields the three relationships

$$\tilde{\mathfrak{z}}_0\tilde{\mathfrak{z}}_1\tilde{\mathfrak{z}}_2 = \mathfrak{z}_0^2(\mathfrak{z}_1^2\mathfrak{z}_2^2 + \mathfrak{z}_2^{-2}), \tag{6.62}$$

$$\tilde{\mathfrak{z}}_0\tilde{\mathfrak{z}}_1^{-1}\tilde{\mathfrak{z}}_2 = \mathfrak{z}_0^2(\mathfrak{z}_1^{-2}\mathfrak{z}_2^2 + \mathfrak{z}_2^{-2}), \tag{6.63}$$

$$\tilde{\mathfrak{z}}_0\tilde{\mathfrak{z}}_2^{-1} = \mathfrak{z}_0^2(\mathfrak{z}_1 + \mathfrak{z}_1^{-1}), \tag{6.64}$$

which give

$$\tilde{\mathfrak{z}}_0^4 = \mathfrak{z}_0^8(\mathfrak{z}_1^2\mathfrak{z}_2^2 + \mathfrak{z}_2^{-2})(\mathfrak{z}_1^{-2}\mathfrak{z}_2^2 + \mathfrak{z}_2^{-2})(\mathfrak{z}_1 + \mathfrak{z}_1^{-1})^2 , \tag{6.65}$$

$$\tilde{\mathfrak{z}}_1^4 = \frac{\mathfrak{z}_1^2\mathfrak{z}_2^2 + \mathfrak{z}_2^{-2}}{\mathfrak{z}_1^{-2}\mathfrak{z}_2^2 + \mathfrak{z}_2^{-2}} , \tag{6.66}$$

$$\tilde{\mathfrak{z}}_2^4 = \frac{(\mathfrak{z}_1^2\mathfrak{z}_2^2 + \mathfrak{z}_2^{-2})(\mathfrak{z}_1^{-2}\mathfrak{z}_2^2 + \mathfrak{z}_2^{-2})}{(\mathfrak{z}_1 + \mathfrak{z}_1^{-1})^2} . \tag{6.67}$$

It will be observed that the formulae (6.65)–(6.67) are invariant under the transformation $\mathfrak{z}_1 \to 1/\mathfrak{z}_1$, which corresponds to reversing the direction of the magnetic field. In order to obtain recurrence relationships in a form appropriate to the one-dimensional Ising model we must bear in mind that, according to analysis of Volume 1, Sect. 2.4, the Curie point lies at zero temperature. This means that we must expect a fixed point on the zero-field axis at $T = 0$. The appropriate variables to choose are ones which are finite at zero temperature. If we concentrate on the region $\mathcal{H} \geq 0$ and $J \geq 0$ ($L \geq 0$, $K \geq 0$) then the variables

$$\mathfrak{u} = \mathfrak{z}_1^{-2} = \exp(-2L) , \tag{6.68}$$

$$\mathfrak{t} = \frac{\mathfrak{z}_2^2\mathfrak{z}_1 - 1}{\mathfrak{z}_2^2\mathfrak{z}_1 + 1} = \tanh\left(\tfrac{1}{2}L + K\right) \tag{6.69}$$

lie in the interval $[0,1]$. From (6.66) and (6.67),

$$\tilde{\mathfrak{u}} = \frac{\mathfrak{u}^2(1+\mathfrak{t})^2 + (1-\mathfrak{t})^2}{2(1+\mathfrak{t}^2)} , \tag{6.70}$$

$$\tilde{\mathfrak{t}} = \frac{4\mathfrak{t}^2 - (1-\mathfrak{u})(\mathfrak{t}^2-1)}{4 + (1-\mathfrak{u})(\mathfrak{t}^2-1)} . \tag{6.71}$$

The first step in the analysis of the recurrence relationships (6.70) and (6.71) is to determine the invariant lines. There are three of these:

(a) The line $\mathfrak{u} = 1$, corresponding to $\mathcal{H} = 0$. On this line there is the one recurrence relationship

$$\tilde{\mathfrak{t}} = \mathfrak{t}^2 \tag{6.72}$$

with *fixed points* $\mathfrak{t} = 1$, $\mathfrak{u} = 1$ and $\mathfrak{t} = 0$, $\mathfrak{u} = 1$.

(b) The line $\mathfrak{t} = 1$, corresponding to $T = 0$. On this line there is the one recurrence relationship

$$\tilde{\mathfrak{u}} = \mathfrak{u}^2 \tag{6.73}$$

with the *fixed points* $\mathfrak{t} = 1$, $\mathfrak{u} = 1$ and $\mathfrak{t} = 1$, $\mathfrak{u} = 0$.

(c) Every point on the line

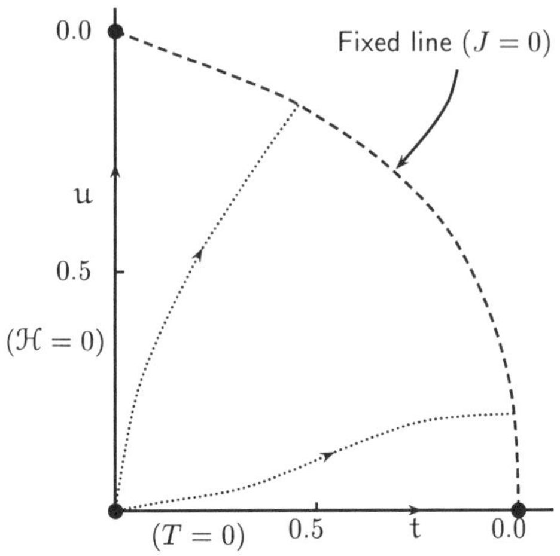

Fig. 6.3. The trajectory flows for the renormalization group transformation of the one-dimensional Ising model. The origin corresponds to $u = t = 1$ and fixed points are denoted by •.

$$u = \left(\frac{1 - t}{1 + t}\right)^2 \tag{6.74}$$

satisfies (6.70) and (6.71). This is a *line of fixed points* corresponding to $J = 0$.

The invariant lines, fixed points and the fixed line are shown in Fig. 6.3, together with examples of trajectory flows. We now consider the linearized recurrence relationships in the neighbourhood of each fixed point.

• $t = 1$, $u = 1$ is the *ferromagnetic fixed point* at $T = 0$, $\mathcal{H} = 0$. For this

$$\begin{pmatrix} \Delta\tilde{u} \\ \Delta\tilde{t} \end{pmatrix} = \begin{pmatrix} 2 & 0 \\ 0 & 2 \end{pmatrix} \begin{pmatrix} \Delta u \\ \Delta t \end{pmatrix} . \tag{6.75}$$

The eigenvalues and eigenvectors and the exponents for this fixed point are given in Table 6.1(a). That the magnetic field exponent $y_1 = y_{\mathcal{H}} = d$ has already been seen in Sect. 6.3 to be a result of the use of the decimation weight function. In this particular case it can also be interpreted as a consequence of the first-order transition at $T = 0$ (see Sect. 2.4) when the magnetization reverses direction as the magnetic field passes through zero.

• $t = 0$, $u = 1$ is the infinite temperature fixed point on $\mathcal{H} = 0$. For this we have

$$\begin{pmatrix} \Delta\tilde{u} \\ \Delta\tilde{t} \end{pmatrix} = \begin{pmatrix} 1 & 0 \\ -\frac{1}{4} & 2 \end{pmatrix} \begin{pmatrix} \Delta u \\ \Delta t \end{pmatrix} . \tag{6.76}$$

The eigenvalues and eigenvectors and the exponents for this fixed point are given in Table 6.1(b). The fixed point is infinitely attractive along the zero-field axis. The marginal exponent $y_1 = 0$ corresponds to the fact that this

Table 6.1. The eigenvalues Λ_j, exponents y_j, and left and right eigenvectors $\boldsymbol{w}_j$ and $\boldsymbol{x}_j$ for (a) the ferromagnetic fixed point $\mathfrak{t} = 1, \mathfrak{u} = 1$, (b) the infinite temperature fixed point $\mathfrak{t} = 0$, $\mathfrak{u} = 1$, (c) the infinite field fixed point $\mathfrak{t} = 1$, $\mathfrak{u} = 0$, (d) the line of fixed points (6.74).

Λ_j	y_j	$\boldsymbol{w}_j$	$\boldsymbol{x}_j$	
2	1	(1,0)	(1,0)	(a)
2	1	(0,1)	(0,1)	
1	0	(1,0)	$(1,-\frac{1}{4})$	(b)
0	$-\infty$	$(\frac{1}{4},1)$	(0,1)	
0	$-\infty$	(1,0)	(1,0)	(c)
1	0	(0,1)	(0,1)	
0	$-\infty$	$\mathfrak{h}(\mathfrak{t})(1,\mathfrak{x}(\mathfrak{t}))$	$\mathfrak{h}(\mathfrak{t})(1,\mathfrak{y}(\mathfrak{t}))$	(d)
1	0	$\mathfrak{h}(\mathfrak{t})(-\mathfrak{y}(\mathfrak{t}),1)$	$\mathfrak{h}(\mathfrak{t})(-\mathfrak{x}(\mathfrak{t}),1)$	

fixed point in on the line of fixed points (6.74). The right eigenvector $\boldsymbol{x}_1$ is tangential to the line of fixed points at this fixed point, where $d\mathfrak{u}/d\mathfrak{t} = -4$.

• $\mathfrak{t} = 1$, $\mathfrak{u} = 0$ is the infinite field fixed point on $T = 0$. For this we have

$$\begin{pmatrix} \Delta\tilde{\mathfrak{u}} \\ \Delta\tilde{\mathfrak{t}} \end{pmatrix} = \begin{pmatrix} 0 & 0 \\ 0 & 1 \end{pmatrix} \begin{pmatrix} \Delta\mathfrak{u} \\ \Delta\mathfrak{t} \end{pmatrix} . \tag{6.77}$$

The eigenvalues and eigenvectors and the exponents for this fixed point are given in Table 6.1(c). The fixed point is infinitely attractive along the zero-temperature axis. The marginal exponent $y_2 = 0$ again corresponds to the fact that this fixed point is on the line of fixed points (6.74). The right eigenvector $\boldsymbol{x}_2$ is tangential to the fixed line at the fixed point where $d\mathfrak{u}/d\mathfrak{t} = 0$.

• The line of fixed points (6.74) at some $0 \le \mathfrak{t} \le 1$ has linearized recurrence relationships

$$\begin{pmatrix} \Delta\tilde{\mathfrak{u}} \\ \\ \Delta\tilde{\mathfrak{t}} \end{pmatrix} = \begin{pmatrix} \dfrac{(1-\mathfrak{t})^2}{1+\mathfrak{t}^2} & \dfrac{-8\mathfrak{t}(1-\mathfrak{t})}{(1+\mathfrak{t}^2)(1+\mathfrak{t})^3} \\ \\ \dfrac{(\mathfrak{t}^2-1)(1+\mathfrak{t})^2}{4(1+\mathfrak{t}^2)} & \dfrac{2\mathfrak{t}}{1+\mathfrak{t}^2} \end{pmatrix} \begin{pmatrix} \Delta\mathfrak{u} \\ \\ \Delta\mathfrak{t} \end{pmatrix} . \tag{6.78}$$

The eigenvalues and eigenvectors and the exponents as functions of $\mathfrak{t}$ are given in Table 6.1(d), where

$$h(t) = \sqrt{\frac{2t}{1+t^2}}, \qquad x(t) = \frac{4(1-t)}{(1+t)^3}, \qquad y(t) = \frac{(1-t)(1+t)^3}{8t}. \tag{6.79}$$

The line of fixed points is infinitely attractive to trajectories which approach it. The marginal exponent $y_2 = 0$ has a corresponding right eigenvector $\boldsymbol{x}_2$ which is tangential to the fixed line, where $du/dt = -4(1-t)/(1+t)^3$.

We now calculate the free energy per lattice site. From (6.17), (6.59), (6.65), (6.68) and (6.69)

$$\mathcal{K}_0(u,t) = \frac{1}{4}\ln\left\{\frac{2(1+u^{-1})^2(1+t^2)[u^2(1+t)^2+(1-t)^2]}{(1-t^2)^2}\right\}. \tag{6.80}$$

The free energy per lattice site at some initial point$(u^{(0)}, t^{(0)})$ is now obtained by inserting this expression into the series (6.50), with $\lambda = 2$, $d = 1$. An interesting problem is to show that this series converges to the closed form expression obtained by more conventional methods. The calculations of Volume 1, Sect. 2.4, when translated into the present notation, yield

$$\phi(u,t) = -\ln\left\{\frac{1}{2u^{3/4}}\left(\frac{1+t}{1-t}\right)^{1/2}\left[1+u+\sqrt{(1-u)^2+4\left(\frac{1-t}{1+t}\right)^2}\right]\right\}. \tag{6.81}$$

It is quite simple to write a computer program, using the recurrence relationships (6.70) and (6.71) to compute $\phi(u^{(0)}, t^{(0)})$, from (6.80) and (6.50). This will give convincing evidence that the series converges to the value given by (6.81). In general it would appear to be rather difficult to sum the series analytically. In the case $\mathcal{H} = 0$, $(u = 1)$, however, the problem has been solved by Nauenberg (1975). His argument, translated into our notation, is as follows. From (6.80),

$$\mathcal{K}_0(1,t) = \frac{1}{2}\ln\left[4\left(\frac{1+t^2}{1-t^2}\right)\right] \tag{6.82}$$

and hence, from (6.71) and (6.50),

$$\begin{aligned}\phi(1,t) &= -\sum_{s=1}^{\infty}\frac{1}{2^s}\left[\ln(2)+\frac{1}{2}\ln\left(\frac{1+t^{2^s}}{1-t^{2^s}}\right)\right],\\ &= -\ln(2)-\ln\left[\prod_{s=1}^{\infty}\left(\frac{1+t^{2^s}}{1-t^{2^s}}\right)^{1/2^{s+1}}\right].\end{aligned} \tag{6.83}$$

Using the identity

$$\frac{1}{\sqrt{1-q^2}} = \prod_{s=1}^{\infty} \left(\frac{1+q^{2^s}}{1-q^{2^s}} \right)^{1/2^{s+1}} , \tag{6.84}$$

gives

$$\phi(1, \mathfrak{t}) = -\ln(2) + \tfrac{1}{2}\ln(1-\mathfrak{t}^2) , \tag{6.85}$$

which also follows from (6.81).

For the ferromagnetic fixed point the exponents given in Table 6.1 can be used together with formulae (2.162) and (2.167), to give the critical exponents

$$\alpha = 1 , \qquad \beta = 0 , \qquad \gamma = 1 , \qquad \delta = \infty . \tag{6.86}$$

The values of β and δ both arise from the first-order phase transition condition $y_1 = d$. Because the matrix in (6.75) is diagonal the scaling fields can be identified as

$$\theta_1 = -\Delta\mathfrak{u} \simeq 2L , \tag{6.87}$$

$$\theta_2 = -\Delta\mathfrak{t} \simeq 2\exp(-2K) . \tag{6.88}$$

From the general analysis of scaling at a critical point described in Sect. 2.9 $\theta_1 \sim \Delta L$ and $\theta_2 \sim \Delta K$. It follows from (6.87) that the first of these two formulae still holds in this case since $L_c = 0$. However, the fact that the critical temperature is $T_c = 0$ implies that $K_c = \infty$, leading to the different choice of scaling field θ_2 given by (6.88). The formula

$$\frac{\partial m}{\partial \mathcal{H}} = \frac{1}{T}\exp\left(\frac{2J}{T}\right) , \tag{6.89}$$

given in Volume 1, Sect. 2.4 for the response of the magnetization m to change of magnetic field, is now consistent with that derived from scaling using the value of γ in (6.86) and (6.88).

A similar analysis is applicable to other one-dimensional models and in Example 6.2 the reader is encouraged to apply the method to the 3-state Potts model. For the case $\lambda = 2$ we have considered only the ferromagnetic situation $J > 0$. This is because application of this weight function to the lattice, by removing alternate sites, maps the antiferromagnetic ground state of alternating spins into the ferromagnetic ground state of aligned spins. In general, from (6.67), a point in phase space with $\mathfrak{z}_2 < 1$ is mapped in one step into a point with $\mathfrak{z}_2 > 1$. Although this does not affect the evaluation of the free energy which must be correct for either sign of J, (since the method is exact), it does mean that this method does not give fixed points associated with the antiferromagnetic model. In particular it is not possible to obtain the fixed point on the zero-temperature axis associated with the critical field (Volume 1, Sect. 4.1). On way of achieving this (Nelson and Fisher 1975) is to take $\lambda = 3$. In fact it is obvious that any odd value of λ would suffice. In the special case of zero magnetic field, it is not difficult to show that

$$\boldsymbol{V} = \mathfrak{Z}_0 \begin{pmatrix} 1 & 1 \\ 1 & -1 \end{pmatrix} \begin{pmatrix} \mathrm{c} & 0 \\ 0 & \mathrm{s} \end{pmatrix} \begin{pmatrix} 1 & 1 \\ 1 & -1 \end{pmatrix}, \tag{6.90}$$

where $\mathrm{c} = \cosh(J/T)$ and $\mathrm{s} = \sinh(J/T)$. Then (6.61) gives

$$2\tilde{\mathfrak{Z}}_0\tilde{\mathrm{c}} = (2\mathfrak{Z}_0\mathrm{c})^\lambda, \tag{6.91}$$

$$2\tilde{\mathfrak{Z}}_0\tilde{\mathrm{s}} = (2\mathfrak{Z}_0\mathrm{s})^\lambda \tag{6.92}$$

and thus the generalization

$$\tilde{\mathfrak{t}} = \mathfrak{t}^\lambda \tag{6.93}$$

of (6.72). For λ odd this equation has the ferromagnetic fixed point $\mathfrak{t} = 1$ and the antiferromagnetic fixed point $\mathfrak{t} = -1$.

6.7 Decimation in Two Dimensions

Before describing some of the approximation methods which have been used to construct renormalization group transformations it is worthwhile exploring the possibility of using the decimation procedure described in the previous section for a two-dimensional lattice. Suppose we make an attempt to do this for the nearest-neighbour Ising model on a square lattice with zero magnetic field. The lattice is divided into black and white sites as shown in Fig. 6.4 and the summation in (6.16) is over the states on all the white sites, leaving the spins on the black sites fixed. Summing the spin states on site G at the centre of the square ABCD gives

$$2\exp(2C)\cosh[K_1(\sigma_\mathsf{A} + \sigma_\mathsf{B} + \sigma_\mathsf{C} + \sigma_\mathsf{D})], \tag{6.94}$$

where C and $K_1 = J/T$ are the trivial and nearest-neighbour couplings respectively. The left-hand side of (6.16) consists of a product of $N/2$ terms of this form, one for each of the squares of type ABCD as shown in Fig. 6.4. It was pointed out by Wilson (1975) that this expression can be rewritten in the form

$$\exp\Big[2C_0 + \mathcal{K}_0(K_1) + \tfrac{1}{2}\mathcal{K}_1(K_1)(\sigma_\mathsf{A}\sigma_\mathsf{B} + \sigma_\mathsf{B}\sigma_\mathsf{C} + \sigma_\mathsf{C}\sigma_\mathsf{D} + \sigma_\mathsf{D}\sigma_\mathsf{A})$$
$$+\mathcal{K}_2(K_1)(\sigma_\mathsf{A}\sigma_\mathsf{C} + \sigma_\mathsf{B}\sigma_\mathsf{D}) + \mathcal{K}_3(K_1)\sigma_\mathsf{A}\sigma_\mathsf{B}\sigma_\mathsf{C}\sigma_\mathsf{D}\Big], \tag{6.95}$$

where

$$\mathcal{K}_0(K_1) = \ln(2) + \tfrac{1}{8}\ln\cosh(4K_1) + \tfrac{1}{2}\ln\cosh(2K_1), \tag{6.96}$$

$$\mathcal{K}_1(K_1) = \tfrac{1}{4}\ln\cosh(4K_1), \tag{6.97}$$

$$\mathcal{K}_2(K_1) = \tfrac{1}{8}\ln\cosh(4K_1), \tag{6.98}$$

$$\mathcal{K}_3(K_1) = \tfrac{1}{8}\ln\cosh(4K_1) - \tfrac{1}{2}\ln\cosh(2K_1). \tag{6.99}$$

By setting

$$\widetilde{C} = 2C + \mathcal{K}_0(K_1)\,, \tag{6.100}$$

$$\widetilde{K}_i = \mathcal{K}_i(K_1)\,, \qquad i = 1,2,3\,, \tag{6.101}$$

we have recurrence relationships of the form of (6.17) and (6.18), with $d = 2$ and $\lambda = \sqrt{2}$, where $\widetilde{K}_1$ is the nearest-neighbour coupling on the lattice of black sites and $\widetilde{K}_2$ and $\widetilde{K}_3$ are respectively a second-neighbour coupling and a four-site coupling on this new lattice. The effect of decimation has been to generate new couplings. We can consider the strategy of going back to the original lattice and including second-neighbour and four-site couplings at this stage. These, however, have the effect of coupling white sites together and it is not difficult to see that this will lead to further couplings. This is the beginning of the proliferation of couplings referred to in Sect. 6.3. One possible approximation is to include a second-neighbour interaction only between black sites in the original lattice and to neglect the four-site coupling. This leads to the modification

$$\widetilde{K}_1 = K_2 + \tfrac{1}{4}\ln\cosh(4K_1)\,, \tag{6.102}$$

$$\widetilde{K}_2 = \tfrac{1}{8}\ln\cosh(4K_1) \tag{6.103}$$

of (6.96)–(6.99). We now follow Wilson (1975) and approximate (6.102) and (6.103) by expanding to quadratic terms to give

$$\widetilde{K}_1 = K_2 + 2K_1^2\,, \tag{6.104}$$

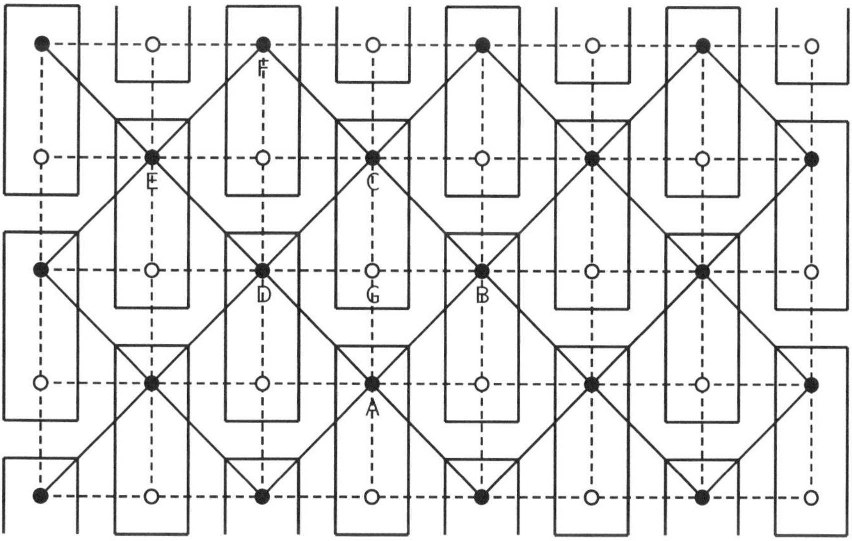

Fig. 6.4. Two site blocks for the nearest-neighbour Ising model on a square lattice.

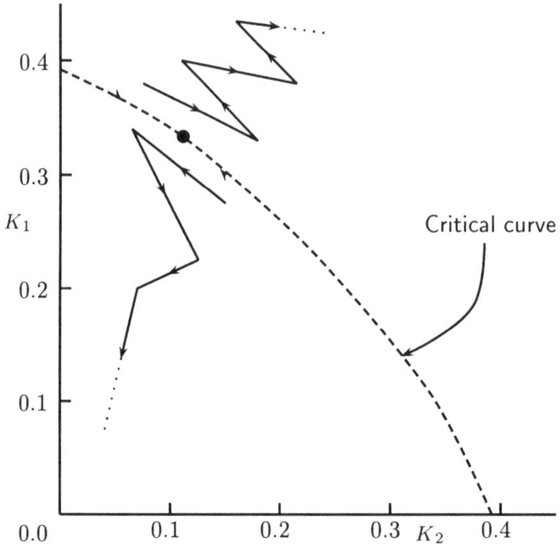

Fig. 6.5. The phase diagram for the approximate treatment of the two-dimensional Ising model using the recurrence relations (6.104) and (6.105).

$$\widetilde{K}_2 = K_1^2 \,. \tag{6.105}$$

These approximate recurrence relations for an Ising model with nearest- and second-neighbour interactions have the fixed points $K_1 = K_2 = 0$ and $K_1 = K_2 = \infty$, which are both sinks and the fixed point $K_1^* = \frac{1}{3}$, $K_2^* = \frac{1}{9}$. Linearizing (6.104) and (6.105) about this fixed point we have

$$\begin{pmatrix} \Delta\widetilde{K}_1 \\ \Delta\widetilde{K}_2 \end{pmatrix} = \begin{pmatrix} \frac{4}{3} & 1 \\ \frac{2}{3} & 0 \end{pmatrix} \begin{pmatrix} \Delta K_1 \\ \Delta K_2 \end{pmatrix} . \tag{6.106}$$

The eigenvalues of the matrix are $\Lambda_1 = \frac{1}{3}(2+\sqrt{10}) = 1.721$ and $\Lambda_2 = \frac{1}{3}(2-\sqrt{10}) = -0.387$. With $\lambda = \sqrt{2}$ it follows from (6.42) that Λ_1 gives the relevant (thermal) exponent $y_1 = y_T = 1.566$. The fixed point (K_1^*, K_2^*) lies on a critical curve, which cuts the axes in the plane of couplings at $(0, 0.392)$ and $(0.392, 0)$ (see Fig. 6.5). We expected these two values to be the same, since an Ising model with only second-neighbour couplings is equivalent to two disconnected Ising models with nearest-neighbour couplings. Trajectories which begin at points enclosed within the region bounded by the axes and the critical curve iterate ultimately to the sink at the origin. Trajectories not beginning on the critical curve and outside this region iterate to the sink at infinity. The fixed point (K_1^*, K_2^*) on the critical curve is attractive to trajectories which begin on the curve. The irrelevant eigenvalue Λ_2 controls the attraction to the fixed point for such trajectories. That it is negative, giving an imaginary exponent y_2, conflicts with the assumption we have made in Sect. 6.4. However, this is not a serious problem. It simply means that the iteration process will jump back forth from one side of the fixed point to the other as shown in Fig. 6.5. The crossing points of the critical curve with the axes give an approximation to the critical temperature of the two-dimensional Ising model. We have $T_c/J = 1/0.392 = 2.550$. This should be compared with the exact value of $T_c/J = 2/\text{arcsinh}(1) = 2.269$ (Sect. 5.10). Also, from (2.198) and the value for the exponent given above, we have the approximate value for $\nu = 1/y_T = 0.639$, which should be compared with the exact value of unity (see Sect. 5.10). Given the crudeness of the approximation the estimate of the critical temperature is surprisingly good although the value of ν is poor. It is interesting to see that if (6.102) and (6.103) are used rather than (6.104) and (6.105) the estimates for the critical temperature and ν are worse. The phase diagram again has the same structure but now $T_c/J = 1.582$ and $\nu = 0.556$. In Example 6.3 alternatives to (6.102) and (6.103) are given, which again produce (6.104) and (6.105) on expansion to quadratic terms. In this case the full equations register a slight improvement. These examples of *ad hoc* approximations applied to a model for which exact results are known serve to illustrate the rather arbitrary nature of the improvements and deterioration in results which can occur. They stand as a warning, when such methods are applied to models for which exact results are not available.

The problem with proliferating interactions which we have encountered in this section arises also for the other standard two and three-dimensional lattices, although it can be avoided in hierarchical lattices, where decimation amounts to a reversal of the procedure by which the hierarchical lattice was created (Kaufman and Griffiths 1981). A type of exact decimation procedure for the Ising model on a triangular lattice has been developed by Hilhorst et al. (1978, 1979). This is based on the star–triangle transformation. It avoids the problem of proliferation of interactions, but leads to spatially dependent couplings. Nevertheless the fixed point can be located analytically and the thermal exponent determined exactly. With these exceptions most real-space renormalization group work has concentrated on approximation methods.

6.8 Lower-Bound and Upper-Bound Approximations

The programme for real-space renormalization group methods is to use the formula (6.16) to obtain the recurrence relationships (6.17) and (6.18) and then to use the methods described in Sects. 6.4 and 6.5 to obtain the critical exponents and free energy respectively. As we have seen in Sect. 6.6, this scheme can be carried out exactly for some one-dimensional models, but Sect. 6.7 describes the problems encountered when a similar approach is attempted in two dimensions. The approximation methods which will be described in the remainder of this chapter are of two broad types. In the methods of Sects. 6.11 and 6.13 the lattice for the system is replaced by a finite subset of sites or a lattice of lower dimensionality. In the methods described in Sects. 6.9, 6.10 and 6.12 an approximation is made to the right-hand side of (6.16) so that the summation can be performed. In these latter methods the modified form of (6.16), when summed over the microstates on $\widetilde{\mathcal{N}}$, will not necessarily reproduce the invariance condition (6.15) for the partition function. There are, however, a number of methods for which this equality can be replaced by an inequality which remains valid throughout the iteration process. Suppose that the approximate recurrence relationships replacing (6.17) and (6.18) are

$$\tilde{\zeta}'_0 = \lambda^d \zeta_0 + \mathcal{U}_0(\zeta_j) , \tag{6.107}$$

$$\tilde{\zeta}'_i = \mathcal{U}_i(\zeta_j) , \qquad i = 1, 2, \ldots, n , \tag{6.108}$$

with (6.15) replaced by

$$Z(\widetilde{N}, \tilde{\zeta}'_i) \le Z(N, \zeta_i) . \tag{6.109}$$

Then, in place of (6.43) and (6.46), we have, respectively,

$$\Phi(\widetilde{N}, \tilde{\zeta}'_i) - \widetilde{N} \tilde{\zeta}'_0 \ge \Phi(N, \zeta_i) - N\zeta_0 , \tag{6.110}$$

$$\phi(\tilde{\zeta}'_i) \ge \lambda^d \phi(\zeta_i) + \mathcal{U}_0(\zeta_i) . \tag{6.111}$$

Now suppose that (6.107) and (6.108) are used to generate a trajectory $\zeta_i^{(0)} \to \zeta_i^{(1)\prime} \to \zeta_i^{(2)\prime} \cdots$ in phase space. By an procedure similar to that of Sect. 6.5 it follows that

$$\phi(\zeta_i^{(0)}) \le \phi^{(\mathcal{U})}(\zeta_i^{(0)}) = -\frac{1}{\lambda^d} \sum_{s=0}^{\infty} \frac{1}{\lambda^{sd}} \mathcal{U}_0(\zeta_i^{(s)\prime}) . \tag{6.112}$$

The transformation generates an upper-bound to the free energy. By a similar argument a set of recurrence relationships

$$\tilde{\zeta}'_0 = \lambda^d \zeta_0 + \mathcal{L}_0(\zeta_j) , \tag{6.113}$$

$$\tilde{\zeta}'_i = \mathcal{L}_i(\zeta_j) , \qquad i = 1, 2, \ldots, n , \tag{6.114}$$

with

$$Z(\widetilde{N}, \tilde{\zeta}'_i) \ge Z(N, \zeta_i) , \tag{6.115}$$

will generate a lower bound

$$\phi^{(\mathcal{L})}(\zeta_i^{(0)}) = -\frac{1}{\lambda^d} \sum_{s=0}^{\infty} \frac{1}{\lambda^{sd}} \mathcal{L}_0(\zeta_i^{(s)\prime}) \tag{6.116}$$

to the free energy.

6.8.1 An Upper-Bound Method

An approach which leads to an upper bound to the free energy was introduced by Niemeijer and van Leeuwen (1974) (see also Niemeijer and van Leeuwen 1976). It is based on a splitting of the Hamiltonian on the right-hand side of (6.16) into two parts

$$\widehat{H}(N,\zeta_i;\sigma(\boldsymbol{r})) = \widehat{H}_0(N,\zeta_i;\sigma(\boldsymbol{r})) + \triangle\widehat{H}(N,\zeta_i;\sigma(\boldsymbol{r}))\,, \tag{6.117}$$

where $\widehat{H}_0$ is referred to as the *zeroth-order part*, compared to which $\triangle\widehat{H}$ is considered to be small. A *zeroth-order transformation* is now defined by

$$\frac{\exp[-\widehat{H}_0(\widetilde{N},\tilde{\zeta}_i;\sigma(\tilde{\boldsymbol{r}}))]}{Z_0(\widetilde{N},\tilde{\zeta}_i)}$$
$$= \frac{\sum_{\{\sigma(\boldsymbol{r})\}}\left[\prod_{\{\tilde{\boldsymbol{r}}\}} p(\sigma(\tilde{\boldsymbol{r}})|\sigma(\boldsymbol{r}),\boldsymbol{r}\in\mathcal{B}(\tilde{\boldsymbol{r}}))\right]\exp[-\widehat{H}_0(N,\zeta_i;\sigma(\boldsymbol{r}))]}{Z_0(N,\zeta_i)}\,, \tag{6.118}$$

where Z_0 is the zeroth-order partition function obtained by replacing $\widehat{H}$ by $\widehat{H}_0$ in (6.4). From (6.13) and (6.14) we see that this transformation is of the form of (6.16) except that it is not assumed that the partition function Z_0 is invariant under the transformation. Defining the zeroth-order average $\langle A\rangle_0$ for any function $A(\sigma(\boldsymbol{r}))$ of the spin states by

$$\langle A\rangle_0 =$$
$$\frac{\sum_{\{\sigma(\boldsymbol{r})\}}\left[\prod_{\{\tilde{\boldsymbol{r}}\}} p(\sigma(\tilde{\boldsymbol{r}})|\sigma(\boldsymbol{r}),\boldsymbol{r}\in\mathcal{B}(\tilde{\boldsymbol{r}}))\right]A(\sigma(\boldsymbol{r}))\exp[-\widehat{H}_0(N,\zeta_i;\sigma(\boldsymbol{r}))]}{\sum_{\{\sigma(\boldsymbol{r})\}}\left[\prod_{\{\tilde{\boldsymbol{r}}\}} p(\sigma(\tilde{\boldsymbol{r}})|\sigma(\boldsymbol{r}),\boldsymbol{r}\in\mathcal{B}(\tilde{\boldsymbol{r}}))\right]\exp[-\widehat{H}_0(N,\zeta_i;\sigma(\boldsymbol{r}))]}\,, \tag{6.119}$$

the recurrence relations (6.16) can now be expressed in the form

$$\exp[-\widehat{H}(\widetilde{N},\tilde{\zeta}_i;\sigma(\tilde{\boldsymbol{r}}))] = \langle\exp[-\triangle\widehat{H}(N,\zeta_i;\sigma(\boldsymbol{r}))]\rangle_0$$
$$\times\sum_{\{\sigma(\boldsymbol{r})\}}\left[\prod_{\{\tilde{\boldsymbol{r}}\}} p(\sigma(\tilde{\boldsymbol{r}})|\sigma(\boldsymbol{r}),\boldsymbol{r}\in\mathcal{B}(\tilde{\boldsymbol{r}}))\right]\exp[-\widehat{H}_0(N,\zeta_i;\sigma(\boldsymbol{r}))]\,. \tag{6.120}$$

Summing over all the spin states $\sigma(\tilde{\boldsymbol{r}})$ gives

$$Z(\widetilde{N},\tilde{\zeta}_i) = Z(N,\zeta_i) = Z_0(N,\zeta_i)\langle\exp[-\triangle\widehat{H}(N,\zeta_i;\sigma(\boldsymbol{r}))]\rangle_0\,. \tag{6.121}$$

These expressions are, of course, exact. The approximation arises when the final factor on the left is replaced by a finite number of terms of the *cumulant expansion*

$$\langle \exp\{-\Delta\widehat{H}(N,\zeta_i;\sigma(\boldsymbol{r}))\}\rangle_0 = \exp\{-\langle\Delta\widehat{H}(N,\zeta_i;\sigma(\boldsymbol{r}))\rangle_0$$
$$+\frac{1}{2!}\langle[\Delta\widehat{H}(N,\zeta_i;\sigma(\boldsymbol{r})) - \langle\Delta\widehat{H}(N,\zeta_i;\sigma(\boldsymbol{r}))\rangle_0]^2\rangle_0$$
$$-\frac{1}{3!}\langle[\Delta\widehat{H}(N,\zeta_i;\sigma(\boldsymbol{r})) - \langle\Delta\widehat{H}(N,\zeta_i;\sigma(\boldsymbol{r}))\rangle_0]^3\rangle_0 + \cdots\}. \tag{6.122}$$

The *first-order cumulant approximation* consists in retaining only the first term in the exponential on the right-hand side of equation (6.122). Substituting into (6.120) and replacing $\tilde{\zeta}_i$ by $\tilde{\zeta}'_i$ gives

$$\exp[-\widehat{H}(\widetilde{N},\tilde{\zeta}'_i,;\sigma(\tilde{\boldsymbol{r}}))] = \langle -\Delta\widehat{H}(N,\zeta_i;\sigma(\boldsymbol{r}))\rangle_0$$
$$\times \sum_{\{\sigma(\boldsymbol{r})\}}\Big[\prod_{\{\tilde{\boldsymbol{r}}\}} p(\sigma(\tilde{\boldsymbol{r}})|\sigma(\boldsymbol{r}),\boldsymbol{r}\in\mathcal{B}(\tilde{\boldsymbol{r}}))\Big]\exp[-\widehat{H}_0(N,\zeta_i;\sigma(\boldsymbol{r}))]. \tag{6.123}$$

The corresponding partition function is again obtained by summing over the spin states $\sigma(\tilde{\boldsymbol{r}})$ to give

$$Z(\widetilde{N},\tilde{\zeta}'_i) = Z_0(N,\zeta_i)\exp\langle -\Delta\widehat{H}(N,\zeta_i;\sigma(\boldsymbol{r}))\rangle_0. \tag{6.124}$$

For any $A(\sigma(\boldsymbol{r}))$

$$\begin{aligned}\langle\exp(A)\rangle_0 &= \exp\langle A\rangle_0\langle\exp(A-\langle A\rangle_0)\rangle_0\\ &= \exp\langle A\rangle_0\langle 1+(A-\langle A\rangle_0)+\frac{1}{2}(A-\langle A\rangle_0)^2+\cdots\rangle_0\\ &\geq \exp\langle A\rangle_0\langle 1+(A-\langle A\rangle_0)\rangle_0\\ &= \exp\langle A\rangle_0.\end{aligned} \tag{6.125}$$

The upper-bound condition (6.109) then follows from (6.121) and (6.124).

6.8.2 A Lower-Bound Method

A starting point for developing a class of lower-bound transformations can be achieved if the right-hand side of (6.16) is modified by replacing $\widehat{H}(N,\zeta_i;\sigma(\boldsymbol{r}))$ by $\widehat{H}(N,\zeta_i;\sigma(\boldsymbol{r}))+\widehat{V}(N,\zeta_i;\sigma(\boldsymbol{r}))$. Then we have

$$\exp[-\widehat{H}(\widetilde{N},\tilde{\zeta}'_i;\sigma(\tilde{\boldsymbol{r}}))] = \sum_{\{\sigma(\boldsymbol{r})\}}\Big[\prod_{\{\tilde{\boldsymbol{r}}\}} p(\sigma(\tilde{\boldsymbol{r}})|\sigma(\boldsymbol{r}),\boldsymbol{r}\in\mathcal{B}(\tilde{\boldsymbol{r}}))\Big]$$
$$\times\exp[-\widehat{H}(N,\zeta_i;\sigma(\boldsymbol{r}))-\widehat{V}(N,\zeta_i;\sigma(\boldsymbol{r}))], \tag{6.126}$$

where the functional form of the Hamiltonian on the left-hand side is the same as that in (6.16), but the renormalized couplings take different values denoted by $\tilde{\zeta}'_i$. Summing (6.126) over the microstates on $\widetilde{\mathcal{N}}$ gives

$$Z(\widetilde{N},\tilde{\zeta}'_i) = Z(N,\zeta_i)\langle \exp[-\widehat{V}(N,\zeta_i;\sigma(\boldsymbol{r}))]\rangle$$
$$\geq Z(N,\zeta_i) - \langle \widehat{V}(N,\zeta_i;\sigma(\boldsymbol{r}))\rangle\,. \tag{6.127}$$

If $\widehat{V}(N,\zeta_i;\sigma(\boldsymbol{r}))$ is chosen so that

$$\langle \widehat{V}(N,\zeta_i;\sigma(\boldsymbol{r}))\rangle = 0\,, \tag{6.128}$$

the transformation satisfies the lower-bound condition (6.115).

6.9 The Cumulant Approximation

We now return to the problem, discussed in Sect. 6.7, of using decimation for the two-dimensional Ising model In this case we adopt the more systematic first-order cumulant approximation described in Sect. 6.8.1 to construct an upper-bound transformation. The zeroth-order Hamiltonian $\widehat{H}_0$ is taken to include the trivial coupling C and the nearest-neighbour coupling K_1. Terms corresponding to a second-neighbour coupling K_2 and a four-site coupling K_3 are included in $\Delta\widehat{H}$. Because the four-site term links together pairs of white sites in Fig. 6.4 the method must be based on pairs of neighbouring squares like ABCD and CDEF. If the Hamiltonian is taken to be made up of overlapping pairs of squares, each square is counted four times and thus factors arising from a single square should be reduced by a factor of four. Denoting the contribution to any factor from the pair of squares by the symbol $[..]_{\square\square}$

$$\left[\sum_{\{\sigma(\boldsymbol{r})\}}\left\{\prod_{\{\tilde{\boldsymbol{r}}\}} p(\sigma(\tilde{\boldsymbol{r}})|\sigma(\boldsymbol{r}),\boldsymbol{r}\in\mathcal{B}(\tilde{\boldsymbol{r}}))\right\}\exp[-\widehat{H}_0(N,C,K_i;\sigma(\boldsymbol{r}))]\right]_{\square\square}$$
$$= \exp(C)\{4\cosh[K_1(\sigma_\mathsf{A}+\sigma_\mathsf{B}+\sigma_\mathsf{C}+\sigma_\mathsf{D})]\}^{1/4}$$
$$\times\{\cosh[K_1(\sigma_\mathsf{E}+\sigma_\mathsf{F}+\sigma_\mathsf{C}+\sigma_\mathsf{D})]\}^{1/4}\,, \tag{6.129}$$

$$[\exp\langle -\Delta\widehat{H}(N,C,K_i;\sigma(\boldsymbol{r}))\rangle_0]_{\square\square} = \exp\Big\{K_2^{(1)}(\sigma_{\mathrm{A}}\sigma_{\mathrm{B}}+\sigma_{\mathrm{E}}\sigma_{\mathrm{F}})$$

$$+K_2^{(2)}\sigma_{\mathrm{C}}\sigma_{\mathrm{D}}+K_2^{(3)}(\sigma_{\mathrm{A}}\sigma_{\mathrm{D}}+\sigma_{\mathrm{B}}\sigma_{\mathrm{C}}+\sigma_{\mathrm{E}}\sigma_{\mathrm{D}}+\sigma_{\mathrm{F}}\sigma_{\mathrm{C}})$$

$$+(K_2+\sigma_{\mathrm{C}}\sigma_{\mathrm{D}}K_3)\tanh[K_1(\sigma_{\mathrm{A}}+\sigma_{\mathrm{B}}+\sigma_{\mathrm{C}}+\sigma_{\mathrm{D}})]$$

$$\times\tanh[K_1(\sigma_{\mathrm{E}}+\sigma_{\mathrm{F}}+\sigma_{\mathrm{C}}+\sigma_{\mathrm{D}})]\Big\}\,, \tag{6.130}$$

where

$$2K_2^{(1)}+K_2^{(2)}+4K_2^{(3)}=K_2\,. \tag{6.131}$$

As in the simpler situation described in Sect. 6.7 the product of the right-hand sides of (6.129) and (6.130) can be re-expressed in the form

$$\begin{aligned}
\exp\Big\{&C+\tfrac{1}{2}\mathcal{K}_0+\mathcal{K}_1^{(1)}(\sigma_{\mathrm{A}}\sigma_{\mathrm{B}}+\sigma_{\mathrm{E}}\sigma_{\mathrm{F}})+\mathcal{K}_1^{(2)}\sigma_{\mathrm{C}}\sigma_{\mathrm{D}}\\
&+\mathcal{K}_1^{(3)}(\sigma_{\mathrm{A}}\sigma_{\mathrm{D}}+\sigma_{\mathrm{B}}\sigma_{\mathrm{C}}+\sigma_{\mathrm{E}}\sigma_{\mathrm{D}}+\sigma_{\mathrm{F}}\sigma_{\mathrm{C}})\\
&+\tfrac{1}{4}\mathcal{K}_2(\sigma_{\mathrm{A}}\sigma_{\mathrm{C}}+\sigma_{\mathrm{B}}\sigma_{\mathrm{D}}+\sigma_{\mathrm{C}}\sigma_{\mathrm{E}}+\sigma_{\mathrm{D}}\sigma_{\mathrm{F}})\\
&+\tfrac{1}{4}\mathcal{K}_3\sigma_{\mathrm{C}}\sigma_{\mathrm{D}}(\sigma_{\mathrm{A}}\sigma_{\mathrm{B}}+\sigma_{\mathrm{E}}\sigma_{\mathrm{F}})\\
&+\mathcal{K}_4(\sigma_{\mathrm{A}}\sigma_{\mathrm{F}}+\sigma_{\mathrm{B}}\sigma_{\mathrm{E}})+\mathcal{K}_5(\sigma_{\mathrm{A}}\sigma_{\mathrm{E}}+\sigma_{\mathrm{B}}\sigma_{\mathrm{F}})+\mathcal{K}_6\sigma_{\mathrm{A}}\sigma_{\mathrm{B}}\sigma_{\mathrm{E}}\sigma_{\mathrm{F}}\\
&+\mathcal{K}_7\sigma_{\mathrm{C}}\sigma_{\mathrm{D}}(\sigma_{\mathrm{A}}\sigma_{\mathrm{E}}+\sigma_{\mathrm{B}}\sigma_{\mathrm{F}})+\mathcal{K}_8\sigma_{\mathrm{C}}\sigma_{\mathrm{D}}(\sigma_{\mathrm{A}}\sigma_{\mathrm{F}}+\sigma_{\mathrm{B}}\sigma_{\mathrm{E}})\\
&+\mathcal{K}_9[\sigma_{\mathrm{A}}\sigma_{\mathrm{B}}(\sigma_{\mathrm{C}}\sigma_{\mathrm{F}}+\sigma_{\mathrm{D}}\sigma_{\mathrm{E}})+\sigma_{\mathrm{E}}\sigma_{\mathrm{F}}(\sigma_{\mathrm{A}}\sigma_{\mathrm{D}}+\sigma_{\mathrm{B}}\sigma_{\mathrm{C}})]\\
&+\mathcal{K}_{10}[\sigma_{\mathrm{A}}\sigma_{\mathrm{B}}(\sigma_{\mathrm{D}}\sigma_{\mathrm{F}}+\sigma_{\mathrm{C}}\sigma_{\mathrm{E}})+\sigma_{\mathrm{E}}\sigma_{\mathrm{F}}(\sigma_{\mathrm{A}}\sigma_{\mathrm{C}}+\sigma_{\mathrm{B}}\sigma_{\mathrm{D}})]\\
&+\mathcal{K}_{11}\sigma_{\mathrm{A}}\sigma_{\mathrm{B}}\sigma_{\mathrm{C}}\sigma_{\mathrm{D}}\sigma_{\mathrm{E}}\sigma_{\mathrm{F}}\Big\}\,,
\end{aligned} \tag{6.132}$$

where

$$\mathcal{K}_0=\tfrac{1}{16}(K_2+K_3)[\tanh(4K_1)+2\tanh(2K_1)]^2$$

$$+\ln(2)+\tfrac{1}{8}\ln\cosh(4K_1)+\tfrac{1}{2}\ln\cosh(2K_1)\,, \tag{6.133}$$

$$\mathcal{K}_1^{(1)}=\tfrac{1}{32}(K_2+K_3)[\tanh^2(4K_1)-4\tanh^2(2K_1)]$$

$$+K_2^{(1)}+\tfrac{1}{32}\ln\cosh(4K_1)\,, \tag{6.134}$$

$$\mathcal{K}_1^{(2)}=\tfrac{1}{32}(K_2+K_3)[\tanh(4K_1)+2\tanh(2K_1)]^2$$

$$+K_2^{(2)}+\tfrac{1}{16}\ln\cosh(4K_1)\,, \tag{6.135}$$

$$\mathcal{K}_1^{(3)} = \tfrac{1}{32}(K_2 + K_3)\tanh(4K_1)[\tanh(4K_1) + 2\tanh(2K_1)]$$
$$+ K_2^{(3)} + \tfrac{1}{32}\ln\cosh(4K_1)\,, \tag{6.136}$$

$$\mathcal{K}_2 = \tfrac{1}{8}(K_2 + K_3)\tanh(4K_1)[\tanh(4K_1) + 2\tanh(2K_1)]$$
$$+ \tfrac{1}{8}\ln\cosh(4K_1)\,, \tag{6.137}$$

$$\mathcal{K}_3 = \tfrac{1}{8}[\tanh^2(4K_1) - 4\tanh^2(2K_1)]$$
$$+ \tfrac{1}{8}\ln\cosh(4K_1) - \tfrac{1}{2}\ln\cosh(2K_1)\,, \tag{6.138}$$

$$\mathcal{K}_4 = \tfrac{1}{32}[(K_2 + K_3)\tanh(4K_1) + 4(K_2 - K_3)\tanh(2K_1)]\,, \tag{6.139}$$

$$\mathcal{K}_5 = \tfrac{1}{32}[(K_2 + K_3)\tanh(4K_1) + 4(K_2 - K_3)\tanh(2K_1)]\,, \tag{6.140}$$

$$\mathcal{K}_6 = \tfrac{1}{32}(K_2 + K_3)[\tanh(4K_1) - 2\tanh(2K_1)]^2\,, \tag{6.141}$$

$$\mathcal{K}_7 = \tfrac{1}{32}[(K_2 + K_3)\tanh(4K_1) - 4(K_2 - K_3)\tanh(2K_1)]\,, \tag{6.142}$$

$$\mathcal{K}_8 = \tfrac{1}{32}[(K_2 + K_3)\tanh(4K_1) - 4(K_2 - K_3)\tanh(2K_1)]\,, \tag{6.143}$$

$$\mathcal{K}_9 = \tfrac{1}{32}(K_2 + K_3)\tanh(4K_1)[\tanh(4K_1) - 2\tanh(2K_1)]\,, \tag{6.144}$$

$$\mathcal{K}_{10} = \tfrac{1}{32}(K_2 + K_3)\tanh(4K_1)[\tanh(4K_1) - 2\tanh(2K_1)]\,, \tag{6.145}$$

$$\mathcal{K}_{11} = \tfrac{1}{32}(K_2 + K_3)[\tanh(4K_1) - 2\tanh(2K_1)]^2\,. \tag{6.146}$$

With

$$\mathcal{K}_1(K_1, K_2, K_3) = 2\mathcal{K}_1^{(1)}(K_1, K_2^{(1)}, K_2, K_3) + \mathcal{K}_1^{(2)}(K_1, K_2^{(2)}, K_2, K_3)$$
$$+ 4\mathcal{K}_1^{(3)}(K_1, K_2^{(3)}, K_2, K_3)\,, \tag{6.147}$$

recurrence relationships of the form (6.17) and (6.18) with $d = 2$, $\lambda = \sqrt{2}$ and $n = 10$ can now be constructed. They reduce to the recurrence relationships given by (6.96)–(6.101), when $K_2 = K_3 = 0$. $\widetilde{K}_1$ and $\widetilde{K}_2$ are the nearest- and second-neighbour couplings on a square like ABCD and $\widetilde{K}_3$ is a four-site coupling on such a square. The couplings $\widetilde{K}_i$, $i = 4, 5, \ldots, 10$ have been generated by the transformation and link sites. Sneddon and

Barber (1977) have investigated the approximation where these proliferating couplings are neglected and only nearest-neighbour, second-neighbour and the four-site coupling on the square are retained. They show that the only fixed point not corresponding to either zero or infinite values of the couplings is $K_1^* = 0.3683$, $K_2^* = 0.1275$, $K_3^* = -0.0303$. Linearizing about this fixed point yields the one relevant eigenvalue $\Lambda_1 = 1.4287$, giving the thermal exponent $y_1 = y_T = 1.029$ with $\nu = 1/y_T = 0.9714$. This represent a considerable improvement over the results of the simpler approximation described in Sect. 6.7. It is tempting to suppose that cumulant approximations to higher orders would yield further improvement. However this does not seem to be the case. Kadanoff (1976b) reports that unpublished results by Houghton and himself yielded respectively $y_T = 1.002$ and $y_T = 1.7$, for the second- and third-order extension to the method described here. This problem with the cumulant approximation does not appear to be related to the difficulties associated with the decimation weight function. Hsu and Gunton (1977) have used the Kadanoff weight function (6.25) to investigate second- and third-order cumulant expansions for the Ising model on the square and simple cubic lattice. They find that the best results are achieved in the decimation limit $\tau \to \infty$ and that, although reasonable results are achieved for the critical temperature and magnetic field exponent, the results for the thermal exponent are poor.

6.10 Bond Moving Approximations

The possibility of achieving a lower-bound transformation in the way described in Sect. 6.8 by defining a function $\widehat{V}(N, \zeta_i; \sigma(\boldsymbol{r}))$ which moved the interaction between the microsystems of one pair of sites to another pair was first discussed by Kadanoff (1975) (see also, Kadanoff et al. 1976). To illustrate this idea consider the zero-field nearest-neighbour Ising model on a hypercubic lattice with Hamiltonian

$$\widehat{H}(N, C, K; \sigma(\boldsymbol{r})) = -NC - K \sum_{\{\boldsymbol{r},\boldsymbol{r}'\}}^{(\text{n.n.})} \sigma(\boldsymbol{r})\sigma(\boldsymbol{r}') , \tag{6.148}$$

where $K = J/T$. Now choose any two nearest-neighbour pairs of sites $\boldsymbol{r}_1$,$\boldsymbol{r}_1'$ and $\boldsymbol{r}_2$,$\boldsymbol{r}_2'$. Remove the interaction (or *bond*) from the first pair and add it to the bond on the second pair. Thus

$$\widehat{V}(N, K; \sigma(\boldsymbol{r})) = K[\sigma(\boldsymbol{r}_1)\sigma(\boldsymbol{r}_1') - \sigma(\boldsymbol{r}_2)\sigma(\boldsymbol{r}_2')] \tag{6.149}$$

and

$$\langle \widehat{V}(N, K; \sigma(\boldsymbol{r}))\rangle = K[\langle\sigma(\boldsymbol{r}_1)\sigma(\boldsymbol{r}_1')\rangle - \langle\sigma(\boldsymbol{r}_2)\sigma(\boldsymbol{r}_2')\rangle] = 0 . \tag{6.150}$$

Of course, this result does not in itself lead to a tractable renormalization scheme. It is necessary to move bonds in a systematic way throughout the

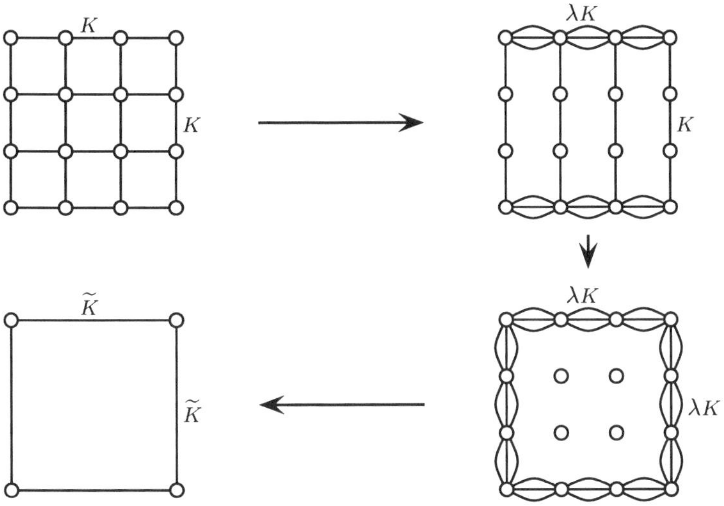

Fig. 6.6. Bond moving and decimation operations on a block of 3 × 3 sites for the Migdal–Kadanoff transformation applied to the zero-field nearest-neighbour Ising model on a square lattice.

lattice so that the summation on the right-hand side of (6.126) can be performed, leading to a Hamiltonian $\widehat{H}(\widetilde{N}, \widetilde{C}, \widetilde{K}; \sigma(\tilde{\boldsymbol{r}}))$ of the same form (6.148) as $\widehat{H}(N, C, K; \sigma(\boldsymbol{r}))$. This can be achieved using the *Migdal–Kadanoff transformation*, first invented by Migdal (1975) and reinterpreted using bond moving ideas by Kadanoff (1976b). In Kadanoff's formulation of this method bonds are moved so that sites can be eliminated by decimation.

As an example of the use of the Migdal–Kadanoff transformation consider the zero-field nearest-neighbour Ising model on a square lattice. Taking a block of $\lambda \times \lambda$ squares, as shown in Fig. 6.6 (for the case $\lambda = 3$) and move the horizontal interior bonds to the upper edge and the vertical interior bonds to the left edge. The interior sites do not now interact with their neighbours, so they will simply each contribute a factor '2' to summation on the left-hand side of (6.126). The coupling on each segment of the boundary will be λK. These sites form a one-dimensional chain connecting the corners to which we can apply the decimation formula (6.93), which together with (6.69) gives

$$\widetilde{K} = \operatorname{arctanh}\{[\tanh(\lambda K)]^{\lambda}\}\,. \tag{6.151}$$

The recurrence relation

$$\widetilde{K} = \operatorname{arctanh}\{[\tanh(\lambda^{d-1} K)]^{\lambda}\}\,, \tag{6.152}$$

obtained by Migdal (1975) for the d-dimensional hypercubic lattice is a simple generalization of (6.151). Equation (6.151) with $\lambda = 2$ can be re-expressed in the form

$$\widetilde{K} = \tfrac{1}{2}\ln[\cosh(4K)]\,, \tag{6.153}$$

giving the critical fixed point $K^* = K_c = 0.304689$ with $T_c/J = 3.282$ and $\nu = 1/y_T = \ln(2)/\ln[2\tanh(4K_c)] = 1.3383$. Neither the critical coupling nor the exponent compare well with the exact values for the two-dimensional Ising model ($K_c = 0.4407$, $\nu = 1$, see Sect. 5.10) and the comparison deteriorates with increasing λ giving for K_c the values 0.2406 and 0.2018 for $\lambda = 3$ and 4 respectively.

An interesting a modification of the method has been proposed by Kadanoff (1976b). He considered the anisotropic Ising model on the d-dimensional hypercubic lattice with different couplings in each axis direction. In the Migdal formulation this would simply lead to the result that each coupling satisfied (6.152) and the fixed points for different values of λ would have all the couplings equal. In the modified method of Kadanoff, instead of the bond moving for all the axes preceding decimation, bond moving and decimation are performed alternately. We illustrate this method for the case $d = 2$ with K_h and K_v denoting horizontal and vertical couplings. First all the internal horizontal bonds are moved to the edges of the $\lambda \times \lambda$ square as shown in the first line of Fig. 6.6. Decimation is then performed in the vertical direction after which the vertical bonds are moved to the edges of the square. Finally decimation is carried out in the horizontal direction. The recurrence relations achieved by this procedure are

$$\begin{aligned} \widetilde{K}_h &= \operatorname{arctanh}\{[\tanh(\lambda K_h)]^\lambda\}\,, \\ \widetilde{K}_v &= \lambda \operatorname{arctanh}\{[\tanh(K_v)]^\lambda\}\,. \end{aligned} \tag{6.154}$$

It is clear that, for any λ, the critical fixed point is now given by

$$K_h^* = \lambda^{-1} K_v^* = K^*\,, \tag{6.155}$$

where K^* is the fixed point value given by (6.151). Substituting for λ from (6.155) into one of the equations (6.154) gives

$$K_v^* \ln[\tanh(K_v^*)] = K_h^* \ln[\tanh(K_h^*)]\,. \tag{6.156}$$

Equation (6.156) has branches[4]

$$K_v^* = K_h^* \tag{6.157}$$

and

$$\sinh(2K_v^*)\sinh(2K_h^*) = 1\,. \tag{6.158}$$

The formula (6.157), which is satisfied by the Migdal critical fixed point, disagrees with (6.155), except when $\lambda = 1$, when (6.158) gives the exact critical temperature for the isotropic Ising model (Sect. 5.10). In general (6.158) gives the critical curve for the anisotropic Ising model (Baxter 1982a). It is, however, a curious property of the method that the individual points on

[4] Equation (6.157) is obvious. Equation (6.158), which is equivalent to $K_v^* = -\frac{1}{2}\ln[\tanh(K_h^*)]$, follows from the fact that this latter equation is invariant under interchange of K_v^* and K_h^*.

this curve are achieved by choosing different (including non-integer) values of λ. This is because, for any chosen value of λ, the couplings renormalize independently and thus can yield only a finite number of discrete points, not a continuous curve. It can be shown that both the Migdal method and the Kadanoff modification lead to the same values for the exponent ν, with the 'best' value being achieve in the limit $\lambda \to 1$ when $\nu = 1.326$.

6.11 Finite-Lattice Approximations

The methods of this type all consist of taking a lattice with a relatively small number of sites and then applying the procedure of Sect. 6.2. In order for the renormalization group transformation to be able to be applied once it must be the case that $\widetilde{N} \geq \lambda^d$ and thus $N \geq \lambda^{2d}$. Within this framework there is still the choice of weight function and of whether to apply periodic boundary conditions to the lattice. Having made these choices the recurrence relationships can be computed exactly. As has been remarked in Sect. 6.3 the approximation here consists of obtaining the fixed points and phase diagram by iterating the recurrence relationships a large number of times. Early applications of this method to the Ising model were by Nauenberg and Nienhuis (1974a, 1974b) who applied the method to four-site blocks on a square lattice and Tjon (1974), who used four three-site blocks on a triangular lattice. The former authors used periodic boundary conditions and the latter did not. Rather than present these calculations, which are described in Niemeijer and van Leeuwen (1976), we shall use as an example the work of Schick et al. (1976, 1977). They used a nine site triangular lattice with periodic boundary conditions to study the antiferromagnetic Ising model in a magnetic field. Given that antiferromagnetism leads to sublattice ordering, in this case on the three sublattices A, B and C shown in Fig. 6.7, the blocks are chosen to reflect this phenomenon. As is shown in Fig. 6.7 the three sites A_1, A_2 and A_3 of sublattice A form a block with block-spin site located at $\widetilde{\mathsf{A}}$, with a similar scheme applying to sublattices B and C. The Hamiltonian for this model has the form

$$\widehat{H}(N, C, L, K, M; \sigma) = -NC - \sum_{\triangle} \Big\{ \tfrac{1}{6} L(\sigma_{\mathrm{A}} + \sigma_{\mathrm{B}} + \sigma_{\mathrm{C}}) + M\sigma_{\mathrm{A}}\sigma_{\mathrm{B}}\sigma_{\mathrm{C}}$$

$$+ \tfrac{1}{2} K(\sigma_{\mathrm{A}}\sigma_{\mathrm{B}} + \sigma_{\mathrm{B}}\sigma_{\mathrm{C}} + \sigma_{\mathrm{C}}\sigma_{\mathrm{A}}) \Big\}, \qquad (6.159)$$

where C, L and K are, as in Sect. 6.6, the trivial, magnetic field and nearest-neighbour couplings and M is a three-spin coupling. For the lattice $\mathcal{N}$, $N = 9$ and the summation is over all twenty-four triangles of the lattice. This yields a lattice $\widetilde{\mathcal{N}}$ with $\widetilde{N} = 3$ and the only triangle in the summation is $\widetilde{\mathsf{A}}\widetilde{\mathsf{B}}\widetilde{\mathsf{C}}$, which must be counted six times because of the periodic boundary conditions. Thus, in this transformation, $\lambda = \sqrt{3}$. Defining

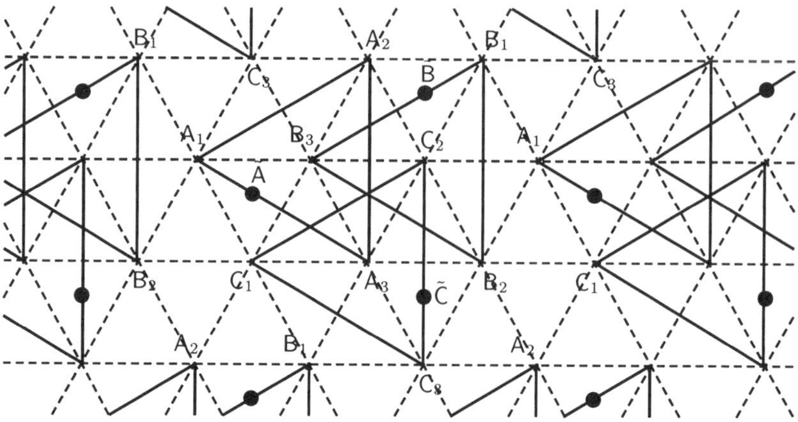

Fig. 6.7. The sublattice and block structure for the triangular lattice showing the form of the periodic boundary conditions.

$$\mathfrak{Z}_0 = \exp(C), \qquad \mathfrak{Z}_1 = \exp(L),$$
$$\mathfrak{Z}_2 = \exp(K), \qquad \mathfrak{Z}_3 = \exp(M), \tag{6.160}$$

(6.16) yields the four independent equations

$$[\tilde{\mathfrak{Z}}_0]^3[\tilde{\mathfrak{Z}}_1]^3[\tilde{\mathfrak{Z}}_2]^9[\tilde{\mathfrak{Z}}_3]^6 = \mathfrak{Z}_0^9 \mathcal{Z}(+1,+1,+1), \tag{6.161}$$

$$[\tilde{\mathfrak{Z}}_0]^3[\tilde{\mathfrak{Z}}_1]^{-3}[\tilde{\mathfrak{Z}}_2]^9[\tilde{\mathfrak{Z}}_3]^{-6} = \mathfrak{Z}_0^9 \mathcal{Z}(-1,-1,-1), \tag{6.162}$$

$$[\tilde{\mathfrak{Z}}_0]^3[\tilde{\mathfrak{Z}}_1][\tilde{\mathfrak{Z}}_2]^3[\tilde{\mathfrak{Z}}_3]^{-6} = \mathfrak{Z}_0^9 \mathcal{Z}(+1,+1,-1), \tag{6.163}$$

$$[\tilde{\mathfrak{Z}}_0]^3[\tilde{\mathfrak{Z}}_1]^{-1}[\tilde{\mathfrak{Z}}_2]^{-3}[\tilde{\mathfrak{Z}}_3]^6 = \mathfrak{Z}_0^9 \mathcal{Z}(-1,-1,+1), \tag{6.164}$$

where $\mathcal{Z}(\sigma_{\tilde{A}}, \sigma_{\tilde{B}}, \sigma_{\tilde{C}})$ is the expression for the right-hand side of (6.16), excluding the trivial coupling, for the designated values of the block spins. These are given (Schick et al. 1977) by

$$\mathcal{Z}(+1,+1,+1) = \mathfrak{Z}_1^9 \mathfrak{Z}_2^{27} \mathfrak{Z}_3^{18} + 9\mathfrak{Z}_1^7 \mathfrak{Z}_2^{15} \mathfrak{Z}_3^6 + 27\mathfrak{Z}_1^5 \mathfrak{Z}_2^7 \mathfrak{Z}_3^2$$
$$+ 18\mathfrak{Z}_1^3 \mathfrak{Z}_2^3 \mathfrak{Z}_3^{-2} + 9\mathfrak{Z}_1^3 \mathfrak{Z}_2^3 \mathfrak{Z}_3^6, \tag{6.165}$$

$$\mathcal{Z}(-1,-1,-1) = \mathfrak{Z}_1^{-9} \mathfrak{Z}_2^{27} \mathfrak{Z}_3^{-18} + 9\mathfrak{Z}_1^{-7} \mathfrak{Z}_2^{15} \mathfrak{Z}_3^{-6} + 27\mathfrak{Z}_1^{-5} \mathfrak{Z}_2^7 \mathfrak{Z}_3^{-2}$$
$$+ 18\mathfrak{Z}_1^{-3} \mathfrak{Z}_2^3 \mathfrak{Z}_3^2 + 9\mathfrak{Z}_1^{-3} \mathfrak{Z}_2^3 \mathfrak{Z}_3^{-6}, \tag{6.166}$$

$$\mathcal{Z}(+1,+1,-1) = \mathfrak{Z}_1^3 \mathfrak{Z}_2^{-9} \mathfrak{Z}_3^{-18} + 6\mathfrak{Z}_1 \mathfrak{Z}_2^{-9} \mathfrak{Z}_3^{-6} + 3\mathfrak{Z}_1^5 \mathfrak{Z}_2^3 \mathfrak{Z}_3^{-6}$$
$$+ 9\mathfrak{Z}_1^{-1} \mathfrak{Z}_2^{-5} \mathfrak{Z}_3^{-2} + 18\mathfrak{Z}_1^3 \mathfrak{Z}_2^{-1} \mathfrak{Z}_3^{-2}$$
$$+ 18\mathfrak{Z}_1 \mathfrak{Z}_2^{-1} \mathfrak{Z}_3^2 + 9\mathfrak{Z}_1 \mathfrak{Z}_2^{-1} \mathfrak{Z}_3^{-6}, \tag{6.167}$$

$$\begin{aligned}\mathcal{Z}(-1,-1,+1) = {} & \mathfrak{z}_1^{-3}\mathfrak{z}_2^{-9}\mathfrak{z}_3^{18} + 6\mathfrak{z}_1^{-1}\mathfrak{z}_2^{-9}\mathfrak{z}_3^{6} + 3\mathfrak{z}_1^{-5}\mathfrak{z}_2^{3}\mathfrak{z}_3^{6} \\ & + 9\mathfrak{z}_1^{1}\mathfrak{z}_2^{-5}\mathfrak{z}_3^{2} + 18\mathfrak{z}_1^{-3}\mathfrak{z}_2^{-1}\mathfrak{z}_3^{2} \\ & + 18\mathfrak{z}_1^{-1}\mathfrak{z}_2^{-1}\mathfrak{z}_3^{-2} + 9\mathfrak{z}_1^{-1}\mathfrak{z}_2^{-1}\mathfrak{z}_3^{6}\,.\end{aligned} \tag{6.168}$$

From (6.161)–(6.168),

$$\begin{aligned}\tilde{\mathfrak{z}}_0 = {} & \mathfrak{z}_0^3\,[\mathcal{Z}(+1,+1,+1)\mathcal{Z}(-1,-1,-1)]^{1/24} \\ & \times\,[\mathcal{Z}(+1,+1,-1)\mathcal{Z}(-1,-1,+1)]^{1/8}\,,\end{aligned} \tag{6.169}$$

$$\tilde{\mathfrak{z}}_1 = \left[\frac{\mathcal{Z}(+1,+1,+1)\mathcal{Z}(+1,+1,-1)}{\mathcal{Z}(-1,-1,-1)\mathcal{Z}(-1,-1,+1)}\right]^{1/8}, \tag{6.170}$$

$$\tilde{\mathfrak{z}}_2 = \left[\frac{\mathcal{Z}(+1,+1,+1)\mathcal{Z}(-1,-1,-1)}{\mathcal{Z}(+1,+1,-1)\mathcal{Z}(-1,-1,+1)}\right]^{1/24}, \tag{6.171}$$

$$\tilde{\mathfrak{z}}_3 = \left[\frac{\mathcal{Z}(+1,+1,+1)}{\mathcal{Z}(-1,-1,-1)}\right]^{1/48}\left[\frac{\mathcal{Z}(-1,-1,+1)}{\mathcal{Z}(+1,+1,-1)}\right]^{1/16}. \tag{6.172}$$

These are the recurrence relationships for the model. It can be seen that there is not a consistent solution given simply by setting $\mathfrak{z}_3 = 1$, ($M = 0$). However, there is a solution for $\mathfrak{z}_2$ alone when $\mathfrak{z}_1 = \mathfrak{z}_3 = 1$. This is because of the symmetry of the terms of the Hamiltonian. For this type of transformation, once a basic set of sites (a triangle in this case) is chosen, all or none of the possible terms in the Hamiltonian corresponding to a particular symmetry must be included. The term corresponding to K is invariant under spin reversal and it is the only possible function of the spins on a triangle (apart from the trivial coupling C) with this property. The terms corresponding to L and M are the only possible combinations of spins which change sign with spin reversal and either both or neither must be included in the Hamiltonian. When they are present the recurrence relationships are invariant under the simultaneous application of $\mathfrak{z}_1 \leftrightarrow 1/\mathfrak{z}_1$, $\mathfrak{z}_3 \leftrightarrow 1/\mathfrak{z}_3$.

The results derived for this model for the antiferromagnetic case ($K < 0$)in the plane $M = 0$ are presented by Schick et al. (1976) and shown in Fig. 6.8. The transition for $L > 0$ is to the antiferromagnetic phase with ground state $(+1,+1,-1)$ and that for $L < 0$ to the antiferromagnetic phase with ground state $(-1,-1,+1)$. The results show good agreement with the Monte Carlo results of Metcalf (1973). As is well-known there is no non-zero Néel temperature on the zero-field axis and the outer termination points of the curves are at the critical field values $\mathcal{H}_c = \pm 6|J|$. The factor six arises as the coordination number of the lattice (see, for example, Volume 1, Sect. 4.2). The zero-point entropy per lattice site at $\mathcal{H} = 0$ is obtained by the method described in Sect. 6.5 and yields the value 0.324, which is very close to the exact value 0.32306 (see Volume 1, Sect. 8.12). The transitions for $L >, < 0$ are con-

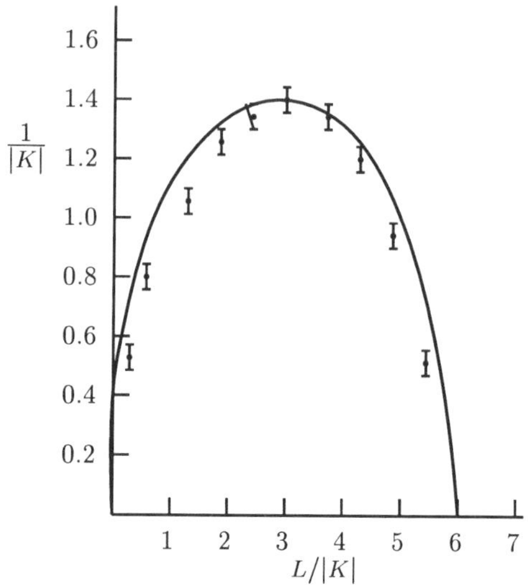

Fig. 6.8. Phase diagram for the continuous phase transition in the plane $M = 0$. The data points are the Monte Carlo results of Metcalf (1973). (Reprinted from Schick et al. (1976), by permission of the publisher Elsevier Science.)

trolled respectively by the fixed points $L^* = \pm\infty$, $K^* = -\infty$, $M^* = \pm 0.113$, $|L^*| + 6K^* + 6|M^*| \sim -0.91$. The two antiferromagnetic phases are each three-fold degenerate corresponding to the three choices for the alignment of spins on the sublattices. This set of six ground states is the same as that of the 3-state antiferromagnetic Potts model (Schick and Griffiths 1977). Alexander (1975) has shown that this leads to a cubic invariant in the Landau theory formulation of exactly the form exhibited by the 3-state Potts model. This implies a first-order transition, a result which for the Potts model is confirmed by other methods (see the discussion in Sect. 3.5).

The physical motivation for the work of Schick et al. (1977) is to formulate a model for the order-disorder transition which occurs for gases adsorbed on monolayer films; in particular the case of helium adsorbed on graphite. While the general fit of the curves in Fig. 6.8 with experimental data for this model is good the work by Bretz (1977) for this system yielded a value $\alpha = 0.34$. The thermal exponent derived from these model calculations has the value 0.956, which from equation (2.167) gives the value $\alpha = -0.093$ for the heat capacity exponent. One might speculate that the exponent -0.26 for this fixed point is associated with a dangerous irrelevant variable (see Sect. 2.14). To achieve the value $\alpha = 0.34$ it follows from (2.167) that the effective dimension of the system would need to be $\bar{d} = 1.587$. This in turn would imply from equations (2.267) and (2.270) that the divergence associated with the correction to scaling has exponent 1.5885. This, however, is mere speculation and further calculations would be needed to settle this point. The global phase diagrams for $M \neq 0$ for both signs of K are described in detail by Schick et al. (1977).

They show that a very rich phase structure is exhibited with first-order phase transition surfaces, and lines of critical, tricritical and critical end points. This renormalization scheme has been used by Schick and Griffiths (1977) for the 3-state Potts model and by Young and Lavis (1979) for a type of bonded lattice fluid which is an extension of the 3-state Potts model. This latter work was extended by Southern and Lavis (1979, 1980) to include the whole class of terms corresponding to the spin-1 model. A spin-1 model renormalized according to this method was used by Lavis et al. (1982) to model phase transitions in monolayers at air/water and oil/water interfaces and Lavis and Quinn (1983) used a spin-$\frac{1}{2}$ model to investigate ferrimagnetic ordering on a triangular lattice.

In general real-space renormalization group procedures on finite lattices give qualitatively satisfactory phase diagrams, although the values of critical exponents tend to be rather disappointing particularly for models exhibiting sublattice ordering. Although one might expect the results to improve with the choice of larger lattices, the method then becomes much more difficult to handle. In the work of Lavis and Southern (1984), which treated a spin-1 model with directional bonding on a body-centred cubic lattice of sixteen sites the recurrence relations have thirty-three thousand terms, which placed the computations near the limit of tractability.

6.12 Variational Approximations

As we indicated in Sect. 6.3 a crucial factor in renormalization group calculations is the choice of weight function. Although an exact transformation would not be affected by this choice, the results of approximate transformations can change quite dramatically when the weight function is changed. In Sects. 6.6, 6.7, 6.9 and 6.10 examples have been given of the use of the decimation weight function and in Sect. 6.11 the majority-rule weight function was employed. The possibility of using a weight function containing a parameter which can be varied to improve the approximation was considered by Kadanoff and Houghton (1975). They used the weight function (6.25), with parameter τ, for transformations of the spin-$\frac{1}{2}$ Ising model on the square lattice. The parameter τ was chosen by using the criterion that two different evaluations of the magnetic field exponent agreed with each other.

A form of variational approximation, designed to optimize the lower-bound condition (6.115) was developed by Kadanoff (1975) (see also Kadanoff et al. 1976). Known as the *one-hypercube approximation*, it is applicable to a hypercubic lattice of any dimension d. Suppose that the centres of the hypercubes of the lattice are denoted, in terms of the set of independent unit vectors $\hat{\boldsymbol{r}}_1, \hat{\boldsymbol{r}}_2, \ldots, \hat{\boldsymbol{r}}_d$, by vectors of the form

$$\boldsymbol{\rho} = a(m_1\hat{\boldsymbol{r}}_1 + m_2\hat{\boldsymbol{r}}_2 + \ldots + m_d\hat{\boldsymbol{r}}_d)\,, \qquad (6.173)$$

where m_ℓ is an integer modulo N_ℓ, with the set of *even* integers $N_1, N_2, \ldots, N_d$ satisfying (6.2). Let the Hamiltonian be such that all the couplings can be expressed in terms of interactions between microsystems on the sites of a single hypercube. Then

$$\widehat{H}(N, \zeta_i; \sigma(\boldsymbol{r})) = -N\zeta_0 - \sum_{i=1}^{n} \zeta_i \sum_{\{\boldsymbol{\rho}\}} \mathfrak{q}_i(\boldsymbol{\rho}), \tag{6.174}$$

where $\mathfrak{q}_i(\boldsymbol{\rho})$ is a function of the $\sigma(\boldsymbol{r})$ at the sites of the cube with centre $\boldsymbol{\rho}$. We assume that the system satisfies the homogeneity condition that $\langle \mathfrak{q}_i(\boldsymbol{\rho})\rangle$ is independent of $\boldsymbol{\rho}$. Then, if in equation (6.126) we choose the form

$$\widehat{V}(N, \mathrm{X}_i(\boldsymbol{\rho}); \sigma(\boldsymbol{r})) = \sum_{i=1}^{n} \sum_{\{\boldsymbol{\rho}\}} \mathrm{X}_i(\boldsymbol{\rho}) \mathfrak{q}_i(\boldsymbol{\rho}), \tag{6.175}$$

where

$$\sum_{\{\boldsymbol{\rho}\}} \mathrm{X}_i(\boldsymbol{\rho}) = 0, \tag{6.176}$$

the condition (6.128) is satisfied, leading to the lower-bound condition (6.115). The formula (6.149) is a particular case of (6.175). The quantities $\mathrm{X}_i(\boldsymbol{\rho})$ must now be chosen to achieve a tractable summation on the right-hand side of (6.126) with the renormalized Hamiltonian, now denoted by $\widehat{H}(\widetilde{N}, \widetilde{\zeta}_i; \sigma(\tilde{\boldsymbol{r}}))$, of the same form as $\widehat{H}(N, \zeta_i, N; \sigma(\boldsymbol{r}))$. In the one-hypercube approximation this is achieved by dividing the hypercubes of the lattice into subsets. The 'blue' subset, with centres denoted by $\boldsymbol{\rho}_\mathrm{b}$ consists of all hypercubes for which all the integers m_ℓ in equation (6.173) are odd and the 'red' subset with centres denoted by $\boldsymbol{\rho}_\mathrm{r}$, consists of all hypercubes for which all the integers m_ℓ are even. There are $N/2^d$ hypercubes in each of the blue and red subsets and the remaining $N(1-2^{1-d})$ hypercubes, with centres denoted by $\boldsymbol{\rho}_\mathrm{g}$ will be called 'green'. It can now be easily checked that the choice

$$\mathrm{X}_i(\boldsymbol{\rho}) = \begin{cases} \zeta_i - \mathrm{Y}_i\,, & \text{if } \boldsymbol{\rho} = \boldsymbol{\rho}_\mathrm{r}\,, \\ \zeta_i\,, & \text{if } \boldsymbol{\rho} = \boldsymbol{\rho}_\mathrm{g}\,, \\ -(2^d - 1)\zeta_i + \mathrm{Y}_i\,, & \text{if } \boldsymbol{\rho} = \boldsymbol{\rho}_\mathrm{b} \end{cases} \tag{6.177}$$

satisfies (6.176) for any choice of the new parameters Y_i, which are chosen to ensure that the summation on the right-hand side of (6.126) is tractable. For the one-hypercube approximation applied to a spin-$\frac{1}{2}$ model, the renormalization scheme is constructed with the sites $\tilde{\boldsymbol{r}}$ of $\widetilde{\mathcal{N}}$ coinciding with the points $\boldsymbol{\rho}_\mathrm{r}$. The scaling parameter $\lambda = 2$, and the Kadanoff–Houghton weight function (6.25) is used with $\mathcal{X}(\tilde{\boldsymbol{r}}) = \mathcal{B}(\tilde{\boldsymbol{r}})$ taken to be the sites on the red hypercube with centre $\tilde{\boldsymbol{r}} = \boldsymbol{\rho}_\mathrm{r}$. Then

$$\prod_{\{\tilde{r}\}} p(\sigma(\tilde{r})|\sigma(r), r \in \mathcal{B}(\tilde{r}))$$

$$= \exp\left\{ \sum_{\{\tilde{r}\}} [\tau\,\sigma(\tilde{r})\mathfrak{p}(\tilde{r}) - \ln(2\cosh(\tau\,\mathfrak{p}(\tilde{r})))] \right\}, \tag{6.178}$$

where $\mathfrak{p}(\tilde{r})$ is the sum of the variables $\sigma(r)$ on all the sites of the red hypercube with centre $\tilde{r}$. The aim is to reduce the dependence of the argument of the summation on the right-hand side of (6.126) to only those terms involving the interactions between the microstates on blue hypercubes. The first term in the exponential on the right-hand side of (6.178) is no problem since each site at the corner of one red hypercube is also at the corner of one blue hypercube. The second term can be 'moved' from red to blue hypercubes by imposing the condition

$$\sum_{i=1}^{n} Y_i \mathfrak{q}_i(\tilde{r}) = \ln[2\cosh(\tau\,\mathfrak{p}(\tilde{r}))], \tag{6.179}$$

which, when all choices are made for the states on the corners of the cube, is equivalent to 2^d linear equations for the n variables. This means, as was the case of the finite-lattice methods of Sect. 6.11, the number n of non-trivial coupling included in the Hamiltonian must be enough to ensure an consistent solution. A detailed analysis of this method for the square lattice was performed by Burkhardt (1976a) for a Hamiltonian with no magnetic field but with nearest-neighbour, second-neighbour and four-site couplings. The value of the parameter τ which maximized the free energy for an initial Hamiltonian equal to its critical fixed point value was 0.761. For this he obtained $T_c/J = 2.066$, $\nu = 0.983$ and $\delta = 15.36$. The recurrence relationships for this system have an invariant subspace with equal nearest- and second-neighbour interactions. Slightly improved results can be achieved by beginning with an initial Hamiltonian with no second-neighbour interactions and performing an preliminary exact decimation to produce a Hamiltonian within this subspace. This was the method employed by Kadanoff (1975) and Kadanoff et al. (1976). A modification of this method applicable to both square and triangular lattices was developed by Southern (1978) and for the body-centred cubic lattice by Burkhardt and Eisenriegler (1978). It has also been generalized to spin-1 models by Burkhardt (1976b) and Burkhardt and Knops (1977) and to the 3-state Potts model by Burkhardt et al. (1976).

6.13 Phenomenological Renormalization

The idea of finite-size scaling was introduced in Sect. 2.16 as an extension of the scaling methods for infinite systems treated in the rest of Chap. 2. An renormalization group justification for this has been given by Brézin (1982) and Suzuki (1977) using momentum-space methods and by

Barber (1983) using real-space methods. We now discuss a real-space renormalization group method based on finite-size scaling which has been developed by Nightingale (1976).

In Sect. 2.16 we described scaling for a d dimensional system $\mathcal{L}_{\mathfrak{d}}(\aleph)$, which was infinite in $\mathfrak{d}$ dimensions and for which the size of the system for the remaining $d - \mathfrak{d}$ dimensions was represented by a parameter $\aleph$ called the thickness. The essential feature of finite-size scaling is that, for such a system with scaling fields $\theta_1, \theta_2, \ldots, \theta_n$, $1/\aleph$ is treated as a scaling field θ_{n+1}. We now adopt a slight change of notation and express the correlation length $\mathfrak{r}(\theta_1, \ldots, \theta_n, \theta_{n+1})$ for $\mathcal{L}_{\mathfrak{d}}(\aleph)$ as $\mathfrak{r}^{(\aleph)}(\theta_1, \ldots, \theta_n)$. Equation (2.301) can now be used to relate the correlation lengths of two similar systems $\mathcal{L}_{\mathfrak{d}}(\aleph)$ and $\mathcal{L}_{\mathfrak{d}}(\widetilde{\aleph})$ with couplings ζ_i and $\tilde{\zeta}_i$ and thicknesses $\aleph$ and $\widetilde{\aleph}$ respectively. Taking $\aleph > \widetilde{\aleph}$ and making the substitution

$$\lambda = \aleph/\widetilde{\aleph}\,, \tag{6.180}$$

we have

$$\mathfrak{r}^{(\aleph)}(\theta_1, \ldots, \theta_n) = \lambda \mathfrak{r}^{(\widetilde{\aleph})}(\tilde{\theta}_1\,, \ldots, \tilde{\theta}_n), \tag{6.181}$$

where

$$\tilde{\theta}_j = \theta_j(\tilde{\zeta}_i) = \lambda^{y_j} \theta_j(\zeta_i) \tag{6.182}$$

relates the scaling fields θ_j and $\tilde{\theta}_j$ for $\mathcal{L}_{\mathfrak{d}}(\aleph)$ and $\mathcal{L}_{\mathfrak{d}}(\widetilde{\aleph})$. Equation (6.181) forms the basis of Nightingale's phenomenological renormalization method. For the infinite system $\mathcal{L}$, defined by the thermodynamic limit $\mathcal{L}_{\mathfrak{d}}(\aleph) \to \mathcal{L}$ as $\aleph \to \infty$ the correlation length $\mathfrak{r}$ satisfies an equation similar to (6.181) except that the same function appears on both sides of the equation in place of the correlation functions for two systems of different thickness. If, however, we suppose that in the limit $\aleph, \widetilde{\aleph} \to \infty$ with λ fixed the correlation lengths for the finite systems approach that of the infinite system, then (6.181) can be taken as an approximation to the scaling form for $\mathcal{L}$. This has the advantage, as we shall see below, that the correlation length for $\mathcal{L}_{\mathfrak{d}}(\aleph)$ can be calculated explicitly as functions of the couplings ζ_i, $i = 1, 2, \ldots, n$, at least for fairly small values of $\aleph$, and we can reasonably suppose that the approximation will increase in accuracy as $\aleph$ is increased. Equation (6.181) can be rewritten as an expression relating the couplings for $\mathcal{L}_{\mathfrak{d}}(\aleph)$ and $\mathcal{L}_{\mathfrak{d}}(\widetilde{\aleph})$. Thus

$$\mathfrak{r}^{(\aleph)}(\zeta_1, \ldots, \zeta_n) = \lambda \mathfrak{r}^{(\widetilde{\aleph})}(\tilde{\zeta}_1, \ldots, \tilde{\zeta}_n)\,. \tag{6.183}$$

If we suppose that there is a renormalization transformation of the form of (6.18) relating the coupling of $\mathcal{L}_{\mathfrak{d}}(\aleph)$ and $\mathcal{L}_{\mathfrak{d}}(\widetilde{\aleph})$ then a fixed point will satisfy the equation

$$\mathfrak{r}^{(\aleph)}(\zeta_1^*, \ldots, \zeta_n^*) = \lambda \mathfrak{r}^{(\widetilde{\aleph})}(\zeta_1^*, \ldots, \zeta_n^*)\,. \tag{6.184}$$

It is, however, clear that for $n > 1$ this transformation is not completely specified by (6.183).

A number of methods for dealing with systems with more than one coupling have been developed and will be described below. We first concentrate on the case $n = 1$, with $\zeta_1 = K = J/T$, where J is some interaction energy between microsystems. The renormalization transformation $K \to \widetilde{K} = \mathcal{K}(K)$ is now defined by

$$\mathfrak{r}^{(\aleph)}(K) = \lambda \mathfrak{r}^{(\tilde{\aleph})}(\widetilde{K}) \tag{6.185}$$

and the fixed points by

$$\mathfrak{r}^{(\aleph)}(K^*) = \lambda \mathfrak{r}^{(\tilde{\aleph})}(K^*) . \tag{6.186}$$

If this transformation is a satisfactory approximation to a system with a non-zero critical temperature T_c, then we expect a finite non-zero critical fixed point solution $K^* = K_c = J/T_c$ to equation (6.186). The scaling field can now be chosen as

$$\theta_1 = K - K^* . \tag{6.187}$$

From (6.182) the thermal exponent $y_K = 1/\nu$ is given by

$$\lambda^{y_K} = \left(\frac{\partial \widetilde{K}}{\partial K} \right)^* \tag{6.188}$$

and, from (6.185),

$$y_K = \frac{\ln \left\{ \left(\frac{\partial \mathfrak{r}^{(\aleph)}}{\partial K} \right)^* \left[\left(\frac{\partial \mathfrak{r}^{(\tilde{\aleph})}}{\partial \widetilde{K}} \right)^* \right]^{-1} \right\}}{\ln(\lambda)} - 1 . \tag{6.189}$$

In Sect. 4.9 we considered the application of the transfer matrix method to a d-dimensional system in which the thermodynamic limit is taken in one dimension. In the terminology of finite-size scaling this corresponds to $\mathfrak{d} = 1$ and, subject to the assumptions described in Sect. 4.9, the correlation length for a system, for which the ground state of the ordered phase has a periodicity η, is given in terms of the ratio of the two largest eigenvalues of the transfer matrix by (4.156). This formula provides the starting point for most phenomenological renormalization calculations. It is clearly of importance to check the accuracy of the method for models for which the eigenvalues of the transfer method are known exactly for any width of system. This has been done for the isotropic and anisotropic square lattice spin-$\frac{1}{2}$ Ising models by Nightingale (1976) and Sneddon (1978), respectively, for the triangular lattice spin-$\frac{1}{2}$ Ising model by Kinzel and Schick (1981) and for the Baxter model (or symmetric eight-vertex model with $J_{21} = J_{22}$ in equation (5.6)) by Nightingale (1977). As an example we consider the simplest case of the isotropic square lattice spin-$\frac{1}{2}$ Ising model.

6.13.1 The Square Lattice Ising Model

We consider a lattice of $N_1 N_2 = N$ sites with the thermodynamic limit being taken for $N_2 \to \infty$ and the thickness of the lattice $\aleph = N_1$. Expressed in lattice fluid form the complete set of eigenvalues for this model is given by equations (4.181)–(4.183). The largest and next largest eigenvalues correspond to taking all plus signs in (4.182) and (4.181) respectively. In spin form this gives

$$\Lambda_1(\aleph; K) = [2\sinh(2K)]^{\frac{1}{2}\aleph} \exp\left\{\tfrac{1}{2}[\gamma_1^{(\aleph)}(K) + \cdots + \gamma_{2\aleph-1}^{(\aleph)}(K)]\right\} \qquad (6.190)$$

and

$$\Lambda_2(\aleph; K) = [2\sinh(2K)]^{\frac{1}{2}\aleph} \exp\left\{\tfrac{1}{2}[\gamma_0^{(\aleph)}(K) + \cdots + \gamma_{2\aleph-2}^{(\aleph)}(K)]\right\} \qquad (6.191)$$

respectively, where the $\gamma_r^{(\aleph)}(K)$, $r = 0, 1, \ldots, 2\aleph - 1$ are given by

$$\cosh[\gamma_r^{(\aleph)}(K)] = \cosh(2K)\coth(2K) - \cos\left(\frac{r\pi}{\aleph}\right). \qquad (6.192)$$

As indicated in Sect. 4.10, the sign ambiguity of the $\gamma_r^{(\aleph)}(K)$ is resolved by imposing the condition that they are all smooth functions of K which are positive at low temperatures (K large). From (4.156) the correlation length is now given by

$$\mathfrak{r}^{(\aleph)}(K) = -\frac{a}{\sqrt{2}}\left[\ln\left|\frac{\Lambda_2(\aleph; K)}{\Lambda_1(\aleph; K)}\right|\right]^{-1}, \qquad (6.193)$$

where we have used the result $c(2) = \frac{1}{4}$, obtained in Sect. 3.3.2. Equations (6.186) and (6.189) can now be used, together with (6.193), to obtain K_c and y_K. It follows from these equations that K_c is given by the crossing point of the curves for $g^{(\aleph)}(K)$ and $g^{(\tilde{\aleph})}(K)$ plotted as functions of K, where

$$g^{(\aleph)}(K) = \left|\frac{\Lambda_2(\aleph; K)}{\Lambda_1(\aleph; K)}\right|^{\aleph}. \qquad (6.194)$$

Nightingale (1976) considered various values of $\aleph$ and $\tilde{\aleph}$ and showed that the best results were obtained when $\tilde{\aleph} = \aleph - 1$. As may be expected the results improved with increasing $\aleph$, but even with quite small lattice thicknesses the results are remarkably good. Plots for $g^{(10)}(K)$ and $g^{(9)}(K)$ are shown in Fig. 6.9. The crossing point of the curves is at $K_c = 0.440286$ and using (6.189) the result $y_K = 1.0067$ is obtained for the critical exponent. The corresponding results for $\tilde{\aleph} = 49$ and $\aleph = 50$ are $K_c = 0.44068412$, $y_K = 1.000233$, the exact results being $K_c = 0.44068679$ and $y_K = 1$. A similar high level of accuracy was also achieved for the other exactly solved models in the work listed above.

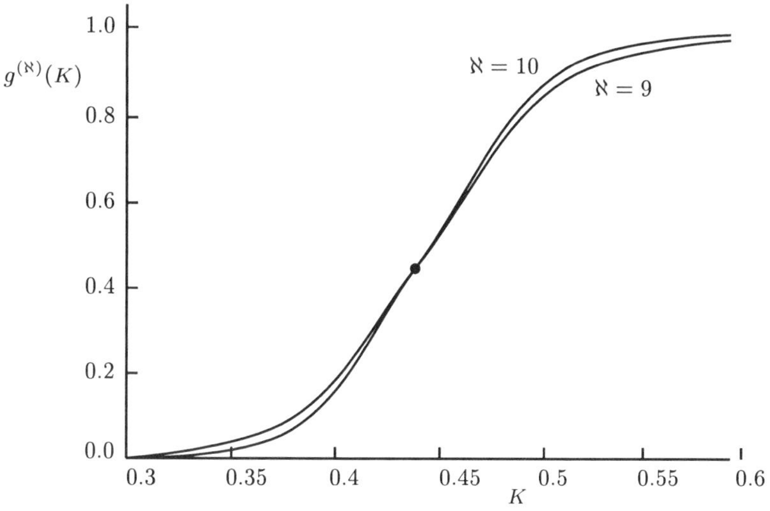

Fig. 6.9. Plots for $g^{(\aleph)}(K)$ with $\aleph = 9, 10$ for the spin-$\frac{1}{2}$ isotropic Ising model on a square lattice.

6.13.2 Other Models

The checks on phenomenological renormalization are, of course, more limited for systems which have not been exactly solved. In some cases the critical temperature is known from transformation methods (Volume 1, Chap. 8). For the square lattice Potts model the critical temperature can be derived from a mapping to a staggered six-vertex model (Volume 1, Sect. 10.14), with value of the exponent α given by exploiting the relationship to the staggered F model. The paper of Sneddon (1978) contains, in addition to the calculations for the spin-$\frac{1}{2}$ anisotropic Ising model, results for spin values $s = 1, \frac{3}{2}, 2, \frac{5}{2}$. For quite small lattice thicknesses ($\aleph = 5, 4, 3$, with $\widetilde{\aleph} = \aleph - 1$) values of the critical temperature correspond quite closely with high-temperature series and lower-bound renormalization group results obtained by Burkhardt and Swendsen (1976). A series of phenomenological renormalization calculations by Nightingale and Blöte (1980), Blöte et al. (1981) and Blöte and Nightingale (1982) for the ν-state Potts model again give encouraging agreement with the exact results available. The method was applied to a triangular Ising ferrimagnetic model by Lavis and Quinn (1987). In that case the sublattice structure gives a periodicity of three rows in the ground state of the ordered phase. By taking $\aleph = 6$, $\widetilde{\aleph} = 3$ they were able to obtain results for the critical temperature in various cases within 1% of the exact values obtained by Lavis and Quinn (1983), using transformation methods.

6.13.3 More Than One Coupling

We now consider the case $n > 1$. Here, as was observed above, the transformation cannot be completely specified by the one equation (6.183) and the fixed point equation (6.184) will yield an $n-1$-dimensional surface in the space of couplings. The simplest method to adopt in such cases is to set $\zeta_i = \omega_i \zeta_1$ for $i = 2, 3, \dots, n$ and then carry out the renormalization procedure for ζ_1 at fixed values of $\omega_2, \omega_3, \dots, \omega_n$. This was the approach adopted by Sneddon (1978) for the anisotropic Ising model and by Kinzel and Schick (1981) for the triangular Ising antiferromagnet in a field. In the latter case (6.183) becomes

$$\mathfrak{r}^{(\aleph)}(K, L) = \lambda \mathfrak{r}^{(\tilde{\aleph})}(\widetilde{K}, \widetilde{L}), \tag{6.195}$$

where, as for the zero-field model, $K = J/T$ and $L = \mathcal{H}/T$. If an equation of the form of (6.195) is used for a model with a critical temperature on the zero-field axis then the magnetic field exponent y_L can be evaluated using a procedure similar to that contained in equations (6.188) and (6.189). Since however, in most situations, the correlation length is an even function of $\mathcal{H}$ (or L) it is necessary to use the second derivative and we have

$$y_L = \frac{\ln\left\{\left(\dfrac{\partial^2 \mathfrak{r}^{(\aleph)}}{\partial L^2}\right)^* \left[\left(\dfrac{\partial^2 \mathfrak{r}^{(\tilde{\aleph})}}{\partial \widetilde{L}^2}\right)^*\right]^{-1}\right\}}{2\ln(\lambda)} - \frac{1}{2}. \tag{6.196}$$

If, as in the case of the antiferromagnetic model of Kinzel and Schick (1981), (6.195) is applied by renormalizing K at fixed values of $L/K = \mathcal{H}/J$ then a fixed point corresponds to any point on the critical curve defined by

$$\mathfrak{r}^{(\aleph)}(K, L) = \lambda \mathfrak{r}^{(\tilde{\aleph})}(K, L)\,, \tag{6.197}$$

in the plane of couplings. The difference between this method and those based on other procedures, which allow renormalization in more than one coupling, is that here each point of the curve is a fixed point, whereas other methods produce curves containing one fixed point, which controls the whole curve. This has the consequence that the single relevant thermal exponent y_K will be a function of $\mathcal{H}/J$. There is also the problem that if y_K is evaluated using (6.189) then it corresponds to the exponent parallel to the L axis, which is not in general the relevant direction with respect to the critical curve. Following Kinzel and Schick (1981), (6.189) can be generalized to define the exponent

$$y(\hat{\boldsymbol{v}}) = \frac{\ln\left\{(\boldsymbol{\nabla}\mathfrak{r}^{(\aleph)}.\hat{\boldsymbol{v}})^* [(\boldsymbol{\nabla}\mathfrak{r}^{(\tilde{\aleph})}.\hat{\boldsymbol{v}})^*]^{-1}\right\}}{\ln(\lambda)} - 1 \tag{6.198}$$

for some arbitrary unit vector $\hat{\boldsymbol{v}}$ in the plane, where '$*$' now simply indicates evaluation at some particularly chosen point (K^*, L^*) on the critical curve. If $\hat{\boldsymbol{v}} = (1, 0)$, the unit vector parallel to the L axis, then (6.198) reduces to

(6.189). The vector $\boldsymbol{u}^* = (\boldsymbol{\nabla}\mathfrak{r}^{(\aleph)} - \lambda\boldsymbol{\nabla}\mathfrak{r}^{(\tilde{\aleph})})^*$ is orthogonal to the critical curve at (K^*, L^*) and if $\hat{\boldsymbol{v}}$ is tangential to the curve (and thus orthogonal to $\boldsymbol{u}^*$) $y(\hat{\boldsymbol{v}}) = 0$. This marginal exponent is not surprising given that the whole of the critical curve consists of fixed points. For different (K^*, L^*) the exponent $y(\hat{\boldsymbol{v}})$ will vary as the direction of $\hat{\boldsymbol{v}}$ is varied. According to the analysis of the linear renormalization group about a fixed point, described in Sect. 6.4, all scaling directions not along the critical curve should yield the same exponent. For the present method this will be the case only when $\boldsymbol{\nabla}\mathfrak{r}^{(\aleph)}$ and $\boldsymbol{\nabla}\mathfrak{r}^{(\tilde{\aleph})}$ are parallel. For this reason Kinzel and Schick (1981) chose values of (K^*, L^*) which satisfied this condition as yielding the 'best' value of y_K, which is now given by

$$y_K = \frac{\ln\left\{|\boldsymbol{\nabla}\mathfrak{r}^{(\aleph)}|^*\left[|\boldsymbol{\nabla}\mathfrak{r}^{(\tilde{\aleph})}|^*\right]^{-1}\right\}}{\ln(\lambda)} - 1\,. \qquad (6.199)$$

Using this method with $\aleph = 9$, $\tilde{\aleph} = 6$ Kinzel and Schick (1981) obtained the value $y_K = 1.2002$ for the thermal exponent on the antiferromagnetic critical curve for the triangular Ising model. It is supposed that this transition is in the universality class of the 3-state Potts model, (Alexander 1975) for which (Volume 1, Sect. 10.14) $\alpha = \frac{1}{3}$. From (2.167) $y_K = 2/(2-\alpha) = \frac{6}{5}$.

Most phenomenological renormalization calculations have concentrated on cases where the phase transitions were known, or expected to be, continuous. Exceptions to this are the ν-state Potts model calculations of Blöte et al. (1981) and Roomany and Wyld (1981). The latter authors reported that the method failed to differentiate between the second-order transition expected for $\nu \leq 4$ and the first-order transition expected for $\nu \geq 5$. Wood and Osbaldstin (1982, 1983) have argued that this inability to distinguish between continuous and first-order transitions is indeed a general characteristic of the method. They developed a general procedure based on phenomenological renormalization for determining an approximated form for phase diagrams including the location of regions of multiphase coexistence. They applied the procedure to the ν-state Potts model and it was applied to a triangular Ising ferrimagnet by Lavis and Quinn (1987).

6.14 Other Renormalization Group Methods

In this chapter we have attempted to outline the basic ideas of the real-space renormalization group and to provide a brief account of some of the main methods used. In keeping with the general approach of this book we have omitted applications to dynamic phenomena and quantum system, for which the reader is referred respectively to the articles by Mazenko and Valls and Pfeuty *et al.* in Burkhardt and van Leeuwen (1982). One approach to the statistical mechanics of lattice systems, which limitations of space has prevent us from including in this volume, is that of *Monte Carlo simulation* (see, for example, Binder 1986). The possibility of improving the results of a Monte Carlo calculations by combining them with a renormalization group procedure was first suggested by Ma (1976b). In the form developed by Swendsen (1979a, 1979b) this *Monte Carlo renormalization method* uses a standard Monte Carlo simulation to obtain a large number of configurations near to criticality. From these correlation functions can be calculated. A renormalization group transformation can now be applied directly to each configuration as many times as is feasible. The last iteration should leave a lattice still large compared to the range of the fixed point Hamiltonian. Approximations to the matrix $\boldsymbol{L}^*$, defined by equation (6.30), can now be calculated from correlation functions between the lattices corresponding to each stage of the renormalization procedure and a sequence of estimates of the critical exponents obtained. Swendsen (1979a) applied the method to the two-dimensional spin-$\frac{1}{2}$ Ising model and the 3-state Potts model using an real-space block-spin procedure with the majority-rule weight function. He obtained results which compared well with other renormalization group methods and were better than those obtained by Monte Carlo simulations alone. In a subsequent paper (Swendsen 1979b) he used different weight functions. This demonstrated that the accuracy and reliability of the results can be strongly affected by the choice made. This problem is, however, one which is shared by other renormalization group procedures (see Sect. 6.12). The method has also been applied to a variety of other models. In particular the application to the three-dimensional Ising model by Pawley et al. (1984) gives results in good agreement with those achieved by series methods.

As indicated in Sect. 6.1, real-space methods represent one of the two ways in which the renormalization group can be applied, the other being the use of renormalization in wave-vector space. This latter method, which involves the elimination of the short-range details of the system by integration over large wave vectors, has led to the *ϵ-expansion method*, (for detailed accounts see, for example, Wilson and Kogut 1974, Ma 1976a), where $\epsilon = 4 - d$. For the Ising model the ϵ-expansion confirms the prediction of Landau–Ginzberg theory in Sect. 3.3.2 that the upper borderline dimension is at $\epsilon = 0$ ($d_{\mathrm{u}} = 4$). The changeover in critical behaviour in this context can be identified with the change between relevance and irrelevance of one of the parameters at the Gaussian fixed point. Critical exponent are here obtained as expansions in powers of ϵ. For the three-dimensional Ising model the expansion to order ϵ^2 gives good agreement with other methods.

Examples

6.1 It is clear that, for the spin-$\frac{1}{2}$ model, the weight function

$$p(\sigma(\tilde{\boldsymbol{r}})|\sigma(\boldsymbol{r}), \boldsymbol{r} \in \mathcal{B}(\tilde{\boldsymbol{r}})) = \tfrac{1}{2}[1 + \sigma(\tilde{\boldsymbol{r}})q(\sigma(\boldsymbol{r}), \boldsymbol{r} \in \mathcal{B}(\tilde{\boldsymbol{r}}))]$$

satisfies condition (6.11), where $q(\sigma(\boldsymbol{r}), \boldsymbol{r} \in \mathcal{B}(\tilde{\boldsymbol{r}}))$ is any function of the microstates of $\mathcal{B}(\tilde{\boldsymbol{r}})$. Show that the Kadanoff–Houghton weight function (6.25) can be expressed in this form with

$$q(\sigma(\boldsymbol{r}), \boldsymbol{r} \in \mathcal{B}(\tilde{\boldsymbol{r}})) = \tanh\left[\tau \sum_{\{\boldsymbol{r} \in \mathcal{X}(\tilde{\boldsymbol{r}})\}} \sigma(\boldsymbol{r})\right].$$

Hence, show that the decimation weight function (6.22) and the majority-rule weight function (6.23) are obtained in the limit $\tau \to \infty$ when $\mathcal{X}(\tilde{\boldsymbol{r}})$ is the single site $\mathring{\boldsymbol{r}} \in \mathcal{B}(\tilde{\boldsymbol{r}})$ and $\mathcal{X}(\tilde{\boldsymbol{r}}) = \mathcal{B}(\tilde{\boldsymbol{r}})$ respectively.

6.2 A one-dimensional 3-state Potts model is in the form of a ring with N lattice sites and nearest-neighbour interaction $-R$ ($R > 0$) between sites with microsystems in the same state and zero for microsystems in different states. In terms of the variable

$$\mathrm{t} = 2\left[\frac{\exp(R/T) - 1}{\exp(R/T) + 2}\right]$$

derive, for $\lambda = 2$, the decimation recurrence relationship

$$\tilde{\mathrm{t}} = \tfrac{1}{2}\mathrm{t}^2.$$

Obtain the fixed points and exponents for this relationship and consider the applicability of this transformation to the case $R < 0$. Show that

$$\phi(\mathrm{t}) = -\sum_{s=1}^{\infty} \frac{1}{2^{s+1}} \ln\left\{\frac{1}{3}\left[\frac{2^{2^s} + \mathrm{t}^{2^s}}{2^{2^s} - \mathrm{t}^{2^s}}\right]\right\}.$$

6.3 Another approach to the problem of decimation for the model described in Sect. 6.7 is to calculate the renormalized nearest-neighbour coupling $\widetilde{K}_1$ by considering an isolated square of sites and the renormalized second-neighbour coupling $\widetilde{K}_2$ by considering a line of three sites. Show that this procedure yields the recurrence relationships

$$\begin{aligned} \widetilde{K}_1 &= K_2 + \ln\cosh(2K_1), \\ \widetilde{K}_2 &= \tfrac{1}{2}\ln\cosh(2K_1). \end{aligned}$$

It will be seen that, as for equations (6.102) and (6.103), the relationships (6.104) and (6.105) are recovered on expansion to quadratic terms. The interested reader with access to computing facilities may like to verify that these equations yield a phase diagram like that shown in Fig. 6.5 with $T_c/J = 2.325$ and $\nu = 0.720$.

7. Series Methods

7.1 Introduction

As we saw in Sect. 1.3, the complete solution of a statistical mechanical model is known once an explicit expression has been obtained for the partition function. In this chapter we shall discuss methods which involve the derivation of expansions of partition functions and free energies, or their derivatives, in series of powers of temperature-dependent variables. These series are called *low temperature*, or *high temperature*, respectively according to whether the expansion variable tends to zero as $T \to 0$ or $T \to \infty$. If an expression for the coefficient of the general term were known then a series would constitute an exact solution. However, the phrase 'series methods' refers primarily to the term-by-term calculation of as many coefficients as possible and the deduction of results for the model from the behaviour of a limited number of coefficient values. Low- and high-temperature series expansions provide approximations to the thermodynamic functions of a model in their respective ranges of temperature. They are, however, more often used to study critical properties, specifically to obtain critical points and exponents. Suppose that

$$f(\zeta) = \sum_{n=0}^{\infty} a_n \zeta^n, \tag{7.1}$$

is some thermodynamic function of the model of interest, with ζ a function of temperature appropriate to high-temperature or low-temperature series. If

$$\lim_{n\to\infty} |a_n|^{1/n} \tag{7.2}$$

exists and is equal to $1/R$, the series converges in the disc $|z| < R$ in the complex plane of ζ. There is always a singularity of $f(\zeta)$ on the circle $|\zeta| = R$, which is the *radius of convergence* of the series. If this singularity is on the positive real axis in the complex plane of ζ then it can be identified with the critical temperature. When this is not the case the analysis is rather more complicated, (Guttmann 1989). In any event, as has been indicated above, the methods to be described in this chapter are those for which only a finite number of terms in a high- or low-temperature series is known. It is thus possible to obtain only an approximation to the location of the singularities exhibited by the full series. These methods provide nevertheless accurate values of critical parameters and exponents for many systems, where no exact solution is known. Of course, where exact solutions are known, comparison with deductions from series methods provides a useful test for the latter.

There are three steps needed in the use of a series approximation for the investigation of a statistical mechanical model:

(i) Defining suitable expansion variables.

(ii) Selecting (or inventing) a method for obtaining the series coefficients and finding as many as skill and computing power allows.

(iii) Analyzing the series.

The methods of series analysis are common to all types of series and models and will be discussed in the next section.

For simplicity we consider spin models, including the ν-state Potts model, which can be expressed in spin variable form (Sect. 3.5), and including a magnetic field $\mathcal{H}$. For low-temperature expansions suitable variables are $\exp(-\Delta E/T)$ and $\exp(-|\mathcal{H}|/T)$, (or powers of these) where ΔE is the positive energy difference between the maximum and minimum energy states of a nearest-neighbour pair of microsystems. These variables both tend to zero as $T \to 0$ and $\exp(-|\mathcal{H}|/T) \to 0$ as $\mathcal{H} \to \infty$ so it can also be used for high-field expansions (Enting and Baxter 1980). For the ferromagnetic spin-$\frac{1}{2}$ Ising model with Hamiltonian (1.49) $\Delta E = 2J$ and zero-field low-temperature expansions in terms of the variable $\exp(-2J/T)$ have already be mentioned briefly in Volume 1, Sect. 8.2.

For high-temperature expansions the obvious variables are $\Delta E/T$ and $|\mathcal{H}|/T$. However, for systems with one internal coupling, corresponding to the nearest-neighbour interaction, and exactly two energy levels $E_0 \pm \Delta E$ for a nearest-neighbour pair there is a better choice. If the upper energy level has degeneracy $\nu - 1$ and the lower energy level is not degenerate. Let

$$\exp\left[\left(E_0 \pm \tfrac{1}{2}\Delta E\right)/T\right] = w\left[1 + \tfrac{1}{2}\tau(\nu - 2 \pm \nu)\right] . \tag{7.3}$$

Solving these equations for w and τ in terms of $\exp[E_0/T]$ and $\exp[\Delta E/T]$ gives

$$\tau = \frac{\exp(\Delta E/T) - 1}{\exp(\Delta E/T) + \nu - 1}, \qquad w = \exp(E_0/T)\frac{\exp(\Delta E/T) + \nu - 1}{\nu \exp(\Delta E/2T)} . \tag{7.4}$$

It is clear that τ is a suitable high-temperature variable in the sense that $\tau \to 0$ as $T \to \infty$. For the spin-$\frac{1}{2}$ Ising model $E_0 = 0$, $\Delta E = 2J$ and $\nu = 2$ and (7.4) become

$$\tau = \tanh(J/T) , \qquad w = \cosh(J/T) . \tag{7.5}$$

This change of variables is the one used in the brief discussion of high-temperature series for the spin-$\frac{1}{2}$ in Volume 1, Sect. 8.3. It will also be used in the more extensive treatment in Sect. 7.4. For the ν-state Potts model, with Hamiltonian (3.87), $\Delta E = R$ and $E_0 = \frac{1}{2}R$ giving

$$\tau = \frac{\exp(R/T) - 1}{\exp(R/T) + \nu - 1}, \qquad w = \frac{\exp(R/T) + \nu - 1}{\nu}. \tag{7.6}$$

These variables are the ones used by Domb (1974d) to transform the partition function of the ν-state Potts model to high-temperature form. It should also be noted that for each model the mapping

$$\exp(-\triangle E/T) \quad \longleftrightarrow \quad \tau\,, \tag{7.7}$$

gives the duality transformation on the square lattice. For the Ising model with coupling $K = J/T$ we obtain from (7.5) $\sinh(K_{\text{L.T.}})\sinh(K_{\text{H.T.}}) = 1$, which is the duality transformation (see Volume 1, Sect. 8.3) between high-temperature coupling $K_{\text{H.T.}}$ and low-temperature coupling $K_{\text{L.T.}}$. With the coupling $K = R/T$ for the Potts model (7.6) gives

$$[\exp(K_{\text{L.T.}}) - 1][\exp(K_{\text{H.T.}}) - 1] = \nu\,, \tag{7.8}$$

which is the duality relationship for the Potts model first discovered by Potts (1952). It follows from these observations that for both the spin-$\frac{1}{2}$ Ising model and the Potts model a high-temperature series in the variable τ can be mapped into a low-temperature series in $\exp(-\triangle E/T)$.

The terms in a low-temperature series correspond to perturbations from a chosen ground state. In Sect. 7.3 we consider the low-temperature series expansion for the ferromagnetic spin-$\frac{1}{2}$ Ising model which arises from perturbations from the ground state where $\sigma(\boldsymbol{r}) = -1$ for all sites of the lattice. The general spin ferromagnetic Ising model and Potts model can be handled in a similar way. However, the spatial isotropy of the exchange interaction terms in classical and quantum Heisenberg Hamiltonians lead to important differences in these cases. For two-dimensional lattices Mermin and Wagner (1966) showed that, for the classical Heisenberg model, there is no spontaneous magnetization at $T > 0$. There is, therefore, no Curie temperature, although the possibility of other types of singularity is not excluded. For three-dimensional lattices there is a Curie temperature, but the low-temperature behaviour is radically different from that of the Ising model (see, for example, Ashcroft and Mermin 1976). There are no small localized perturbations from the ground state and, for the quantum Heisenberg model, the excitations are spin waves. It follows that there can be no low-temperature series of the type discussed in Sect. 7.3 for these models. In fact, the changes in the spontaneous magnetization and heat capacity from their zero-temperature values are proportional to $T^{\frac{3}{2}}$. For both the quantum and classical Heisenberg models at $T < T_{\text{c}}$, there is a divergence in the magnetic susceptibility as $\mathcal{H} \to 0$. In the classical Heisenberg model the low-temperature heat capacity has the form $Nk_{\text{B}} + aT$, where a is a constant. For the Heisenberg antiferromagnet on a loose-packed lattice the Ising model-type ground state, with alternate spins oppositely directed, is not an eigenstate of the Hamiltonian and, apart from the $d = 1$, $s = \frac{1}{2}$ case which was solved by Bethe (1931), determination of the ground state remains an unsolved problem. (For further discussion of these problems see Rushbrooke et al. 1974.) In Sect. 7.4 we describe the high-temperature expansion of the spin-$\frac{1}{2}$ Ising model using the variable τ defined in (7.5) and in Sect. 7.5 we develop the high-temperature *linked-cluster* or *connected-graph expansion* for general spin models for which the procedure of Sect. 7.4 is not applicable. In Sect. 7.6 the method is applied to general-s Ising models and the Heisenberg models. The related finite-cluster and finite-lattice methods are discussed in Sect. 7.7 and the chapter concludes with a summary of the results derived from all the preceding methods.

7.2 The Analysis of Series

Methods for analyzing power series to extract critical parameters and exponents are described in detail by Gaunt and Guttmann (1974) and Guttmann (1989). In relation to this work the latter makes the important point that in series analysis it is a question of "fitting series coefficients to an assumed functional form – an assumed form may be implicit or explicit, but is invariably present".

Suppose that a system has a critical point given by $\zeta = \zeta_c$ in terms of some high- or low-temperature (real) variable ζ. This critical point is assumed to give a leading power-law singularity of the form

$$f(\zeta) \simeq A_0\left(1 - \frac{\zeta}{\zeta_c}\right)^{-\lambda}, \qquad \zeta \to \zeta_c - , \tag{7.9}$$

in some response function $f(\zeta)$. The problem now is to extract the values of ζ_c and λ from a finite number of the coefficients of the power-series expansion (7.1). In the complex plane of ζ the radius of convergence R of the series (7.1) must satisfy the condition $R \leq \zeta_c$. It may, however, be shown (Hille 1959) that if $a_n \geq 0$ for all n greater than some $n_0 \geq 0$ then $\zeta = R$ is a singular point of $f(\zeta)$. If $f(\zeta)$ has only one singularity for real values of ζ, it follows that $\zeta_c = R$. In this case, as indicated in Sect. 7.1, the critical point is the nearest singularity to the origin and $1/\zeta_c$ is given by the limit (7.2). This then provides us with a 'simple' method for obtaining the ζ_c. However, as we shall see in Sect. 7.4, in the case of the high-temperature susceptibility series for the square Ising model, convergence is very slow and still leaves an error of about 5% compared to the exact result when a series of 54 terms is used. It is clear that more sophisticated methods of analysis are required.

7.2.1 The Ratio Method

This, the first widely used technique, was introduced by Domb and Sykes (1956, 1957a, 1957b). Its mathematical basis is in a theorem of Darboux (Szegö 1975):

Theorem 7.2.1. *Let $g(w)$ be regular in the complex plane of w for $|w| < 1$ with power-series expansion*

$$g(w) = \sum_{n=0}^{\infty} b_n w^n. \tag{7.10}$$

If $g(w)$ has a finite number of singularities at the distinct points $\exp(\mathrm{i}\theta_1)$, $\exp(\mathrm{i}\theta_2)$, …, $\exp(\mathrm{i}\theta_p)$ *on the unit circle $|w| = 1$ and*

$$g(w) \simeq \sum_{m=0}^{\infty} A_m^{(s)} [1 - w\exp(-\mathrm{i}\theta_s)]^{m\mu_s - \lambda_s}, \qquad s = 1, 2, \ldots, p, \tag{7.11}$$

in a neighbourhood of $w = \exp(\mathrm{i}\theta_s)$, where $\mu_s > 0$, then

$$b_n \simeq \sum_{m=0}^{\infty}\sum_{s=1}^{p} A_m^{(s)} \frac{\Gamma(m\mu_s - \lambda_s + 1)}{\Gamma(n+1)\Gamma(m\mu_s - \lambda_s - n + 1)}[-\exp(\mathrm{i}\theta_s)]^n, \tag{7.12}$$

in the following sense: if M is an arbitrary positive number, and if a sufficiently large number of terms in the series for m in (7.12) is taken, then an expression is obtained which approximates b_n with an error of $\mathrm{O}(n^M)$.

Suppose this theorem is applied to the case of the function $f(\zeta)$ with a single singularity on the circle of convergence at the point $\zeta = \zeta_\mathrm{c}$ on the positive real axis. With the identifications $w = \zeta/\zeta_\mathrm{c}$, $g(\zeta/\zeta_\mathrm{c}) = f(\zeta)$, $b_n = a_n\zeta_\mathrm{c}^n$, $\lambda_1 = \lambda$, $\mu_1 = \mu$ and $A_m^{(1)} = A_m$, equation (7.10) becomes identical to (7.1) and (7.11) and (7.12) give

$$f(\zeta) \simeq \sum_{m=0}^{\infty} A_m \left(1 - \frac{\zeta}{\zeta_\mathrm{c}}\right)^{m\mu - \lambda}, \tag{7.13}$$

$$a_n \simeq \frac{(-1)^n}{\zeta_\mathrm{c}^n} \sum_{m=0}^{\infty} A_m \frac{\Gamma(m\mu - \lambda + 1)}{\Gamma(n+1)\Gamma(m\mu - \lambda - n + 1)}, \tag{7.14}$$

respectively. The leading term in (7.13) corresponds to the scaling form (7.9) for $f(\zeta)$. Subsequent terms give corrections to scaling. Neglecting all but the leading term equation (7.14) gives the ratio formula

$$\frac{a_n}{a_{n-1}} = \frac{\lambda + n - 1}{n\zeta_\mathrm{c}}. \tag{7.15}$$

This is an expression for the ratio a_n/a_{n-1} as a linear function of $1/n$ with gradient $(\lambda - 1)/\zeta_\mathrm{c}$ and intercept $1/\zeta_\mathrm{c}$ on the '$1/n = 0$'-axis. Graphical interpolation of the data could therefore be used to determine ζ_c and λ.

A number of variants of the ratio method have been developed, to improve the approximations to critical parameters and exponents (see Guttmann 1989). However, unless the singularity of interest dominates the series expansion, the ratios still do not give good estimates for critical parameters and exponents. Smooth monotonic behaviour of the coefficients is required and, although oscillations arising from another known singularity on the circle of convergence can sometimes be catered for (Sect. 7.4 and Example 7.1), applications to cases where the critical singularity is not at least as close as any other singularity to the origin do not yield good results.

7.2.2 Padé Approximants

This method was initiated by Baker (1961) (see also, Baker 1965). The $[N/M]$ Padé approximant to a power series in the variable ζ is the quotient

$$[N/M] = \frac{P_N(\zeta)}{Q_M(\zeta)} = \frac{p_0 + p_1\zeta + \cdots + p_N\zeta^L}{1 + q_1\zeta + \cdots + q_M\zeta^M} \tag{7.16}$$

of two polynomials $P_N(\zeta)$ and $Q_M(\zeta)$ of degree N and M respectively. The coefficients of the polynomials can be chosen (uniquely) to agree with the first $N + M + 1$ coefficients of the series. These coefficients can be obtained by solving two systems of linear equations.

Rather than determining the Padé approximant of $f(\zeta)$, given by (7.1) a more rapid convergence can often be achieved by using a related function. From (7.9),

$$\frac{\mathrm{d}\ln[f(\zeta)]}{\mathrm{d}\zeta} \simeq \frac{\lambda}{\zeta - \zeta_c}, \qquad \zeta \to \zeta_c - 0 \tag{7.17}$$

has a critical point corresponds to a simple pole, with residue λ. Since Padé approximants can represent simple poles exactly, approximants of the form

$$\frac{\mathrm{d}\ln[f(\zeta)]}{\mathrm{d}\zeta} = \frac{P_N(\zeta)}{Q_M(\zeta)} \tag{7.18}$$

to the *Dlog series* of $f(\zeta)$ should converge much faster than approximants to the series for $f(\zeta)$.

With suitable adaption Padé approximant methods can be applied to more complicated singularity structures. The case

$$f(\zeta) \simeq \sum_{k=0}^{m} A_k \left(1 - \frac{\zeta}{\zeta_k}\right)^{-\lambda_k} \tag{7.19}$$

was considered by Baker and Hunter (1973). In general the analysis is rather complicated but two particular cases are of interest:

All the exponents $\lambda_j = \lambda$, *which is known.* For this Baker and Hunter (1973) apply a procedure which they call *exponent renormalization.*[1] The following theorem is not difficult to prove:

[1] This terminology is not related to the exponent renormalization described in Volume 1, Sect. 9.3.

Theorem 7.2.2. *If $f(\zeta)$ with the power-series expansion (7.1) has leading power-law form*

$$f(\zeta) \simeq \sum_{k=0}^{m} A_k \left(1 - \frac{\zeta}{\zeta_k}\right)^{-\lambda} , \tag{7.20}$$

then $g(\zeta)$ with the power-series expansion

$$g(\zeta) = \sum_{n=0}^{\infty} b_n \zeta^n , \tag{7.21}$$

with

$$b_n = a_n \begin{pmatrix} -\lambda \\ n \end{pmatrix} , \tag{7.22}$$

has leading power-law behaviour

$$g(\zeta) \simeq \sum_{k=0}^{m} A_k \left(1 + \frac{\zeta}{\zeta_k}\right)^{-1} . \tag{7.23}$$

Thus, if λ is known, the derived coefficient of (7.1) can be transformed using (7.22) to give a series to the same degree for $g(\zeta)$.

All the singularities are confluent at $\zeta_j = \zeta_c$, which is known. In this case we assume that $\lambda_0 = \lambda$ is known and $\lambda_k = \lambda - \Delta_k$, where $0 < \Delta_1 \le \Delta_1 \le \cdots \le \Delta_m < \lambda$. Thus

$$f(\zeta) \simeq \left(1 - \frac{\zeta}{\zeta_c}\right)^{-\lambda} \left[A_0 + A_1 \left(1 - \frac{\zeta}{\zeta_c}\right)^{\Delta_1} + A_2 \left(1 - \frac{\zeta}{\zeta_c}\right)^{\Delta_2} \cdots \right] . \tag{7.24}$$

For this case Baker and Hunter (1973) propose a method very similar to that for the exponent renormalization case. An alternative suggested by Adler et al. (1981) is to use the function

$$F(\zeta) = \lambda f(\zeta) - (\zeta_c - \zeta) \frac{\mathrm{d}f(\zeta)}{\mathrm{d}\zeta} . \tag{7.25}$$

Substituting into (7.25) from (7.24) gives

$$F(\zeta) \simeq \left(1 - \frac{\zeta}{\zeta_c}\right)^{-\lambda} \left[A_1\Delta_1 \left(1 - \frac{\zeta}{\zeta_c}\right)^{\Delta_1} + A_2\Delta_2 \left(1 - \frac{\zeta}{\zeta_c}\right)^{\Delta_2} \cdots \right] . \tag{7.26}$$

The logarithmic derivative of $F(\zeta)$ has a simple pole at ζ_c with residue $\lambda - \Delta_1$. In principle one can derive successive confluent exponents $\Delta_1, \Delta_2, \ldots$ by repeated application of this transformation.

7.2.3 The Differential Approximant Method

This method, which is due to Guttmann and Joyce (1972),[2] can be seen as a natural extension of the use (7.18) of Dlog Padé approximants. That equation can be expressed in the form

$$Q_M(\zeta)\frac{\mathrm{d}f(\zeta)}{\mathrm{d}\zeta} - P_N(\zeta)f(\zeta) = 0\,. \tag{7.27}$$

The determined coefficients of $f(\zeta)$ are used to obtain the polynomials $Q_M(\zeta)$ and $P_N(\zeta)$ and then the differential equation is integrated. A natural generalization of (7.27) is

$$\sum_{k=0}^{\kappa} Q^{(k)}(\zeta)\frac{\mathrm{d}^k f(\zeta)}{\mathrm{d}\zeta^k} + R(\zeta)f(\zeta) = 0\,. \tag{7.28}$$

Again the known coefficients are used to determine the polynomials $Q^{(k)}(\zeta)$, $k = 1, 2, \ldots, \kappa$, and $R(\zeta)$ and the differential equation is integrated.

7.3 Low-Temperature Series for the Spin-$\frac{1}{2}$ Ising Model

We consider a lattice $\mathcal{N}$ of N sites with each site $\boldsymbol{r}$ occupied by an Ising spin $\sigma(\boldsymbol{r}) = \pm 1$. A constant-field partition function is used and it is assumed initially that $\mathcal{H} < 0$. The terms in a low-temperature expansion correspond to perturbations of the ground state, in which $\sigma(\boldsymbol{r}) = -1$, for all sites of $\mathcal{N}$. For the perturbed state in which n spins are reversed the Hamiltonian (1.49) can be expressed in the form

$$\widehat{H}(\mathcal{H}, J; N_\mathrm{u}, \mathrm{n}) = \widehat{H}_0(\mathcal{H}, J) + 2JN_\mathrm{u} + 2\mathrm{n}|\mathcal{H}|\,, \tag{7.29}$$

where N_u is the number of unlike $(+-)$ nearest-neighbour pairs and

$$\widehat{H}_0(\mathcal{H}, J) = -N\left(\tfrac{1}{2}\mathrm{z}J + |\mathcal{H}|\right) \tag{7.30}$$

is the Hamiltonian for the ground state where $\mathrm{n} = N_\mathrm{u} = 0$. Using an index k to distinguish between different configurations with the same number of $+$ spins we have

$$N_\mathrm{u} = \mathrm{zn} - 2\mathrm{q}[\mathrm{n}, \mathrm{k}]\,, \tag{7.31}$$

where $\mathrm{q}[\mathrm{n}, \mathrm{k}]$ is the number of nearest-neighbour $++$ spin pairs. With the usual notation, $\widehat{\mathsf{H}} = \widehat{H}/T$, $\widehat{\mathsf{H}}_0 = \widehat{H}_0/T$, (7.31) and (7.29) give

$$\widehat{\mathsf{H}}(L, K; \mathrm{n}, \mathrm{k}) = \widehat{\mathsf{H}}_0(L, K) + 2K(\mathrm{zn} - 2\mathrm{q}[\mathrm{n}, \mathrm{k}]) + 2\mathrm{n}|L|\,, \tag{7.32}$$

where $K = J/T$ and $L = \mathcal{H}/T$. The constant field partition function $Z(L, K)$ is given by

[2] See also the article by Joyce and Guttmann in Graves-Morris (1973) and a discussion of this method in Guttmann (1989).

$$\exp[\widehat{H}_0(L,K)]Z(L,K) = \sum_{\mathfrak{n}=1}^{\infty}\sum_{\{\mathfrak{k}\}} \exp[-\widehat{H}(L,K;\mathfrak{n},\mathfrak{k}) + \widehat{H}_0(L,K)]$$

$$= 1 + \sum_{\mathfrak{n}=1}^{\infty}\sum_{\{\mathfrak{k}\}} \Omega[\mathfrak{n},\mathfrak{k}]\mathfrak{Z}^{\mathfrak{n}}\mathfrak{X}^{\frac{1}{2}z\mathfrak{n}-\mathfrak{q}[\mathfrak{n},\mathfrak{k}]}, \tag{7.33}$$

where

$$\mathfrak{Z} = \exp(-2|L|), \qquad \mathfrak{X} = \exp(-4K), \tag{7.34}$$

both tend to zero with T. We take the + spins to be the vertices and ++ nearest-neighbour pairs to be the lines of graphs on $\mathcal{N}$. Since every nearest-neighbour ++ pair is joined by a line, the subgraphs of $\mathcal{N}$ are all *section graphs* (Appendix A.7) and the topology of a graph $\mathfrak{g}$ is denoted by $[\mathfrak{n},\mathfrak{k}]$. The configuration number $\Omega[\mathfrak{n},\mathfrak{k}]$ which denotes the number of ways of placing the graph $[\mathfrak{n},\mathfrak{k}]$ on the lattice, satisfies (A.218). To obtain terms of the low-temperature series expansion (7.33) it is necessary to compute $\Omega[\mathfrak{n},\mathfrak{k}]$ for successive values of $\mathfrak{n}$ and all the consequent values of $\mathfrak{k}$. Values of $\overline{\omega}[\mathfrak{n},\mathfrak{k}] = \Omega[\mathfrak{n},\mathfrak{k}]/N$ for graphs on the simple cubic lattice are listed in Table A.1. The terms in equation (7.33) can be grouped to yield

$$\exp[\widehat{H}_0(L,K)]Z(L,K) = 1 + \sum_{\mathfrak{n}=1}^{\infty} G(\mathfrak{n};\mathfrak{X})\mathfrak{Z}^{\mathfrak{n}}, \tag{7.35}$$

where

$$G(\mathfrak{n};\mathfrak{X}) = N\sum_{\{\mathfrak{k}\}} \overline{\omega}[\mathfrak{n},\mathfrak{k}]\mathfrak{X}^{\frac{1}{2}z\mathfrak{n}-\mathfrak{q}[\mathfrak{n},\mathfrak{k}]}. \tag{7.36}$$

The dimensionless free energy $\Phi(N,L,K)$ is then given by

$$\Phi(N,L,K) - \widehat{H}_0(L,K) = -\ln\left[1 + \sum_{\mathfrak{n}=1}^{\infty} G(\mathfrak{n};\mathfrak{X})\mathfrak{Z}^{\mathfrak{n}}\right]$$

$$= -N\sum_{\mathfrak{n}=1}^{\infty} g(\mathfrak{n};\mathfrak{X})\mathfrak{Z}^{\mathfrak{n}}, \tag{7.37}$$

where the expansion of the logarithm yields

$$\begin{aligned} Ng(1;\mathfrak{X}) &= G(1;\mathfrak{X}), \\ Ng(2;\mathfrak{X}) &= G(2;\mathfrak{X}) - \tfrac{1}{2}[G(1;\mathfrak{X})]^2, \\ Ng(3;\mathfrak{X}) &= G(3;\mathfrak{X}) - G(1;\mathfrak{X})G(2;\mathfrak{X}) + \tfrac{1}{3}[G(1;\mathfrak{X})]^3. \end{aligned} \tag{7.38}$$

Using the values of $\overline{\omega}[\mathfrak{n},\mathfrak{k}]$ and $\mathfrak{q}[\mathfrak{n},\mathfrak{k}]$ given in Table A.1, the first few $G(\mathfrak{n};\mathfrak{X})$ in (7.36), for the simple cubic lattice, are

$$\begin{aligned}
G(1;\mathfrak{X}) &= N\mathfrak{X}^3\,,\\
G(2;\mathfrak{X}) &= 3N\mathfrak{X}^5 + \tfrac{1}{2}(N^2-7N)\mathfrak{X}^6\,,\\
G(3;\mathfrak{X}) &= 15N\mathfrak{X}^7 + 3(N^2-12N)\mathfrak{X}^8\\
&\qquad + \tfrac{1}{6}(N^3-21N^2+128N)\mathfrak{X}^9\,,\\
G(4;\mathfrak{X}) &= 3N\mathfrak{X}^8 + 83N\mathfrak{X}^9 + \cdots\,.
\end{aligned} \tag{7.39}$$

From (7.30) and (7.37)–(7.39), the free energy, up to third degree terms in $\mathfrak{Z}$ and ninth degree in $\mathfrak{X}$, can be expressed in the form

$$\begin{aligned}
\Phi(N,L,K) + N(3K+|L|) = -N\Big[&\mathfrak{X}^3\mathfrak{Z} + \left(3\mathfrak{X}^5 - \tfrac{7}{2}\mathfrak{X}^6\right)\mathfrak{Z}^2\\
&+\left(15\mathfrak{X}^7 - 36\mathfrak{X}^8 + \tfrac{64}{3}\mathfrak{X}^9\right)\mathfrak{Z}^3\\
&+(3\mathfrak{X}^8 + 83\mathfrak{X}^9 + \cdots)\mathfrak{Z}^4 + \cdots\Big]\,.
\end{aligned} \tag{7.40}$$

The coefficient of $\mathfrak{Z}^{\mathfrak{n}}$ in (7.40) is $g(\mathfrak{n};\mathfrak{X})$. It will be seen that the $g(\mathfrak{n};\mathfrak{X})$ for $\mathfrak{n} \geq 3$ are independent of N and, by comparison with (7.39), each $g(\mathfrak{n};\mathfrak{X})$ is the part of $G(\mathfrak{n};\mathfrak{X})$ which is linear in N. In fact, this proves to be true for any value of $\mathfrak{n}$ and for any lattice. Expressing the logarithm as a double series in $\mathfrak{Z}$ and $\mathfrak{X}$ gives

$$\Phi(N,L,K) - \widehat{H}_0(L,K) = -N\sum_{\mathfrak{n}=1}^{\infty}\sum_{\{\mathfrak{k}\}}\omega[\mathfrak{n},\mathfrak{k}]\mathfrak{Z}^{\mathfrak{n}}\mathfrak{X}^{\frac{1}{2}z\mathfrak{n}-\mathfrak{q}[\mathfrak{n},\mathfrak{k}]}, \tag{7.41}$$

where $\omega[\mathfrak{n},\mathfrak{k}]$ is a lattice constant, which is the N-independent term in $\overline{\omega}[\mathfrak{n},\mathfrak{k}]$, (Appendix A.7). Equation (7.41) together with (7.30) then implies that $\Phi(N,L,K) = N\phi(L,K)$, where the dimensionless free energy ϕ per lattice site is independent of N. This is the extensive property of Φ which, when N, as has been assumed, is large, must result from statistical mechanical theory, if the latter is to agree with classical thermodynamics. Any other outcome would mean that the model is physically unrealistic. In fact, it can be shown rigorously from the form of the Hamiltonian for an Ising model or equivalent lattice gas that $N^{-1}\ln Z$ tends to a limit independent of N as the size of the lattice tends to infinity in such a way that the ratio of the number of boundary sites to N tends to zero.[3] The magnetization density m is given by

$$m = \frac{1}{N}\frac{\partial \ln Z}{\partial L} = -\frac{1}{N}\frac{\partial \Phi}{\partial L}\,, \tag{7.42}$$

and the reduced susceptibility is

$$\tilde{\chi} = \frac{\partial m}{\partial \mathcal{H}}\,. \tag{7.43}$$

[3] See Griffiths (1972) for this and more general cases and Sect. 4.2 for a discussion of the thermodynamic limit for a one-component lattice gas.

It is also convenient to define the dimensionless susceptibility

$$\chi = \frac{\partial m}{\partial L} = T\tilde{\chi}. \tag{7.44}$$

Since we have assumed that $\mathcal{H} < 0$, we have $|L| = -L$, $\mathfrak{Z} = \exp(2L)$ and it can be deduced from (7.29) and (7.37) that

$$\begin{aligned} m &= -1 + 2\sum_{\mathfrak{n}=1}^{\infty} \mathfrak{n}g(\mathfrak{n};\mathfrak{X})\mathfrak{Z}^{\mathfrak{n}}, \\ \chi &= 4\sum_{\mathfrak{n}=1}^{\infty} \mathfrak{n}^2 g(\mathfrak{n};\mathfrak{X})\mathfrak{Z}^{\mathfrak{n}}. \end{aligned} \tag{7.45}$$

The zero-field free energy and susceptibility can be obtained by putting $\mathfrak{Z} = 1$. For the simple cubic lattice (7.40) gives

$$\chi_0 = 4\mathfrak{X}^3 + 48\mathfrak{X}^5 - 56\mathfrak{X}^6 + 548\mathfrak{X}^7 - 1104\mathfrak{X}^8 + 6080\mathfrak{X}^9 + \cdots . \tag{7.46}$$

As $T \to 0$, both $\mathfrak{Z}^{\mathfrak{n}}$ and the coefficient $G(\mathfrak{n};\mathfrak{X})$ tend to zero in (7.35). However, the ground state with $\mathcal{H} < 0$, in which $\sigma(\boldsymbol{r}) = -1$ for all $\boldsymbol{r}$, is also the limiting state as $|\mathcal{H}| \to \infty$ at constant $T > 0$. In this limit $\mathfrak{Z}^{\mathfrak{n}} \to 0$, but $\mathfrak{X}$, and hence $G(\mathfrak{n};\mathfrak{X})$ and $g(\mathfrak{n};\mathfrak{X})$, are constant. The series (7.35) or (7.37) can therefore be regarded as *high-field expansions* as well as low-temperature expansions. In the lattice fluid transcription of the Ising model (Sect. 4.4) the spin states $\sigma(\boldsymbol{r}) = +1$ and $\sigma(\boldsymbol{r}) = -1$ correspond respectively to molecules and vacant lattice sites. Thus a high-field expansion becomes a fugacity or density expansion (see Domb 1974b). The unity term in (7.35) represents the vacuum state of chemical potential $\mu = -\infty$. The terms in $\mathfrak{Z}^{\mathfrak{n}}$ correspond to the formation of clusters of molecules as μ increases from $-\infty$.

A brief indication of the reasoning required to obtain the lattice constants $\omega[\mathfrak{n},\mathfrak{k}]$ is given in Appendix A.7. A procedure, variously referred to as the *partial-sum*, *shadow* or *code* method was developed by Sykes et al. (1965, 1973a). This was used to obtain high-field groupings (Sykes et al. 1973b) and low-temperature groupings (Sykes et al. 1973c) for two-dimensional lattices and for three-dimensional lattice (Sykes et al. 1973d and Sykes et al. 1973e respectively). The expansion (7.46) for χ_0 on the simple cubic lattice was extended to $\mathfrak{X}^{20}$ with expansions to $\mathfrak{X}^{15}$, $\mathfrak{X}^{28}$ and $\mathfrak{X}^{40}$ for the diamond, body-centred cubic and face-centred cubic respectively. In two dimensions the expansion of χ_0 on the square lattice was obtained to $\mathfrak{X}^{11}$ and in further papers (Sykes et al. 1973e, 1975a, 1975b, 1975c) on the triangular and honeycomb lattices to $\mathfrak{X}^{21}$. The procedure has also been applied to the Ising model on the triangular lattice with a pure triplet interaction (Sykes and Watts 1975d) and to the Ising model with $s > \frac{1}{2}$ (Fox and Guttmann 1973). Recent work by Vohwinkel (1993), using a modification of the shadow method, extended the magnetization series for the spin-$\frac{1}{2}$ Ising model on a simple cubic lattice to $\mathfrak{X}^{32}$. He also obtain results on the four-dimensional hypercubic lattice to $\mathfrak{X}^{48}$ and for the 3-state Potts model on the simple cubic lattice to degree 56 in the corresponding variable.

7.4 High-Temperature Series for the Spin-$\frac{1}{2}$ Ising Model

7.4.1 The Free Energy

The constant-field partition function for the spin-$\frac{1}{2}$ Ising model with Hamiltonian (1.49) can be expressed in the form

$$Z(L,K) = \sum_{\{\sigma(\boldsymbol{r})\}} \prod_{\{\boldsymbol{r},\boldsymbol{r}'\}}^{(\text{n.n.})} \exp[K\sigma(\boldsymbol{r})\sigma(\boldsymbol{r}')] \prod_{\{\boldsymbol{r}\}} \exp[L\sigma(\boldsymbol{r})] . \tag{7.47}$$

The terms in each product can be transformed in the manner of (7.3) with

$$\begin{aligned} \exp[K\sigma(\boldsymbol{r})\sigma(\boldsymbol{r}')] &= \cosh(K)[1+\sigma(\boldsymbol{r})\sigma(\boldsymbol{r}')\tau] , \\ \exp[L\sigma(\boldsymbol{r})] &= \cosh(L)[1+\sigma(\boldsymbol{r})\theta] , \end{aligned} \tag{7.48}$$

where

$$\tau = \tanh(K) , \qquad \theta = \tanh(L) \tag{7.49}$$

and (7.47) becomes

$$\begin{aligned} Z(L,K) = {} & [\cosh(K)]^{\frac{1}{2}zN}[\cosh(L)]^N \\ & \times \sum_{\{\sigma(\boldsymbol{r})\}} \prod_{\{\boldsymbol{r},\boldsymbol{r}'\}}^{(\text{n.n.})} [1+\sigma(\boldsymbol{r})\sigma(\boldsymbol{r}')\tau] \prod_{\{\boldsymbol{r}\}} [1+\sigma(\boldsymbol{r})\theta] . \end{aligned} \tag{7.50}$$

A binomial expansion of the first product gives a sum of $2^{\frac{1}{2}zN}$ terms each of the form τ^{q} multiplied by q factors $\sigma(\boldsymbol{r})\sigma(\boldsymbol{r}')$, each corresponding to a different nearest-neighbour pair. A subgraph corresponding to a given term can be obtained by drawing a line between nearest-neighbour sites $\boldsymbol{r}$ and $\boldsymbol{r}'$ for each factor $\sigma(\boldsymbol{r})\sigma(\boldsymbol{r}')$ present in the term. The spin variable $\sigma(\boldsymbol{r})$ is raised to a power equal to the valency of the vertex $\boldsymbol{r}$ in the subgraph. When all the spin variables are summed over the values $+1$ and -1, any term whose subgraph has vertices of odd valency disappears and the series is restricted to terms whose subgraphs have all vertices of even valency. These are called *zero-field or polygon graphs* (see Volume 1, Sect. 8.5 and Appendix A.7). For even n, $\sigma^{\mathsf{n}}(\boldsymbol{r}) = 1$ for $\sigma(\boldsymbol{r}) = \pm 1$, so, in the summation over all $\sigma(\boldsymbol{r})$, each vertex yields a factor 2, as does each lattice site not occupied by a vertex. Thus

$$\prod_{\{\boldsymbol{r},\boldsymbol{r}'\}}^{(\text{n.n.})} [1+\sigma(\boldsymbol{r})\sigma(\boldsymbol{r}')\tau] = 2^N[1+\mathsf{X}(0|\tau)] , \tag{7.51}$$

where

$$\mathsf{X}(0|\tau) = \sum_{\{\mathsf{q},\mathsf{k}\}} \Omega(0|\mathsf{q},\mathsf{k})\tau^{\mathsf{q}} , \tag{7.52}$$

the summation being over all q and k corresponding to zero-field graphs, denoted by their topologies $(0|\mathsf{q},\mathsf{k})$, $\Omega(0|\mathsf{q},\mathsf{k})$ being the number of ways in which $(0|\mathsf{q},\mathsf{k})$ can be placed on the lattice (Appendix A.7).

A binomial expansion of the second product in (7.50) yields terms of the form $\sigma(\boldsymbol{r}_1)\sigma(\boldsymbol{r}_2)\ldots\sigma(\boldsymbol{r}_{\mathsf{n}})\theta^{\mathsf{n}}$, each of which must be multiplied into each term in the expansion of the first product. However, unless the latter term has a subgraph with vertices of odd valency at the sites $\boldsymbol{r}_1, \boldsymbol{r}_2, \ldots, \boldsymbol{r}_{\mathsf{n}}$ and all other vertices of even valency, the product will disappear after all the spin variables are summed over the values ± 1. Since, as is pointed out in Appendix A.7, the number of odd vertices in a subgraph must be even, the coefficient for any odd n is zero. Thus

$$Z(L,K) = [\cosh(K)]^{\frac{1}{2}zN}[2\cosh(L)]^{N} Z^*(\theta,\tau) = Z(-L,K)\,, \qquad (7.53)$$

where

$$Z^*(\theta,\tau) = 1 + \sum_{\mathsf{s}=0}^{\infty} \mathsf{X}(2\mathsf{s}|\tau)\theta^{2\mathsf{s}}\,, \qquad \mathsf{X}(2\mathsf{s}|\tau) = \sum_{\{\mathsf{q},\mathsf{k}\}} \Omega(2\mathsf{s}|\mathsf{q},\mathsf{k})\tau^{\mathsf{q}}\,, \quad (7.54)$$

$\Omega(2\mathsf{s}|\mathsf{q},\mathsf{k})$ being the number of ways in which the graph with topology $(2\mathsf{s}|\mathsf{q},\mathsf{k})$ can be placed on the lattice. The first index $2\mathsf{s}$ denotes the number of odd vertices in the topology. Subgraphs contributing to the θ^2 term in (7.54), which have two odd vertices ($\mathsf{s}=1$), are called *magnetic graphs*, while those contributing to the θ^4 term, which have four odd vertices ($\mathsf{s}=2$), are called *hypermagnetic graphs*. Subgraphs of the simple cubic lattice are shown in Fig. A.1.

From (7.50) and (7.51) the zero-field high-temperature expansion of the partition function is given by

$$Z(0,K) = [\cosh(K)]^{\frac{1}{2}zN} 2^N [1 + \mathsf{X}(0|\tau)]\,, \qquad (7.55)$$

where $\mathsf{X}(0|\mathsf{q},\mathsf{k})$ is given by (7.52). The problem now is to compute the numbers $\Omega(0|\mathsf{q},\mathsf{k})$. In the case of a linear chain of N sites, in the form of a ring, with the N-th site as a nearest neighbour of the first, this can be done easily for all q. Then the only zero-field graph is the entire ring of N sites, so that $\Omega(0|\mathsf{q},\mathsf{k}) = 1$ if $\mathsf{q} = N$ and zero otherwise. This gives $\mathsf{X}(0|\tau) = \tau^N$. Since $z=2$, the total number of nearest-neighbour pairs is N and (7.55) becomes

$$Z(0,K) = [2\cosh(K)]^N [1 + \tau^N]\,, \qquad (7.56)$$

which agrees with the result obtained by the transfer matrix method of Volume 1, Sect. 2.4.1.

For the square lattice the zero-field values of $\varpi(0|\mathsf{q},\mathsf{k}) = \Omega(0|\mathsf{q},\mathsf{k})/N$ for up to ten lines are given in Table A.2. Using these (7.55), for the square lattice with terms up to τ^{10}, becomes

$$\begin{aligned} Z(0,K) = [\cosh(K)]^{2N} 2^N \big[&1 + N\tau^4 + 2N\tau^6 \\ &+ \tfrac{1}{2}N(N+9)\tau^8 + 2N(N+6)\tau^{10} + \cdots\big]\,. \end{aligned} \qquad (7.57)$$

The zero-field free energy can now be calculated by taking the logarithm of the partition function. A little manipulation establishes that powers of N greater than one cancel in the coefficients of powers of τ as far as the terms that have been calculated, indicating (up to this order), as for low-temperature series, that lattice constants $\omega(0|\mathsf{q},\mathsf{q})$ can be used in place the full forms for $\overline{\omega}(0|\mathsf{q},\mathsf{q})$. The dimensionless free energy per lattice site then has the series expansion

$$\begin{aligned}\phi(0,K) &= -N^{-1}\ln Z(0,K)\\ &= -\left\{\ln[2\cosh^2(K)] + \tau^4 + 2\tau^6 + \tfrac{9}{2}\tau^8 + 12\tau^{10} + \cdots\right\}. \end{aligned} \tag{7.58}$$

Using an identity between graphs and loops on a finite square lattice established by Sherman (1960, 1962),[4] Burgoyne (1963) developed a combinatorial method which, if the finite character of the lattice is ignored, allows an effective summing of the high-temperature series (7.58) to give Onsager's formula (see, for example, Volume 1, (8.137)) for the free energy.

7.4.2 Susceptibility Series

The absence of higher powers of N in the series (7.58) is the physically essential extensive property of $\Phi(0,K)$, which is found to apply for expansions up to any power of τ. Assuming this property of the free energy holds for all T and $\mathcal{H}$ it follows, from (7.53) and (7.54), that

$$\begin{aligned}\phi(L,K) &= -N^{-1}\ln Z(L,K)\\ &= -\left\{\tfrac{1}{2}\mathsf{z}\ln[\cosh(K)] + \ln[2\cosh(L)] + N^{-1}\ln Z^*(\theta,\tau)\right\}, \end{aligned} \tag{7.59}$$

where

$$\begin{aligned}N^{-1}\ln Z^*(\theta,\tau) &= \textstyle\sum_{\mathsf{s}=1}^{\infty}\mathsf{x}(2\mathsf{s}|\tau)\theta^{2\mathsf{s}}\,,\\ \mathsf{x}(2\mathsf{s}|\tau) &= \textstyle\sum_{\{\mathsf{q},\mathsf{k}\}}\omega(2\mathsf{s}|\mathsf{q},\mathsf{k})\tau^{\mathsf{q}}\,. \end{aligned} \tag{7.60}$$

For $\mathcal{H}=0$, it follows from (7.53) and (7.60) that

$$\begin{aligned}N^{-1}\ln Z(0,K) &= \tfrac{1}{2}\mathsf{z}\ln[\cosh(K)] + N^{-1}\ln Z^*(0,\tau)\\ &= \tfrac{1}{2}\mathsf{z}\ln[\cosh(K)] + \mathsf{x}(0|\tau)\,. \end{aligned} \tag{7.61}$$

From (1.41) the dimensionless configurational energy $u(0,K)$ per site is then given by

$$u(0,K) = \frac{\partial\phi}{\partial K} = -\tfrac{1}{2}\mathsf{z}\tau + u^*(\tau)\,, \tag{7.62}$$

where

[4] This identity was conjectured by Feynmann in some unpublished lecture notes where he used it to develop a variation of the method of Kac and Ward (1952) for solving the Ising model.

$$u^*(\tau) = -\frac{dx(0;\tau)}{d\tau}\frac{d\tau}{dK} = -(1-\tau^2)\sum_{\{q,k\}} q\omega(0|q,k)\tau^{q-1} \tag{7.63}$$

and, from (7.42), (7.49), (7.53) and (7.60), the magnetization density is

$$m = -\frac{\partial \phi(L,K)}{\partial L} = \theta + (1-\theta^2)\sum_{s=1}^{\infty} 2s x(2s|\tau)\theta^{2s-1}. \tag{7.64}$$

Then, from (7.44), the dimensionless susceptibility is

$$\chi = 1-\theta^2 - \frac{2\theta}{N}\frac{\partial \ln Z^*(\theta,\tau)}{\partial\theta} + \frac{(1-\theta^2)^2}{N}\frac{\partial^2 \ln Z^*(\theta,\tau)}{\partial\theta^2}. \tag{7.65}$$

Since $\theta = 0$ when $L = \mathcal{H} = 0$, the zero-field susceptibility is given by $\tilde{\chi}_0 = \chi_0/T$, where

$$\chi_0 = 1 + \frac{1}{N}\left(\frac{\partial^2 \ln Z^*(\theta,\tau)}{\partial\theta^2}\right)_{\theta=0}. \tag{7.66}$$

By differentiating (7.64) with respect to θ, it can be seen that the only series term which is non-zero, after setting $\theta = 0$, is the one corresponding to $s = 1$. It follows that

$$\chi_0 = 1 + 2\sum_{\{q,k\}} \omega(2|q,k)\tau^q. \tag{7.67}$$

The zero-field susceptibility thus depends only on subgraphs with two odd vertices (magnetic graphs).

We now consider some illustrations of (7.67). The first of these is the ring lattice of N sites, for which $\omega(2|q,1) = 1$ for any q, (Appendix A.7). The formula (7.67) gives a binomial series, yielding

$$\chi_0 = \frac{1+2\tau(1-\tau^{N-1})}{(1-\tau)}, \tag{7.68}$$

which is true for finite N. Since, $\tau < 1$, $\tau^{N-1} \to 0$ as $N \to \infty$, and in the limit of infinite N

$$\chi_0 = \frac{1+\tau}{1-\tau} = \exp(2K). \tag{7.69}$$

This is equivalent to Volume 1, (2.101), which was derived using the transfer matrix method.

For a general lattice $\omega(2|q,1)$ is given by (A.221). Thus, for a Bethe lattice of coordination number z, (7.69) gives

$$\begin{aligned}\chi_0 &= 1 + \sum_{q=1}^{\infty} z(z-1)^{q-1}\tau^q \\ &= \frac{1+\tau}{1-(z-1)\tau} = \frac{2}{z[\exp(-2K)-(z-2)/z]}.\end{aligned} \tag{7.70}$$

This is equivalent to the Bethe-pair approximation expression for χ_0 (Volume 1, (7.41)), confirming that the Bethe-pair formalism gives exact results for the interior of a Bethe lattice (Volume 1, Sect. 7.7, Baxter 1982a). The radius of convergence of the series in (7.70) is given by

$$\tau = (\mathrm{z}-1)^{-1}, \qquad T = \frac{2J}{\ln[\mathrm{z}/(\mathrm{z}-2)]}, \tag{7.71}$$

which is the Bethe-pair critical temperature.

For all regular lattices the only magnetic graphs for $\mathrm{q}=1$ and $\mathrm{q}=2$ are one and two-link chains, with lattice constants given by (A.221). Thus, from (7.49) and (7.67),

$$\chi_0 = 1 + \mathrm{z}\tau + \mathrm{z}(\mathrm{z}-1)\tau^2 + \mathrm{O}(\tau^3), \tag{7.72}$$

where

$$\tau = \frac{J}{T} + \mathrm{O}(T^{-3}). \tag{7.73}$$

From (7.72) and (7.73), it follows that

$$\tilde{\chi}_0^{-1} = \frac{T}{\chi_0} = T - \mathrm{z}J + \frac{\mathrm{z}J^2}{T} + \mathrm{O}(T^{-2}). \tag{7.74}$$

Experimental plots of this form, of the inverse susceptibility against temperature, are asymptotic at high temperatures to a line meeting the temperature axis at a value called the *paramagnetic Curie temperature*. It follows from (7.74) that the paramagnetic Curie temperature is $\mathrm{z}J$, which is higher than the critical Curie temperature for both regular and Bethe lattice Ising ferromagnets. Since the analysis does not depend on the sign of J the paramagnetic Curie temperature for the Ising antiferromagnet has the negative value $-\mathrm{z}|J|$. For both ferromagnets and antiferromagnets it can be seen from the $\mathrm{z}J^2/T$ in (7.74) that the curve of χ_0/T will lie above its asymptote at high temperatures.

Tables A.2 and A.3 give the data on the square lattice for zero-field graphs for $\mathrm{q} \le 10$ and magnetic graphs for $\mathrm{q} \le 7$. Using this information and summing the lattice constants over the subgraph topologies for each value of q we have, from (7.67),

$$\begin{aligned}\chi_0 = {} & 1 + 4\tau + 12\tau^2 + 36\tau^3 + 100\tau^4 + 276\tau^5 \\ & + 740\tau^6 + 1972\tau^7 + \cdots .\end{aligned} \tag{7.75}$$

7.4.3 Coefficient Relations for Susceptibility Series

A *closed graph* is defined as one with no vertices of valency one. For instance, zero-field graphs are closed but chains and tadpoles are not and, as we see in Appendix A.7, calculation of lattice constants for non-closed graphs is in general more difficult than for closed graphs. The elimination of contributions to series from non-closed graphs is therefore desirable. From the chain recurrence relationship (A.223), Sykes (1961) derived a different equation of the form

$$\mathfrak{S}(\mathrm{q}) = \omega(2;\mathrm{q},1) - 2(\mathrm{z}-1)\omega(2;\mathrm{q}-1,1) + (\mathrm{z}-1)^2\omega(2;\mathrm{q}-2,1) \tag{7.76}$$

for chain lattice constants, where $\mathfrak{S}(\mathrm{q})$ is the sum of *closed* subgraph lattice constants of order q or less.[5] Sykes (1961) eliminated non-closed graph contributions in the zero-field susceptibility series by using a difference equation, similar in form to (7.76), for the coefficients of powers of τ in χ_0. This is equivalent to the relation

$$\chi_0 = \Theta(\tau)\Big[1 - (\mathrm{z}-2)\tau - (\mathrm{z}-1)\tau^2 + \sum_{\mathrm{q}=3}^{\infty} \mathrm{d}(\mathrm{q})\tau^{\mathrm{q}}\Big], \tag{7.77}$$

where

$$\Theta(\tau) = [1 - \tau(\mathrm{z}-1)]^{-2} \tag{7.78}$$

corresponds to the left-hand side of the difference equation and the $\mathrm{d}(\mathrm{q})$ are linearly dependent on the closed subgraph lattice constants. Sykes (1961) gave explicit expressions for $\mathrm{d}(\mathrm{q})$, but has never published his derivation. We shall use an alternative approach, due to Nagle and Temperley (1968), based on the weak-graph transformation (Appendix A.8).

The partition function factor $Z^*(\theta,\tau)$, defined by (7.54) can be put into the 'vertex-weight' form (A.240) by using (A.241). Since the vertex weight then depends only on the valency γ, the weak-graph transformation takes the form (A.258). In this case (A.261) is the explicit formula for $\mathfrak{N}_{\gamma'}$. The essential step is to adjust the value of y to make $\mathfrak{N}_1 = 0$, thus giving zero weight to all non-closed subgraphs. From (A.261), y is given by

$$\tfrac{1}{2}(1+\theta)(y-\tau^{\frac{1}{2}})(1+y\tau^{\frac{1}{2}})^{\mathrm{z}-1} + \tfrac{1}{2}(1-\theta)(y+\tau^{\frac{1}{2}})(1-y\tau^{\frac{1}{2}})^{\mathrm{z}-1} = 0\,. \tag{7.79}$$

This equation cannot be solved to give y as a closed form in θ, but y can be regarded as a series in odd powers of θ, since, if a pair of values (y,θ) satisfies (7.79), then so does $(-y,-\theta)$. This implies that $y(\theta)$ is an odd function. It then follows, from (A.261), that $\mathfrak{N}_{\gamma'}$ can be expressed as a series in odd or even powers of θ according to whether γ' is odd or even. Since χ_0 is given by (7.66), explicit values are not needed for the coefficients of θ^3 and higher powers. It is straightforward to show that

$$y = \theta\sqrt{\tau\Theta(\tau)} + \mathrm{O}(\theta^3) \tag{7.80}$$

is the form of the solution of (7.79). From (A.261) and (7.78)

---

[5] For the form of $\mathfrak{S}(\mathrm{q})$ see Sykes (1961) or Domb (1974a).

$$\mathfrak{N}_0 = 1 + \theta^2\tau\Theta(\tau)z\left[1 - \frac{1}{2}\tau(z-1)\right] + O(\theta^4)\,. \tag{7.81}$$

It is convenient to define the new normalized vertex weights

$$\mathfrak{M}_{\gamma'} = \mathfrak{N}_{\gamma'}/\mathfrak{N}_0\,, \qquad \gamma' = 1,\ldots,z\,. \tag{7.82}$$

The form of $\mathfrak{N}_{\gamma'}$, and thus $\mathfrak{M}_{\gamma'}$, differs according to whether γ' is even or odd. For even and odd values we set $\gamma' = 2\nu$ and $\gamma' = 2\nu + 1$ respectively. From (A.261), (7.81) and (7.82),

$$\begin{aligned} \mathfrak{M}_{2\nu} &= \tau^\nu + \theta^2\tau^\nu\nu\Theta(\tau)[2(\nu-1)(1+\tau)^2 - 1 + \tau^2] + O(\theta^4)\,,\\ \mathfrak{M}_{2\nu+1} &= 2\theta\nu\tau^\nu(1+\tau)\sqrt{\tau\Theta(\tau)} + O(\theta^3)\,. \end{aligned} \tag{7.83}$$

For the Ising model it is the factor $Z^*(\theta,\tau)$ of the partition function, defined in (7.53) and (7.54), which is of the form (A.240). From (A.256), in this special case where the vertex weights depend only on the site valencies,

$$\begin{aligned} N^{-1}\ln Z^*(\theta,\tau) &= -\tfrac{1}{2}z\ln(1+y^2) + N^{-1}\ln Z^*(\mathfrak{N}_0,\ldots,\mathfrak{N}_z)\\ &= \ln\mathfrak{N}_0 - \tfrac{1}{2}z\ln(1+y^2)\\ &\quad + N^{-1}\ln Z^*(\mathfrak{M}_0,\ldots,\mathfrak{M}_z)\,. \end{aligned} \tag{7.84}$$

From (7.60), $N^{-1}\ln Z^*(\mathfrak{M}_0,\ldots,\mathfrak{M}_z)$ can be expressed as a sum of lattice constants multiplied by appropriate products of the $\mathfrak{M}_{\gamma'}$, for $\gamma' = 2,\ldots,z$.[6] In fact, we are concerned in deriving an expression only for the zero-field susceptibility, which is obtained from $Z^*(\theta,\tau)$ using (7.66). This means, since $\mathfrak{M}_{2\nu+1}$ is linear in θ, that we need consider contributions only from zero-field graphs and from magnetic subgraphs with exactly two odd vertices. Thus

$$N^{-1}\ln Z^*(\mathfrak{M}_0,\ldots,\mathfrak{M}_z) = \Psi(0|\tau,\theta) + \Psi(2|\tau,\theta) + O(\theta^3)\,, \tag{7.85}$$

where

$$\Psi(2s|\tau,\theta) = \sum_{\{q,k\}} \omega(2s|q,k) \prod^{(2s;q,k)} \mathfrak{M}_{\gamma'}\,, \qquad s = 0,1, \tag{7.86}$$

the product in each case being over the vertices of the appropriate graph. From (7.63), (7.83) and (A.250),

$$\begin{aligned} \left(\frac{d^2\Psi(0|\tau,\theta)}{d\theta^2}\right)_{\theta=0} &= 2\Theta(\tau)\sum_{\{q,k\}}\tau^q \sum_{\{i\}}^{(0;q,k)} \nu_i[2(\nu_i-1)\\ &\quad + (1+\tau)^2 - 1 + \tau^2]\\ &= 2\Theta(\tau)[\tau u^*(\tau) + 2(1+\tau)^2\Sigma(0|\tau)]\,, \end{aligned} \tag{7.87}$$

where

---

[6] $\gamma' = 0$, with $\mathfrak{M}_0 = 1$, corresponds to a lattice site not present in the subgraph and $\mathfrak{M}_1 = 0$.

$$\Sigma(0|\tau) = \sum_{\{q,k\}} \omega(0|q,k)\tau^q \sum_{\{i\}}^{(0;q,k)} \nu_i(\nu_i - 1), \tag{7.88}$$

in each case the inner summation being over all the vertices of the appropriate subgraph. For the example of the honeycomb lattice, discussed below, it is important to note that even vertices of valency two ($\nu_i = 1$) make no contribution to the sum in (7.88). In a similar way

$$\left(\frac{d^2\Psi(2|\tau,\theta)}{d\theta^2}\right)_{\theta=0} = 8\Theta(\tau)(1+\tau)^2\Sigma(2|\tau), \tag{7.89}$$

where

$$\Sigma(2|\tau) = \sum_{\{q,k\}} \omega(2|q,k)\tau^q \nu_1\nu_2, \tag{7.90}$$

$(2\nu_1 + 1)$ and $(2\nu_2 + 1)$ being the vertex weights of the two odd vertices in the appropriate subgraph. Now, from (7.66), (7.78), (7.80), (7.81), (7.85), (7.87) and (7.89),

$$\begin{aligned} \chi_0 = \Theta(\tau)[1 - (z-2)\tau - (z-1)\tau^2 + 2\tau u^*(\tau) \\ + 4(1+\tau)^2\{\Sigma(0|\tau) + 2\Sigma(2|\tau)\}], \end{aligned} \tag{7.91}$$

which is of the form of (7.77). The first three terms in the square bracket in (7.91) give the Bethe lattice susceptibility (7.70). It should be noted that the series for $u^*(\tau)$ (given by (7.63)), $\Sigma(0|\tau)$ and $\Sigma(2|\tau)$ contain terms from separated closed subgraphs as well as from connected ones.

As an example of the use of (7.91), terms in the series for the honeycomb lattice susceptibility will be now be derived. Since, for this lattice, $z = 3$, there are no even vertices of valency greater than two and $\Sigma(0|\tau) = 0$. Data for zero-field and closed magnetic subgraphs for $q \leq 14$ are given in Table A.4. From the zero-field subgraphs and (7.63),

$$u^*(\tau) = -(3\tau^5 - 3\tau^7 + 15\tau^9 - 24\tau^{11} + 93\tau^{13} + \cdots), \tag{7.92}$$

and from the magnetic subgraphs and (7.90),

$$(1+\tau)^2\Sigma(2|\tau) = \tfrac{3}{2}\tau^{11} + 3\tau^{12} + 3\tau^{13} + 9\tau^{14} + \cdots. \tag{7.93}$$

Using the expansion of the Bethe lattice terms, given in (7.70), with $z = 3$ and expressing $\Theta(\tau)$ as a series, equation (7.91) yields

$$\begin{aligned} \chi_0 = 1 + 3\tau + 6\tau^2 + 12\tau^3 + 24\tau^4 + 48\tau^5 + 90\tau^6 \\ + 168\tau^7 + 318\tau^8 + 600\tau^9 + 1098\tau^{10} + 2004\tau^{11} \\ + 3696\tau^{12} + 6792\tau^{13} + 12270\tau^{14} + \cdots. \end{aligned} \tag{7.94}$$

From this result for the honeycomb lattice it is now possible to derive the corresponding expansion for the triangular lattice. The star–triangle transformation, (Volume 1, Sect. 8.4) gives the relationship

$$\tau_{\rm H}^2 = \frac{\tau_{\rm T}(1+\tau_{\rm T})}{1+\tau_{\rm T}^3} \tag{7.95}$$

between the variables $\tau_{\rm H}$ and $\tau_{\rm T}$ for the honeycomb and triangular lattices respectively and it was shown by Fisher (1959) (see also, Syozi 1972, Appendix D). that

$$\chi_{0\rm T}(\tau_{\rm T}) = \tfrac{1}{2}[\chi_{0\rm H}(\tau_{\rm H}) + \chi_{0\rm H}(-\tau_{\rm H})] \,. \tag{7.96}$$

Substituting from (7.94) gives the form

$$\begin{aligned}\chi_0 = {} & 1 + 6\tau + 30\tau^2 + 138\tau^3 + 606\tau^4 + 2586\tau^5 + 10818\tau^6 + 44564\tau^7 \\ & + 141318\tau^8 + 280236\tau^9 + 211920\tau^{10} - 507708\tau^{11} \\ & - 1834800\tau^{12} - 2280864\tau^{13} + 543660\tau^{14} + \cdots \end{aligned} \tag{7.97}$$

for the triangular lattice susceptibility series.

7.5 The Linked-Cluster Expansion

We now consider high-temperature series for more general magnetic models, particularly the general-s Ising model and the classical and quantum Heisenberg models. Since the number of energy levels of a nearest-neighbour pair is greater that two, the type of transform of variables used in Sect. 7.4 is no longer possible and expansions will be in the coupling $K = J/T$.

7.5.1 Multi-Bonded Graphs

At each site $\boldsymbol{r}$ of the lattice $\mathcal{N}$ there is a spin operator $\hat{\boldsymbol{s}}(\boldsymbol{r})$ which can have a number of components, which are themselves either classical scalar quantities or quantum operators. We define the usual field coupling $L = \mathcal{H}/T$, but for later convenience we introduce distinct interactions $-J(\boldsymbol{r},\boldsymbol{r}')$ between nearest-neighbour pairs $(\boldsymbol{r},\boldsymbol{r}')$ with corresponding couplings $K(\boldsymbol{r},\boldsymbol{r}') = J(\boldsymbol{r},\boldsymbol{r}')/T$. The dimensionless Hamiltonian has the form

$$\widehat{H}(L, K(\boldsymbol{r},\boldsymbol{r}'); \hat{\boldsymbol{s}}(\boldsymbol{r})) = \widehat{H}_0(L; \hat{s}_z(\boldsymbol{r})) + \widehat{H}_1(K(\boldsymbol{r},\boldsymbol{r}'); \hat{\mathfrak{e}}(\boldsymbol{r})) \,, \tag{7.98}$$

where

$$\widehat{H}_0(L; \hat{s}_z(\boldsymbol{r})) = -(L/s)\sum_{\{\boldsymbol{r}\}} \hat{s}_z(\boldsymbol{r}) \,, \tag{7.99}$$

$$\widehat{H}_1(K(\boldsymbol{r},\boldsymbol{r}'); \hat{\mathfrak{e}}(\boldsymbol{r},\boldsymbol{r}')) = -\sum_{\{\boldsymbol{r},\boldsymbol{r}'\}}^{(\rm n.n.)} K(\boldsymbol{r},\boldsymbol{r}')\hat{\mathfrak{e}}(\boldsymbol{r},\boldsymbol{r}') \,. \tag{7.100}$$

The operator $\hat{\mathfrak{e}}(\boldsymbol{r},\boldsymbol{r}')$ is some function of the spin operators $\hat{\boldsymbol{s}}(\boldsymbol{r})$ and $\hat{\boldsymbol{s}}(\boldsymbol{r}')$ on the nearest-neighbour pair of sites $\boldsymbol{r}$ and $\boldsymbol{r}'$ and the magnetic field is in the z-direction, in which the component of $\hat{\boldsymbol{s}}(\boldsymbol{r})$ is $\hat{s}_z(\boldsymbol{r})$.[7]

[7] In the treatment of quantum systems will assume a set of units in terms of which $\hbar = 1$.

The probability function defined in (1.12) is replaced by the density operator defined in (A.265), with the partition function given by (A.266). For the Hamiltonian defined by equations (7.98)–(7.100)

$$\begin{aligned} Z(L,K(\boldsymbol{r},\boldsymbol{r}')) &= \mathrm{Trace}\{\exp[-\widehat{H}(L,K(\boldsymbol{r},\boldsymbol{r}');\hat{\boldsymbol{s}}(\boldsymbol{r}))]\} \\ &= \mathrm{Trace}\left\{\exp[-\widehat{H}_1(K(\boldsymbol{r},\boldsymbol{r}');\hat{\mathfrak{e}}(\boldsymbol{r},\boldsymbol{r}'))]\prod_{\{\boldsymbol{r}\}}\exp\left(\frac{L\hat{s}_z(\boldsymbol{r})}{s}\right)\right\}. \end{aligned} \tag{7.101}$$

For the Ising model the trace reduces to the usual summation, but for the quantum Heisenberg model it denotes the algebraic trace of the appropriate spin-operator matrices and for the classical Heisenberg model it corresponds to integration over polar coordinates. Given that the spin operators on all the lattice sites are of the same type,

$$z_0(L) = \mathrm{Trace}\{\exp(L\hat{s}_z(\boldsymbol{r})/s)\} \tag{7.102}$$

is independent of the lattice site and, as in Appendix A.9, it is convenient to define

$$\begin{aligned} Z^+(L,K(\boldsymbol{r},\boldsymbol{r}')) &= [z_0(L)]^{-N} Z(L,K(\boldsymbol{r},\boldsymbol{r}')) \\ &= \mathrm{Trace}\left\{\exp[-\widehat{H}_1(K(\boldsymbol{r},\boldsymbol{r}');\hat{\mathfrak{e}}(\boldsymbol{r},\boldsymbol{r}'))]\hat{p}_0(L)\right\}, \end{aligned} \tag{7.103}$$

where

$$\hat{p}_0(L) = [z_0(L)]^{-N}\prod_{\{\boldsymbol{r}\}}\exp[L\hat{s}_z(\boldsymbol{r})/s] \tag{7.104}$$

is the density operator for the non-interacting system ($K=0$).

To obtain an expansion for $\ln Z^+$ in powers of K the Maclaurin method is used.[8] This gives

$$\ln Z^+(L,K) = \left[\exp\left(K\sum_{\{\boldsymbol{r},\boldsymbol{r}'\}}^{(\mathrm{n.n.})}\mathcal{D}(\boldsymbol{r},\boldsymbol{r}')\right)\ln Z^+(L,K(\boldsymbol{r},\boldsymbol{r}'))\right]_0, \tag{7.105}$$

where

$$\mathcal{D}(\boldsymbol{r},\boldsymbol{r}')) = \frac{\partial}{\partial K(\boldsymbol{r},\boldsymbol{r}')} \tag{7.106}$$

and the subscript '0' indicates that all the couplings $K(\boldsymbol{r},\boldsymbol{r}')$ are set equal to zero. Using the expansion

[8] The function $f(\boldsymbol{x})$ of the vector $\boldsymbol{x}$ can be expressed in the form

$$f(\boldsymbol{\delta x}) = [\exp(\boldsymbol{\delta x}\cdot\boldsymbol{\nabla})f(\boldsymbol{x})]_{\boldsymbol{x}=0}$$

at a point $\boldsymbol{\delta x}$ near the origin. The expansion in powers of the components of $\boldsymbol{\delta x}$ is then obtained by using the power-series expansion of the exponential.

$$\exp\left(K\sum_{\{\boldsymbol{r},\boldsymbol{r}'\}}^{(\text{n.n.})}\mathcal{D}(\boldsymbol{r},\boldsymbol{r}')\right)=1+\sum_{\mathring{q}=1}^{\infty}\frac{K^{\mathring{q}}}{\mathring{q}!}\left[\sum_{\{\boldsymbol{r},\boldsymbol{r}'\}}^{(\text{n.n.})}\mathcal{D}(\boldsymbol{r},\boldsymbol{r}')\right]^{\mathring{q}} \tag{7.107}$$

in (7.105) gives a power-series expansion in K. The expression in square brackets can be expanded multinomially into a sum of terms each containing a product of $\mathring{q}$ nearest-neighbour pair factors $\mathcal{D}(\boldsymbol{r},\boldsymbol{r}')$. Two operators corresponding to nearest-neighbour pairs with a site in common will not in general commute and terms should, therefore, be grouped together according to the set of operators involved, in all their permutations. A subgraph of $\mathcal{N}$ corresponding to each group of products is obtained by placing a line between the nearest-neighbour sites $\boldsymbol{r}$ and $\boldsymbol{r}'$ for each factor $\mathcal{D}(\boldsymbol{r},\boldsymbol{r}')$ present. However, the class of graphs obtained differs from those of Sect. 7.4 since a factor $\mathcal{D}(\boldsymbol{r},\boldsymbol{r}')$ may occur more than once in each product of the group, giving several lines linking a pair of vertices $\boldsymbol{r}$ and $\boldsymbol{r}'$. We thus have *multi-bonded graphs* (Appendix A.7).

As an elementary example, consider the $\mathring{q}=2$ term in the K-expansion of (7.107) for the square lattice. We concentrate on one square group of sites labelled 1, 2, 3 and 4 and replacing $\boldsymbol{r}$ and $\boldsymbol{r}'$ by the site numbers for these four sites

$$\begin{aligned}\left[\sum_{\{\boldsymbol{r},\boldsymbol{r}'\}}^{(\text{n.n.})}\mathcal{D}(\boldsymbol{r},\boldsymbol{r}')\right]^{2} &= [\mathcal{D}(1,2)]^2+[\mathcal{D}(2,3)]^2+[\mathcal{D}(3,4)]^2+[\mathcal{D}(4,1)]^2\\ &+2\mathcal{D}(1,2)\mathcal{D}(3,4)+2\mathcal{D}(2,3)\mathcal{D}(4,1)\\ &+[\mathcal{D}(1,2)\mathcal{D}(2,3)+\mathcal{D}(2,3)\mathcal{D}(1,2)]+[\mathcal{D}(4,1)\mathcal{D}(1,2)+\mathcal{D}(1,2)\mathcal{D}(4,1)]\\ &+[\mathcal{D}(2,3)\mathcal{D}(3,4)+\mathcal{D}(3,4)\mathcal{D}(2,3)]+[\mathcal{D}(3,4)\mathcal{D}(4,1)+\mathcal{D}(4,1)\mathcal{D}(3,4)]\\ &+\cdots.\end{aligned} \tag{7.108}$$

The first four terms on the right-hand side of (7.108) have the multi-bonded topology $(2,1)$ given in the first line of Table A.5. Terms five and six are disconnected pairs with commuting operators with topology $(2,2)$ and terms seven to ten are connected, with two terms each corresponding to the possible non-commutativity of the operators and with topology $(2,3)$.

If we now perform a multinomial expansion on each term in (7.107) and retain as distinct all orderings of distinct operators there will be $\mathring{q}!/\prod_{t=1}^{q}\alpha_t!$ terms corresponding to the multi-bonded graph $(\mathring{q},\mathring{k})$. Thus, substituting into (7.105)

$$\ln Z^{+}(L,K)=\sum_{\mathring{q}=1}^{\infty}K^{\mathring{q}}\sum_{\{\mathring{k}\}}\mathring{\Omega}(\mathring{q},\mathring{k})C(L;\mathring{q},\mathring{k})\left(\prod_{t=1}^{q}\alpha_t!\right)^{-1}, \tag{7.109}$$

where[9]

[9] It is important to note that we can set $K(\boldsymbol{r},\boldsymbol{r}')=0$ for all lattice edges apart from those of the graph $(\mathring{q},\mathring{k})$ *before* differentiating.

$$\mathrm{C}(L;\mathring{\mathrm{q}},\mathring{\mathrm{k}}) = \left[\widehat{D}(\mathring{\mathrm{q}},\mathring{\mathrm{k}}) \ln Z^{+}(L, K(\boldsymbol{r},\boldsymbol{r}'))\right]_0 \tag{7.110}$$

and

$$\widehat{D}(\mathring{\mathrm{q}},\mathring{\mathrm{k}}) = \frac{\prod_{\mathrm{t}=1}^{\mathrm{q}} \alpha_{\mathrm{t}}!}{\mathring{\mathrm{q}}!} \sum_{\{\nu(\mathring{\mathrm{q}},\mathring{\mathrm{k}})\}} \mathcal{D}(\boldsymbol{r}_1,\boldsymbol{r}'_1)\cdots\mathcal{D}(\boldsymbol{r}_{\mathring{\mathrm{q}}},\boldsymbol{r}'_{\mathring{\mathrm{q}}}), \tag{7.111}$$

the sum being over all permutations $\nu(\mathring{\mathrm{q}},\mathring{\mathrm{k}})$ of the operators for $(\mathring{\mathrm{q}},\mathring{\mathrm{k}})$. A partial trace can then be performed over all the lattice sites not occupied by the n graph vertices. This operation involves only $\hat{p}_0(L)$, which can then be replaced by

$$\hat{p}'_0(L) = [z_0(L)]^{-\mathrm{n}} \prod_{\{\boldsymbol{r}\}}^{(\mathrm{k})} \exp(L\hat{s}_z(\boldsymbol{r})/s). \tag{7.112}$$

The remaining part of the trace operation is confined to the n graph vertices. Since there is only a single term $K(\boldsymbol{r},\boldsymbol{r}')\hat{\mathfrak{e}}(\boldsymbol{r},\boldsymbol{r}')$ in $\widehat{H}$ for each nearest-neighbour link, the trace can be regarded as taken over the silhouette graph with topology k. This will be denoted by the symbol $(\mathrm{Trace})_{\mathrm{k}}$. From (7.110), $\mathrm{C}(L;\mathring{\mathrm{q}},\mathring{\mathrm{k}})$ depends only on $(\mathrm{Trace})_{\mathrm{k}}$ and the multiplicities $\alpha_1,\alpha_2,\ldots,\alpha_{\mathrm{q}}$ and is therefore independent of the lattice type.

7.5.2 Connected Graphs and Stars

Suppose that the graph with topology k is disjoint with the two parts denoted by k_1 and k_2. In this case the operators for k_1 and k_2 commute and, from (A.264), it follows that

$$(\mathrm{Trace})_{\mathrm{k}} = (\mathrm{Trace})_{\mathrm{k}_1}(\mathrm{Trace})_{\mathrm{k}_2}, \tag{7.113}$$

$$\ln(\mathrm{Trace})_{\mathrm{k}} = \ln(\mathrm{Trace})_{\mathrm{k}_1} + \ln(\mathrm{Trace})_{\mathrm{k}_2}. \tag{7.114}$$

Since the formula (7.110) for $\mathrm{C}(L;\mathring{\mathrm{q}},\mathring{\mathrm{k}})$ involves partial differentiation with respect to the couplings for all the bonds of the graph, this process will yield a result of zero for each part of the decomposition. It follows that $\mathrm{C}(L;\mathring{\mathrm{q}},\mathring{\mathrm{k}})$ in non-zero *only for connected graphs* and the expansion in (7.109) can be restricted to such graphs.[10] Since only connected graphs are involved $\Omega(\mathrm{q},\mathrm{k}) = N\omega(\mathrm{q},\mathrm{k})$, where $\omega(\mathrm{q},\mathrm{k})$ is the lattice constant (Appendix A.7). From (7.103) and (7.109), the dimensionless free energy per lattice site is given by

[10] Hence the alternative name 'connected graph expansion'. Note that a multi-bonded graph is connected if and only if its silhouette is connected.

$$\begin{aligned}\phi(L,K) &= -N^{-1}\ln Z(L,K)\\ &= -\ln z_0(L)\\ &\quad -\sum_{\mathring{\mathrm{q}}=1}^{\infty} K^{\mathring{\mathrm{q}}} \sum_{\{\mathring{\mathrm{k}}\}} \omega(\mathrm{q},\mathrm{k})\mathrm{S}(\mathrm{q},\mathrm{k};\mathring{\mathrm{q}},\mathring{\mathrm{k}})\mathrm{C}(L;\mathring{\mathrm{q}},\mathring{\mathrm{k}})\left(\prod_{\mathrm{t}=1}^{\mathrm{q}}\alpha_{\mathrm{t}}!\right)^{-1},\end{aligned} \tag{7.115}$$

where $\mathrm{S}(\mathrm{q},\mathrm{k};\mathring{\mathrm{q}},\mathring{\mathrm{k}})$ is given by (A.231).

In Sects. 7.3 and 7.4 we deduced the extensive property of the free energy for certain Ising model low-temperature and high-temperature series by extrapolation from the first few terms. In (7.115) we now have a more rigorously derived result which justifies the replacement of $\overline{\omega}(\mathrm{q},\mathrm{k})$ by $\omega(\mathrm{q},\mathrm{k})$. Nevertheless, (7.109) (with the expansion restricted to connected graphs) is not merely a stepping stone to (7.115). Since no assumption was made in the derivation of (7.109) about the number N of lattice sites it is valid for a finite lattice and indeed for any connected group of sites.

We have shown that, in general, the expansion in (7.109) can be restricted to connected graphs. In certain models this restriction can be strengthened to also exclude graphs with articulation points (see Sect. A.7). This is the case for models in which the spectrum of interaction energies of a graph in zero field, with the spin state on one vertex fixed, is independent of that spin state. We can then write

$$(\mathrm{Trace})_{\mathrm{k}} = z_0(0)(\mathrm{Trace})^*_{\mathrm{k}}\,, \tag{7.116}$$

where $(\mathrm{Trace})^*_{\mathrm{k}}$ denotes the partial trace with one spin state fixed. This applies to the spin-$\frac{1}{2}$ Ising model, where the only possible change of state for a spin $\sigma(\boldsymbol{r})$ is a reversal from $\sigma(\boldsymbol{r}) = 1$ to $\sigma(\boldsymbol{r}) = -1$ or vice-versa and the zero-field Hamiltonian is invariant under reversal of all spins. It does not apply to the $s > \frac{1}{2}$ Ising model, as can easily be seen by considering a single pair of nearest-neighbour spins. It is, however, applicable to the D-vector model. A graph of topology k with at least one articulation point can be regarded as consisting of two graphs k_1 and k_2 with only one vertex (the articulation point) in common. Fixing the spin state at this vertex we then have

$$\begin{aligned}(\mathrm{Trace})_{\mathrm{k}} &= z_0(0)(\mathrm{Trace})^*_{\mathrm{k}_1}(\mathrm{Trace})^*_{\mathrm{k}_2}\\ &= [z_0(0)]^{-1}(\mathrm{Trace})_{\mathrm{k}_1}(\mathrm{Trace})_{\mathrm{k}_2}\,,\end{aligned} \tag{7.117}$$

$$\ln(\mathrm{Trace})_{\mathrm{k}} = -\ln z_0(0) + \ln(\mathrm{Trace})_{\mathrm{k}_1} + \ln(\mathrm{Trace})_{\mathrm{k}_2}\,. \tag{7.118}$$

Since $z_0(0)$ is independent of $K(\boldsymbol{r},\boldsymbol{r}')$ for every nearest-neighbour pair a similar argument to that used for disconnected graphs establishes the result that $\mathrm{C}(L;\mathring{\mathrm{q}},\mathring{\mathrm{k}}) = 0$ when the silhouette graph has an articulation point. The summations in (7.109) and (7.115) can thus be restricted to stars (i.e. connected graphs without articulation points). Other applications of star–graph expansions are given by Domb (1974a, 1974c).

7.5.3 Moments, Cumulants and Finite Clusters

When any functions $f(y)$ and $\ln f(y)$ are expanded in powers of y the coefficients in the series for $f(y)$ are called *moments* and those for $\ln f(y)$ are called *cumulants*. For the case $f(y) = \exp(\widehat{H}_0)Z$ and $y = \exp(-2|L|)$, equations (7.38) give the first few cumulants in terms of moments. Equations (7.109)–(7.111) give the cumulant expansion of $Z^+(L, K)$ in powers of K. It is not difficult to show that the corresponding moment expansion is given by

$$Z^+(L,K) = 1 + \sum_{\mathring{q}=1}^{\infty} K^{\mathring{q}} \sum_{\{\mathring{k}\}} \Omega(\mathring{q},\mathring{k})\mathrm{B}(L;\mathring{q},\mathring{k})\left(\prod_{t=1}^{q}\alpha_t!\right)^{-1}, \tag{7.119}$$

where

$$\mathrm{B}(L;\mathring{q},\mathring{k}) = \mathrm{Trace}\left\{\widehat{E}(\mathring{q},\mathring{k})\hat{p}_0(L)\right\} \tag{7.120}$$

and

$$\widehat{E}(\mathring{q},\mathring{k}) = \frac{\prod_{t=1}^{q}\alpha_t!}{\mathring{q}!}\sum_{\{\nu(\mathring{q},\mathring{k})\}} \hat{\mathfrak{e}}(\boldsymbol{r}_1,\boldsymbol{r}'_1)\cdots\hat{\mathfrak{e}}(\boldsymbol{r}_{\mathring{q}},\boldsymbol{r}'_{\mathring{q}}) \,. \tag{7.121}$$

Too sharp a distinction should not be drawn between moment and cumulant methods since, to derive the thermodynamic properties of a model, it is always necessary to obtain a expression for $\ln Z(L, K)$. In Sect. 7.3, assuming that $\ln Z(L, K)$ is a thermodynamically extensive quantity for large N, we identified the cumulants as the terms linear in N in the expressions for the moments.

In Sect. 1.3 we introduced the use of the angle brackets $\langle\cdots\rangle$ to indicate the expectation value (or ensemble average) of a statistical mechanical quantity. The corresponding situation for quantum statistics is presented in Appendix A.9. Using the Hamiltonian defined in (7.98) the corresponding formula for the expectation value of a quantum operator $\hat{\mathcal{O}}$ is

$$\begin{aligned}\langle\hat{\mathcal{O}}\rangle &= \frac{\mathrm{Trace}\{\hat{\mathcal{O}}\exp[-\widehat{H}(L,K(\boldsymbol{r},\boldsymbol{r}');\hat{\mathfrak{s}}(\boldsymbol{r}))]\}}{Z(L,K(\boldsymbol{r},\boldsymbol{r}'))}, \\ &= \frac{\mathrm{Trace}\left\{\hat{\mathcal{O}}\exp\left[\sum_{\{\boldsymbol{r},\boldsymbol{r}'\}}^{(\mathrm{n.n.})} K(\boldsymbol{r},\boldsymbol{r}')\hat{\mathfrak{e}}(\boldsymbol{r},\boldsymbol{r}')\right]\hat{p}_0(L)\right\}}{Z^+(L,K(\boldsymbol{r},\boldsymbol{r}'))}.\end{aligned} \tag{7.122}$$

We now denote the (not necessarily distinct) quantum operators $\hat{\mathfrak{e}}(\boldsymbol{r},\boldsymbol{r}')$ for the graph $(\mathring{q},\mathring{k})$ by $\hat{\phi}_1, \hat{\phi}_2, \ldots, \hat{\phi}_{\mathring{q}}$. The corresponding couplings are denoted by $\kappa_1, \kappa_2, \ldots, \kappa_{\mathring{q}}$, with $\mathcal{D}_i$ denoting the partial derivative with respect to κ_i. The notation $\langle\langle\cdots\rangle\rangle$ is defined by (A.277) and (A.288) gives the relationship between single bracket terms $\langle\cdots\rangle_0$ and the double bracket expression $\langle\langle\cdots\rangle\rangle_0$, when both are evaluated with all the couplings κ_i set to zero. In fact it is shown in Appendix A.9 that double brackets reduce to single brackets for commuting operators. Using the notation[11]

$$\lceil\hat{\phi}_{\mathring{q}} \cdots \hat{\phi}_1\rfloor = \mathcal{D}_{\mathring{q}}\mathcal{D}_2 \cdots \mathcal{D}_1 \ln Z^+(L, \kappa(\boldsymbol{r}, \boldsymbol{r}')) . \tag{7.123}$$

it was shown in Appendix A.9 that

$$\lceil\hat{\phi}_{\mathring{q}} \cdots \hat{\phi}_1\rfloor_0 = \sum_{\ell=1}^{\mathring{q}} (-1)^{\ell-1}(\ell-1)! \sum_{\{\pi_{\mathring{q}}(\ell)\}} \langle\langle\hat{\phi}^{(\mathring{q})} \cdots \hat{\phi}^{(j)}\rangle\rangle_0 \cdots \\ \cdots \langle\langle\hat{\phi}^{(t)} \cdots \hat{\phi}^{(r)}\rangle\rangle_0 \cdots \langle\langle\cdots \hat{\phi}^{(1)}\rangle\rangle_0 , \tag{7.124}$$

$$\langle\langle\hat{\phi}_{\mathring{q}} \cdots \hat{\phi}_1\rangle\rangle_0 = \sum_{\ell=1}^{\mathring{q}} \sum_{\{\pi_{\mathring{q}}(\ell)\}} \lceil\hat{\phi}^{(\mathring{q})} \cdots \hat{\phi}^{(j)}\rfloor_0 \cdots \lceil\hat{\phi}^{(t)} \cdots \hat{\phi}^{(r)}\rfloor_0 \cdots \\ \cdots \lceil\cdots \hat{\phi}^{(1)}\rfloor_0 , \tag{7.125}$$

where $\pi_{\mathring{q}}(\ell)$ denotes a partition of $\hat{\phi}_1, \hat{\phi}_2, \ldots, \hat{\phi}_{\mathring{q}}$ into ℓ subsets. From (7.120) and (7.121) it can be seen that each factor $\langle\langle\cdots\rangle\rangle_0$ in (7.124) corresponds to a coefficient $\mathrm{B}(L; \mathring{q}', \mathring{k}')$ for some graph $(\mathring{q}', \mathring{k}')$ in the moment expansion (7.119) with

$$\mathrm{B}(L; \mathring{q}, \mathring{k}) = \langle\langle\hat{\phi}_{\mathring{q}} \cdots \hat{\phi}_1\rangle\rangle_0 . \tag{7.126}$$

From (7.124) and (A.288) it can be seen that $\lceil\hat{\phi}_{\mathring{q}} \cdots \hat{\phi}_1\rfloor_0$ is invariant under any permutation of the operators *even when the operators do not commute.* The right-hand side of (7.111) could be replaced by any single term in the summation and it follows from (7.110) and (7.123) that the coefficients $\mathrm{C}(L; \mathring{q}, \mathring{k})$ in the cumulant expansion (7.109) are given by

$$\mathrm{C}(L; \mathring{q}, \mathring{k}) = \lceil\hat{\phi}_{\mathring{q}} \cdots \hat{\phi}_1\rfloor_0 . \tag{7.127}$$

For this reason equations (7.124) and (7.125) are called *generalized moment-cumulant relations.* A rearrangement of (7.125) gives the iterative relationship

$$\lceil\hat{\phi}_{\mathring{q}} \cdots \hat{\phi}_1\rfloor_0 = \langle\langle\hat{\phi}_{\mathring{q}} \cdots \hat{\phi}_1\rangle\rangle_0 - \sum_{\ell=2}^{\mathring{q}} \sum_{\{\pi_{\mathring{q}}(\ell)\}} \lceil\hat{\phi}^{(\mathring{q})} \cdots \hat{\phi}^{(j)}\rfloor_0 \cdots \\ \cdots \lceil\hat{\phi}^{(t)} \cdots \hat{\phi}^{(r)}\rfloor_0 \cdots \lceil\cdots \hat{\phi}^{(1)}\rfloor_0 . \tag{7.128}$$

From (7.125) and (7.128) it can be shown that

[11] Other authors (for example Rushbrooke et al. 1974) use the standard square bracket pair [···] for this quantity. We have used the special symbol ⌈···⌋ to avoid confusion with the more general use of square brackets.

$$\lceil \hat{\phi}_1 \cdots \hat{\phi}_{\mathring{q}} \rfloor_0 = \langle\!\langle \hat{\phi}_1 \cdots \hat{\phi}_{\mathring{q}} \rangle\!\rangle_0 - \sum_{\{\pi_{\mathring{q}}(2)\}} \lceil \hat{\phi}^{(1)} \cdots \hat{\phi}^{(\ell)} \rfloor_0 \langle\!\langle \hat{\phi}^{(\ell+1)} \cdots \hat{\phi}^{(\mathring{q})} \rangle\!\rangle_0 , \tag{7.129}$$

where one operator ($\hat{\phi}_1$ say) is present in all the $\lceil \cdots \rfloor_0$ in the summation. Although the proof is rather lengthy, it can be shown directly from (7.125) that, if the $\langle\!\langle \cdots \rangle\!\rangle_0$ are the coefficients in a moment expansion then the $\lceil \cdots \rfloor_0$ are the corresponding cumulant coefficients (Rushbrooke et al. 1974). It can also be shown directly from (7.128) that if $\langle\!\langle \cdots \rangle\!\rangle_0$ and $\lceil \cdots \rfloor_0$ are related to graphs and if the all the terms in $\langle\!\langle \cdots \rangle\!\rangle_0$ satisfy factorization relations like (7.113) then the $\lceil \cdots \rfloor_0$ corresponding to separated graphs are zero.

7.6 Applications of the Linked-Cluster Expansion

7.6.1 The General-s Ising Model

The spin magnitude can now take any positive integer or half-integer value and

$$\hat{\mathfrak{e}}(\boldsymbol{r}, \boldsymbol{r}') = \frac{s_z(\boldsymbol{r}) s_z(\boldsymbol{r}')}{s^2} \tag{7.130}$$

are commuting operators with the $s_z(\boldsymbol{r})$ regarded as classical scalars with the range of values $s, s-1, \ldots, 1-s, -s$. It is useful to define

$$\sigma(\boldsymbol{r}) = s_z(\boldsymbol{r})/s \tag{7.131}$$

and (7.98), with $K(\boldsymbol{r}, \boldsymbol{r}') = K$, becomes

$$\widehat{H}(L, K; \sigma(\boldsymbol{r})) = -K \sum_{\{\boldsymbol{r}, \boldsymbol{r}'\}}^{(\text{n.n.})} \sigma(\boldsymbol{r})\sigma(\boldsymbol{r}') - L \sum_{\{\boldsymbol{r}\}} \sigma(\boldsymbol{r}) . \tag{7.132}$$

This equation is exactly of the same form as (1.49) except that there $\sigma(\boldsymbol{r}) = \pm 1$, which is the special case $s = \frac{1}{2}$ of the present treatment. From (7.102)

$$\begin{aligned} z_0(L) &= \sum_{\{\sigma\}} \exp(L\sigma) = \exp(L) \sum_{j=0}^{2s} \exp(-jL/s) \\ &= \frac{\sinh\left[\left(s + \frac{1}{2}\right) L/s\right]}{\sinh\left[\frac{1}{2} L/s\right]} , \end{aligned} \tag{7.133}$$

which reduces to $2s + 1$ when $L = 0$. We shall also need to define

$$z_\gamma(L) = \sum_{\{\sigma\}} \sigma^\gamma \exp(L\sigma) = \exp(L) \sum_{j=0}^{2s} \left[\frac{s-j}{s}\right]^\gamma \exp(-jL/s) \tag{7.134}$$

and

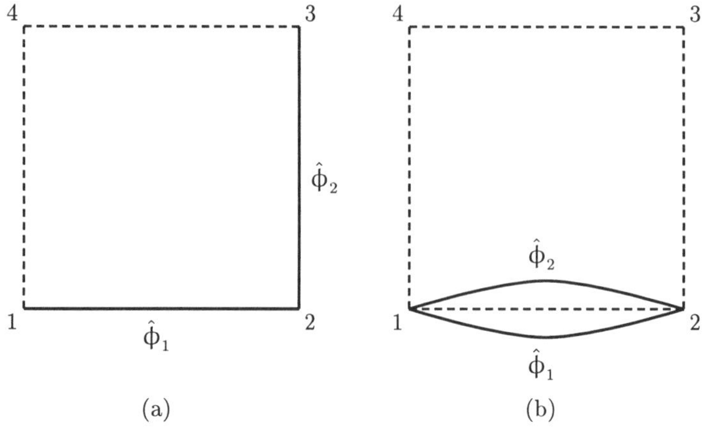

Fig. 7.1. $\mathring{q} = 2$ graphs topologies with **(a)** $\mathring{k} = 1$ and **(b)** $\mathring{k} = 2$.

$$\bar{z}_\gamma(L) = \frac{z_\gamma(L)}{z_0(L)} . \tag{7.135}$$

Since $-\sigma$ has the same range as σ, it follows from (7.134) and (7.135) that $z_0(L)$ is an even function and $\bar{z}_\gamma(L)$ is even or odd according as γ is even or odd. For the spin-s Ising model all the operators commute and $\langle\langle\cdots\rangle\rangle_0 = \langle\cdots\rangle_0$. For $\mathring{q} = 2$ there are two connected graph topologies, shown as (a) and (b) in Fig. 7.1. From (7.124), (7.127) and (7.134), we have

$$\begin{aligned}
\mathrm{C}(L;2,1) &= \langle\hat{\phi}_1\hat{\phi}_2\rangle_0 - \langle\hat{\phi}_1\rangle_0\langle\hat{\phi}_2\rangle_0 \\
&= \langle[\sigma(\boldsymbol{r}_1)\sigma(\boldsymbol{r}_2)][\sigma(\boldsymbol{r}_2)\sigma(\boldsymbol{r}_3)]\rangle_0 - \langle\sigma(\boldsymbol{r}_1)\sigma(\boldsymbol{r}_2)\rangle_0\langle\sigma(\boldsymbol{r}_2)\sigma(\boldsymbol{r}_3)\rangle_0 \\
&= \langle\sigma(\boldsymbol{r}_1)[\sigma(\boldsymbol{r}_2)]^2\sigma(\boldsymbol{r}_3)\rangle_0 - \langle\sigma(\boldsymbol{r}_1)\sigma(\boldsymbol{r}_2)\rangle_0\langle\sigma(\boldsymbol{r}_2)\sigma(\boldsymbol{r}_3)\rangle_0 \\
&= \bar{z}_2(L)[\bar{z}_1(L)]^2 - [\bar{z}_1(L)]^4 ,
\end{aligned} \tag{7.136}$$

$$\begin{aligned}
\mathrm{C}(L;2,2) &= \langle\hat{\phi}_1\hat{\phi}_2\rangle_0 - \langle\hat{\phi}_1\rangle_0\langle\hat{\phi}_2\rangle_0 \\
&= \langle[\sigma(\boldsymbol{r}_1)\sigma(\boldsymbol{r}_2)]^2\rangle_0 - [\langle\sigma(\boldsymbol{r}_1)\sigma(\boldsymbol{r}_2)\rangle_0]^2 \\
&= \langle[\sigma(\boldsymbol{r}_1)]^2[\sigma(\boldsymbol{r}_2)]^2\rangle_0 - [\langle\sigma(\boldsymbol{r}_1)\sigma(\boldsymbol{r}_2)\rangle_0]^2 \\
&= [\bar{z}_2(L)]^2 - [\bar{z}_1(L)]^4 .
\end{aligned} \tag{7.137}$$

When $\mathcal{H} = L = 0$, $\bar{z}_1(0) = 0$ but $\bar{z}_2(0) \neq 0$, so that $\mathrm{C}(0;2,1) = 0$ but $\mathrm{C}(0;2,2) \neq 0$. This illustrates the point that, although the structure of the $\mathrm{C}(L;\mathring{q},\mathring{k})$ in terms of the $\langle\cdots\rangle_0$, as given by (7.124) and (7.127), are the same for a given $\mathring{q}$, there is a dependence on the graph topology $\mathring{k}$ through the $\langle\cdots\rangle_0$.

It is not difficult to generalize the result $\mathrm{C}(0;2,1) = 0$ to any graph where some of the vertex valencies $\gamma_1, \gamma_2, \ldots, \gamma_n$ are odd. Since each $\hat{\phi}_i$ corresponds to a line of the graph $(\mathring{q}, \mathring{k})$, a division of $\hat{\phi}_1, \hat{\phi}_2, \ldots, \hat{\phi}_{\mathring{q}}$ into subsets represents a division of the graph into subgraphs, with each line of the original graph belonging to exactly one subgraph. If some of the valencies are odd then at least one subgraph must have some odd valencies so that the corresponding $\langle \cdots \rangle_0$ expressed as a product of $\bar{z}_\gamma(L)$, has a factor with γ odd. Since $\bar{z}_\gamma(0) = 0$ when γ is odd the result follows.

7.6.2 *D*-Vector Models

In this case the spin operator $\hat{\boldsymbol{s}}(\boldsymbol{r})$ at each of the sites $\boldsymbol{r}$ of the lattice $\mathcal{N}$ is a D-dimensional[12] unit vector $\boldsymbol{s}(\boldsymbol{r})$ with

$$\hat{\mathfrak{e}}(\boldsymbol{r},\boldsymbol{r}') = \boldsymbol{s}(\boldsymbol{r}) \cdot \boldsymbol{s}(\boldsymbol{r}') . \tag{7.138}$$

For $D = 1$ the vector has a single component, usually taken to be in the direction of the z axis with values ± 1; this is the spin-$\frac{1}{2}$ Ising model. For $D = 2$ the two components can be taken to be in the directions of the x and y axes and the model is called the *plane rotator model.* The $D = 3$ case is the *classical Heisenberg model*

Another member of this taxonomy is the *spherical model* first proposed by Kac and solved by Berlin and Kac (1952) (see also Joyce 1972). This is derived by first modifying the interaction part of the Ising Hamiltonian to

$$\widehat{H}_1(\kappa(\boldsymbol{r}-\boldsymbol{r}');\sigma(\boldsymbol{r})) = -\sum_{\{\boldsymbol{r},\boldsymbol{r}'\}} \kappa(\boldsymbol{r}-\boldsymbol{r}')\sigma(\boldsymbol{r})\sigma(\boldsymbol{r}') , \tag{7.139}$$

where the sum is now over all pairs of sites, and then replacing the Ising constraint $|\sigma(\boldsymbol{r})| = 1$, for all lattice sites, with the condition

$$\sum_{\{\boldsymbol{r}\}} |\sigma(\boldsymbol{r})|^2 = N , \tag{7.140}$$

subject to which each individual variable $\sigma(\boldsymbol{r})$ is allowed to range over all real values. This model has been solved for $d = 1, 2, 3$ and has a phase transition for $d = 3$. For $d \leq 2$ it has a phase transition if $\kappa(\boldsymbol{r}-\boldsymbol{r}') \sim |\boldsymbol{r}-\boldsymbol{r}'|^{-d-\delta}$, where $0 < \delta < d$, (see Joyce (1972) and references therein). Although this model was introduced by reference to the Ising model ($D = 1$) it can be seen that the constraint (7.140), if applied to vectors of any dimension, would effectively combines the roles of N and D. In fact the properties of the model are more similar to those of the Heisenberg model than the Ising model and it can be shown that it is equivalent to the $D \to \infty$ limit of the D-vector model (Stanley 1968b, Kac and Thompson 1971, Pearce and Thompson 1977).

For the classical Heisenberg model $\boldsymbol{s}(\boldsymbol{r})$ can be taken to specify the direction of a magnetic dipole in three dimensions with

$$\hat{\mathfrak{e}}(\boldsymbol{r},\boldsymbol{r}') = s_x(\boldsymbol{r})s_x(\boldsymbol{r}') + s_y(\boldsymbol{r})s_y(\boldsymbol{r}') + s_z(\boldsymbol{r})s_z(\boldsymbol{r}') . \tag{7.141}$$

---

12 The name 'D-vector model' is due to Stanley (1968a, 1974). The names 'n-vector model' and '$O(n)$ model' are also used, the latter because, in the absence of symmetry breaking fields, there is invariance under rotations in the n-dimensional space of $\hat{\boldsymbol{s}}(\boldsymbol{r})$. We use D rather than n to avoid confusion with other usages.

The direction of any $\boldsymbol{s}(\boldsymbol{r})$ can be specified by the polar angles (θ, ψ). With the field along the z-axis identified as the $\theta = 0$ direction, (7.133) is replaced by

$$z_0(L) = \int_{-\pi}^{\pi} \mathrm{d}\psi \int_0^{\pi} \exp(L\cos\theta)\mathrm{d}\theta = \frac{4\pi \sinh(L)}{L}, \tag{7.142}$$

so that $z_0(0) = 4\pi$.

Equation (7.116) applies for D-vector models since $\hat{\mathfrak{e}}(\boldsymbol{r}, \boldsymbol{r}')$ is obviously independent of the choice of the polar axis $\theta = 0$, which can be chosen as the direction of the fixed spin. The connected graph expansion can, therefore, be restricted to star graphs.

The most general, anisotropic, form of the *quantum Heisenberg model* is given by

$$\hat{\mathfrak{e}}(\boldsymbol{r}, \boldsymbol{r}') = [\lambda_x \hat{s}_x(\boldsymbol{r})\hat{s}_x(\boldsymbol{r}') + \lambda_y \hat{s}_y(\boldsymbol{r})\hat{s}_y(\boldsymbol{r}') + \hat{s}_z(\boldsymbol{r})\hat{s}_z(\boldsymbol{r}')]/s^2, \tag{7.143}$$

with s taking integer or (integer + half) values. In the extreme anisotropic case $\lambda_x = \lambda_y = 0$ the general-s Ising model is recovered. The case $\lambda_x = 0$, $\lambda_y = 1$ is the planar or *X–Y model.*[13] The name 'Heisenberg model' is usually taken to refer to the isotropic case $\lambda_x = \lambda_y = 1$ when

$$\begin{aligned} \hat{\mathfrak{e}}(\boldsymbol{r}, \boldsymbol{r}') &= [\hat{s}_x(\boldsymbol{r})\hat{s}_x(\boldsymbol{r}') + \hat{s}_y(\boldsymbol{r})\hat{s}_y(\boldsymbol{r}') + \hat{s}_z(\boldsymbol{r})\hat{s}_z(\boldsymbol{r}')]/s^2 \\ &= \hat{\boldsymbol{s}}(\boldsymbol{r}).\hat{\boldsymbol{s}}(\boldsymbol{r}')/s^2. \end{aligned} \tag{7.144}$$

The limits of $\hat{s}_x(\boldsymbol{r})/s$, $\hat{s}_y(\boldsymbol{r})/s$ and $\hat{s}_z(\boldsymbol{r})/s$, as $s \to \infty$, can be regarded as the scalar components of a classical unit vector so that the classical Heisenberg model is recovered. Normally a representation is chosen so that the matrix of the field component $\hat{s}_z(\boldsymbol{r})$ is diagonal and $z_0(L)$ has the same form, (7.133), as for the Ising model, with $z_0(0) = 2s + 1$.

7.6.3 The Classical Heisenberg Model

We now consider the derivation of high-temperature expansion coefficients for the classical zero-field Heisenberg model. This involves evaluating expressions of the type

$$\langle \hat{\phi}_1 \cdots \hat{\phi}_{\mathring{q}} \rangle_0 = (4\pi)^{-\mathfrak{n}} \mathrm{Trace}\{\hat{\phi}_1 \cdots \hat{\phi}_{\mathring{q}}\}, \tag{7.145}$$

where $\mathfrak{n}$ is the number of vertices in the multi-bonded graph with $\mathring{q}$ lines. Since each $\hat{\phi}$-factor has the form (7.141) the product $\hat{\phi}_1 \cdots \hat{\phi}_{\mathring{q}}$ can be expanded into $3^{\mathring{q}}$ terms, each a product of $2\mathring{q}$ vector components. Each such term can be regarded as the product of $\mathfrak{n}$ factors, one for each vertex, and the trace can be taken over each factor separately. We thus need to evaluate expressions of the type

$$\frac{1}{4\pi}\mathrm{Trace}\{s_x^{\eta_x} s_y^{\eta_y} s_z^{\eta_z}\} = \frac{1}{4\pi}\int_{-\pi}^{\pi} \mathrm{d}\psi \int_0^{\pi} s_x^{\eta_x} s_y^{\eta_y} s_z^{\eta_z} \sin\theta \mathrm{d}\theta, \tag{7.146}$$

---

[13] In our notation it should be called the 'Y–Z model'.

where η_x, η_y and η_z are integers and

$$s_x = \sin\theta\cos\psi\,, \qquad s_y = \sin\theta\sin\psi\,, \qquad s_z = \cos\theta\,. \tag{7.147}$$

It follows that

$$\frac{1}{4\pi}\mathrm{Trace}\{s_x^{\eta_x} s_y^{\eta_y} s_z^{\eta_z}\} = \frac{1}{4\pi}\mathrm{I}(\eta_x,\eta_y,\eta_z)\mathrm{J}(\eta_x,\eta_y), \tag{7.148}$$

where

$$\begin{aligned}
\mathrm{I}(\eta_x,\eta_y,\eta_z) &= \int_0^{\pi} (\cos\theta)^{\eta_z}(\sin\theta)^{(\eta_x+\eta_y+1)}\mathrm{d}\theta \\
&= [1+(-1)^{\eta_z}]\int_0^{\frac{1}{2}\pi} (\cos\theta)^{\eta_z}(\sin\theta)^{(\eta_x+\eta_y+1)}\mathrm{d}\theta \\
&= \tfrac{1}{2}[1+(-1)^{\eta_z}]\mathrm{B}\left(\tfrac{1}{2}\eta_x+\tfrac{1}{2}\eta_y+1,\tfrac{1}{2}\eta_z+1\right),
\end{aligned} \tag{7.149}$$

$$\begin{aligned}
\mathrm{J}(\eta_x,\eta_y) &= \int_{-\pi}^{\pi} (\cos\psi)^{\eta_x}(\sin\psi)^{\eta_y}\mathrm{d}\psi \\
&= [1+(-1)^{\eta_x}][1+(-1)^{\eta_y}]\int_0^{\frac{1}{2}\pi} (\cos\psi)^{\eta_x}(\sin\psi)^{\eta_y}\mathrm{d}\psi \\
&= \tfrac{1}{2}[1+(-1)^{\eta_x}][1+(-1)^{\eta_y}]\mathrm{B}\left(\tfrac{1}{2}\eta_x+1,\tfrac{1}{2}\eta_y+1\right),
\end{aligned} \tag{7.150}$$

$\mathrm{B}(a,b)$ being the *beta function*. It follows that the right-hand side of (7.148) is non-zero only if η_x, η_y and η_z are all even integers. Using the standard formula expressing the beta function in terms of gamma functions (see, for example, Abramowitz and Segun 1965), (7.148)–(7.150) yield

$$\frac{1}{4\pi}\mathrm{Trace}\{s_x^{\eta_x} s_y^{\eta_y} s_z^{\eta_z}\} = \begin{cases} \dfrac{\Gamma\left(\frac{1}{2}\eta_x+\frac{1}{2}\right)\Gamma\left(\frac{1}{2}\eta_y+\frac{1}{2}\right)\Gamma\left(\frac{1}{2}\eta_z+\frac{1}{2}\right)}{2\pi\Gamma\left(\frac{1}{2}\eta_x+\frac{1}{2}\eta_y+\frac{1}{2}\eta_z+\frac{3}{2}\right)}\,, & \eta_x,\eta_y,\eta_z \text{ even}, \\ 0, & \text{otherwise}. \end{cases} \tag{7.151}$$

The gamma functions can be evaluated using the formulae (Abramowitz and Segun 1965) $\Gamma(a+1) = a\Gamma(a)$ and $\Gamma(\frac{1}{2}) = \sqrt{\pi}$. It can now be shown that $\mathrm{C}(0;\mathring{\mathrm{q}},\mathring{\mathrm{k}})$, given by (7.124) and (7.127), is zero unless all the vertex valencies of the graph $(\mathring{\mathrm{q}},\mathring{\mathrm{k}})$ are even. The argument for this is similar to that used for the Ising model in Sect. 7.6.1. The terms in the summation on the right-hand side of (7.124) represent divisions of the graph $(\mathring{\mathrm{q}},\mathring{\mathrm{k}})$ into subgraphs with each line appearing exactly once. If there is a vertex of $(\mathring{\mathrm{q}},\mathring{\mathrm{k}})$ with odd valency then one of the subgraphs in the division has a vertex with odd valency. When the $\langle\cdots\rangle_0$ expression for this factor is expanded into a product of vertex contributions the valency γ of any vertex is related to its corresponding η_x, η_y and η_y by

$$\eta_x + \eta_y + \eta_z = \gamma\,. \tag{7.152}$$

If γ is odd then at least one of η_x, η_y and η_z is odd and the result follows. The types of graphs for which $\mathrm{C}(0; \mathring{q}, \mathring{k})$ needs to be evaluated can be further reduced by using Theorem 2, p. 519 of Stanley (1974). Suppose that a graph $(\mathring{q}+1, \mathring{k}')$ is formed from $(\mathring{q}, \mathring{k})$ by placing a new vertex on one of its lines. Then, in our notation, the theorem establishes the result

$$\mathrm{C}(0; \mathring{q}+1, \mathring{k}') = \tfrac{1}{3}\mathrm{C}(0; \mathring{q}, \mathring{k}). \tag{7.153}$$

The corresponding result for the D-vector model is given by replacing '3' by 'D' and for the general-s Ising model there is a similar result obtained by replacing '3' by $\bar{z}_2(0)$, given by (7.134) and (7.135). When $s = \frac{1}{2}$ these two become equivalent with $\bar{z}_2(0) = D = 1$.

As a simple illustration we use (7.151) to derive the coefficients for the graphs of Fig. 7.1. Since for graph (a) vertices 1 and 3 have odd valencies $\mathrm{C}(0; 2, 1) = 0$. From (7.124), (7.127) and (7.138)

$$\mathrm{C}(0; 2, 2) = \langle [\boldsymbol{s}(\boldsymbol{r}_1) \cdot \boldsymbol{s}(\boldsymbol{r}_2)]^2 \rangle_0 - \langle \boldsymbol{s}(\boldsymbol{r}_1) \cdot \boldsymbol{s}(\boldsymbol{r}_2) \rangle_0^2\,, \tag{7.154}$$

where, from (7.145) and (7.151),

$$\begin{aligned} \langle \boldsymbol{s}(\boldsymbol{r}_1).\boldsymbol{s}(\boldsymbol{r}_2) \rangle_0 &= (4\pi)^{-2}\mathrm{Trace}\{s_x(\boldsymbol{r}_1)s_x(\boldsymbol{r}_2) \\ &\quad + s_y(\boldsymbol{r}_1)s_y(\boldsymbol{r}_2) + s_z(\boldsymbol{r}_1)s_z(\boldsymbol{r}_2)\} \\ &= 0\,. \end{aligned} \tag{7.155}$$

It follows that

$$\begin{aligned} \mathrm{C}(0; 2, 2) &= \langle [\boldsymbol{s}(\boldsymbol{r}_1) \cdot \boldsymbol{s}(\boldsymbol{r}_2)]^2 \rangle_0 \\ &= (4\pi)^{-2}\mathrm{Trace}\{s_x^2(\boldsymbol{r}_1)s_x^2(\boldsymbol{r}_2) + s_y^2(\boldsymbol{r}_1)s_y^2(\boldsymbol{r}_2) + s_z^2(\boldsymbol{r}_1)s_z^2(\boldsymbol{r}_2) \\ &\quad + 2s_x(\boldsymbol{r}_1)s_x(\boldsymbol{r}_2)s_y(\boldsymbol{r}_1)s_y(\boldsymbol{r}_2) + 2s_y(\boldsymbol{r}_1)s_y(\boldsymbol{r}_2)s_z(\boldsymbol{r}_1)s_z(\boldsymbol{r}_2) \\ &\quad + 2s_z(\boldsymbol{r}_1)s_z(\boldsymbol{r}_2)s_x(\boldsymbol{r}_1)s_x(\boldsymbol{r}_2)\} \\ &= 3(4\pi)^{-2}[\mathrm{Trace}\{s_x^2(\boldsymbol{r}_1)\}]^2 \\ &= \left[\frac{\Gamma(\frac{1}{3})\Gamma^2(\frac{1}{2})}{2\pi\Gamma(\frac{5}{2})}\right]^2 = \frac{1}{3}\,. \end{aligned} \tag{7.156}$$

Using the notation $\mathrm{C}(0; \mathring{q}, 2)$ for the coefficient of a single-bonded polygon of $\mathring{q}$ lines it follows from (7.153) and (7.156) that

$$\mathrm{C}(0; \mathring{q}, 2) = \frac{1}{3^{(\mathring{q}-1)}}. \tag{7.157}$$

In deriving tables of coefficients like those in Stanley (1974) considerable use is made of iteration relations based on (7.128) and (7.129).

7.6.4 The Quantum Heisenberg Model

In this case, because of the non-commutativity of the operators (7.144), we must, according to equation (A.288), take

$$\langle\langle \hat{\phi}_{\mathring{q}} \cdots \hat{\phi}_1 \rangle\rangle_0 = \frac{\prod_{t=1}^{q} \alpha_t!}{\mathring{q}!} \sum_{\{\nu(\mathring{q},\mathring{k})\}} \langle \hat{\phi}^{(\mathring{q})} \cdots \hat{\phi}^{(1)} \rangle_0 , \tag{7.158}$$

when the summation is over all distinguishable permutations of the operators for the graph $(\mathring{q}, \mathring{k})$. From (7.122)

$$\langle \hat{\phi}^{(\mathring{q})} \cdots \hat{\phi}^{(1)} \rangle_0 = \mathrm{Trace}\{\hat{\phi}^{(\mathring{q})} \cdots \hat{\phi}^{(1)} \hat{\rho}_0(L)\} , \tag{7.159}$$

where $\hat{\rho}_0(L)$ is given by (7.104). The trace of a product of operators is invariant under cyclic permutation. In general this does not help in reducing the number of terms in the summation in (7.158) because of the presence of the operator $\hat{\rho}_0(L)$. However, when $L = \mathcal{H} = 0$, $\hat{\rho}_0(0) = 1/(1+2s)^N$ and

$$\langle\langle \hat{\phi}_{\mathring{q}} \cdots \hat{\phi}_1 \rangle\rangle_0 = \frac{\mathring{q} \prod_{t=1}^{q} \alpha_t!}{(1+2s)^{n} \mathring{q}!} \sum_{\{\nu'(\mathring{q},\mathring{k})\}} \mathrm{Trace}\{\hat{\phi}^{(\mathring{q})} \cdots \hat{\phi}^{(1)}\} , \tag{7.160}$$

where $\{\nu'(\mathring{q}, \mathring{k})\}$ is the set of permutations which are not cyclically related and the trace is restricted to the sites occupied by the graph $(\mathring{q}, \mathring{k})$.

We denote the $(2s+1)$-dimensional matrices representing the operators $\hat{s}_x(\boldsymbol{r})$, $\hat{s}_y(\boldsymbol{r})$ and $\hat{s}_z(\boldsymbol{r})$ at any site of the lattice by $\boldsymbol{S}_x$, $\boldsymbol{S}_y$ and $\boldsymbol{S}_z$ respectively. Consider now a graph $(\mathring{q}, \mathring{k})$ of n sites which are labelled $1, 2, \ldots, n$. Then the $(2s+1)^n$-dimensional matrix representation $\boldsymbol{E}(i,j)$ of the operator $\hat{\mathfrak{e}}(i,j)$ defined in (7.144) is

$$\boldsymbol{E}(i,j) = \boldsymbol{E}_x(i,j) + \boldsymbol{E}_y(i,j) + \boldsymbol{E}_z(i,j) , \tag{7.161}$$

where

$$\boldsymbol{E}_w(i,j) = \frac{1}{s^2} \boldsymbol{I} \otimes \cdots \otimes \boldsymbol{I} \otimes \boldsymbol{S}_w \otimes \boldsymbol{I} \otimes \cdots \otimes \boldsymbol{I} \otimes \boldsymbol{S}_w \otimes \boldsymbol{I} \otimes \cdots \otimes \boldsymbol{I}, \quad w = x, y, z , \tag{7.162}$$

in which $\boldsymbol{I}$ is the $(2s+1)$-dimensional unit matrix and the matrix $\boldsymbol{S}_w$ is at the i-th and j-th places in the direct matrix product. The problem now is to evaluate expressions of the form $\mathrm{Trace}\{\boldsymbol{E}(i_1,j_1)\boldsymbol{E}(i_2,j_2) \cdots \boldsymbol{E}(i_{\mathring{q}}, j_{\mathring{q}})\}$, where $(i_1,j_1), (i_2,j_2), \ldots, (i_{\mathring{q}}, j_{\mathring{q}})$ represents some arbitrary ordering of the bonds of $(\mathring{q}, \mathring{k})$. From (7.161)

$$\mathrm{Trace}\{\boldsymbol{E}(i_1,j_1) \cdots \boldsymbol{E}(i_{\mathring{q}}, j_{\mathring{q}})\} = \sum_{\{u,v,\ldots,w\}} \mathrm{Trace}\{\boldsymbol{E}_u(i_1,j_1) \cdots \cdots \boldsymbol{E}_w(i_{\mathring{q}}, j_{\mathring{q}})\}, \tag{7.163}$$

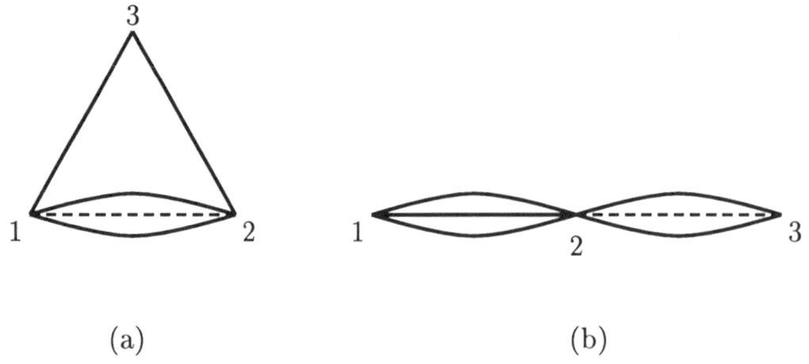

Fig. 7.2. Two graphs with $\mathfrak{n} = 3$.

where $\{u, v, \ldots, w\}$ denotes all choices of $\mathring{\mathfrak{q}}$ x's, y's and z's. From (7.162) and the properties of the direct matrix product (see, for example, Lloyd 1953)

$$\mathrm{Trace}\{\boldsymbol{E}_u(\mathfrak{i}_1, \mathfrak{j}_1) \cdots \boldsymbol{E}_w(\mathfrak{i}_{\mathring{\mathfrak{q}}}, \mathfrak{j}_{\mathring{\mathfrak{q}}})\} = \frac{1}{s^{2\mathring{\mathfrak{q}}}} \mathrm{Trace}\{\boldsymbol{\Pi}_1\} \cdots \cdots \mathrm{Trace}\{\boldsymbol{\Pi}_\mathfrak{n}\}, \tag{7.164}$$

where $\boldsymbol{\Pi}_\mathfrak{j}$ corresponds to vertex $\mathfrak{j}$ and has the form $\boldsymbol{S}_u \boldsymbol{S}_v \cdots \boldsymbol{S}_w$ with $\gamma_\mathfrak{j}$ factors, the identifications of $u, v, \ldots, w$ in terms of x, y and z being given by the left-hand side of (7.164).

It may be helpful at this point to use the two graphs of Fig. 7.2 as examples. For graph (a), from (7.160)–(7.164),

$$\begin{aligned}\langle\langle \hat{\phi}_4 \cdots \hat{\phi}_1 \rangle\rangle_0 = \frac{1}{3s^8(2s+1)^3} \sum_{\{t,u,v,w\}} & [\mathrm{Trace}\{\boldsymbol{S}_t \boldsymbol{S}_u \boldsymbol{S}_v\} \mathrm{Trace}\{\boldsymbol{S}_t \boldsymbol{S}_u \boldsymbol{S}_w\} \\ & \times \mathrm{Trace}\{\boldsymbol{S}_v \boldsymbol{S}_w\} \\ & + \mathrm{Trace}\{\boldsymbol{S}_t \boldsymbol{S}_u \boldsymbol{S}_v\} \mathrm{Trace}\{\boldsymbol{S}_t \boldsymbol{S}_u \boldsymbol{S}_w\} \mathrm{Trace}\{\boldsymbol{S}_w \boldsymbol{S}_v\} \\ & + \mathrm{Trace}\{\boldsymbol{S}_t \boldsymbol{S}_v \boldsymbol{S}_u\} \mathrm{Trace}\{\boldsymbol{S}_t \boldsymbol{S}_u \boldsymbol{S}_w\} \mathrm{Trace}\{\boldsymbol{S}_w \boldsymbol{S}_v\}],\end{aligned} \tag{7.165}$$

where $\{t, u, v, w\}$ indicates that the summation is over all terms where t, u, v and w take all choices of x, y and z. In fact the first two terms on the right can be combined since $\mathrm{Trace}\{\boldsymbol{S}_v \boldsymbol{S}_w\} = \mathrm{Trace}\{\boldsymbol{S}_w \boldsymbol{S}_v\}$. For graph (b)

$$\begin{aligned}\langle\langle \hat{\phi}_5 \cdots \hat{\phi}_1 \rangle\rangle_0 = \frac{1}{2s^{10}(2s+1)^3} \sum_{\{t,u,v,w,r\}} & [\mathrm{Trace}\{\boldsymbol{S}_t \boldsymbol{S}_u \boldsymbol{S}_v\} \\ & \times \mathrm{Trace}\{\boldsymbol{S}_t \boldsymbol{S}_u \boldsymbol{S}_v \boldsymbol{S}_w \boldsymbol{S}_r\} \mathrm{Trace}\{\boldsymbol{S}_w \boldsymbol{S}_r\} \\ & + \mathrm{Trace}\{\boldsymbol{S}_t \boldsymbol{S}_u \boldsymbol{S}_v\} \mathrm{Trace}\{\boldsymbol{S}_t \boldsymbol{S}_u \boldsymbol{S}_w \boldsymbol{S}_v \boldsymbol{S}_r\} \mathrm{Trace}\{\boldsymbol{S}_w \boldsymbol{S}_r\}].\end{aligned} \tag{7.166}$$

To evaluate the expressions in (7.165) and (7.166) we must now use explicit forms for the spin matrices $\boldsymbol{S}_x$, $\boldsymbol{S}_y$ and $\boldsymbol{S}_z$. In terms of the orthonormal basis kets $|s, m\rangle$ a particular vertex contribution is given by

$$\mathrm{Trace}\{\boldsymbol{S}_u\boldsymbol{S}_v\cdots\boldsymbol{S}_w\} = \sum_{m=-s}^{s} \langle s,m|\hat{s}_u\hat{s}_v\cdots\hat{s}_w|s,m\rangle\,, \tag{7.167}$$

where

$$\hat{s}_x|s,m\rangle = f(s,m)|s,m-1\rangle + f(s,-m)|s,m+1\rangle\,, \tag{7.168}$$

$$\hat{s}_y|s,m\rangle = \mathrm{i}f(s,m)|s,m-1\rangle - \mathrm{i}f(s,-m)|s,m+1\rangle\,, \tag{7.169}$$

$$\hat{s}_z|s,m\rangle = m|s,m\rangle, \tag{7.170}$$

with $m = -s, -s+1, \ldots, s-1, s$ and

$$f(s,m) = \tfrac{1}{2}\sqrt{(s+m)(s+1-m)} \tag{7.171}$$

(see, for example, Edmonds 1957). It is not difficult to show that

$$\mathrm{Trace}\{\boldsymbol{S}_u\boldsymbol{S}_v\} = \tfrac{1}{3}\delta^{\mathrm{Kr}}(u,v)s(s+1)(2s+1)\,, \tag{7.172}$$

$$\mathrm{Trace}\{\boldsymbol{S}_u\boldsymbol{S}_v\boldsymbol{S}_w\} = \tfrac{1}{6}\mathrm{i}\epsilon(u,v,w)s(s+1)(2s+1)\,, \tag{7.173}$$

where $\epsilon(u,v,w)$ is the Levi-Civita tensor, which is zero unless u, v and w are all different, when it has the value $+1$ or -1 according as they are an even or odd permutation of the three symbols. The operator $\hat{\boldsymbol{s}} = (\hat{s}_x, \hat{s}_y, \hat{s}_z)$ satisfies the commutation relation

$$\hat{\boldsymbol{s}} \wedge \hat{\boldsymbol{s}} = \mathrm{i}\hat{\boldsymbol{s}}\,, \tag{7.174}$$

which is invariant under the operations $\mathcal{C} = \{x \to y \to z \to x \to \cdots\}$ and $\mathcal{P} = \{x \leftrightarrow y, z \to -z\}$. It follows that $\mathrm{Trace}\{\boldsymbol{S}_u\boldsymbol{S}_v\cdots\boldsymbol{S}_w\}$ is invariant under the operations $\mathcal{C}$ and $\mathcal{P}$. Suppose now that $\langle s,m|\hat{s}_u\hat{s}_v\cdots\hat{s}_w|s,m\rangle$ contains η_u operators $\hat{s}_u$ for $u = x, y, z$. As for the classical Heisenberg model the numbers η_x, η_y and η_z are related to the vertex valency γ by (7.152). It follows from (7.168) and (7.169) that $\langle s,m|\hat{s}_u\hat{s}_v\cdots\hat{s}_w|s,m\rangle$ will be zero unless η_x and η_y are both either even or odd. Because of the invariance under $\mathcal{C}$ it then follows that the trace is zero unless all of η_x, η_y and η_z are either even or odd. This is a weaker condition than (7.151) for the classical Heisenberg model. These results are exemplified by (7.172) and (7.173) and, applying them to (7.165), it follows that, for non-zero terms in the summation, t, u, v must be some ordering of x, y, z with $w = v$. Then

$$\begin{aligned}
\langle\!\langle\hat{\phi}_4\cdots\hat{\phi}_1\rangle\!\rangle_0 &= \frac{1}{3s^8(2s+1)^3}\sum_{\{t,u,v\}} \mathrm{Trace}\{\boldsymbol{S}_t\boldsymbol{S}_u\boldsymbol{S}_v\}\mathrm{Trace}\{\boldsymbol{S}_v^2\} \\
&\quad [2\,\mathrm{Trace}\{\boldsymbol{S}_t\boldsymbol{S}_u\boldsymbol{S}_v\} + \mathrm{Trace}\{\boldsymbol{S}_t\boldsymbol{S}_v\boldsymbol{S}_u\}] \\
&= \frac{1}{3s^8(2s+1)^3}\sum_{\{t,u,v\}} [\mathrm{Trace}\{\boldsymbol{S}_t\boldsymbol{S}_u\boldsymbol{S}_v\}]^2\mathrm{Trace}\{\boldsymbol{S}_v^2\} \\
&= \frac{2}{s^8(2s+1)^3}[\mathrm{Trace}\{\boldsymbol{S}_x\boldsymbol{S}_y\boldsymbol{S}_z\}]^2\mathrm{Trace}\{\boldsymbol{S}_z^2\} \\
&= -\frac{(s+1)^3}{54s^5}\,,
\end{aligned} \tag{7.175}$$

where the summations are now over all permutations of x, y, z. For graph (b) we need expressions of the form $\mathrm{Trace}\{\boldsymbol{S}_t\boldsymbol{S}_u\boldsymbol{S}_v\boldsymbol{S}_w\boldsymbol{S}_r\}$. From the arguments given above it must be the case that, for a non-zero term, η_x, η_y and η_z must be 1, 1 and 3 in some order. Using the invariance under the operation $\mathcal{C}$ and $\mathcal{P}$ there are only two traces for which we need explicit expressions. These are given by Rushbrooke and Wood (1958). In our notation

$$\mathrm{Trace}\{\boldsymbol{S}_u\boldsymbol{S}_v\boldsymbol{S}_w\boldsymbol{S}_w\boldsymbol{S}_w\} = \tfrac{1}{30}\mathrm{i}\epsilon(u,v,w)s(s+1)(2s+1)(3s^2+3s-1)\,, \tag{7.176}$$

$$\mathrm{Trace}\{\boldsymbol{S}_u\boldsymbol{S}_w\boldsymbol{S}_v\boldsymbol{S}_w\boldsymbol{S}_w\} = \tfrac{1}{30}\mathrm{i}\epsilon(u,v,w)s(s+1)(2s+1)(s^2+s-2)\,, \tag{7.177}$$

where u, v, w is any cyclic permutation of x, y, z. In (7.166) the only non-zero terms occur for t, u, v all different and $w = r$. With this constraint applying to t, u, v

$$\begin{aligned}
\langle\!\langle \hat{\phi}_5 \cdots \hat{\phi}_1 \rangle\!\rangle_0 &= \frac{1}{2s^{10}(2s+1)^3} \sum_{\{t,u,v,w\}} \mathrm{Trace}\{\boldsymbol{S}_t\boldsymbol{S}_u\boldsymbol{S}_v\}\mathrm{Trace}\{\boldsymbol{S}_w^2\} \\
&\quad \times [\mathrm{Trace}\{\boldsymbol{S}_t\boldsymbol{S}_u\boldsymbol{S}_v\boldsymbol{S}_w\boldsymbol{S}_w\} + \mathrm{Trace}\{\boldsymbol{S}_t\boldsymbol{S}_u\boldsymbol{S}_w\boldsymbol{S}_v\boldsymbol{S}_w\}] \\
&= \frac{1}{2s^{10}(2s+1)^3} \sum_{\{t,u,v\}} \mathrm{Trace}\{\boldsymbol{S}_t\boldsymbol{S}_u\boldsymbol{S}_v\}\mathrm{Trace}\{\boldsymbol{S}_t^2\} \\
&\quad \times [3\,\mathrm{Trace}\{\boldsymbol{S}_t\boldsymbol{S}_u\boldsymbol{S}_v\boldsymbol{S}_v\boldsymbol{S}_v\} + \mathrm{Trace}\{\boldsymbol{S}_u\boldsymbol{S}_t\boldsymbol{S}_v\boldsymbol{S}_t\boldsymbol{S}_t\}] \\
&= -\frac{(s+1)^3}{36s^7}[2s(s+1)-1]\,.
\end{aligned} \tag{7.178}$$

It can be seen that, with all possible simplifications, the evaluation of $\langle\!\langle \hat{\phi}_{\mathring{q}} \cdots \hat{\phi}_1 \rangle\!\rangle_0$ can be quite intricate, even for small values of $\mathring{q}$. Fortunately much of the complicated sorting process can be computerized. The expressions $s^{2q}\langle\!\langle \hat{\phi}_{\mathring{q}} \cdots \hat{\phi}_1 \rangle\!\rangle_0$ are called *reduced mean traces* by Rushbrooke et al. (1974) and they are given in a table by these authors up to $\mathring{q} = 8$.

7.6.5 Correlations and Susceptibility

Any quantum operator $\hat{\mathcal{O}}$, which is an arbitrary function of the spin operators on the lattice sites, has expectation value $\langle\hat{\mathcal{O}}\rangle$ with respect to the general Hamiltonian $\widehat{H}(L, K(\boldsymbol{r},\boldsymbol{r}'); \hat{\boldsymbol{s}}(\boldsymbol{r}))$ given by (7.122). We are now interested in determining the high-temperature series expansion for $\langle\hat{\mathcal{O}}\rangle$ in the case $K(\boldsymbol{r},\boldsymbol{r}') = K$ for all nearest neighbours. The simplest way to do this is to modify (7.98) by replacing $\widehat{H}(L, K(\boldsymbol{r},\boldsymbol{r}'); \hat{\boldsymbol{s}}(\boldsymbol{r}))$ with

$$\widehat{H}_{\mathrm{M}}(L, K(\boldsymbol{r},\boldsymbol{r}'), M; \hat{\boldsymbol{s}}(\boldsymbol{r})) = \widehat{H}(L, K(\boldsymbol{r},\boldsymbol{r}'); \hat{\boldsymbol{s}}(\boldsymbol{r})) - M\mathcal{O}\,. \tag{7.179}$$

Then it is not difficult to see,[14] using (A.277) and (A.279), that for the expectation value of $\mathcal{O}$ with respect to this modified Hamiltonian

$$\begin{aligned}\langle\hat{\mathcal{O}}\rangle &= \frac{\mathrm{Trace}\{\hat{\mathcal{O}}\exp[-\widehat{H}_{\mathrm{M}}(L,K(\boldsymbol{r},\boldsymbol{r}'),M;\hat{\boldsymbol{s}}(\boldsymbol{r}))]\}}{Z_{\mathrm{M}}(L,K(\boldsymbol{r},\boldsymbol{r}'),M)}\\ &= \frac{\mathrm{Trace}\{\mathcal{D}_M\exp[-\widehat{H}_{\mathrm{M}}(L,K(\boldsymbol{r},\boldsymbol{r}'),M;\hat{\boldsymbol{s}}(\boldsymbol{r}))]\}}{Z_{\mathrm{M}}(L,K(\boldsymbol{r},\boldsymbol{r}'),M)}\\ &= \langle\!\langle\hat{\mathcal{O}}\rangle\!\rangle = \lceil\hat{\mathcal{O}}\rfloor\\ &= \mathcal{D}_M\ln Z_{\mathrm{M}}^{+}(L,K(\boldsymbol{r},\boldsymbol{r}'),M),\end{aligned} \tag{7.180}$$

where $\mathcal{D}_M$ denotes partial differentiation with respect to M and the symbols $\langle\!\langle\cdots\rangle\!\rangle$ and $\lceil\cdots\rfloor$ are used in relation to the modified Hamiltonian. Using the argument of Sect. 7.5.1 the expectation value when $M=0$ and all $K(\boldsymbol{r},\boldsymbol{r}')=K$ is now given by

$$\begin{aligned}\langle\hat{\mathcal{O}}\rangle &= \left[\exp\left(K\sum_{\{\boldsymbol{r},\boldsymbol{r}'\}}^{(\mathrm{n.n.})}\mathcal{D}(\boldsymbol{r},\boldsymbol{r}')\right)\{\mathcal{D}_M\ln Z_{\mathrm{M}}^{+}(L,K(\boldsymbol{r},\boldsymbol{r}'),M)\}_{M=0}\right]_0\\ &= \langle\hat{\mathcal{O}}\rangle_0+\sum_{\mathring{\mathrm{q}}=1}^{\infty}K^{\mathring{\mathrm{q}}}\sum_{\{\mathring{\mathrm{k}}\}}\mathring{\Omega}(\mathring{\mathrm{q}},\mathring{\mathrm{k}})\mathrm{G}(L;\mathring{\mathrm{q}},\mathring{\mathrm{k}})\left[\prod_{\mathrm{t}=1}^{\mathrm{q}}\alpha_{\mathrm{t}}!\right]^{-1},\end{aligned} \tag{7.181}$$

where

$$\mathrm{G}(L;\mathring{\mathrm{q}},\mathring{\mathrm{k}}) = \left[\widehat{D}(\mathring{\mathrm{q}},\mathring{\mathrm{k}})\mathcal{D}_M\ln Z_{\mathrm{M}}^{+}(L,K(\boldsymbol{r},\boldsymbol{r}'),M)\right]_0, \tag{7.182}$$

in which the subscript '0' now indicates that all $K(\boldsymbol{r},\boldsymbol{r}')$ *and* M are set to zero. We have already seen in Sect. 7.5.3 that, when used in a formula like (7.110), $\widehat{D}(\mathring{\mathrm{q}},\mathring{\mathrm{k}})$ can be replaced by any one term in the summation in (7.111) and it follows that

$$\mathrm{G}(L;\mathring{\mathrm{q}},\mathring{\mathrm{k}}) = \lceil\hat{\mathcal{O}}\hat{\phi}_{\mathring{\mathrm{q}}}\cdots\hat{\phi}_1\rfloor_0\,. \tag{7.183}$$

By the argument used in Sect. 7.5.1 it is clear that $\mathrm{G}(L;\mathring{\mathrm{q}},\mathring{\mathrm{k}})=0$ unless $\hat{\mathcal{O}}$ is a function of some of the spin operators on the sites of $(\mathring{\mathrm{q}},\mathring{\mathrm{k}})$.

We now consider some particular examples, mainly concentrating on the case $L=0$. First suppose that $\hat{\mathcal{O}}=[\hat{s}_z(\boldsymbol{r}_0)]^{\ell}$, where $\boldsymbol{r}_0$ is some arbitrarily chosen site of the lattice and ℓ is a positive integer. The only non-zero coefficients $\mathrm{G}(0;\mathring{\mathrm{q}},\mathring{\mathrm{k}})$ will be ones where $\boldsymbol{r}_0$ is a vertex of $(\mathring{\mathrm{q}},\mathring{\mathrm{k}})$ and each term in the expansion of such coefficients, using the generalized moment-cumulant relation (7.124), will contain a factor $\langle\!\langle[\hat{s}_z(\boldsymbol{r}_0)]^{\ell}\hat{\phi}^{(t)}\cdots\hat{\phi}^{(r)}\rangle\!\rangle_0$ for some set of operators representing a connected subgraph of $(\mathring{\mathrm{q}},\mathring{\mathrm{k}})$ with $\boldsymbol{r}_0$ as a vertex. For the general-s Ising model the factor $[\hat{s}_z(\boldsymbol{r}_0)]^{\ell}$ is equivalent to increasing

[14] When $\mathcal{O}$ and $\widehat{H}_{\mathrm{M}}$ do not commute it is necessary to use the invariance of the trace under cyclic permutation.

the valency of the vertex $\boldsymbol{r}_0$ by ℓ. Since any graph must have an even number of odd vertices, it follows that

$$\mathrm{G}(0; \mathring{\mathrm{q}}, \mathring{\mathrm{k}}) = 0\,, \qquad \text{if } \ell \text{ is odd}\ . \tag{7.184}$$

For the classical and quantum Heisenberg models the effect of $[\hat{s}_z(\boldsymbol{r}_0)]^\ell$ is to increase η_z for the vertex contribution at $\boldsymbol{r}_0$ to $\eta_z + \ell$ and again the result (7.184) applies. Since $\langle[\hat{s}_z(\boldsymbol{r}_0)]^\ell\rangle_0 = 0$ when $L = 0$ and ℓ is odd we have the result

$$\langle[\hat{s}_z(\boldsymbol{r}_0)]^\ell\rangle = 0\,, \qquad \text{if } \ell \text{ is odd}\ . \tag{7.185}$$

A similar argument leads to the result

$$\langle[\hat{s}_x(\boldsymbol{r}_0)]^\ell\rangle = \langle[\hat{s}_y(\boldsymbol{r}_0)]^\ell\rangle = 0\,, \qquad \text{if } \ell \text{ is odd}\ . \tag{7.186}$$

Given that all sites of the lattice are equivalent, the dimensionless magnetization per lattice site $m = \langle\hat{s}_z(\boldsymbol{r}_0)\rangle/s$ and it follows from (7.185) that $m = 0$ when $L = 0$.

We have already seen in Sect. 7.4.2 that, for the $s = \frac{1}{2}$ Ising model, the high-temperature series expansion (7.64) for the magnetization yields the result $m = 0$ when $L = 0$. The range of validity of the expansion is limited to the region above the Curie temperature where there is no spontaneous magnetization. We now see that the same is the case for the general-s Ising model and for the classical and quantum Heisenberg models. The operator $\hat{\boldsymbol{s}}^2(\boldsymbol{r}) = \hat{s}_x^2(\boldsymbol{r}) + \hat{s}_y^2(\boldsymbol{r}) + \hat{s}_z^2(\boldsymbol{r})$ is equal to one for the classical Heisenberg model and has the single eigenvalue $s(s+1)$ for the general-s Ising model and the quantum Heisenberg model. When $L = 0$, both the Heisenberg models have symmetry under cyclic permutation of the x, y and z axes and in that case

$$\langle\hat{s}_z^2(\boldsymbol{r})\rangle = \tfrac{1}{3}\langle\hat{\boldsymbol{s}}^2(\boldsymbol{r})\rangle = \begin{cases} \frac{1}{3}\,, & \text{classical H. M.}\ , \\[2ex] \frac{1}{3}s(s+1)\,, & \text{quantum H. M.}\ . \end{cases} \tag{7.187}$$

The expectation value is interaction independent and the only contribution to the series expansion (7.181) is the first term. For all three models

$$\langle\hat{s}_z^2(\boldsymbol{r})\rangle_0 = s^2\left(\frac{\partial^2 \ln z_0(L)}{\partial L^2}\right)_{L=0} \tag{7.188}$$

with $s = 1$ for the classical Heisenberg model. From (7.133) and (7.142) this gives results agreeing with (7.187) for the Heisenberg models. For the general-s Ising model the situation is a little more complicated as the Hamiltonian does not have rotational symmetry. Equation (7.188) gives the same value for $\langle\hat{s}_z^2(\boldsymbol{r})\rangle_0$ as for the quantum Heisenberg model, but in this case the lack of symmetry under interchange of the axes prevents us concluding that $\langle\hat{s}_z^2(\boldsymbol{r})\rangle = \langle\hat{s}_z^2(\boldsymbol{r})\rangle_0$.

Now consider the case

$$\hat{\mathcal{O}} = \begin{cases} \hat{s}_z(\boldsymbol{r}_1)\hat{s}_z(\boldsymbol{r}_2)\,, & \text{general-}s\text{ I. M.}\ , \\[2ex] \hat{\boldsymbol{s}}(\boldsymbol{r}_1)\boldsymbol{.}\,\hat{\boldsymbol{s}}(\boldsymbol{r}_2)\,, & \text{classical and quantum H. M.s}\ , \end{cases} \tag{7.189}$$

where $\boldsymbol{r}_1 \neq \boldsymbol{r}_2$ are two arbitrarily chosen (not necessarily nearest-neighbour) sites of the lattice. It follows from (7.130), (7.127), (7.141), (7.144) and (7.183) that

$$\mathsf{G}(L;\mathring{\mathsf{q}},\mathring{\mathsf{k}}) = s^2\mathsf{C}(L;\mathring{\mathsf{q}}',\mathring{\mathsf{k}}'), \tag{7.190}$$

where $s = 1$ for the classical Heisenberg model and $(\mathring{\mathsf{q}}',\mathring{\mathsf{k}}')$ is the graph obtained from $(\mathring{\mathsf{q}},\mathring{\mathsf{k}})$ by placing an extra line between the sites $\boldsymbol{r}_1$ and $\boldsymbol{r}_2$. When $L = 0$ the only non-zero coefficients are those when both $\boldsymbol{r}_1$ and $\boldsymbol{r}_2$ are vertices of $(\mathring{\mathsf{q}},\mathring{\mathsf{k}})$ and thus $\mathring{\mathsf{q}}' = \mathring{\mathsf{q}} + 1$. If $(\mathring{\mathsf{q}},\mathring{\mathsf{k}})$ is disjoint in two parts one containing $\boldsymbol{r}_1$ and the other $\boldsymbol{r}_2$ then $(\mathring{\mathsf{q}}',\mathring{\mathsf{k}}')$ is connected. The sites $\boldsymbol{r}_1$ and $\boldsymbol{r}_2$ are, however, articulation points of the graph and so for the classical Heisenberg model, as was shown in Sect. 7.5.2, $\mathsf{C}(0;\mathring{\mathsf{q}}+1,\mathring{\mathsf{k}}') = 0$. This can also be shown to be true for the quantum Heisenberg model for any graph which can be separated into disjoint parts by the removal of a single line (Rushbrooke and Wood 1958, Rushbrooke et al. 1974). In the case of the Heisenberg models, therefore, the expansion is over connected graphs.

According to the discussion of Sect. 1.5 the (fluctuation) pair correlation function between the spins states on sites $\boldsymbol{r}_1$ and $\boldsymbol{r}_2$ is given by

$$\Gamma_2(\sigma(\boldsymbol{r}_1),\sigma(\boldsymbol{r}_2)) = \begin{cases} \langle\sigma(\boldsymbol{r}_1)\sigma(\boldsymbol{r}_2)\rangle - \langle\sigma(\boldsymbol{r}_1)\rangle\langle\sigma(\boldsymbol{r}_2)\rangle\,, & \text{general-}s\text{ I. M. ,} \\ \langle\hat{\boldsymbol{s}}(\boldsymbol{r}_1).\hat{\boldsymbol{s}}(\boldsymbol{r}_2)\rangle - \langle\hat{\boldsymbol{s}}(\boldsymbol{r}_1)\rangle\boldsymbol{.}\langle\hat{\boldsymbol{s}}(\boldsymbol{r}_2)\rangle\,, & \begin{array}{l}\text{classical and}\\ \text{quantum H. M.s}\,.\end{array} \end{cases} \tag{7.191}$$

From (7.185) and (7.186), when $L = 0$, the second term in each of these forms is zero and the fluctuation function is the same as the total correlation function. The evaluation of the pair correlation function series is that for $\hat{\mathsf{O}}$ given by (7.189). The formula (7.190), which has been used by Stanley (1974) to derive the series expansion for the zero-field pair correlation function of the classical Heisenberg model, shows that any coefficient in the correlation function series is a coefficient in the series of $\ln Z^+$, but not necessarily that for the same lattice. Consider, for example, the case where $\boldsymbol{r}_1$ and $\boldsymbol{r}_2$ are second-neighbour sites on the square lattice. Then a graph contributing to the correlation series would be one of topology $(6,2)$ given in the last line of Table A.5 with $\boldsymbol{r}_1$ and $\boldsymbol{r}_2$ coincident with two vertices of valency three, which are now connected by a line. The silhouette of the resulting graph could be embedded in a triangular lattice but not the square lattice.

The pair correlation function $\Gamma_2(\sigma(\boldsymbol{r}_1),\sigma(\boldsymbol{r}_2))$ yields most quantities of interest. In particular it has been used by Stanley (1974) to derive the zero-field susceptibility series for the classical Heisenberg model. The susceptibility is given by the fluctuation-response function formula (1.62), which in this case has the form

$$\chi = N^{-1} \sum_{\{\boldsymbol{r}_1,\boldsymbol{r}_2\}} \Gamma_2(\sigma(\boldsymbol{r}_1),\sigma(\boldsymbol{r}_2))\,. \tag{7.192}$$

Since in zero field the magnetization is zero the zero-field susceptibility is

$$\chi_0 = (Ns^2)^{-1}\Big[\sum_{\{r\}}\Big\langle\{\hat{s}_z(r)\}^2\Big\rangle + 2\sum_{\{r_1,r_2\}}\langle\hat{s}_z(r_1)\hat{s}_z(r_2)\rangle\Big], \qquad (7.193)$$

where each pair $r_1 \neq r_2$ is counted only once. If we now concentrate on the Heisenberg models and use equation (7.187) and the symmetry of the models under permutation of the axes this equation gives

$$\chi_0 = \tfrac{1}{3}\Big[\Delta(s) + 2(Ns^2)^{-1}\sum_{\{r_1,r_2\}}\langle\hat{s}(r_1)\cdot\hat{s}(r_2)\rangle\Big], \qquad (7.194)$$

where for the quantum Heisenberg model $\Delta(s) = (s+1)/s$, and for the classical Heisenberg model the value corresponds to the limit $s \to \infty$ with $\Delta(s) \to 1$. Since $\langle\hat{s}(r_1).\hat{s}(r_2)\rangle_0 = \langle\hat{s}(r_1)\rangle_0.\langle\hat{s}(r_2)\rangle_0 = 0$ it follows from (7.181) and (7.190) that in the limit of large N, when $\mathring{\Omega}(\mathring{\mathrm{q}},\mathring{\mathrm{k}})$ can be replaced by $N\mathring{\omega}(\mathring{\mathrm{q}},\mathring{\mathrm{k}})$,

$$(Ns^2)^{-1}\sum_{\{r_1,r_2\}}\langle\hat{s}(r_1)\cdot\hat{s}(r_2)\rangle = \sum_{\mathring{\mathrm{q}}=1}^{\infty}K^{\mathring{\mathrm{q}}}\sum_{\{\mathring{\mathrm{k}}\}}\mathring{\omega}(\mathring{\mathrm{q}},\mathring{\mathrm{k}})\left(\prod_{\mathrm{t}=1}^{\mathrm{q}}\alpha_{\mathrm{t}}!\right)^{-1}$$
$$\times\sum_{\{r_1,r_2\}}\mathrm{C}(0;\mathring{\mathrm{q}}+1,\mathring{\mathrm{k}}'(r_1,r_2)), \qquad (7.195)$$

where $\mathring{\mathrm{k}}'(r_1,r_2)$ denotes the topology obtained from $\mathring{\mathrm{k}}$ by adding a line between r_1 and r_2.

An alternative approach, like that of Sect. 7.4, is to find the terms in L^2 in the moment expansion. These are found to involve sums over sets of graphs similar to those just derived in the correlation expansion, in each of which an additional line is inserted between a pair of vertices of a connected graph $(\mathring{\mathrm{q}},\mathring{\mathrm{k}})$. The susceptibility χ_0 depends on the corresponding terms in the cumulant expansion which can be derived by the 'coefficient of N' method.

7.7 Finite Methods

In this section we consider two related methods, the finite-cluster method introduced by Domb (1960, 1974a) and the finite-lattice method introduced by de Neef (1975) and developed by de Neef and Enting (1977). Although both these methods are primarily techniques for obtaining high-temperature expansions, the latter has also been used for low-temperature expansions for $s \geq 1$ Ising models (Enting et al. 1994, Jensen et al. 1996, Jensen and Guttmann 1996) and for the Potts model (Guttmann and Enting 1993, 1994, Briggs et al. 1994, Arisue and Tabata 1997).

7.7.1 The Finite-Cluster Method

A starting point for this method (Rushbrooke 1964, Rushbrooke et al. 1974) is the observation that (7.109) applies to any finite connected group of sites $\mathfrak{g}$. From (7.103), (A.231) and (7.109) the dimensionless free energy of a graph $\mathfrak{g}$ of N sites and $\mathrm{q}_{\mathfrak{g}}$ bonds is

$$\Phi^{(\mathfrak{g})}(N;L,K) = -N\ln z_0(L)$$
$$-\sum_{\mathring{\mathrm{q}}=1}^{\infty} K^{\mathring{\mathrm{q}}} \sum_{\{\mathring{\mathrm{k}}\}} \Omega^{(\mathfrak{g})}(\mathrm{q},\mathrm{k})\mathrm{S}(\mathrm{q},\mathrm{k};\mathring{\mathrm{q}},\mathring{\mathrm{k}})\mathrm{C}(L;\mathring{\mathrm{q}},\mathring{\mathrm{k}})\left(\prod_{\mathrm{t}=1}^{\mathrm{q}} \alpha_{\mathrm{t}}!\right)^{-1} . \tag{7.196}$$

The right-hand side of (7.196) is a linear function of the $\Omega^{(\mathfrak{g})}(\mathrm{q},\mathrm{k})$, which give the number of weak embeddings of the graphs (q,k) in $\mathfrak{g}$ and thus

$$\Phi^{(\mathfrak{g})}(N;L,K) = -\sum_{\mathrm{q}=0}^{\mathrm{q}_{\mathfrak{g}}}\sum_{\{\mathrm{k}\}} \Omega^{(\mathfrak{g})}(\mathrm{q},\mathrm{k})\Psi(L,K;\mathrm{q},\mathrm{k}) , \tag{7.197}$$

where the first term on the right-hand side of (7.196) has been included by taking

$$\Omega^{(\mathfrak{g})}(0,1) = N , \qquad \Psi(L,K;0,1) = \ln z_0(L) , \tag{7.198}$$

which just corresponds to the weak embedding of a single site in $\mathfrak{g}$. At the other extreme there is only one topology and one way for weakly embedding $\mathfrak{g}$ into itself and so $\Omega^{(\mathfrak{g})}(\mathrm{q}_{\mathfrak{g}},1) = 1$. Since (q,k) can be the silhouette of multi-bonded graphs $(\mathring{\mathrm{q}},\mathring{\mathrm{k}})$ with any number $\mathring{\mathrm{q}} \geq \mathrm{q}$ of lines, $\Psi(L,K;\mathrm{q},\mathrm{k})$ is an infinite series in powers of K, the lowest power being K^{q}. It can be seen from (7.196) that $\Psi(L,K;\mathrm{q},\mathrm{k})$ does not depend on $\mathfrak{g}$ and, to determine this quantity, we use (7.197) itself and the set of similar equations obtained by replacing $\mathfrak{g}$ by its connected subgraphs.

As an illustration consider the case where $\mathfrak{g}$ is a square of sites as shown in Fig. 7.3. Now there is only one topology for each connected subgraph and it can easily be seen that

$$\Phi^{(0,1)}(1;L,K) = -\Psi(L,K;0,1) , \tag{7.199}$$

$$\Phi^{(1,1)}(2;L,K) = 2\Psi(L,K;0,1) + \Psi(L,K;1,1) , \tag{7.200}$$

$$\Phi^{(2,1)}(3;L,K) = -3\Psi(L,K;0,1) - 2\Psi(L,K;1,1)$$
$$- \Psi(L,K;2,1) , \tag{7.201}$$

$$\Phi^{(3,1)}(4;L,K) = -4\Psi(L,K;0,1) - 3\Psi(L,K;1,1)$$
$$- 2\Psi(L,K;2,1) - \Psi(L,K;3,1) , \tag{7.202}$$

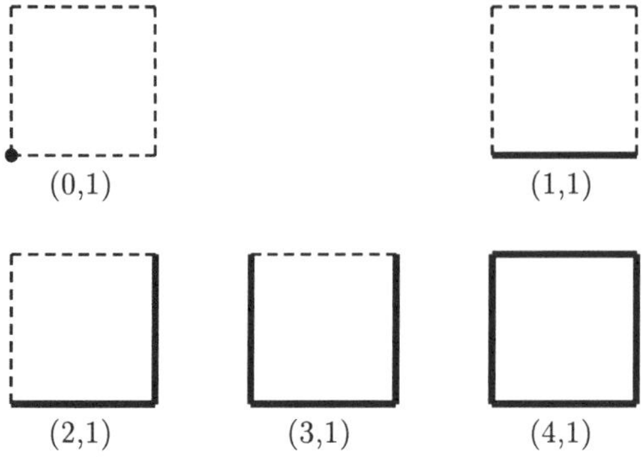

Fig. 7.3. Subgraphs of a square in the finite-cluster method.

$$\begin{aligned}\Phi^{(4,1)}(4;L,K) = -4\Psi(L,K;0,1) - 4\Psi(L,K;1,1) \\ - 4\Psi(L,K;2,1) - 4\Psi(L,K;3,1) \\ - \Psi(L,K;4,1)\,.\end{aligned} \tag{7.203}$$

The free energy $\Phi^{(0,1)}(1;L,K)$ for a single vertex is $-\ln[z_0(L)]$ and (7.199) is equivalent to the second of equations (7.198). For any particular type of Hamiltonian the quantities $\Psi(L,K;\mathfrak{q},1)$, for $\mathfrak{q} = 1,\ldots,4$, can be obtained from (7.199)–(7.203) once the free energies for the corresponding clusters of sites have been calculated.

As an example, for the spin-$\frac{1}{2}$ Ising model in zero field

$$\begin{aligned}\Phi^{(0,1)}(1;0,K) &= -\ln(2)\,, \\ \Phi^{(1,1)}(2;0,K) &= -\ln(4\cosh K)\,, \\ \Phi^{(2,1)}(3;0,K) &= -\ln(8\cosh^2 K)\,, \\ \Phi^{(3,1)}(4;0,K) &= -\ln(16\cosh^3 K)\,, \\ \Phi^{(4,1)}(4;0,K) &= -\ln\{4[3+\cosh(4K)]\}\,.\end{aligned} \tag{7.204}$$

On substituting into (7.199)–(7.203) this gives

$$\begin{aligned}&\Psi(0,K;0,1) = \ln(2)\,, &&\Psi(0,K;1,1) = \ln(\cosh K)\,, \\ &\Psi(0,K;2,1) = 0\,, &&\Psi(0,K;3,1) = 0\,, \\ &\Psi(0,K;4,1) = \ln\left[\frac{3+\cosh(4K)}{4\cosh^4 K}\right]\,.\end{aligned} \tag{7.205}$$

As $(2,1)$ and $(3,1)$ are graphs with articulation points the zero values obtained for $\Psi(0,K;2,1)$ and $\Psi(0,K;3,1)$ have been predicted in Sect. 7.5.2.

Where $\mathfrak{g}$ is a lattice with a large number N of sites (7.197) gives the free energy per lattice site

$$\phi(L,K) = -\sum_{\mathfrak{q}=0}^{\infty}\sum_{\{\mathfrak{k}\}} \omega(\mathfrak{q},\mathfrak{k})\Psi(L,K;\mathfrak{q},\mathfrak{k}), \tag{7.206}$$

where the sum is over all connected subgraphs $(\mathfrak{q},\mathfrak{k})$ of the lattice graph.

Now suppose that the aim is to obtain a high-temperature expansion up to and including the term in K^m. Recalling that $\Psi(L,K;\mathfrak{q},\mathfrak{k})$, which depends only on the graph $(\mathfrak{q},\mathfrak{k})$, contains no powers of K with index less that $\mathfrak{q}$, we can drop all terms with $\mathfrak{q} > m$, from the right-hand side of (7.206). This will involve both terminating the outer summation at $\mathfrak{q} = m$ and truncating the infinite power-series forms of the remaining $\Psi(L,K;\mathfrak{q},\mathfrak{k})$. Indicating this truncation by using a subscript m we have

$$\phi_m(L,K) = -\sum_{\mathfrak{q}=0}^{m}\sum_{\{\mathfrak{k}\}} \omega(\mathfrak{q},\mathfrak{k})\Psi_m(L,K;\mathfrak{q},\mathfrak{k}). \tag{7.207}$$

A set on $m+1$ linear equations for $\Psi_m(L,K;\mathfrak{q},\mathfrak{k})$, $\mathfrak{q} = 0,1,\ldots,m$, are, from (7.197),

$$\Phi^{(\mathfrak{q}',\mathfrak{k}')}(\mathfrak{n}';L,K) = -\sum_{\mathfrak{q}=0}^{\mathfrak{q}'}\sum_{\{\mathfrak{k}\}} \Omega^{(\mathfrak{q}',\mathfrak{k}')}(\mathfrak{q},\mathfrak{k})\Psi_m(L,K;\mathfrak{q},\mathfrak{k}),$$

$$\mathfrak{q}' = 0,1,\ldots,m, \tag{7.208}$$

where $\mathfrak{n}'$ is the number of sites of the graph with topology $(\mathfrak{q};,\mathfrak{k}')$. The graphs $(\mathfrak{q}',\mathfrak{k}')$ are given a 'dictionary ordering', first being divided into sets of given $\mathfrak{q}'$, with $\mathfrak{q}'$ increasing from 0 to m and then arranged in each set according to the index $\mathfrak{k}'$. If the terms on the right-hand side of (7.208) are similarly arranged the $\Omega^{(\mathfrak{q}',\mathfrak{k}')}(\mathfrak{q},\mathfrak{k})$ form the elements of a triangular matrix called the *T matrix* (see Appendix A.12). With a change of sign the coefficients on the right-hand side of (7.203) are a simple example of this. Because of the triangular form of the matrix equations (7.208) are easily solved for $\Psi_m(L,K;\mathfrak{q},\mathfrak{k})$ (Example 7.6), which can then be substituted into (7.207).

A method for deriving partition functions of finite clusters of spins in the classical Heisenberg model was developed by Joyce (1967) and used by Joyce and Bowers (1966) to calculate high-temperature expansions on various lattices. Joyce's method was extended by Domb (1976) to give a systematic procedure for calculating the partition functions of star topologies in the D-vector model. This work was used by English et al. (1979) to obtain high-temperature series to K^{12} for general D and to K^{13} for the classical Heisenberg model.

7.7.2 The Finite-Lattice Method

In a sequence of papers Hijmans and de Boer (1955) developed a systematic method for generating a hierarchy of closed-form cluster methods for order-disorder problems. The lowest-order approximations in this hierarchy correspond to the zeroth-order single-site and first-order site-pair (quasi-chemical) approximations referred to in Volume 1, Chap. 7. At a higher level the approximation is more accurate that of Volume 1, Sect. 7.2 as it takes into account all subclusters of sites between the maximum group chosen and the single sites. A recent resurgence of interest in this work has occurred because of the success of the finite-lattice method of series expansions of de Neef and Enting (1977). The combinatorial procedures needed to effect the Hijmans and de Boer programme turn out to be exactly those needed to begin a finite-lattice calculation and in Appendix A.12 we give a brief account of this development.

The starting point for the finite-lattice method is a formula,

$$\Phi(\mathcal{N}) = \sum_{\mathfrak{g} \in \mathcal{S}} \chi(\mathcal{S}; \mathfrak{g}) \Phi(\mathfrak{g}) , \tag{7.209}$$

similar to (A.331), which expresses the free energy for a lattice $\mathcal{N}$ of N sites in the form of a weighted sum of the free energies of a set $\mathcal{S}$ of subgraphs of $\mathcal{N}$. The coefficients $\chi(\mathcal{S}; \mathfrak{g})$ are dependent on the set $\mathcal{S}$ of graphs chosen and the number of terms of a power series which can be given correctly by this procedure is determined by the lowest power of the expansion variable which occurs in a connected graph which cannot be embedded in a $\mathfrak{g} \in \mathcal{S}$.

The upper limit on size of the members of $\mathcal{S}$ can be determined in different ways. The simplest procedure, used by Hijmans and de Boer (1955) and in the related method of Enting and Baxter (1977), is to choose a particular graph, or block of sites, and then the generate the set $\mathcal{S}$ by successive overlapping (see Appendix A.12). The method favoured by de Neef and Enting (1977) on the square lattice is the impose an upper limit on the circumference of the rectangular blocks, which are, as before, generated by an overlapping procedure. The method for determining the coefficients $\chi(\mathcal{S}; \mathfrak{g})$ for this constraint is discussed by Enting (1978). Once the coefficients have been obtained the remaining task is to obtain the free energy or partition functions for the graphs of $\mathcal{S}$. For the square lattice the members of $\mathcal{S}$ are rectangular blocks and transfer matrix methods can be used.

The finite-lattice method is not in essence a high- or low-temperature method. However, the cases where it has been applied as a high-temperature expansion seem to be confined to the spin-$\frac{1}{2}$ Ising model and Potts model where, in the case of the square lattice, duality allows a low-temperature interpretation of the results. De Neef (1975) obtained the high-temperature free energy expansion for the 3-state Potts model on the square lattice up to τ^{22} (see de Neef and Enting 1977). Guttmann and Enting (1993) obtained the magnetization and susceptibility for the spin-$\frac{1}{2}$ Ising model on the simple cubic lattice up to τ^{26} and Guttmann and Enting (1994) obtained corresponding results for the 3-state Potts model up to τ^{21}.

Low-temperature expansions, as we saw in Sect. 7.3, correspond to accounting for successive perturbations from a chosen ground state. In the same way the difference in the application of the finite-lattice method to high- and low-temperature series is that in the former free boundary conditions are used in the application of the transfer matrix procedure to the members of $\mathcal{S}$, whereas for the latter the boundary sites are fixed in the ground state. On the simple cubic lattice Guttmann and Enting (1993) extended the low-temperature spin-$\frac{1}{2}$ Ising model free energy expansion to degree 26 and Guttmann and Enting (1994) obtain corresponding results for the 3-state Potts model to degree 43. Briggs et al. (1994) use the method to obtain low-temperature series for the ν-state Potts model on the square lattice for various values of ν. Recent extensions of this work have been carried out by Arisue and Tabata (1997). Spin-s Ising models for $s > \frac{1}{2}$, do not, even on the square lattice, have duality properties, and it is of particular interest to obtain series approximations for such models. Adler and Enting (1984) obtained 45-term expansions for the magnetization, heat capacity and susceptibility of the spin-1 Ising model on the square lattice. The results were extended to the 79th degree by Enting et al. (1994) and higher-order spin values were investigated by Jensen et al. (1996).

7.8 Results and Analysis

In this section we bring together the results for the three types of temperature expansions discussed in this chapter. These results take the form of values for critical parameters and critical exponents. According to the ideas of universality (Sect. 2.3) the critical exponents will change with the symmetry group of the order parameter of the model and the dimension of the lattice. In addition to these factors critical parameters, like the critical temperature also differ between different lattice structures.

In two dimensions. The critical temperatures for the spin-$\frac{1}{2}$ Ising model on lattices are know from duality theory and the critical exponents $\alpha = 0$ (logarithmic singularity), $\beta = \frac{1}{8}$ and $\gamma = \frac{7}{4}$ are given by Onsager's solution (see, for example, Volume 1, Chap. 8) with $\delta = 1 + \gamma/\beta = 15$ being a consequence of the Widom scaling law (2.169). For the ν-state Potts model the critical temperature on the square lattice is known from duality or from the equivalence to a staggered six-vertex model (Volume 1, Sect. 10.14). correspondence. For $0 < \nu \leq 4$ it was conjectured by den Nijs (1979) that

$$\alpha = \frac{2\pi - 8\mu}{3\pi - 6\mu}, \tag{7.210}$$

and independently by Nienhuis et al. (1980) and Pearson (1980) that

$$\beta = \frac{\pi + 2\mu}{12\pi}, \tag{7.211}$$

where $\mu = \arccos(\sqrt{\mu}/2)$. Using the scaling laws (2.169) and (2.170) this gives

$$\gamma = \frac{7\pi^2 - 8\mu\pi + 4\mu^2}{6\pi(\pi - 2\mu)}, \qquad \delta = \frac{15\pi^2 - 16\mu\pi + 4\mu^2}{\pi^2 - 4\mu^2}. \tag{7.212}$$

The formula (7.210) for α has now been establish by Black and Emery (1981) by further exploiting the relationship to the staggered F model.

In three dimensions. Estimates for the critical exponents for both the Ising and Heisenberg[15] models can be obtain from field theory results for the n-vector model, for which $n = 1$ is equivalent to the Ising model and $n = 3$ the Ising model. These are given by Le Guillou and Zinn-Justin (1980) as $\alpha = 0.110(5)$, $\beta = 0.325(2)$, $\gamma = 1.241(2)$ for the Ising model and $\alpha = -0.115(9)$, $\beta = 0.3645(3)$, $\gamma = 1.386(4)$ for the Heisenberg model. Nickel (1991) examined the ϕ^4 field theory exponents and showed, that by including a second confluence exponent, estimates for the Ising model could be obtained which agreed well with those of high-temperature series. The rational values for the Ising exponents in two dimensions are supported by conformal field theory. Conformal field theories in three dimensions are more problematic, but it is tempting to suppose that the Ising exponents are still rational. Mojumder (1991) has proposed a theory based on partial renormalization of superconformal dimensions of matter fields on a $(2,0)$ supersymmetric string world sheet and has obtained the values $\alpha = \frac{1}{8}$, $\beta = \frac{3}{8}$.

7.8.1 High-Temperature Series for Spin-$\frac{1}{2}$ Ising and Potts Models

As we saw in Sect. 7.1 the particular feature of these models is that there exists a high-temperature variable τ, defined in the Ising model by (7.5) and the Potts model by (7.6), in terms of which the series methods of Sect. 7.4 and the finite-lattice methods of Sect. 7.7 can be applied. Using these methods large numbers of coefficients for high-temperature spin-$\frac{1}{2}$ Ising model series have now been obtained. For the square lattice the zero-field susceptibility series given in (7.75) has been extended to 54 terms (quoted by Guttmann 1989 and obtained by Nickel 1985[16]). The last few of these are

$$\begin{aligned}\chi_0 = \cdots &+ 5429147554971826760 20\tau^{51} \\ &+ 13294400774247124354 76\tau^{52} \\ &+ 32546159798488760642 44\tau^{53} \\ &+ 79654880659404621053 80\tau^{54} + \cdots .\end{aligned} \tag{7.213}$$

For the body-centred cubic lattice

$$\begin{aligned}\chi_0 = 1 &+ 8\tau + 56\tau^2 + 392\tau^3 \\ &+ \cdots + 1995182186382338 96\tau^{21} + \cdots ,\end{aligned} \tag{7.214}$$

[15] It is now generally believed (Stanley 1987) that the exponents of the quantum Heisenberg model are independent of the spin s and, since the classical Heisenberg model is given by the limit $s \to \infty$, this too will have the same exponents.

[16] Personal communication to Professor Guttmann.

(Nickel 1982), and for the simple cubic Guttmann and Enting (1993) used the finite-lattice method to obtain the expansion

$$\begin{aligned}\Phi = & 1 + 3\tau^4 + 22\tau^6 + 192\tau^8 \\ & + \cdots + 16809862992\tau^{22} + \cdots .\end{aligned} \tag{7.215}$$

The problem now is to extract the values of the critical parameter τ_c and the appropriate critical exponents. Methods for doing this are discussed briefly in Sect. 7.2 and have been applied to the series (7.213)–(7.215). Before discussing the conclusions which are drawn it is perhaps worthwhile to revert briefly to a simpler approach.

For a series of positive coefficients a_n, if the series had only one singularity for real values of τ, $\tau_c = R$ the radius of convergence for which $|a_n|^{-1/n}$ are successive approximations. This could be tested against the high-temperature susceptibility series (7.75) and (7.213). With $\tau_c^{(n)} = |a_n|^{-1/n}$, we have $\tau_c^{(5)} = 0.3299$, $\tau_c^{(6)} = 0.3325$, $\tau_c^{(7)} = 0.3383$, ..., $\tau_c^{(52)} = 0.3618$ $\tau_c^{(53)} = 0.3857$, $\tau_c^{(54)} = 0.3930$. Comparing these results with the exact value $\tau_c = 1/(1+\sqrt{2}) \simeq 0.41421$, it will be seen that convergence is very slow. The error in the final estimate is about 5% and extrapolation is unlikely to lead to much improvement. If we suppose that the series has only one singularity on the circle of convergence, we could consider the use of the ratio formula (7.15) (with $\zeta_c = \tau_c$ and $\lambda = \gamma$). This is an expression for the ratio a_n/a_{n-1} as a linear function of $1/n$ with gradient $(\gamma - 1)/\tau_c$ and intercept $1/\tau_c$ on the '$1/n = 0$'-axis. Graphical interpolation of the data could therefore be used to determine τ_c and γ. Plots of $a_n/(za_{n-1})$ against $1/n$ are shown in Fig. 7.4 on three different lattices. An approximation to the parameters for the interpolated line can be obtained by solving the linear equations given by (7.15) for two different values of n. Using the coefficients for the square lattice given in (7.213) the reader can easily verify that better results are obtained by using $n = 52$ and $n = 54$ than by using a consecutive pair of terms ($n = 52$ and $n = 53$ or $n = 53$ or $n = 54$). This is because the oscillations in the values of the coefficients at small values of n, shown in Fig. 7.4, still persist at these higher values of n. This contrasts with the rather more linear appearance of the plots for the triangular and face-centred cubic lattices shown in Fig. 7.4. To understand the reason for this difference in behaviour we must remember that, in deriving the ratio formula (7.15), it was assumed that the critical point was the only singularity on the circle of convergence. In fact this assumption is false for the Ising model on any loose-packed lattice.[17] In all these cases perfect antiferromagnetic ordering is possible with a singularity on the negative real axis at $-\tau_c$, corresponding to the antiferromagnetic critical point. This leads to oscillations is the series coefficients. Theorem 7.2.1 can be applied to the case where there are two singularities at $\zeta = \pm\zeta_c$ to give

[17] Square, honeycomb, simple cubic, body-centre cubic and diamond (see Volume 1, Appendix A.1).

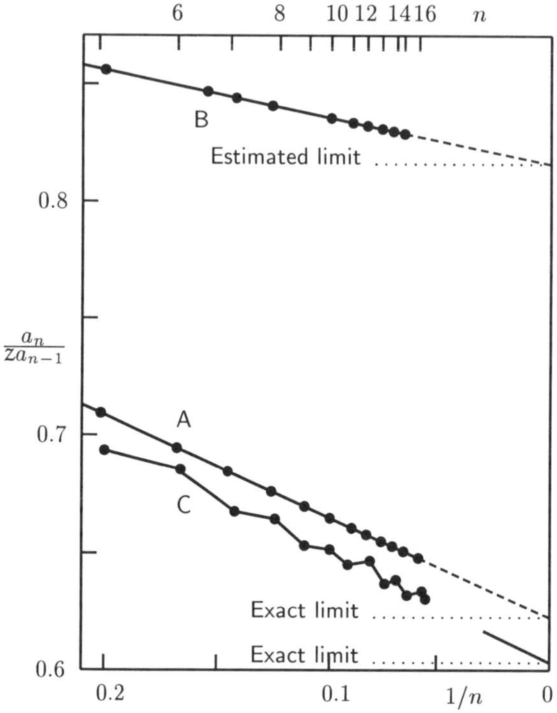

Fig. 7.4. Interpolated plots of $a_n/(za_{n-1})$ against $1/n$ for high-temperature susceptibility series of the Ising model on the square (A, $z = 4$), triangular (B, $z = 6$) and face-centred cubic (C, $z = 12$) lattices. (Reprinted from Guttmann (1989), p. 18, by permission of the publisher Academic Press Ltd.)

$$\frac{a_n}{a_{n-2}} = \frac{(\lambda + n - 1)(\lambda + n - 2)}{n(n-1)\zeta_c^2} \tag{7.216}$$

(Example 7.1). Applying this formula to the series (7.213), for $n = 53$ and $n = 54$, gives $\tau_c = 0.41415$, $\gamma = 1.7358$ (see equation (5.126)).

An extensive analysis of high-temperature Ising series on simple-cubic, body-centred cubic and face-centred cubic lattices was carried out by Graves-Morris (1988). Using a generalized inverse Padé approximant method it was concluded that $\gamma = 1.2406(17)$ is the critical exponent value common to all these lattices. This result was achieved by assuming a leading confluent exponent $\Delta = \frac{1}{2}$. An analysis using partial differential approximants by Chen et al. (1982) and Fisher and Chen (1985) lead to the result $\gamma = 1.2385(15)$ with the confluent exponent $\Delta = 0.54(5)$ and by George and Rehr (1984) and Nickel and Rehr (1990), using second-order differential approximants, to $\gamma = 1.2378(6)$ with $\Delta = 0.52(3)$.

Apart from the calculation of critical exponents, it is of interest to use series expansions to obtain a more detailed form for the leading singular behaviour of thermodynamic functions. Although Onsager's solution of the square lattice Ising model gives information along the zero-field axis it does not provide details of the singular form of the zero-field susceptibility. For this it is known that

$$\chi_0 = C_0^{(+)}\mu^{-7/4} + C_1^{(+)}\mu^{-3/4} + D_0 + C_2^{(+)}\mu^{1/4} + D_1\mu + E_0^{(+)}\mu\ln(\mu) + C_3^{(+)}\mu^{5/4} + \cdots , \qquad (7.217)$$

where $\mu = (1 - T_c/T)$, $C_0^{(+)} = 0.9625817322$ (Barouch et al. 1973, Tracy and McCoy 1973), $C_1^{(+)} = C_0^{(+)}(\sqrt{(2)}/16)\ln(1+\sqrt{2})$, (Wu et al. 1976) and $D_0 = -0.10413323411$, (Kong et al. 1986).[18] The remaining terms in (7.217) were obtained by Gartenhaus and Scott McCullough (1988) by analyzing Nickel's series (7.214) and (7.213) using Padé approximants and second-order homogeneous differential approximants.

For the ν-state Potts model Kim and Joseph (1975) obtained free energy series in the presence of arbitrary external fields to τ^8 for two and three-dimensional lattices and analyzed their results using Dlog Padé approximants. In two dimensions they obtained $\gamma = 1.42(5)$ for $\nu = 3$ and $\gamma = 1.20(5)$ for $\nu = 4$. These compare with the exact results $\gamma = \frac{13}{9}$ and $\gamma = \frac{7}{6}$ respectively derived from (7.212). They also presented evidence that the transition is first-order in two dimensions for $\nu > 4$, in agreement with the known exact results (Baxter 1973, 1978) and in three dimensions for $\nu > 2$, in agreement with Monte Carlo (Hermann 1979, Knak Jensen and Mouritsen 1979) and Monte Carlo renormalization (Blöte and Swendsen 1979) calculations. For the two-dimensional 3-state Potts model de Neef and Enting (1977) used finite-lattice methods to obtain a free energy series to τ^{22}. Using Padé approximants they estimate a value for $\alpha = 0.42(5)$.

7.8.2 Low-Temperature Series

While high-temperature series are, in general, well behaved, with uncertainty in the derived exponents being at the third or fourth decimal place the situation for low-temperature series is less satisfactory because of the presence of non-physical singularities closer to the origin that the physical singularity. A case of this behaviour is the body-centred cubic series for magnetization and susceptibility derived by Sykes et al. (1973e). The coefficients in these series show an apparently random distribution of signs which Gaunt and Guttmann (1974) show is due to a complex pair of singularities closer to the origin than the physical singularity. The application of Dlog Padé approximants to the magnetization series yielded a best estimate of $\beta = 0.31605$ for the critical exponent. Gaunt and Sykes (1973) analyzed the low-temperature magnetization and susceptibility series of Sykes et al. (1973e) for the diamond and face-centred cubic lattices. Using both ratio and Dlog Padé approximant methods they concluded that the data is 'quite consistent with $\beta = \frac{5}{16}$ and

[18] To this order a similar expression is known for $T < T_c$, with $C_0^{(+)} = 0.0255369719$ (Barouch et al. 1973, Tracy and McCoy 1973), $C_1^{(-)} = -C_0^{(-)}(\sqrt{(2)}/16)\ln(1+\sqrt{2})$ (Wu et al. 1976) and D_0 with the same value (Kong et al. 1986).

not inconsistent with $\gamma' = \frac{5}{4}$'. A recent application of the shadow method to the spin-$\frac{1}{2}$ Ising model on the simple cubic lattice (Vohwinkel 1993) extended the magnetization series to $\mathfrak{X}^{32}$. The shadow method has also been applied to $s = 1$ and $s = \frac{3}{2}$ Ising models on a number of two- and three-dimensional lattices by Fox and Guttmann (1973). They conclude that there is no evidence that critical exponents differ from their $s = \frac{1}{2}$ values and give the estimates $\beta = 0.324(15)$ and $\gamma' = 1.26(8)$ in three dimensions. Using a corner matrix formalism Baxter and Enting (1979) extended the square lattice spin-$\frac{1}{2}$ expansion for χ_0 to $\mathfrak{X}^{23}$.

The major development in low-temperature expansions in recent years has been in the use of finite-lattice methods. Briggs et al. (1994) derived series for the ν-state Potts model on the square lattice for $\nu = 2, 3, .., 10$. They show how the method can be used to distinguish between second- and first-order transitions with a first-order transition being predicted for $\nu > 4$. Guttmann and Enting (1993) derive a low-temperature series to $\mathfrak{X}^{26}$ for the spin-$\frac{1}{2}$ on the simple cubic lattice. Using a combination of transformation and Padé approximant methods they estimate the critical exponents as $\alpha' = 0.124(6)$, $\beta = 0.329(9)$ and $\gamma' = 1.251(28)$. The method has also been applied to the spin-1 Ising model by Adler and Enting (1984) and Enting et al. (1994). In the latter work the series was obtain to $\mathfrak{X}^{79}$ and accurate estimates are obtained for the critical temperature and for the location of the non-physical singularities. This work has been extended to $s = \frac{1}{3}$, 2, $\frac{5}{2}$ and 3 by Jensen et al. (1996).

7.8.3 High-Temperature K Expansions

Stanley (1967) use the linked-cluster method to obtain zero-field susceptibility series for three two-dimensional and three three-dimensional lattices in most cases up to K^9. He analyzed the results using ratio and Padé approximant methods and estimated that $\gamma = 1.38(2)$ for the three dimensional lattices. Domb (1976) developed a systematic procedure for calculating the finite-cluster partitions functions needed for the method described in Sect. 7.7.1. This was then applied by English et al. (1979) to the problem of finding free energy and heat capacity expansions for the D-vector model for general D on the face-centred cubic lattice. They obtained the series for the Heisenberg model up to K^{13}. Using ratio methods they estimate the critical parameter $K_c^{-1} = 3.167$ and $\alpha = -0.148$.

For the quantum Heisenberg model the critical exponents are expected to be the same as those for the classical case and independent of s. Rushbrooke and Wood (1955) obtained the zero-field susceptibility series for the spin-$\frac{1}{2}$ quantum Heisenberg model up to K^6 for simple cubic, body-centre cubic and face-centred cubic lattice. They estimated the critical temperature using the ratio method and made the interesting observation that their results agree with the formula

$$K_c^{-1} = \tfrac{5}{96}(z-1)[11s(s+1)-1], \tag{7.218}$$

to within 1%.

Examples

7.1 Suppose that the function $f(\zeta)$ has two singularities on the circle of convergence $|\zeta| = \zeta_c$ at $\zeta = \pm\zeta_c$. Given that the leading singular behaviours are

$$f(\zeta) \simeq A_0^{(\pm)}\left(1 \mp \frac{\zeta}{\zeta_c}\right)^{\lambda}, \qquad \zeta \to \zeta_c\mp,$$

show, using Theorem 7.2.1, that

$$a_n \simeq \frac{1}{\zeta_c^n}\left[(-1)^n A^{(+)} + A^{(-)}\right]\frac{\Gamma(1-\lambda)}{\Gamma(n+1)\Gamma(1-n-\lambda)}.$$

Hence derive the ratio formula (7.216). (It will be observed that (7.216) can also be derived directly from (7.15). This means that the same results can be derived for the critical parameters by using the simple ratio formula with some n and $n+2$ rather than $n+1$.)

7.2 Show using the material of Sect. 7.3 that, for the simple cubic lattice, the free energy per site in zero field has a low-temperature expansion of the form

$$\phi(0, K) = \tfrac{3}{4}\ln(\mathfrak{X}) - \mathfrak{X}^3 + \tfrac{7}{2}\mathfrak{X}^6 - 15\mathfrak{X}^7 + 33\mathfrak{X}^8 - \cdots.$$

7.3 Show that the formula (5.128) for the spontaneous magnetization m_s of the square lattice spin-$\frac{1}{2}$ Ising model can be expressed in the form

$$m_s = \frac{(1+\mathfrak{X})^{\frac{1}{4}}(1-6\mathfrak{X}+\mathfrak{X}^2)^{1/8}}{(1-\mathfrak{X})^{\frac{1}{2}}},$$

in terms of the low-temperature variable $\mathfrak{X} = \exp(-4K)$. Expand this expression to give the series

$$m_s = 1 - 2\mathfrak{X}^2 - 8\mathfrak{X}^3 - 34\mathfrak{X}^4 - 152\mathfrak{X}^5 - 714\mathfrak{X}^6 - 3472\mathfrak{X}^7 - \cdots.$$

Use the ratio method to obtain an estimate for the critical parameter value and the exponent β. These can be compared with the exact values $\mathfrak{X}_c = 3 - 2\sqrt{2} \simeq 0.1716$ and $\beta = 0.125$.

7.4 Obtain, from equation (7.40), the low-temperature expansion

$$m = 1 - 2\mathfrak{X}^3 - 12\mathfrak{X}^5 + 14\mathfrak{X}^6 - 90\mathfrak{X}^7 + 192\mathfrak{X}^8 - 792\mathfrak{X}^9 + O(\mathfrak{X}^{10}),$$

for the zero-field magnetization of the Ising model on the simple cubic lattice. Determine the series for $d\ln(m)/d\mathfrak{X}$ and use the Padé approximant method to obtain approximations to $\mathfrak{X}_c$ and β. (The results should compare quite well with the best series results $\mathfrak{X}_c \simeq 0.412$, $\beta \simeq 0.33$.)

7.5 The only contribution to the summation over pairs $\{\boldsymbol{r}_1, \boldsymbol{r}_2\}$ in equation (7.195) when $\mathring{q} = 1$ is when $\boldsymbol{r}_1$ and $\boldsymbol{r}_2$ coincide with the ends of the single bond, giving the graph of Fig. 7.1(b). The first few terms for the zero-field susceptibility for the classical Heisenberg model are

$$\chi_0 = \frac{1}{3} + \frac{4}{3}K + \frac{40}{9}K^2 + \frac{1952}{135}K^3 + \cdots$$

(Stanley 1967) and the corresponding expansion for the spin-$\frac{1}{2}$ quantum Heisenberg model is

$$\chi_0 = 1 + 3K + 6K^2 + 11K^3 + \cdots$$

(Rushbrooke and Wood 1955). For the classical Heisenberg model (7.156) can be used to obtain the linear term in χ_0. Obtain $C(0; 2, 2)$ for the spin-$\frac{1}{2}$ quantum Heisenberg model and verify the linear term in this case. You may also like to use similar methods to verify the quadratic terms in each series.

7.6 Express equations (7.199)–(7.202) in T-matrix form and by inverting the matrix and substituting into (7.206) show that

$$\begin{aligned}\phi(N; L, K) = {} & [\omega(0,1) - 2\omega(1,1) + \omega(2,1)]\Phi^{(0,1)}(1; L, K) \\ & + [\omega(1,1) - 2\omega(2,1) + \omega(3,1)]\Phi^{(1,1)}(2; L, K) \\ & + [\omega(2,1) - 2\omega(3,1)]\Phi^{(2,1)}(3; L, K) \\ & + \omega(3,1)\Phi^{(3,1)}(4; L, K)\end{aligned}$$

correctly reproduces the terms of the cumulant expansion up to $O(K^3)$.

8. Dimer Assemblies

8.1 Introduction

In this volume and in Volume 1 we have been mainly concerned with lattice models where each microsystem occupies one site and where two microsystems can occupy any pair of sites. In such systems there can be no provision for size and shape effects although, as in the water models of Bell and Lavis (1970) and Bell (1972) (see Volume 1, Sect. 7.5) directional bonding can be included. One way to modify this type of lattice fluid model is to introduce an infinite repulsion between microsystems extending as far as a certain range of neighbours; such models are called *exclusion models*. The hard-square and hard-hexagon models (see Sect. 4.8) are of this type.

A mathematically similar approach is to suppose that the microsystem itself is a connected subgraph of the lattice $\mathcal{N}$ (see Appendix A.7). An *arrangement* of microsystems on the lattice then consists of a subgraph each of whose components is a microsystem. A *covering* is an arrangement which is a *spanning graph*, that is one in which every lattice site is a vertex of a microsystem. In this context a microsystem with a single vertex and no lines is called a *monomer* and a microsystem consisting of a single line and two vertices of valency one is a *dimer*. In this chapter we are concerned with *pure dimer assemblies*, for which each arrangement or *configuration* is a covering of the lattice consisting entirely of dimers. For a lattice $\mathcal{N}$ of N sites, it is clear that a dimer covering is possible only when N is even, a situation henceforth assumed, and then the number of dimers in the covering is $M = \frac{1}{2}N$.

Figure 8.1 shows a pure dimer configuration on the square lattice. Suppose there are M_1 horizontal and M_2 vertical dimers and that their respective fugacities are $\mathfrak{z}_1$ and $\mathfrak{z}_2$. Then, if $\Omega(N; M_1, M_2)$ is the number of dimer configurations on the lattice for given values of M_1 and M_2, the partition function is

$$Z(N, ; \mathfrak{z}_1, \mathfrak{z}_2) = \sum_{\{M_1+M_2=\frac{1}{2}N\}} \Omega(N; M_1, M_2)\mathfrak{z}_1^{M_1}\mathfrak{z}_2^{M_2}. \tag{8.1}$$

In the case of the triangular and honeycomb lattices there are three dimer species, corresponding to the three principle lattice directions and

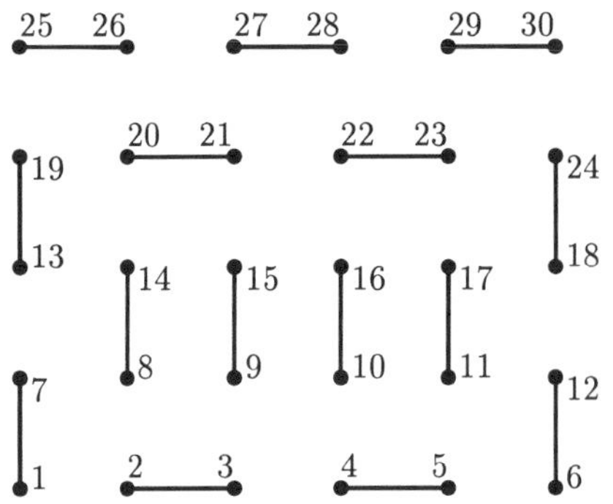

Fig. 8.1. A pure dimer configuration on the square lattice.

$$Z(N;\mathfrak{z}_1,\mathfrak{z}_2,\mathfrak{z}_3)=\sum_{\{M_1+M_2+M_3=\frac{1}{2}N\}}\Omega(N;M_1,M_2,M_3)\mathfrak{z}_1^{M_1}\mathfrak{z}_2^{M_2}\mathfrak{z}_3^{M_3}\,. \tag{8.2}$$

The fugacities can have various physical interpretations which will be discussed later in the chapter.

In Volume 1 exact results for some $d = 2$ partition functions were obtained using transformation methods (Chap. 8) and transfer matrix methods (Chap. 10). Further use of transfer matrix methods for obtaining exact results are given in Chap. 5 of this volume. In this chapter we describe a method for evaluating exactly the partition functions (8.1) and (8.2). The square lattice case was first solved by Temperley and Fisher (1961), Fisher (1961) and Kasteleyn (1961). The method described in this chapter can be applied to any regular two-dimensional lattice. In Sect. 8.4 it will be shown that an alternative solution for the Ising model can be derived by transformation to an equivalent dimer system.

8.2 The Dimer Partition Function

Let $\Pi_0 = (1, 2, \ldots, N)$ be some labelling of the sites of $\mathcal{N}$ so that the partition into pairs $(1, 2), (3, 4), \ldots, (N-1, N)$, with (i, j) being a nearest-neighbour pair occupied by a dimer, represents a possible dimer covering of $\mathcal{N}$. In the case of the square lattice with an even number of sites in each horizontal row, Π_0 could be a labelling from left to right along successive rows, with the corresponding covering having all the dimers horizontal. Let $\Pi = (\pi_1, \pi_2, \ldots, \pi_N)$ be some permutation of Π_0 with the properties

$$\pi_1 < \pi_2\,, \qquad \pi_3 < \pi_4\,, \ldots\ldots, \pi_{N-1} < \pi_N\,, \tag{8.3}$$

$$\pi_1 < \pi_3 < \pi_5 < \cdots < \pi_{N-1}\,. \tag{8.4}$$

Let $\boldsymbol{Q}$ be an $N \times N$ matrix with elements $\langle i|\boldsymbol{Q}|j\rangle = q(i, j)$ satisfying the conditions

$$q(i,j) = \begin{cases} q(j,i) > 0\,, & i \text{ and } j \text{ a nearest-neighbour pair,} \\ 0\,, & i \text{ and } j \text{ not a nearest-neighbour pair.} \end{cases} \tag{8.5}$$

Then $(\pi_1, \pi_2), (\pi_3, \pi_4), \ldots, (\pi_{N-1}, \pi_N)$ denotes a dimer covering of $\mathcal{N}$ if and only if

$$q(\pi_1, \pi_2)q(\pi_3, \pi_4) \cdots q(\pi_{N-1}, \pi_N) \neq 0\,. \tag{8.6}$$

For the matrix $\boldsymbol{Q}$ we define the *dimer partition function*

$$Z(N; \boldsymbol{Q}) = \sum_{\{\Pi\}} q(\pi_1, \pi_2)q(\pi_3, \pi_4) \cdots q(\pi_{N-1}, \pi_N)\,. \tag{8.7}$$

In view of (8.6) a term in the summation in (8.7) not corresponding to a dimer covering is zero and the summation can therefore be restricted to dimer coverings. The dimer partition functions (8.1) and (8.2) are special cases of (8.7) where $q(i,j)$ is equal to the same fugacity value for all parallel dimers in any particular principal lattice direction. In Appendix A.10 we give some mathematical results which can be used to evaluate the dimer partition function (8.7) for any regular two-dimensional lattice. These results can be summarized in two theorems:

Theorem 8.2.1. *There exists an antisymmetric matrix* $\boldsymbol{Z}$ *whose elements are related to those of* $\boldsymbol{Q}$ *by*

$$z(i,j) = \pm q(i,j)\,, \qquad z(j,i) = \mp q(i,j)\,, \tag{8.8}$$

for a particular choice of signs, such that

$$\mathrm{Det}\{\boldsymbol{Z}\} = [Z(N; \boldsymbol{Q})]^2. \tag{8.9}$$

In Appendix A.10 we show that this theorem follows from two results proved by Lieb and Loss (1993). The evaluation of the dimer partition function for a particular choice of fugacities $q(i,j)$ is thus reduced to:

(i) the determination of the choice of signs for the elements of $\boldsymbol{Z}$;

(ii) the evaluation of the determinant of $\boldsymbol{Z}$.

The first step in achieving (i) is to give a direction to every nearest-neighbour pair of the lattice by adding an arrow (see Appendix A.7.7). For any closed polygon on the lattice made up of sites and nearest-neighbour pairs, an arrow on a side is either clockwise or anticlockwise and a polygon is said to be *clockwise odd* if its sides contain an odd number of clockwise arrows. The *elementary polygons* are those enclosing single faces of the lattice. In Appendix A.10 we show that:

Theorem 8.2.2. *If arrows are assigned to the nearest-neighbour pairs of* $\mathcal{N}$ *in such a way that every elementary polygon is clockwise odd then the assignment of signs to the elements of* $z(i,j)$ *such that* $z(i,j) > 0$ *or* $z(i,j) < 0$ *according as the arrow along* (i,j) *is in the direction* $i \to j$ *or* $j \to i$, *is that which satisfies (8.9).*

This result is established using the relationship between the dimer partition function and the Pfaffian of the matrix $\boldsymbol{Z}$.

Step (ii) is now the evaluation of the determinant of the $N \times N$ matrix $\boldsymbol{Z}$. Since N is large this is not an obviously simple task. However, in the case where, as in (8.1) and (8.2), the fugacities are the same along each principle lattice direction and periodic boundary conditions are applied, the signs of the elements of $\boldsymbol{Z}$ can be chosen periodically. Then $\boldsymbol{Z}$ becomes a cyclic matrix and in Appendix A.11 we describe a procedure evaluating the determinant of a cyclic $N \times N$ matrix in the limit $N \to \infty$. We now consider the cases of the square and honeycomb lattices.

8.2.1 The Square Lattice Case

In this case $q(i,j) = \mathfrak{z}_1$ and $q(i,j) = \mathfrak{z}_2$ for all horizontal and vertical nearest-neighbour pairs respectively, with corresponding elements of $\boldsymbol{Z}$, $z(i,j) = \pm\mathfrak{z}_1$ and $z(i,j) = \pm\mathfrak{z}_2$. The signs are chosen to match an arrow configuration which makes every elementary square polygon clockwise odd. For $\text{Det}\{\boldsymbol{Z}\}$ to be evaluated the arrow assignment must also be periodic and a suitable one for the square lattice is shown in Fig. 8.2.

Let the nearest-neighbour lattice distance be a. Then disregarding boundary effects, a translation by a multiple of a in the vertical direction takes the oriented lattice into itself. However, because of the alternating arrow sense on the vertical edges, a translation in the horizontal direction must be by an even multiple of a in order to take the oriented lattice into itself. The unit cell of the oriented lattice can be taken to be a $2a \times a$ rectangle, containing a horizontal nearest-neighbour pair of sites. The two sites in any unit cell can be labelled by the letters L and R (left and right respectively). Such a cell is shown in Fig. 8.3. The aim now, following the general lines of Montroll (1964), is to express $\boldsymbol{Z}$ as a cyclic block matrix, as defined in Appendix A.11. We first adopt a new method of indexing the lattice sites. Suppose that the lattice consists of N_2 rows each of $2N_1$ sites, so that there are N_2 rows each of N_1 unit cells. The two sites of a unit cell are labelled $(\boldsymbol{j}, \text{L})$ and $(\boldsymbol{j}, \text{R})$, where

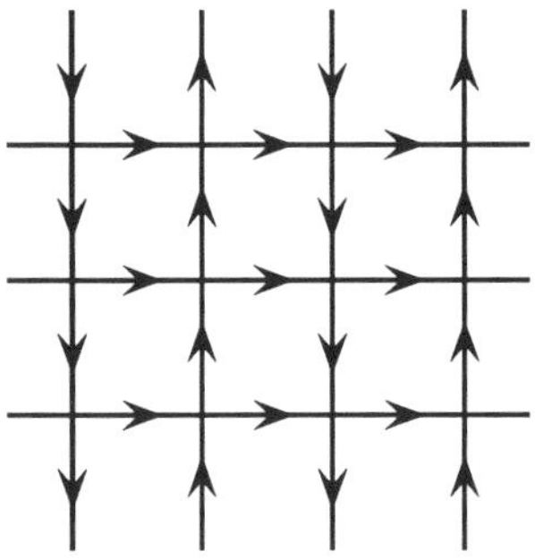

Fig. 8.2. An arrow assignment for the square lattice.

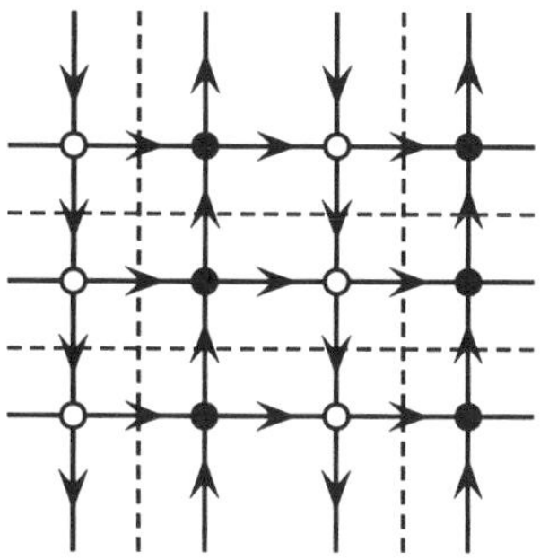

Fig. 8.3. A unit cell and sites in the neighbouring unit cells. The cell boundaries are marked by broken lines and L and R sites are represented by • and ∘ respectively.

$\boldsymbol{j} = (j_1, j_2)$, with j_1 and j_2 the integer coordinates of the cell. Now define the 2×2 block

$$\boldsymbol{z}(\boldsymbol{j}, \boldsymbol{\ell}) = \begin{pmatrix} z(\boldsymbol{j}, \mathrm{L}; \boldsymbol{\ell}, \mathrm{L}) & z(\boldsymbol{j}, \mathrm{L}; \boldsymbol{\ell}, \mathrm{R}) \\ z(\boldsymbol{j}, \mathrm{R}; \boldsymbol{\ell}, \mathrm{L}) & z(\boldsymbol{j}, \mathrm{R}; \boldsymbol{\ell}, \mathrm{R}) \end{pmatrix} . \tag{8.10}$$

Since all cells are identical, $\boldsymbol{z}(\boldsymbol{j}, \boldsymbol{\ell})$ depends only on the relative positions of $\boldsymbol{j}$ and $\boldsymbol{\ell}$ and thus

$$\boldsymbol{z}(\boldsymbol{j}, \boldsymbol{\ell}) = \boldsymbol{z}(\boldsymbol{\ell} - \boldsymbol{j}) = \boldsymbol{z}(\ell_1 - j_1, \ell_2 - j_2) . \tag{8.11}$$

From (8.5) and (8.8), $\boldsymbol{z}(\boldsymbol{j}, \boldsymbol{\ell}) \neq \boldsymbol{0}$ only if there are nearest-neighbour links between sites in the cells $\boldsymbol{j}$ and $\boldsymbol{\ell}$. From Fig. 8.3 it can be seen that the only non-zero blocks are

$$\begin{aligned} &\boldsymbol{z}(0,0) = \begin{pmatrix} 0 & \mathfrak{z}_1 \\ -\mathfrak{z}_1 & 0 \end{pmatrix} , \\ &\boldsymbol{z}(1,0) = \begin{pmatrix} 0 & 0 \\ \mathfrak{z}_1 & 0 \end{pmatrix} , \qquad \boldsymbol{z}(-1,0) = \begin{pmatrix} 0 & -\mathfrak{z}_1 \\ 0 & 0 \end{pmatrix} , \\ &\boldsymbol{z}(0,1) = \begin{pmatrix} \mathfrak{z}_2 & 0 \\ 0 & -\mathfrak{z}_2 \end{pmatrix} , \qquad \boldsymbol{z}(0,-1) = \begin{pmatrix} -\mathfrak{z}_2 & 0 \\ 0 & \mathfrak{z}_2 \end{pmatrix} . \end{aligned} \tag{8.12}$$

Periodic boundary conditions are now imposed on the right and left and top and bottom of the lattice so that

$$\boldsymbol{z}(\boldsymbol{j} + \boldsymbol{n}) = \boldsymbol{z}(\boldsymbol{j}) , \qquad \boldsymbol{n} = (n_1 N_1, n_2 N_2) , \tag{8.13}$$

where n_1 and n_2 are any integers. Equations (8.11) and (8.13) imply that conditions (A.321) are satisfied and that $\boldsymbol{Z}$ is a cyclic block matrix with blocks of dimension 2×2. The determinant of $\boldsymbol{Z}$ can thus be evaluated using the method of Appendix A.11. The disadvantage is that extra 'torus-winding' terms have been introduced into $\boldsymbol{Z}$ and since the discussion in Appendix A.10 depends on the lattice being planar this approach cannot be used for finite lattices. However, it may be assumed that the boundary conditions will not affect densities in the thermodynamic limit $N_1 \to \infty$ and $N_2 \to \infty$.[1] Using (A.323) with $\theta_1 = 2\pi k_1/N_1$ and $\theta_2 = 2\pi k_2/N_2$,

$$\begin{aligned}\boldsymbol{\zeta}(\theta_1,\theta_2) &= \boldsymbol{z}(0,0) + \boldsymbol{z}(1,0)\exp(\mathrm{i}\theta_1) + \boldsymbol{z}(-1,0)\exp(-\mathrm{i}\theta_1) \\ &\qquad + \boldsymbol{z}(0,1)\exp(\mathrm{i}\theta_2) + \boldsymbol{z}(0,-1)\exp(-\mathrm{i}\theta_2) \\ &= \begin{pmatrix} \mathfrak{z}_2\exp(\mathrm{i}\theta_2) - \mathfrak{z}_2\exp(-\mathrm{i}\theta_2) & \mathfrak{z}_1 - \mathfrak{z}_1\exp(-\mathrm{i}\theta_1) \\ -\mathfrak{z}_1 + \mathfrak{z}_1\exp(\mathrm{i}\theta_1) & \mathfrak{z}_2\exp(-\mathrm{i}\theta_2) - \mathfrak{z}_2\exp(\mathrm{i}\theta_2) \end{pmatrix}\end{aligned} \tag{8.14}$$

and thus

$$\mathrm{Det}\{\boldsymbol{\zeta}(\theta_1,\theta_2)\} = 4\left[\mathfrak{z}_1^2\sin^2\left(\tfrac{1}{2}\theta_1\right) + \mathfrak{z}_2^2\sin^2(\theta_2)\right]. \tag{8.15}$$

From (8.9), (8.15) and (A.325) the dimensionless free energy $\phi(\mathfrak{z}_1,\mathfrak{z}_2)$ per lattice site in the limit $N = 2N_1N_2 \to \infty$ is given by

$$\begin{aligned}\phi(\mathfrak{z}_1,\mathfrak{z}_2) &= -\lim_{N_1\to\infty}\lim_{N_2\to\infty}\frac{\ln Z(Z;\mathfrak{z}_1,\mathfrak{z}_2)}{2N_1N_2} \\ &= -\frac{1}{4}\lim_{N_1\to\infty}\lim_{N_2\to\infty}\frac{\ln[\mathrm{Det}\{\boldsymbol{Z}(\mathfrak{z}_1,\mathfrak{z}_2)\}]}{N_1N_2} \\ &= -\frac{1}{(4\pi)^2}\int_0^{2\pi}\mathrm{d}\theta_1\int_0^{2\pi}\mathrm{d}\theta_2\ln\left\{4\left[\mathfrak{z}_1^2\sin^2\left(\tfrac{1}{2}\theta_1\right) + \mathfrak{z}_2^2\sin^2(\theta_2)\right]\right\}.\end{aligned} \tag{8.16}$$

With the change of variable $\theta_2 \to \frac{1}{2}\theta_2$, this result can be expressed in the more symmetric form

$$\begin{aligned}\phi(\mathfrak{z}_1,\mathfrak{z}_2) &= -\tfrac{1}{2}\ln(2) \\ &\qquad + \frac{1}{(4\pi)^2}\int_0^{2\pi}\mathrm{d}\theta_1\int_0^{2\pi}\mathrm{d}\theta_2\ln[g(\mathfrak{z}_1,\mathfrak{z}_2;\cos\theta_1,\cos\theta_2)],\end{aligned} \tag{8.17}$$

where

$$g(\mathfrak{z}_1,\mathfrak{z}_2;x_1,x_2) = [\mathfrak{z}_1^2 + \mathfrak{z}_2^2 - x_1\mathfrak{z}_1^2 - x_2\mathfrak{z}_2^2]^{-1}. \tag{8.18}$$

In the thermodynamic limit the numbers ρ_1 and ρ_2 of horizontal and vertical dimers respectively, per lattice site, are given by

$$\rho_s = -\mathfrak{z}_s\frac{\partial\phi}{\partial\mathfrak{z}_s} \tag{8.19}$$

$$= \frac{1}{2\pi^2}\int_0^{\pi}\mathrm{d}\theta_1\int_0^{\pi}\mathrm{d}\theta_2\, g(\mathfrak{z}_1,\mathfrak{z}_2;\cos\theta_1,\cos\theta_2), \qquad s = 1,2. \tag{8.20}$$

[1] This is proved by Montroll (1964), Sect. 4.8, using Ledermann's theorem.

It can be seen, from these formulae, that $\rho_1 + \rho_2 = \frac{1}{2}$, as is necessary for consistency. Using the standard result

$$\int_0^\pi \frac{\mathrm{d}\theta}{y - x\cos\theta} = \frac{\pi}{\sqrt{y^2 - x^2}}, \tag{8.21}$$

this gives

$$\begin{aligned}\rho_s &= \frac{1}{2\pi}\int_0^\pi \frac{\eta_s^2(1 - cos\theta)}{\sqrt{(\eta_2^2 + 1 - \eta_s^2\cos\theta)^2 - 1}}\mathrm{d}\theta \\ &= \int_0^\pi \frac{\eta_s\sqrt{1-\cos\theta}}{\sqrt{\eta_s^2 + 2 - \eta_s^2\cos\theta}}\mathrm{d}\theta\,, \qquad s = 1, 2\,,\end{aligned} \tag{8.22}$$

where $\eta_1 = \mathfrak{z}_1/\mathfrak{z}_2$ and $\eta_2 = 1/\eta_1 = \mathfrak{z}_2/\mathfrak{z}_1$. Using the substitution $u = 1 - \cos\theta$ this gives

$$\begin{aligned}\rho_s &= \frac{1}{2\pi}\int_0^2 \frac{\mathrm{d}u}{\sqrt{(2-u)(2\eta_s^{-2} + u)}} \\ &= \frac{\arctan(\eta_s)}{\pi}\,.\end{aligned} \tag{8.23}$$

This remarkably simple result is due to Fisher (1961). When $\mathfrak{z}_1 = 0$ all the dimers are vertical $Z(N; 0, \mathfrak{z}_2) = \mathfrak{z}_2^{N/2}$, $\phi(0, \mathfrak{z}_2) = -\frac{1}{2}\ln(\mathfrak{z}_2)$ and similarly, when $\mathfrak{z}_2 = 0$, all the dimers are horizontal and $\phi(\mathfrak{z}_1, 0) = -\frac{1}{2}\ln(\mathfrak{z}_1)$. Using these results with (8.19) and (8.23), we obtain the alternative forms

$$\phi(\mathfrak{z}_1, \mathfrak{z}_2) = -\tfrac{1}{2}\ln(\mathfrak{z}_2) - \frac{1}{\pi}\int_0^{\mathfrak{z}_1/\mathfrak{z}_2} \frac{\arctan(\eta)\mathrm{d}\eta}{\eta}, \tag{8.24}$$

$$\phi(\mathfrak{z}_1, \mathfrak{z}_2) = -\tfrac{1}{2}\ln(\mathfrak{z}_1) - \frac{1}{\pi}\int_0^{\mathfrak{z}_2/\mathfrak{z}_1} \frac{\arctan(\eta)\mathrm{d}\eta}{\eta}. \tag{8.25}$$

It follows from these equations that $\phi(\mathfrak{z}_1, \mathfrak{z}_2)$ is a continuous function of $\mathfrak{z}_1$ and $\mathfrak{z}_2$ with continuous derivatives of all orders except at $\mathfrak{z}_1 = \mathfrak{z}_2 = 0$. There can, therefore, be no phase transitions. From (8.1) the number of ways $\Omega(N)$ of distributing the $\frac{1}{2}N$ dimers on the lattice of N sites is

$$\Omega(N) = \sum_{\{M_1 + M_2 = \frac{1}{2}N\}} \Omega(N, M_1, M_2) = Z(N; 1, 1)\,. \tag{8.26}$$

For large N, from (8.24) or (8.25),

$$\frac{\ln\Omega(N)}{N} = -\phi(1,1) = \frac{1}{\pi}\int_0^1 \frac{\arctan(\eta)\mathrm{d}\eta}{\eta}\,. \tag{8.27}$$

Expanding the integrand in a series, valid for $\eta \le 1$, and integrating gives

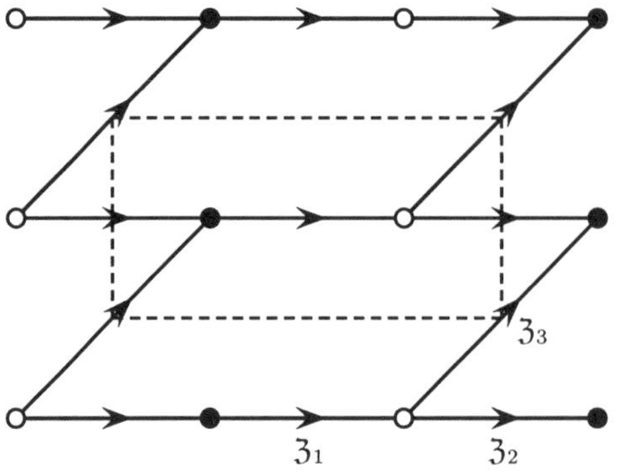

Fig. 8.4. An arrow assignment for the honeycomb lattice. The unit cell boundaries are marked by broken lines and L and R sites are represented by • and ∘ respectively.

$$\frac{\ln \Omega(N)}{N} = \frac{1}{\pi}\sum_{k=0}^{\infty}\frac{(-1)^k}{(2k+1)^2} = \frac{\mathrm{G}}{\pi}, \tag{8.28}$$

where G = 0.915966 is *Catalan's constant* (Abramowitz and Segun 1965).

The partition function for the finite $2N_1 \times N_2$ lattice was obtained by Fisher (1961) and Kasteleyn (1961). For even N_2 this takes the form

$$Z(N;\mathfrak{z}_1,\mathfrak{z}_2) = 2^{\frac{1}{2}N}\prod_{k_1=1}^{N_1}\prod_{k_2=1}^{\frac{1}{2}N_2}\left[\mathfrak{z}_1^2\cos^2\left(\frac{k_1\pi}{2N_1+1}\right)+\mathfrak{z}_2^2\cos^2\left(\frac{k_2\pi}{N_2+1}\right)\right] \tag{8.29}$$

in the present notation. Fisher used an alternative combinatorial method to evaluate the Pfaffian, while Kastelyn, by some ingenious algebra, was able to dispense with the periodic condition (8.13). The formula (8.17) for the free energy per site follows from (8.29) in the thermodynamic limit, confirming that in this limit intensive quantities are independent of the original boundary conditions. By putting $\mathfrak{z}_1 = \mathfrak{z}_2 = 1$ in (8.29), values of $\Omega(N)$ for finite lattices can be derived. The number of ways of arranging 32 symmetric dominos on a chess board corresponds to taking $2N_1 = N_2 = 8$ and gives $\Omega(64) = 12{,}988{,}816$ which is surprisingly large. This gives $\ln\Omega(64)/64 = 0.25593$, to be compared with the value 0.291561 obtained from (8.28) in the thermodynamic limit.

8.2.2 The Honeycomb Lattice Case

An arrow assignment for the honeycomb lattice (Montroll 1964) is shown in Fig. 8.4. There are clockwise arrows on three of the six edges forming an elementary polygon so that the clockwise odd condition is satisfied. The broken lines denote the boundary of a unit cell containing a nearest-neighbour pair of sites. The notation for the representation of L and R sites is the same as in Fig. 8.3. To present the cell structure as clearly as possible the usual geometry of the lattice has been distorted, although parallel edges remain parallel. This does not affect the evaluation of the partition function which depends only on the topology of the vertex connections and the fugacities assigned to the lattice directions. These are shown on the bottom right of the figure.

The lattice topology determines an important conservation property of the honeycomb dimer configurations. It can be seen from Fig. 8.4 that the occupation of any member of a row of horizontal edges by a dimer implies that exactly one of the rows of diagonal edges below is unoccupied. So if there are m_1 and m_2 dimers of species 1 and 2 respectively in a given row of horizontal edges then there will be $m_3 = N_1 - m_1 - m_2$ dimers in the row of diagonal edges above and the same number m_3 in the row of diagonal edges below. The number m_3 is thus the same for all rows of diagonal edges. Since the three sets of parallel edges indexed by the numbers 1, 2 and 3 are topologically equivalent a similar conservation rule applies to dimers of species 1 and 2. In the geometry of Fig. 8.4 there is thus the same number of dimers in each column of type 1 edges and in each column of type 2 edges. It can be seen from Fig. 8.1 that similar conservation rules do not apply to the square lattice. The reason is that there the number of vacant edges in a row of vertical edges depends not only on the number of dimers in the row of horizontal edges immediately below, but also on the number of dimers in the row of vertical edges below that.

The analysis of the honeycomb lattice is similar to that for the square lattice, although $\boldsymbol{z}(0,\pm 1) = \mathbf{0}$, since, from Fig. 8.4, there are no nearest-neighbour links between the sites of a given cell and those of the cells immediately above and below it. However, there are diagonal edges connecting the R site of a given cell with the L site of the cell above and to the right and the L site to the R site of cell below and to the left. The non-zero blocks are, therefore,

$$\begin{aligned}
&\boldsymbol{z}(0,0) = \begin{pmatrix} 0 & \mathfrak{z}_1 \\ -\mathfrak{z}_1 & 0 \end{pmatrix}, \\
&\boldsymbol{z}(1,0) = \begin{pmatrix} 0 & 0 \\ \mathfrak{z}_2 & 0 \end{pmatrix}, \qquad \boldsymbol{z}(-1,0) = \begin{pmatrix} 0 & -\mathfrak{z}_2 \\ 0 & 0 \end{pmatrix}, \\
&\boldsymbol{z}(1,1) = \begin{pmatrix} 0 & 0 \\ \mathfrak{z}_3 & 0 \end{pmatrix}, \qquad \boldsymbol{z}(-1,-1) = \begin{pmatrix} 0 & -\mathfrak{z}_3 \\ 0 & 0 \end{pmatrix}.
\end{aligned} \tag{8.30}$$

From (A.323),

$$\begin{aligned}
\boldsymbol{\zeta}(\theta_1,\theta_2) = {} & \boldsymbol{z}(0,0) + \boldsymbol{z}(1,0)\exp(\mathrm{i}\theta_1) + \boldsymbol{z}(-1,0)\exp(-\mathrm{i}\theta_1) \\
& + \boldsymbol{z}(1,1)\exp(\mathrm{i}[\theta_1+\theta_2]) + \boldsymbol{z}(-1,-1)\exp(-\mathrm{i}[\theta_1+\theta_2]),
\end{aligned} \tag{8.31}$$

and thus

$$\begin{aligned}
\mathrm{Det}\{\boldsymbol{\zeta}(\theta_1,\theta_2)\} = {} & \mathfrak{z}_1^2 + \mathfrak{z}_2^2 + \mathfrak{z}_3^2 - 2\mathfrak{z}_1\mathfrak{z}_2\cos\theta_1 \\
& + 2\mathfrak{z}_2\mathfrak{z}_3\cos\theta_2 - 2\mathfrak{z}_1\mathfrak{z}_3\cos(\theta_1+\theta_2).
\end{aligned} \tag{8.32}$$

As in the case of the square lattice this then yields the dimensionless free energy $\phi(\mathfrak{z}_1,\mathfrak{z}_2,\mathfrak{z}_3)$ per lattice site in the thermodynamic limit. We obtain

$$\phi(\mathfrak{z}_1,\mathfrak{z}_2,\mathfrak{z}_3) = -\frac{1}{(4\pi)^2}\int_0^{2\pi} d\theta_1 \int_0^{2\pi} d\theta_2 \ln[\mathfrak{z}_1^2+\mathfrak{z}_2^2+\mathfrak{z}_3^2 - 2\mathfrak{z}_1\mathfrak{z}_2\cos\theta_1 + 2\mathfrak{z}_2\mathfrak{z}_3\cos\theta_2 - 2\mathfrak{z}_1\mathfrak{z}_3\cos(\theta_1+\theta_2)]\,. \qquad (8.33)$$

With the change of variable $\theta_1 \to \pi - \theta_1$ this becomes

$$\phi(\mathfrak{z}_1,\mathfrak{z}_2,\mathfrak{z}_3) = -\frac{1}{(4\pi)^2}\int_0^{2\pi} d\theta_1 \int_0^{2\pi} d\theta_2 \ln[\mathfrak{z}_1^2+\mathfrak{z}_2^2+\mathfrak{z}_3^2 + 2\mathfrak{z}_1\mathfrak{z}_2\cos\theta_1 + 2\mathfrak{z}_2\mathfrak{z}_3\cos\theta_2 + 2\mathfrak{z}_1\mathfrak{z}_3\cos(\theta_2-\theta_1)]\,, \qquad (8.34)$$

a result given, without derivation, by Wu (1968). As would be expected, this expression is symmetrical in the fugacities, which can be permuted without altering the value of $\phi(\mathfrak{z}_1,\mathfrak{z}_2,\mathfrak{z}_3)$.

So far the analysis has been very similar to that for the square lattice. However, there are important differences in the thermodynamics of the two systems, as appears when (8.34) is transformed to a single integral expression. It is not difficult to show that the argument of the logarithm in the integrand of (8.34) can be expressed in the form

$$Y(\mathfrak{z}_1,\mathfrak{z}_2,\mathfrak{z}_3;\cos\theta_1) + X(\mathfrak{z}_1,\mathfrak{z}_2,\mathfrak{z}_3;\cos\theta_1)\cos[\theta_2 - \gamma(\mathfrak{z}_1,\mathfrak{z}_2;\cos\theta_1)]\,, \qquad (8.35)$$

where

$$Y(\mathfrak{z}_1,\mathfrak{z}_2,\mathfrak{z}_3;x) = W(\mathfrak{z}_1,\mathfrak{z}_2;x) + \mathfrak{z}_3^2\,, \qquad (8.36)$$

$$X(\mathfrak{z}_1,\mathfrak{z}_2,\mathfrak{z}_3;x) = 2\mathfrak{z}_3\sqrt{W(\mathfrak{z}_1,\mathfrak{z}_2;x)}\,, \qquad (8.37)$$

$$\gamma(\mathfrak{z}_1,\mathfrak{z}_2;x) = \arctan\left(\frac{\mathfrak{z}_1\sqrt{1-x^2}}{\mathfrak{z}_2+\mathfrak{z}_1\,x}\right), \qquad (8.38)$$

$$W(\mathfrak{z}_1,\mathfrak{z}_2;x) = \mathfrak{z}_1^2+\mathfrak{z}_2^2+2\mathfrak{z}_1\mathfrak{z}_2\,x\,, \qquad (8.39)$$

and

$$Y^2(\mathfrak{z}_1,\mathfrak{z}_2,\mathfrak{z}_3;x) - X^2(\mathfrak{z}_1,\mathfrak{z}_2,\mathfrak{z}_3;x) = [W(\mathfrak{z}_1,\mathfrak{z}_2;x) - \mathfrak{z}_3^2]^2\,. \qquad (8.40)$$

The integration with respect to θ_2 in (8.34) can now be performed, using the standard formula

$$\int_0^{\pi} \ln(y - x\cos\omega)d\omega = \pi\ln\left[\tfrac{1}{2}\left(y+\sqrt{y^2-x^2}\right)\right], \qquad x < y,$$

quoted in Volume 1, Sect. 8.11, to give

$$\begin{aligned}\int_0^{2\pi} \ln[Y + X\cos(\theta_2 - \gamma)]\mathrm{d}\theta_2 &= \int_{\pi-\gamma}^{3\pi-\gamma} \ln(Y - X\cos\omega)\mathrm{d}\omega \\ &= \int_0^{2\pi} \ln(Y - X\cos\omega)\mathrm{d}\omega \\ &= 2\pi \ln\left[\tfrac{1}{2}\left(Y + \sqrt{Y^2 - X^2}\right)\right] \\ &= 2\pi \ln\left[\tfrac{1}{2}\left(W + \mathfrak{z}_3^2 + |W - \mathfrak{z}_3^2|\right)\right]. \end{aligned} \quad (8.41)$$

This last expression depends on θ_1 through $\cos\theta_1$ and so the range of integration can be halved. Then, making the substitution $\theta_1 = \pi - \theta$, we have

$$\begin{aligned}\phi(\mathfrak{z}_1, \mathfrak{z}_2, \mathfrak{z}_3) = -\frac{1}{(4\pi)} \int_0^{\pi} \ln\{\tfrac{1}{2}[W(\mathfrak{z}_1, \mathfrak{z}_2; -\cos\theta) + \mathfrak{z}_3^2 \\ + |W(\mathfrak{z}_1, \mathfrak{z}_2; -\cos\theta) - \mathfrak{z}_3^2|]\}\mathrm{d}\theta. \end{aligned} \quad (8.42)$$

The fugacities $\mathfrak{z}_1$, $\mathfrak{z}_2$ and $\mathfrak{z}_3$ are non-negative and, in order to understand the behaviour of this model, we need to determine the analytic nature of $\phi(\mathfrak{z}_1, \mathfrak{z}_2, \mathfrak{z}_3)$ in this octant of the $(\mathfrak{z}_1, \mathfrak{z}_2, \mathfrak{z}_3)$ space. This octant can be divided into four regions:

(i) $\mathfrak{z}_1 > \mathfrak{z}_2 + \mathfrak{z}_3$, $\mathfrak{z}_2 < \mathfrak{z}_3 + \mathfrak{z}_1$, $\mathfrak{z}_3 < \mathfrak{z}_1 + \mathfrak{z}_2$.

(ii) $\mathfrak{z}_2 > \mathfrak{z}_3 + \mathfrak{z}_1$, $\mathfrak{z}_3 < \mathfrak{z}_1 + \mathfrak{z}_2$, $\mathfrak{z}_1 < \mathfrak{z}_2 + \mathfrak{z}_3$.

(iii) $\mathfrak{z}_3 > \mathfrak{z}_1 + \mathfrak{z}_2$, $\mathfrak{z}_1 < \mathfrak{z}_2 + \mathfrak{z}_3$, $\mathfrak{z}_2 < \mathfrak{z}_3 + \mathfrak{z}_1$.

(iv) $\mathfrak{z}_1 < \mathfrak{z}_2 + \mathfrak{z}_3$, $\mathfrak{z}_2 < \mathfrak{z}_3 + \mathfrak{z}_1$, $\mathfrak{z}_3 < \mathfrak{z}_1 + \mathfrak{z}_2$.

The boundaries between regions (i), (ii) and (iii) respectively and region (iv) are the surfaces $\mathcal{C}_1$, $\mathcal{C}_2$ and $\mathcal{C}_3$ given by

$$\mathcal{C}_1: \qquad \mathfrak{z}_1 = \mathfrak{z}_2 + \mathfrak{z}_3\,, \quad (8.43)$$

$$\mathcal{C}_2: \qquad \mathfrak{z}_2 = \mathfrak{z}_3 + \mathfrak{z}_1\,, \quad (8.44)$$

$$\mathcal{C}_3: \qquad \mathfrak{z}_3 = \mathfrak{z}_1 + \mathfrak{z}_2\,. \quad (8.45)$$

For all θ in the range $[0, \pi]$, $\mathfrak{z}_3^2 > W(\mathfrak{z}_1, \mathfrak{z}_2; -\cos\theta)$ in region (iii) and, from (8.19) and (8.42),

$$\phi(\mathfrak{z}_1, \mathfrak{z}_2, \mathfrak{z}_3) = -\tfrac{1}{2}\ln\mathfrak{z}_3\,, \qquad \rho_1 = \rho_2 = 0\,, \qquad \rho_3 = \tfrac{1}{2}\,. \quad (8.46)$$

So in region (iii) the system is in a 'frozen' state of perfect order, with all dimers parallel in the direction associated with the fugacity $\mathfrak{z}_3$. By symmetry it follows that states of perfect order also exist with the dimers aligned in the other lattice directions. In region (i)

$$\phi(\mathfrak{z}_1, \mathfrak{z}_2, \mathfrak{z}_3) = -\tfrac{1}{2}\ln\mathfrak{z}_1\,, \qquad \rho_1 = \tfrac{1}{2}\,, \qquad \rho_2 = \rho_3 = 0\,, \quad (8.47)$$

and, in region (ii),

$$\phi(\mathfrak{z}_1, \mathfrak{z}_2, \mathfrak{z}_3) = -\tfrac{1}{2}\ln\mathfrak{z}_2\,, \qquad \rho_1 = \rho_3 = 0\,, \qquad \rho_2 = \tfrac{1}{2}\,. \quad (8.48)$$

These frozen configurations do not admit small (localized) perturbations. Consider, for example, the case of the system in region (iii), where, in the geometry of Fig. 8.4, all horizontal edges are unoccupied, while all edges in each row of diagonal sites are occupied giving $m_3 = N_1$ for all such rows. Now suppose that one dimer of any row of diagonal dimers is displaced to a horizontal edge. By the conservation rule discussed above, there must be a similar displacement from all other rows of diagonal dimers, so that the disturbance is propagated throughout the lattice. The situation is similar to that in an ferroelectric ground state of the six-vertex model (see Volume 1, Sect. 10.6). In fact, in the next section, we shall prove the equivalence between a honeycomb dimer model and a five-vertex ferroelectric model and confirm the absence of small ground-state perturbations. The honeycomb lattice situation contrasts with that on the square lattice, where there can be small disturbances in the basic configuration or in the corresponding configuration with all dimers vertical. Fig. 8.1 can, for instance, be regarded as showing a perturbation in a 'corner' of a larger lattice, with all other dimers on this lattice in their basic configuration positions.

Consider the case where the system is in region (iv). There is now a real value θ_0 is the range $(0, \pi)$ such that

$$\mathfrak{z}_3^2 = W(\mathfrak{z}_1, \mathfrak{z}_2; -\cos\theta_0)\,. \tag{8.49}$$

The range of integration in (8.42) can be split into two parts and, using (8.39)

$$\phi(\mathfrak{z}_1, \mathfrak{z}_2, \mathfrak{z}_3) = -\frac{\theta_0}{2\pi}\ln\mathfrak{z}_3 - \frac{1}{4\pi}\int_{\theta_0}^{\pi}\ln[\mathfrak{z}_1^2 + \mathfrak{z}_2^2 - 2\mathfrak{z}_1\mathfrak{z}_2\cos\theta]\mathrm{d}\theta\,. \tag{8.50}$$

From this result it follows that

$$\rho_3 = \frac{\theta_0}{2\pi} = \frac{1}{2\pi}\arccos\left(\frac{\mathfrak{z}_1^2 + \mathfrak{z}_2^2 - \mathfrak{z}_3^2}{2\mathfrak{z}_1\mathfrak{z}_2}\right), \tag{8.51}$$

and by symmetry

$$\rho_1 = \frac{1}{2\pi}\arccos\left(\frac{\mathfrak{z}_2^2 + \mathfrak{z}_3^2 - \mathfrak{z}_1^2}{2\mathfrak{z}_2\mathfrak{z}_3}\right), \tag{8.52}$$

$$\rho_2 = \frac{1}{2\pi}\arccos\left(\frac{\mathfrak{z}_3^2 + \mathfrak{z}_1^2 - \mathfrak{z}_2^2}{2\mathfrak{z}_3\mathfrak{z}_1}\right). \tag{8.53}$$

It can easily be verified that

$$\rho_1 + \rho_2 + \rho_3 = \tfrac{1}{2}\,. \tag{8.54}$$

From (8.39) and (8.49), as the phase point approaches the surface $\mathcal{C}_3$, $\theta_0 \to \pi$. It then follows from (8.51) that $\rho_3 \to \frac{1}{2}$ and it must be the case that $\rho_2 \to 0$ and $\rho_1 \to 0$. We can also see that $\phi(\mathfrak{z}_1, \mathfrak{z}_2, \mathfrak{z}_3)$, given by (8.50), tends to the value $-\frac{1}{2}\ln\mathfrak{z}_3$. The dimer densities and the free energy density are continuous over the boundary $\mathcal{C}_3$ between the regions (iii) and (iv). They are also continuous over the boundaries $\mathcal{C}_1$ and $\mathcal{C}_2$. From (8.46)–(8.48), in the ordered regions (i)–(iii),

$$\frac{\partial \rho_s}{\partial \mathfrak{z}_j} = 0\,, \qquad s,j = 1,2,3\,, \tag{8.55}$$

and, from (8.51)–(8.53), in the disordered region (iv),

$$\frac{\partial \rho_s}{\partial \mathfrak{z}_j} = \begin{cases} \dfrac{\mathfrak{z}_s}{\tau(\mathfrak{z}_1,\mathfrak{z}_2,\mathfrak{z}_3)}\,, & \text{if } s = j\,, \\[2ex] \dfrac{\mathfrak{z}_i^2 - \mathfrak{z}_s^2 - \mathfrak{z}_j^2}{2\mathfrak{z}_j\tau(\mathfrak{z}_1,\mathfrak{z}_2,\mathfrak{z}_3)}\,, & \text{if } s \neq j \text{ and } i \neq s,j\,, \end{cases} \tag{8.56}$$

where

$$\tau(\mathfrak{z}_1,\mathfrak{z}_2,\mathfrak{z}_3) = \pi\sqrt{(\mathfrak{z}_1+\mathfrak{z}_2-\mathfrak{z}_3)(\mathfrak{z}_2+\mathfrak{z}_3-\mathfrak{z}_1)(\mathfrak{z}_3+\mathfrak{z}_1-\mathfrak{z}_2)(\mathfrak{z}_1+\mathfrak{z}_2+\mathfrak{z}_3)}\,. \tag{8.57}$$

In view of condition (8.54) only six of these nine partial derivatives are independent. From (8.43)–(8.45) and (8.57), it can be seen that, as any one of the boundaries of the disordered region is approached from within that region, the partial derivatives given by (8.56) exhibit singular behaviour. Specifically, as surface $\mathcal{C}_1$ is approached within region (iv),

$$\frac{\partial \rho_s}{\partial \mathfrak{z}_j} \sim \frac{1}{\sqrt{\mathfrak{z}_2+\mathfrak{z}_3-\mathfrak{z}_1}}\,, \qquad s,j = 1,2,3\,. \tag{8.58}$$

The surfaces $\mathcal{C}_1$, $\mathcal{C}_2$ $\mathcal{C}_3$ are surfaces of second-order phase transitions. This result was stated, without derivation, by Kasteleyn (1963).

Consider now the case where the fugacity $\mathfrak{z}_s$ is a smooth function of a coupling $K_s = \varepsilon_s/T$, for $s = 1,2,3$ and the system is controlled by changing the temperature. The critical temperature T_c at which the system passes from the disordered state (iv) to the ordered state (i) is then given as the solution of

$$\mathfrak{z}_2(T_c) + \mathfrak{z}_3(T_c) = \mathfrak{z}_1(T_c)\,, \tag{8.59}$$

and, inside (iv) and close to the surface $\mathcal{C}_1$,

$$\mathfrak{z}_2(T) + \mathfrak{z}_3(T) - \mathfrak{z}_1(T) \simeq [\mathfrak{z}_2'(T_c) + \mathfrak{z}_3'(T_c) - \mathfrak{z}_1'(T_c)](T - T_c)\,, \tag{8.60}$$

unless

$$\mathfrak{z}_2'(T_c) + \mathfrak{z}_3'(T_c) = \mathfrak{z}_1'(T_c)\,. \tag{8.61}$$

From (1.31), (1.41) and (8.19), the internal energy u per lattice site is given by

$$u = \sum_{s=1}^{3} \varepsilon_s \frac{\partial \phi}{\partial K_s} = \sum_{s=1}^{3} \varepsilon_s \frac{\partial \phi}{\partial \mathfrak{z}_s}\frac{\mathrm{d}\mathfrak{z}_s}{\mathrm{d}K_s} = -\sum_{s=1}^{3} \varepsilon_s \rho_s \frac{\mathrm{d}\ln \mathfrak{z}_s}{\mathrm{d}K_s}\,. \tag{8.62}$$

Since the dimer densities ρ_s are continuous across the critical surfaces u must also be so and, from (8.46)–(8.48),

$$u = -\frac{1}{2}\varepsilon_s \frac{\mathrm{d}\ln \mathfrak{Z}_s}{\mathrm{d}K_s}, \qquad s = 1, 2, 3, \tag{8.63}$$

in the ordered regions (i), (ii) and (iii) respectively. The heat capacity c per lattice site, which is given by

$$c = k_\mathrm{B} \frac{\partial u}{\partial T}, \tag{8.64}$$

will involve the partial derivatives of the dimer densities with respect to the fugacities. It follows from (8.58) and (8.60) that

$$c \sim \frac{1}{\sqrt{T - T_\mathrm{c}}}, \tag{8.65}$$

as $\mathcal{C}_1$ is approached from inside region (iv), unless (8.61) is satisfied. This gives a critical exponent $\alpha = \frac{1}{2}$.

We now consider a simple interpretation of the fugacities. Suppose that each dimer carries a small electric dipole aligned with its axis and that an electric field $\mathcal{E}$ acts on the system parallel to the direction labelled 3. Given that $\mathcal{E}$ includes the magnitude of the dipole moment, thereby being measured in units of energy, and that dipole-dipole interactions are negligible in comparison with the dipole-field interaction, this model corresponds to the case $\varepsilon_1 = \varepsilon_2 = \varepsilon_3 = \mathcal{E}$ and we have

$$\mathfrak{Z}_3 = \exp(K) + \exp(-K) = 2\cosh(K). \tag{8.66}$$

In a similar way

$$\mathfrak{Z}_1 = \mathfrak{Z}_2 = 2\cosh\left(\tfrac{1}{2}K\right). \tag{8.67}$$

Since $\mathfrak{Z}_3 > \mathfrak{Z}_2 = \mathfrak{Z}_1$, a transition at a fixed value of $\mathcal{E}$ will occur on the surface $\mathcal{C}_3$ at a critical coupling $K_\mathrm{c} = \mathcal{E}/T_\mathrm{c}$ given by

$$\cosh(K_\mathrm{c}) = 2\cosh\left(\tfrac{1}{2}K_\mathrm{c}\right), \tag{8.68}$$

which can be transformed to

$$\frac{\mathcal{E}}{T_\mathrm{c}} = 2\operatorname{arccosh}\left(\frac{1+\sqrt{3}}{2}\right). \tag{8.69}$$

An equivalent relationship (again without derivation) is given by Kasteleyn (1963), who regards the dimers as polar diatomic molecules adsorbed on to a solid whose surface layer of atoms has the honeycomb form. For this model the internal energy in region (iii) is given, from (8.63), by

$$u = -\tfrac{1}{2}\mathcal{E}\tanh(K). \tag{8.70}$$

A second-order transition, where, as in this model, the heat capacity has a square-root singularity for $T > T_\mathrm{c}$ and is finite or zero for $T < T_\mathrm{c}$ has been given the name *K-type* (K standing for Kasteleyn) by Nagle et al. (1989). These authors also use the name *O-type* (O standing for Onsager) for the kind of transition, which occurs in the two-dimensional Ising model, where the heat capacity has a logarithmic singularity on each side of the critical point. These two types of transition are further discussed in Sect. 8.5.

8.3 The Modified KDP Model Equivalence

Figure 8.5(a) shows a pure dimer distribution on the honeycomb lattice. Suppose that the lattice is contracted by reducing the lengths of the diagonal edges to zero and letting the two vertices at the ends of each such edge coalesce into one. The resulting square lattice is shown in Fig. 8.5(b) and it can be seen that the possible vertex line patterns deriving from the horizontal and vertical dimers in Fig. 8.5(a) are those, in the second line of Fig. 5.1, labelled 1, 3, 4, 5 and 6. The vertex line patterns, 2, 7 and 8 are absent because they would correspond to overlapping dimers on the original honeycomb lattice. With suitable relations between dimer fugacities and vertex energies the honeycomb lattice dimer model is thus equivalent to a five-vertex model. This model can be regarded as a special case of the six-vertex model with the vertex energy $e_2 = \infty$ (see Volume 1, Chap. 10), or of the eight-vertex model with vertex energies $e_2 = e_7 = e_8 = \infty$ (Chap. 5).[2]

To avoid confusion with numerical indices we denote the energies of vertices of types 1, 3 and 4 by $e_{\mathfrak{a}}$, $e_{\mathfrak{b}}$ and $e_{\mathfrak{b}'}$ respectively. Since, from (5.1) the numbers of types 5 and 6 are equal for a large lattice we can give their energies the same value, denoted by $e_{\mathfrak{c}}$. In the honeycomb dimer model shown in Fig. 8.5(a), let the fugacities $\mathfrak{z}_1$, $\mathfrak{z}_2$ and $\mathfrak{z}_3$ be associated respectively with diagonal, vertical and horizontal edges. Since the contraction of a dimer-occupied diagonal edge results in a type 1 vertex on the square lattice we must set

$$\exp(-e_{\mathfrak{a}}/T) = \mathfrak{z}_1 \,. \tag{8.71}$$

For each vertical dimer the fugacity $\mathfrak{z}_2$ can be decomposed into two factors $\sqrt{\mathfrak{z}_2}$ and one of these assigned to each of the two sites occupied by the dimer segments. Treating the fugacity $\mathfrak{z}_3$ in a similar way and comparing Figs. 8.5(a), (b) and 5.1, we have

$$\exp(-e_{\mathfrak{b}}/T) = \mathfrak{z}_2 \,, \qquad \exp(-e_{\mathfrak{b}'}/T) = \mathfrak{z}_3 \,, \qquad \exp(-e_{\mathfrak{c}}/T) = \sqrt{\mathfrak{z}_2\mathfrak{z}_3} \,, \tag{8.72}$$

[2] The properties of this model are identical to those of the model of Wu (1968), in which $e_1 = \infty$. In the $e_1 = \infty$ case the derivation of the equivalence between the dimer model and the five-vertex model is more complicated and involves an intermediate lattice. This is a consequence of the convention for vertex line configurations used in vertex models. It should be noted that the arrows used in vertex models, like those shown in Fig. 5.1, represent polarization along the edges and have no connection to the use of arrows on edges earlier in this chapter.

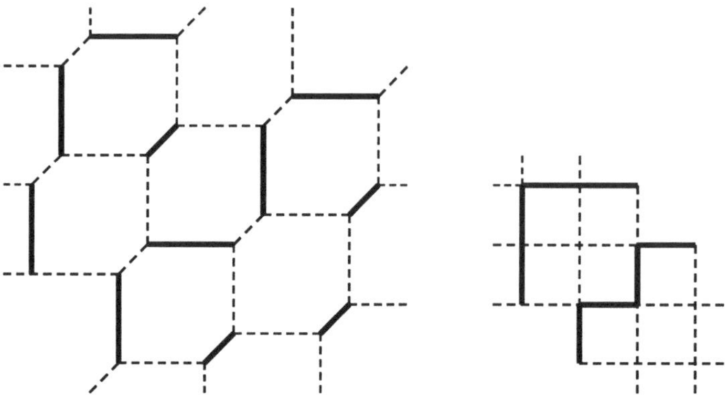

(a) (b)

Fig. 8.5. (a) A pure dimer distribution on the honeycomb lattice. **(b)** The square lattice achieved by contracting the diagonal lines of (a) to zero.

from which it follows that a condition for equivalence between the dimer and five-vertex models is that

$$e_{\mathfrak{b}} + e_{\mathfrak{b}'} = 2e_{\mathfrak{c}} \,. \tag{8.73}$$

As was indicated in Volume 1, Sect. 10.2, the ferroelectric six-vertex model is one where the ground-state vertices are of type 1, 2, 3 or 4 and the antiferroelectric model is one where the ground-state vertices are of type 5 or 6. For the latter to be the case it must be that $e_{\mathfrak{c}} < e_{\mathfrak{b}}$ and $e_{\mathfrak{c}} < e_{\mathfrak{b}'}$, a situation which is excluded by (8.73). When this condition holds the partition function $Z^{(5\mathrm{v})}(\frac{1}{2}N, T)$ of the five-vertex model is related to that of the honeycomb dimer model by

$$Z^{(5\mathrm{v})}\left(\tfrac{1}{2}N, T\right) = Z(N; \mathfrak{z}_1, \mathfrak{z}_2, \mathfrak{z}_3) \,. \tag{8.74}$$

The energy u per lattice site for the five-vertex model on the contracted lattice of $\frac{1}{2}N$ sites is given, from (8.62), (8.71) and (8.72), by

$$u = 2(\rho_1 e_{\mathfrak{a}} + \rho_2 e_{\mathfrak{b}} + \rho_3 e_{\mathfrak{b}'}) \,. \tag{8.75}$$

The case of the six-vertex model for which the only non-zero vertex energies are those of type 1 and 2, which are equal and negative, is called the *KDP model* (see Volume 1, Table 10.1). The case of the present model, where

$$e_{\mathfrak{a}} = -\varepsilon_1 < 0 \,, \qquad e_{\mathfrak{b}} = e_{\mathfrak{b}'} = e_{\mathfrak{c}} = 0 \,, \tag{8.76}$$

differs from the KDP model in that the energy of a type 2 vertex is infinity rather than $-\varepsilon_1$. It is called the *modified KDP model*. From (8.71), (8.72) and (8.75), with $K_1 = \varepsilon_1/T$,

$$\mathfrak{z}_1 = \exp(K_1) \,, \qquad \mathfrak{z}_2 = \mathfrak{z}_3 = 1 \,, \tag{8.77}$$

$$u = -2\rho_1\varepsilon_1 \,. \tag{8.78}$$

The only transition surface relevant to this model is $\mathcal{C}_1$, which, from (8.43) and (8.77), gives

$$\exp(K_{1c}) = 2\,, \qquad T_c = \varepsilon_1/\ln(2)\,, \tag{8.79}$$

for the critical temperature.

In region (i), when $T < T_c$, $\rho_1 = \frac{1}{2}$ and $u = -\varepsilon_1$, so that for the five-vertex model every vertex is of type 1. Like the original KDP model, this modified model is in a frozen perfectly ordered ferroelectric ground state below the critical temperature. The argument used in Volume 1, Sect. 10.3 that the ferroelectric ground state, in which all vertices are of type 1, does not admit small perturbations still applies here. The only difference is that configuration graphs like those shown in Volume 1, Fig. 10.5 cannot any longer 'touch' at a vertex, since such a vertex is of type 2. The equivalent state in the honeycomb dimer model is one in which all the dimers are parallel with fugacity $\mathfrak{z}_1$. It follows that no small perturbations to this state, or the other two fully aligned states, can occur. This confirms the argument of Sect. 8.2.2.

In region (iv), when $T > T_c$, it follows from (8.52) and (8.78) that

$$\rho_1 = \frac{1}{2\pi}\arccos\left[1 - \frac{1}{2}\exp(2K_1)\right]\,. \tag{8.80}$$

The zero-field heat capacity c_0 per vertex in the modified KDP model is thus given by

$$c_0 = k_B\frac{\partial u}{\partial T} = 2k_BK_1^2\frac{\partial \rho_1}{\partial K_1} = \frac{k_BK_1^2\exp(K_1)}{\pi\sqrt{1-\frac{1}{4}\exp(2K_1)}}\,, \tag{8.81}$$

which exhibits a square-root singularity as $T \to T_c$. This same singular behaviour is exhibited by the corresponding six-vertex (KDP) model, except in that case there is a finite discontinuity in u at the critical temperature (see Volume 1, Sect. 10.10).

Now suppose that an electric field with horizontal and vertical components $\mathcal{E}_h$ and $\mathcal{E}_v$ in the plane of the lattice is applied to the modified KDP model. Given that these quantities include the magnitude of the horizontal or vertical component of the dipole moment of a polarized vertex (see Fig. 5.1) the vertex energies given in (8.76) are changed to

$$\begin{aligned} e_{\mathfrak{a}} &= -\varepsilon_1 - \mathcal{E}_v - \mathcal{E}_h\,, \qquad & e_{\mathfrak{b}} &= -\mathcal{E}_v + \mathcal{E}_h\,, \\ e_{\mathfrak{b}'} &= \mathcal{E}_v - \mathcal{E}_h\,, \qquad & e_{\mathfrak{c}} &= 0\,. \end{aligned} \tag{8.82}$$

Since (8.73) is still satisfied, the model remains equivalent to the honeycomb dimer model. When

$$-\tfrac{1}{2}\varepsilon_1 < \mathcal{E}_v\,, \qquad -\tfrac{1}{2}\varepsilon_1 < \mathcal{E}_h\,, \tag{8.83}$$

$e_{\mathfrak{a}} < e_{\mathfrak{b}}$ and $e_{\mathfrak{a}} < e_{\mathfrak{b}'}$ and again the only relevant transition surface is $\mathcal{C}_1$. For fixed values of the electric field components the system will go into a perfectly ordered state with all vertices of type 1 at a critical temperature given by

$$\exp\left(\frac{\varepsilon_1 + \mathcal{E}_{\mathrm{v}} + \mathcal{E}_{\mathrm{h}}}{T_{\mathrm{c}}}\right) = 2\cosh\left(\frac{\mathcal{E}_{\mathrm{v}} - \mathcal{E}_{\mathrm{h}}}{T_{\mathrm{c}}}\right). \tag{8.84}$$

When

$$\mathcal{E}_{\mathrm{h}} < -\tfrac{1}{2}\varepsilon_1, \qquad \mathcal{E}_{\mathrm{h}} < \mathcal{E}_{\mathrm{v}}, \tag{8.85}$$

the transition is to a perfectly ordered state with all vertices of type 3 and when

$$\mathcal{E}_{\mathrm{v}} < -\tfrac{1}{2}\varepsilon_1, \qquad \mathcal{E}_{\mathrm{v}} < \mathcal{E}_{\mathrm{h}}, \tag{8.86}$$

the transition is to a perfectly ordered state with all vertices of type 4. In each of these cases the calculation of the transition temperature is similar to that given for the case (8.83).

8.4 The Ising Model Equivalence

In Sect. 7.4 we described the application of high-temperature series methods to the spin-$\frac{1}{2}$ Ising Model and the form of the zero-field partition function appropriate to these methods is given by (7.52) and (7.55). As a preliminary to the work of this section we need to express these equations in a slightly modified form. From (7.49), (7.52) and (7.55),

$$Z^{(\mathrm{I.M.})}(N;0,K) = 2^N[\sinh(K)]^{\frac{1}{2}zN}\left[v^{-\frac{1}{2}zN} + \sum_{\{\mathsf{q},\mathsf{k}\}}\Omega(0|\mathsf{q},\mathsf{k})v^{\mathsf{q}-\frac{1}{2}zN}\right]. \tag{8.87}$$

Given any zero-field graph $(0|\mathsf{q},\mathsf{k})$ *placed in a particular way on the lattice*, the set of $\frac{1}{2}zN-\mathsf{q}$ edges of the lattice which do not belong to $(0|\mathsf{q},\mathsf{k})$ is called a *complement* of the graph. It is clear that $(0|\mathsf{q},\mathsf{k})$ has $\Omega(0|\mathsf{q},\mathsf{k})$ complements, corresponding to the number of ways of placing the graph on the lattice. The second factor in square brackets in (8.87) corresponds to a sum over all complements of all zero-field graphs, with a factor v^{-1} for every edge in a complement, including the whole lattice $\mathcal{N}$, which is the complement of the zero-field graph with no edges. Let $(\mathsf{q}',\mathsf{k}')$ denote a graph which is a complement of a zero-field graph, where k' is now used to distinguish all the different topologies *and locations* of the graphs with q' edges. Then (8.87) can be expressed in the form

$$Z^{(\mathrm{I.M.})}(N;0,K) = 2^N[\sinh(K)]^{\frac{1}{2}zN}\sum_{\{\mathsf{q}',\mathsf{k}'\}}(v^{-1})^{\mathsf{q}'}. \tag{8.88}$$

This result can be generalized to an anisotropic Ising Model. Suppose the lattice is divided into η classes of edges with s_i edges of class i where $\mathsf{s}_1 +$

$s_2 + \cdots + s_\eta = \frac{1}{2}zN$. Let the nearest-neighbour coupling on the edges of class i be K_i with $v_i = \tanh(K_i)$. Then in place of (8.88) we have

$$Z^{(\text{I.M.})}(N;0,K_i) = 2^N \left\{ \prod_{i=1}^{\eta} [\sinh(K_i)]^{s_i} \right\} \sum_{\{q',k'\}} \left[\prod_{i=1}^{\eta} (v_i^{-1})^{q_i'} \right], \qquad (8.89)$$

where q_i' is the number of edges of class i in the graph (q', k'). The equivalence between pure dimer and Ising models is established by finding a relationship between complementary graphs and dimer graphs, formed, as in Figs. 8.1 and 8.5(a), by placing a line on every edge occupied by a dimer. However, for dimer graphs, exactly one line is incident at each vertex, while for zero-field graphs, or their complements, several lines can be incident. The way to deal with this is indicated by the procedure of Sect. 8.3, where the five-vertex graph of Fig. 8.5(b) is obtained from the dimer graph of Fig. 8.5(a) by deleting edges and coalescing sites. The reverse process would be the expansion of the square lattice of Fig. 8.5(b) by bifurcating each site into two sites connected by a new edge. Fisher (1966) gave a general prescription for splitting sites and inserting edges so as to yield a one-to-one relationship between complementary graphs on the original lattice and dimer graphs on the new lattice. The simplest application of Fisher's method is to the honeycomb lattice Ising Model. An outline of this procedure, including a derivation of the free energy of the anisotropic Ising Model on the honeycomb, triangular and square lattices, will now be given.

Since all vertices of zero-field graphs have even valency, those on the honeycomb lattice must have either no lines or two lines incident at each vertex. A complementary graph must, therefore, have either one or three lines incident at each vertex. Suppose that each honeycomb site is split into three sites connected by a triangle of new edges, each • site in Fig. 8.4 being replaced by a downward pointing triangle and each ∘ site by an upward pointing triangle. This new lattice has two types of elementary polygons, the new triangles and dodecagons with six sides originating from the honeycomb hexagons and six sides from the new triangles. Hence the term *3-12 lattice* used by Nagle et al. (1989). The six sites A, B, C, D, E and F shown in Fig. 8.6 constitute a unit cell in the new lattice, the sites $A^{p}rime$, B′, E′ and F′ being nearest-neighbour sites in adjacent unit cells. Now consider a dimer graph on this lattice. If none of the new edges AB, BC and CA is occupied by a line of the dimer graph then the three edges E′A, F′B and CD originating from the honeycomb lattice will all be so occupied. Otherwise not more than one of the three edges AB, BC and CA can be occupied by a dimer line. If, say, there is a dimer line on AB then there will also be one on CD, but not on E′A or F′B. Similar arguments apply for BC and CA. Hence, if the honeycomb lattice is recovered by replacing each triangle by a single site, then either three lines or one line are incident at this site. It follows that there is a one-to-one correspondence between dimer graphs on the new lattice and complementary graphs on the honeycomb lattice.

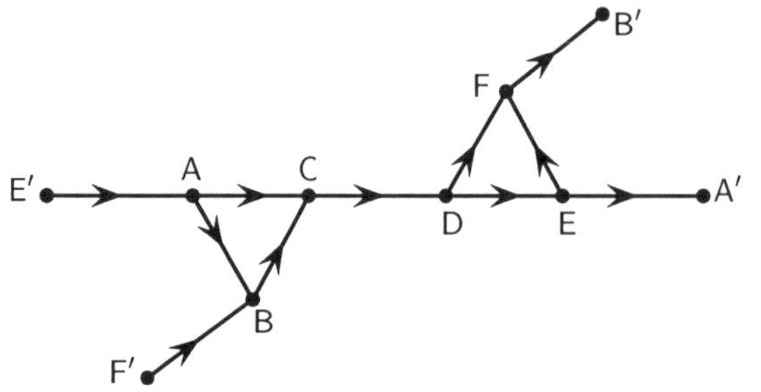

Fig. 8.6. The 3-12 lattice obtained by splitting the sites of a honeycomb lattice.

For the dimer assembly on the 3-12 lattice let the fugacities on the edges of the triangles be unity and those on the honeycomb edges be $\mathfrak{z}_1$, $\mathfrak{z}_2$ and $\mathfrak{z}_3$ assigned as in Fig. 8.4. Let N be the number of sites of the honeycomb lattice. Then the 3-12 lattice will have $3N$ sites. Because of the one-to-one correspondence between dimer graphs on the 3-12 lattice and complementary graphs on the honeycomb lattice, the dimer partition function on the 3-12 lattice is given by

$$Z(3N;\mathfrak{z}_1,\mathfrak{z}_2,\mathfrak{z}_3) = \sum_{\{\mathsf{q}',\mathsf{k}'\}} \mathfrak{z}_1^{\mathsf{q}'_1}\mathfrak{z}_2^{\mathsf{q}'_2}\mathfrak{z}_3^{\mathsf{q}'_3}, \tag{8.90}$$

where the summation on the left is over all complementary graphs $(\mathsf{q}',\mathsf{k}')$ on the honeycomb lattice. In fact this summation is just the $\eta = 3$ case of the summation in (8.89) with $\mathfrak{z}_i = v_i^{-1}$, $i = 1,2,3$. For the honeycomb lattice $\mathsf{s}_1 = \mathsf{s}_2 = \mathsf{s}_3 = \frac{1}{2}N$ and, from (8.89) and (8.90),

$$\begin{aligned} Z^{(\mathrm{I.M.})}(N;0,K_i) = {} & [4\sinh(K_1)\sinh(K_2)\sinh(K_3)]^{N/2} \\ & \times Z(3N;v_1^{-1},v_2^{-1},v_3^{-1})\,. \end{aligned} \tag{8.91}$$

Let the honeycomb lattice consist of a $N_1 \times N_2$ array of unit cells with $N = 2N_1N_2$. It is now necessary to find $Z(3N;v_1^{-1},v_2^{-1},v_3^{-1})$ for the 3-12 lattice whose unit cell is shown in Fig. 8.6. This was derived from the honeycomb lattice and the arrow assignment for the honeycomb edges remains unchanged, while that for the triangle of new edges makes these triangles clockwise odd. The dodecagons are also clockwise odd. Equation (8.31) can still be used for the diagonal block matrices, but these matrices are now 6×6, since there are six sites in the unit cell. Taking the sites of the unit cell in the literal order A, B, C, D, E and F it follows, using Fig. 8.6, that

$$\zeta(\theta_1,\theta_2) = \begin{pmatrix} 0 & 1 & 1 & 0 & -\frac{e^{-i\theta_1}}{v_2} & 0 \\ -1 & 0 & 1 & 0 & 0 & -\frac{e^{-i(\theta_1+\theta_2)}}{v_3} \\ -1 & -1 & 0 & \frac{1}{v_1} & 0 & 0 \\ 0 & 0 & -\frac{1}{v_1} & 0 & 1 & 1 \\ \frac{e^{i\theta_1}}{v_2} & 0 & -1 & 0 & 1 & \\ 0 & \frac{e^{i(\theta_1+\theta_2)}}{v_3} & 0 & -1 & -1 & 0 \end{pmatrix}, \tag{8.92}$$

giving

$$\begin{aligned} \mathrm{Det}\{\zeta(\theta_1,\theta_2)\} = {} & v_1^{-2} + v_2^{-2} + v_3^{-2} + v_1^{-2}v_2^{-2}v_3^{-2} \\ & - 2v_1^{-1}v_2^{-1}(v_3^{-2}-1)\cos\theta_1 - 2v_2^{-1}v_3^{-1}(v_1^{-2}-1)\cos\theta_2 \\ & - 2v_3^{-1}v_1^{-1}(v_2^{-2}-1)\cos(\theta_1+\theta_2)\,. \end{aligned} \tag{8.93}$$

Defining

$$\mathrm{s}_i = \sinh(2K_i) = \frac{2v_i}{1-v_i^2}\,, \qquad \mathrm{c}_i = \cosh(2K_i) = \frac{1+v_i^2}{1-v_i^2}\,, \qquad i = 1,2,3\,, \tag{8.94}$$

and using

$$\sinh(K_i) = \frac{v_i}{\sqrt{1-v_i^2}} \tag{8.95}$$

gives

$$\begin{aligned} \mathrm{Det}\{\zeta(\theta_1,\theta_2)\} = {} & \tfrac{1}{2}[\sinh(K_1)\sinh(K_2)\sinh(K_3)]^{-2}[\mathrm{c}_1\mathrm{c}_2\mathrm{c}_3 + 1 \\ & -\mathrm{s}_1\mathrm{s}_2\cos\theta_1 - \mathrm{s}_2\mathrm{s}_3\cos\theta_2 - \mathrm{s}_3\mathrm{s}_1\cos(\theta_1+\theta_2)]\,. \end{aligned} \tag{8.96}$$

From (8.9), (8.91), (8.96) and (A.325), the dimensionless free energy, per lattice site of the honeycomb lattice anisotropic Ising Model in zero field is

$$\begin{aligned} \phi^{(\mathrm{I.M.})}(0,K_i) = {} & -\tfrac{1}{2}\ln[4\sinh(K_1)\sinh(K_2)\sinh(K_3)] \\ & - \lim_{N_1\to\infty}\lim_{N_2\to\infty} \frac{Z(3N;v_1^{-1},v_2^{-1},v_3^{-1})}{2N_1N_2} \\ = {} & -\ln(2) - \frac{1}{16\pi^2}\int_0^{2\pi} d\theta_1 \int_0^{2\pi} d\theta_2 \ln[\mathrm{c}_1\mathrm{c}_2\mathrm{c}_3 + 1 \\ & - \mathrm{s}_1\mathrm{s}_2\cos\theta_1 - \mathrm{s}_2\mathrm{s}_3\cos\theta_2 - \mathrm{s}_3\mathrm{s}_1\cos(\theta_1+\theta_2)]\,. \end{aligned} \tag{8.97}$$

The star–triangle transformation developed in Volume 1, Sect. 8.5 can now be used to obtain the corresponding expression for the triangular lattice. The result is

$$\phi^{(\mathrm{I.M.})}(0, K_i) = -\ln(2) - \frac{1}{8\pi^2}\int_0^{2\pi} \mathrm{d}\theta_1 \int_0^{2\pi} \mathrm{d}\theta_2 \ln[\mathrm{c}_1\mathrm{c}_2\mathrm{c}_3 + \mathrm{s}_1\mathrm{s}_2\mathrm{s}_3 - \mathrm{s}_1\cos\theta_1 - \mathrm{s}_2\cos\theta_2 - \mathrm{s}_3\cos(\theta_1+\theta_2)], \qquad (8.98)$$

and setting $K_3 = 0$, $\mathrm{s}_3 = 0$, $\mathrm{c}_3 = 1$ yield the result

$$\phi^{(\mathrm{I.M.})}(0, K_i) = -\ln(2) - \frac{1}{8\pi^2}\int_0^{2\pi} \mathrm{d}\theta_1 \int_0^{2\pi} \mathrm{d}\theta_2 \ln[\mathrm{c}_1\mathrm{c}_2 - \mathrm{s}_1\cos\theta_1 - \mathrm{s}_2\cos\theta_2], \qquad (8.99)$$

which applies to the anisotropic Ising Model on the square lattice. The isotropic ($\mathrm{c}_i = \mathrm{c}$, $\mathrm{s}_i = \mathrm{s}$) cases of these formulae were derived in Volume 1, Chap. 8.

8.5 K-Type and O-Type Transitions

Considering for simplicity the isotropic case, we have shown that the 3-12 dimer model, with fugacity $\mathfrak{z} = v^{-1} = \coth(K) > 1$ on the original honeycomb edges and $\mathfrak{z} = 1$ on the new triangle edges, is equivalent to the honeycomb lattice Ising Model ferromagnet. It follows that the transitions in this dimer model are similar to those in a two-dimensional Ising Model ferromagnet, and are thus O-type according to the definition at the end of Sect. 8.2.2. With these fugacities the ground state of the 3-12 dimer model has dimers on the honeycomb edges (for example F′B or CD in Fig. 8.5) but no dimers on the triangle edges. This ground state admits localized perturbations. Dimers can, for example, be moved cyclically from the honeycomb to the triangle edges in any of the dodecagons without affecting the dimer configuration on the rest of the lattice. The dimer on F′B in Fig. 8.5 could be moved to BC, that on CD to DE and so on. We have thus encountered three types of critical behaviour in pure dimer assemblies: K-type in the honeycomb lattice model of Sect. 8.2.2, O-type on the 3-12 lattice and no transition in the square lattice model of Sect. 8.2.1. In the last two cases the ground states admit small perturbations, while for the first case the ground state is frozen with small perturbations excluded, so that the heat capacity is zero below the critical temperature.

K-type critical behaviour is not confined to dimer systems. In Volume 1, Sect. 10.12 it was shown that, in the six-vertex ferroelectric model with non-zero electric field there is a second-order transition at a field-dependent critical temperature, with a square-root singularity in the heat capacity above the critical temperature. Below the critical temperature the state is frozen, fully polarized and the heat capacity is zero. The square-root singular behaviour above the critical temperature still applies in zero field, but the transition to the completely polarized state is first-order with a latent heat (Volume 1, Sect. 10.10, see also Lieb and Wu 1972). From the evidence presented so far it may seem that a square-root singularity in the heat capacity above a critical

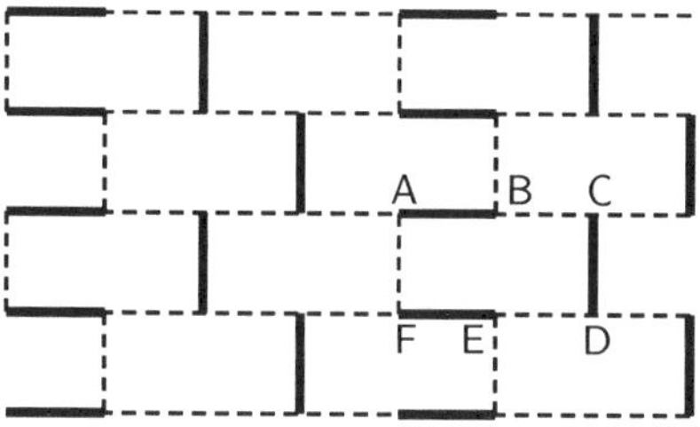

Fig. 8.7. The V_2H_2 dimer model. For the ground state, the positions of the dimers are represented by full lines and the empty edges by broken lines.

temperature is associated with a frozen state below the critical temperature. The situation is, however, not quite so simple, as we shall see.

A sequence of honeycomb lattice dimer models is discussed by Nagle et al. (1989). These models have a less simple assignment of fugacities than that for the models of Sect. 8.2.2, where the fugacities corresponded to the three edge orientations in the usual hexagonal geometry. Using the brick lattice geometry, obtained by rotating the diagonal edges into the vertical position, edge types and fugacity assignments are now made according to the arrangement of dimers in a ground state. One model of this type is called V^2H^2. The ground state for that model is shown in Fig. 8.7, with the full line segments representing the positions of the dimers and the broken lines representing the unoccupied edges. For any arrangement of dimers a fugacity of unity is assigned to a dimer in a ground-state position and a fugacity $\mathfrak{z} = \exp(-\varepsilon/T)$, with $\varepsilon > 0$, to a dimer in any other position. Nagle et al. (1989) also considered dimer models on the 4-8 lattice, formed from a square lattice by replacing each vertex by a quadrangle of edges, and what they called the *SQK model*. In this model extra edges are inserted in the brick lattice to make it into a square lattice, but the dimers on these new vertical edges are given different fugacities from that on the brick lattice edges. Various applications have been made of these types of models, including the chain conformal system described in the next section.

The V^2H^2 model has some interesting critical properties described by Nagle et al. (1989). First, note that the ground state shown in Fig. 8.7 admits local perturbations. In the hexagon ABCDEF the dimers on AB, CD and EF, can, for example, be shifted to BC, DE and FA, without disturbing the configuration elsewhere on the lattice. Thus, as would be expected, the heat capacity below the critical temperature is not identically zero. It does, however, remain finite as the critical temperature is approached from below and has a square-root singularity as the critical temperature is approached from above. The transition is K-type according to the definition at the end of Sect. 8.2.2, where we chose for the reason just seen to provide for the possibility of non-zero subcritical values of the heat capacity. Similar behaviour is found in six-vertex antiferroelectric models (Volume 1, Sect. 10.13), for which there is a critical electric field $\mathcal{E}_c$ such that, for $\mathcal{E} < \mathcal{E}_c$, the ground state is sublattice ordered with zero polarization. This ground state admits small perturbations to the ordering which do not change the overall zero value of the polarization. When the temperature is reduced at a constant non-zero value of $\mathcal{E} < \mathcal{E}_c$ there is a second-order transition to the zero polarization state with a square-root heat capacity singularity. From the possibility of small perturbations we expect the heat capacity to be non-zero below the critical temperature and this is indeed the case. In the zero polarization region the free energy density $f(\mathcal{E},T) = f(0,T)$ (Lieb and Wu 1972) so that the heat capacity $c(\mathcal{E},T) = c(0,T)$, which is finite but non-zero. For $\mathcal{E} > \mathcal{E}_c$, antiferroelectric models[3] undergo a transition similar to that in ferroelectric models, while at $\mathcal{E} = 0$ there is yet another kind of critical behaviour. This is a transition of infinite order in which $c(0,T)$ and all its derivatives are continuous.

As we have seen a frozen state below the transition is not a necessary condition for a K-type transition. However, lattice conservation properties may be relevant. The row conservation rule for dimers established at the beginning of Sect. 8.2.2 depends only on the topology of the brick (honeycomb) lattice and not on the assignment of fugacities. So that it applies to any brick lattice dimer system including the V^2H^2 model. Again, six-vertex models have an ice rule arrow conservation property, which results from the permitted vertex connections and does not depend on the vertex energies. It therefore applies to both ferroelectric and antiferroelectric models. For the eight-vertex models, the ice rule does not apply and, as we showed in Chap. 5, there is then a new range of critical properties, including the O-type transition as a special case. Both conservation rules and anisotropy play a role in determining the properties of dimer models. The six-vertex models do not have K-type second-order transitions when the field is reduced to zero, although FE models do show a square-root singularity in the heat capacity as the first-order transition is approached from above. A non-zero field obviously introduces an element of anisotropy. As shown in Sect. 8.3, the highly anisotropic five-vertex model has a K-type transition to the completely polarized state, even in zero field. However, the introduction of a degree of spatial anisotropy does not necessarily change the character of transitions in two-dimensional lattice systems. As may be deduced from the expressions obtained in Sect. 8.4 the Ising ferromagnets still have O-type transitions even when there are different couplings in different lattice directions. Obviously a similar statement can be made for the equivalent dimer systems.

8.6 The Chain Conformal Transition

In Volume 1, Sect. 7.4 consideration was given to phase transitions occurring in *amphipathic monolayers*. These are monolayers of molecules which have a polar head group with an affinity for water and a hydrocarbon chain which

---

[3] Apart from the F model, which is a special case (see Volume 1, Sect. 10.13).

resist immersion in water. Two-dimensional vapour–liquid first-order transitions occur at very low surface pressures, and at higher surface pressure in the appropriate temperature range a further transition is observed. The states on the lower pressure and higher pressure sides of this transition are usually called *liquid-expanded* and *liquid-condensed*. The transition from the liquid-condensed to liquid-expanded state is thought to be due to disordering of the chain conformations, a process which can take two forms. One of these is the orientation of the carbon chains. A simple model of this kind, due to Bell et al. (1978), is presented in Volume 1, Sect. 7.4. In the other form, which is used in the theory of Nagle (1975a, 1986) (see also Nagle et al. 1989), the disordering mechanism is *isomerism* or chain 'bending'. In the lowest energy state of a saturated carbon chain each successive link (carbon-carbon bond) is at a trans angle with the previous link, implying that all carbon atoms of the chain are coplanar. The higher energy isomers result from some of the successive pairs of links bending from the trans angle to one of two gauche angles. After changes from trans to gauche the carbon atoms of a chain are no longer coplanar and the chain takes up a larger monolayer area. Thus, in a closely packed assembly of erect chains, the isomerism of one chain disturbs the conformation of the others and promotes isomerization, so that the effect is cooperative.

In the model of Nagle (1975a) the interfacial plane is replaced by a line, which is the lower boundary of a vertical plane lattice on which the links of the various chains are placed. Although the model makes all isomers coplanar it is possible to vary the horizontal boundary length $\tilde{\ell}$ per chain (the analogue of the experimental monolayer area per chain) and the chain link density. However, since both dimensions of the lattice are, as usual, regarded as tending to infinity the results are restricted to the limiting case of chains with an infinite number of links. In this section we present a simplified version of the model due to Nagle (1975b).[4] This displays similar critical behaviour to the full model, but has more results obtainable analytically and is related to the dimer assembly on the honeycomb lattice, discussed in Sect. 8.2.2. Most of our notation is different from Nagle's.

In Fig. 8.8 (as in Fig. 8.7) the honeycomb lattice is represented with the brick lattice structure. Let the nearest-neighbour distance in the brick lattice be b. Chain distributions on the lattice are obtained by supposing that each vertical edge can be associated with a vertical link of length $2b$ and each horizontal edge with a chain link of length $\sqrt{2}b$, at an angle 45° to the vertical, the centre of the link being coincident with that of the edge in each case. The constraint is imposed that all links are connected into continuous chains running from the lower to the upper boundary of the lattice and the chains are packed as closely as is possible without two chains touching. It will be found that these requirements are satisfied if exactly one of the three edges meeting at each vertex is associated with a chain link. Thus, if each

[4] Also equivalent to the case $a_{\text{vdw}} = \infty$ of Nagle (1975a).

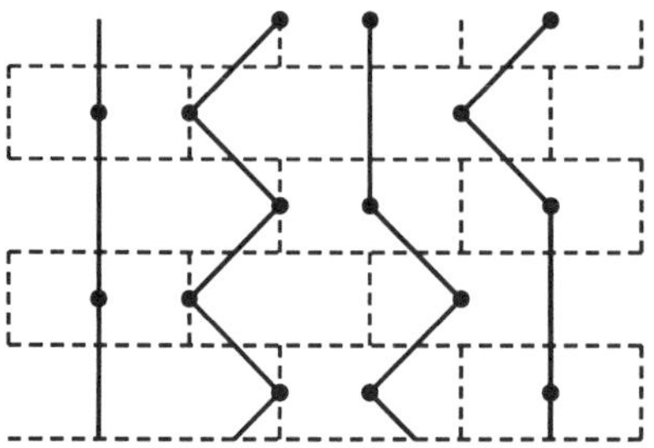

Fig. 8.8. The brick lattice (*broken lines*) and chain conformations (*full lines*). The chain links are connected at the points marked •.

such edge is occupied by a dimer, there is just one dimer segment on each site. There is, therefore, a one-to-one relation between pure dimer distributions on the lattice and chain configurations satisfying the required conditions. Such a chain configuration is represented by the full lines in Fig. 8.8. The points where the chain links are connected are at the centres of the bricks of the lattice.

Chains with every link vertical, like the one on the left in Fig. 8.8 are regarded as in the lowest energy all-trans state, while sloping links represent bending from trans to gauche angles. The number M_v and M_h of vertical and horizontal dimers on the brick lattice are respectively equal to the number of trans and gauche links. The horizontal dimers on the brick lattice correspond to dimers of types 1 and 2 on the honeycomb lattice. So the horizontal dimer density on a lattice of N sites is

$$\rho_\mathrm{h} = \frac{M_\mathrm{h}}{N} = \rho_1 + \rho_2 \,. \tag{8.100}$$

Since all the horizontal edges of the brick lattice are equivalent, we can, without loss of generality, in the dimer model set

$$\mathfrak{z}_1 = \mathfrak{z}_2 = \mathfrak{z}\,, \qquad \mathfrak{z}_3 = 1\,. \tag{8.101}$$

When $\mathfrak{z} > \frac{1}{2}$ the system is in region (iv), as defined in Sect. 8.2.2, and the dimensionless free energy density $\phi(\mathfrak{z}) = \phi(\mathfrak{z}, \mathfrak{z}, 1)$ is given by (8.50). From (8.52) and (8.53), $\rho_1 = \rho_2$ and

$$\rho_\mathrm{h} = \rho_1 + \rho_2 = -\mathfrak{z}\frac{\partial \phi}{\partial \mathfrak{z}} = \frac{1}{\pi}\arccos\left(\frac{1}{2\mathfrak{z}}\right) . \tag{8.102}$$

From (8.2) and (8.101), the honeycomb lattice partition function is given by

$$Z(N;\mathfrak{z}) = \sum_{M_\mathrm{h}=0}^{\frac{1}{2}N} \Omega(N, M_\mathrm{h})\mathfrak{z}^{M_\mathrm{h}}. \tag{8.103}$$

By the method of the maximum term (Volume 1, Sect. 2.8), in the limit of large N,

$$\begin{aligned} \phi(\mathfrak{z}) &\simeq -\frac{1}{N}\ln Z(N,\mathfrak{z}) \\ &\simeq -\rho_\mathrm{h}\ln(\mathfrak{z}) - s(\rho_\mathrm{h})\,, \end{aligned} \tag{8.104}$$

where $\rho_{\rm h}$ and $\mathfrak{z}$ are related by (8.102) and

$$s(\rho_{\rm h}) \simeq \frac{1}{N} \ln \Omega(N, M_{\rm h}) \tag{8.105}$$

is the dimensionless entropy per lattice site. Equations (8.50), (8.102) and (8.104) would give an explicit expression for s as a function of $\rho_{\rm h}$ which can be understood as the entropy of chain mixing per lattice site in the chain model.

Returning to this picture of the system, let $\tilde{\ell}$ be the monolayer length per chain. Then, with every chain in the all-trans state, $\rho_{\rm h} = 0$ and $\tilde{\ell} = b$. With every chain in the all-gauche state, like the second chain from the left in Fig. 8.8, $\rho_{\rm h} = \frac{1}{2}$ and $\tilde{\ell} = 2b$. We now derive $\tilde{\ell}$ for any $\rho_{\rm h}$ in the allowed range $[0, \frac{1}{2}]$. Let the number of sites in each row and column of the brick lattice be S_1 and S_2 respectively. Given that S_1 and S_2 are large enough for edge effects to be neglected

$$N = S_1 S_2 = 2(M_{\rm v} + M_{\rm h})\,. \tag{8.106}$$

Denoting by C the number of chains, then the total length of the assembly is

$$S_1 b = C\tilde{\ell}\,. \tag{8.107}$$

The lengths of the projections onto the vertical of gauche and trans links are respectively b and $2b$ and the sum of the projections of all links in any chain is $S_2 b$. Thus

$$CS_2 = 2M_{\rm v} + M_{\rm h} = N - M_{\rm h} \tag{8.108}$$

and, from (8.100)–(8.108),

$$\tilde{\ell} = \frac{b}{1 - \rho_{\rm h}}\,. \tag{8.109}$$

Suppose that the only variable energy term is $M_{\rm h}\varepsilon$, where $\varepsilon > 0$ is the energy per gauche link, and consider the canonical distribution defined by the variables $S_1 b$, C, N and T. If C and N are kept fixed then in the coupling density formulation the system is specified by the density $\ell = \tilde{\ell}/b = S_1/C$ and coupling $K = \varepsilon/T$. The mean number $N/2C$ of links per chain will then be a fixed parameter although the individual chains will, rather unrealistically, have unequal numbers of links. Thus, for example, an all-trans chain will have $\frac{1}{2}S_2$ links and an all-gauche chain S_2 links. However, Nagle (1975b) argues that, as $S_2 \to \infty$, the probability distribution for the number of links in a chain will contract to a narrow peak. The dimensionless (Helmholtz) free energy density for this model is

$$\psi(\ell, K) = K\rho_{\rm h} - s(\rho_{\rm h})\,, \tag{8.110}$$

where, from (8.109),

$$\rho_{\rm h} = 1 - 1/\ell\,. \tag{8.111}$$

The linear pressure $\widetilde{P}$ is conjugate to the length of the assembly $C\tilde{\ell}$. Since $\widetilde{P}C\tilde{\ell} = (\widetilde{P}Cb/N)(\ell N)$, the pressure field conjugate to ℓ is $P = \widetilde{P}Cb/N$. Then defining the pressure coupling $L = P/T$ we have, from (8.104), (8.110) and (8.111)

$$\begin{aligned} L &= -\frac{\partial \psi}{\partial \ell} \\ &= -\frac{1}{\ell^2}\left(K - \frac{\partial s}{\partial \rho_{\rm h}}\right) \\ &= -\frac{1}{\ell^2}[K + \ln(\mathfrak{Z})]\,, \end{aligned} \tag{8.112}$$

where $\mathfrak{Z}$ is a function of ℓ, given, from (8.102) and (8.111), by

$$\mathfrak{Z} = \left\{2\cos\left[\pi\left(1 - \frac{1}{\ell}\right)\right]\right\}^{-1}. \tag{8.113}$$

To understand the critical properties of the model we analyze the form of the isotherms of L plotted against ℓ for different values of K. To do this we need

$$\left(\frac{\partial L}{\partial \ell}\right)_K = \frac{2[K + \ln(\mathfrak{Z})]}{\ell^3} - \frac{\pi\sqrt{4\mathfrak{Z}^2 - 1}}{\ell^4}\,, \tag{8.114}$$

obtained from (8.112) and (8.113).

As indicated above, when $\mathfrak{Z} > \frac{1}{2}$, the system is in an unfrozen state. As $\mathfrak{Z} \to \frac{1}{2} + 0$, $\rho_{\rm h} \to 0$ and $\ell \to 1$. In this limit the system is in the all-trans state with all the chain links vertical. Along any isotherm, as $\mathfrak{Z} \to \frac{1}{2}$,

$$L \;\to\; L_0(K) = K_{\rm t} - K\,, \tag{8.115}$$

$$\left(\frac{\partial L}{\partial \ell}\right)_K \;\to\; 2[K - K_{\rm t}]\,, \tag{8.116}$$

where

$$K_{\rm t} = \ln(2)\,. \tag{8.117}$$

Although the monolayer length attains its minimum value at $\mathfrak{Z} = \frac{1}{2}$, L_0 is finite for all K.

For $K < K_{\rm t}$, $(T > T_{\rm t} = \varepsilon/\ln 2)$, $L_0 > 0$ and, from (8.114), $(\partial L/\partial \ell)_K < 0$ for all $\mathfrak{Z} > \frac{1}{2}$. Thus as L increases towards L_0, ℓ decreases monotonically to unity. From (8.112), values of $L > L_0$ require values of $\mathfrak{Z} < \frac{1}{2}$. However, we have seen that the system then remains in its frozen state with $\ell = 1$. It follows that, at $L = L_0$, the expansion coefficient $-(\partial \ell/\partial L)_K$ decreases discontinuously from a positive value to zero. Nagle (1975a) calls this a 'weak' second-order transition.

From (8.115), $K = K_{\rm t}$ gives $L_0 = 0$ and, for $K > K_{\rm t}$, $(T < T_{\rm t} = \varepsilon/\ln 2)$, we have $L_0 < 0$ or equivalently $P_0 = TL_0 < 0$. Although negative pressures are not physically realizable it is of interest to consider them in this model. For $K > K_{\rm t}$ it follows from (8.114) that $(\partial L/\partial \ell)_K > 0$ when $\ell = 1$ $(\mathfrak{Z} = \frac{1}{2})$

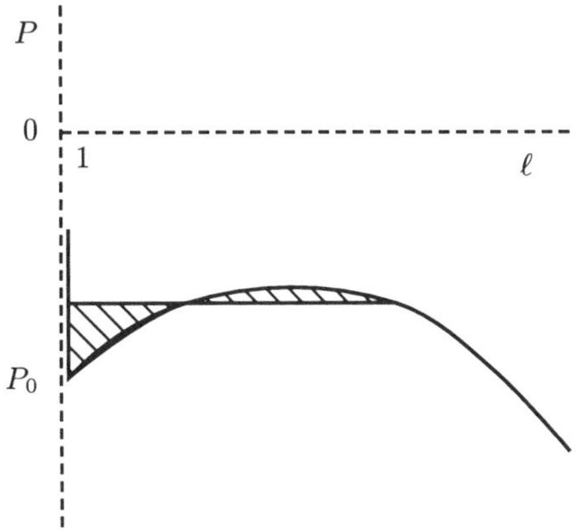

Fig. 8.9. An isotherm with an instability loop and the equal areas construction giving two points on the coexistence curve for the chain model.

implying an instability loop like that shown in Fig. 8.9. One branch of the coexistence curve is given by $\ell = 1$, $L \le 0$. Suppose the other branch of the coexistence curve is the set of points $(\ell^*(K), L^*(K))$, $(K > K_t)$. Then using the equal-areas rule (Volume 1, Sect. 1.6)

$$L^*(K) = L(\ell^*(K), K)\,, \tag{8.118}$$

$$\int_1^{\ell^*} L(\ell, K)\mathrm{d}\ell = (\ell^* - 1)L^*(K)\,. \tag{8.119}$$

In a neighbourhood of the point $(\ell^*(K_t), L^*(K_t)) = (1, 0)$ it follows from (8.112) and (8.113) that

$$L(\ell, K) \simeq -(K - K_t) + 2(K - K_t)(\ell - 1) - \tfrac{1}{2}\pi^2(\ell - 1)^2\,, \tag{8.120}$$

and substituting into equations (8.118) and (8.119) gives

$$\ell^*(K) \simeq 1 + \frac{3}{\pi^2}(K - K_t)\,, \tag{8.121}$$

$$L^*(K) \simeq -(K - K_t) + \frac{3}{2\pi^2}(K - K_t)^2\,. \tag{8.122}$$

Plotted in the $(\tilde{\ell}, T)$ plane, the coexistence curve has a cusp at the point $\tilde{\ell}_t = b$, $T_t = \varepsilon/\ln(2)$ and (8.121) gives the critical exponent $\beta_t = 1$. We have used the subscript 't' in this discussion because the point $L_t = 0$, $\ell_t = 1$, (where $K = K_t = \ln 2$) is the meeting point of a line of (weak) second-order transitions and a line of first-order transitions. It is, therefore, a kind of tricritical point (see Volume 1, Sect. 6.8 and Sect. 2.12 of this volume) although it does have the unusual feature that both the second- and first-order transitions are to a perfectly ordered (all-trans) state.

The more general chain model of Nagle (1975a) allows empty spaces in the lattice. The dimer system equivalent to this model is the SQK model (see Sect. 8.5) with extra edges added to the brick lattice to make it into a square lattice. Dimers occupying these new edges form a third species, corresponding not to chain links but to empty spaces in the chain distribution.

This means that the lattice area a per link is a variable, in contrast to the model discussed above where $a = a_0 = \frac{1}{2}b^2$. A attractive van der Waals energy term $-a_{\text{vdw}}(a_0/a)^c$ is included, a_{vdw} and c being positive parameters with a_0 the minimum value of a. As $a_{\text{vdw}} \to \infty$ the equilibrium value of $a \to a_0$, recovering the simplified model with no empty spaces. The phase diagram is similar to that of the simplified model except that L_{t} is now positive and increases as a_{vdw} decreases. In the (ℓ, L) plane the wedge-shaped coexistence region moves up the L axis. Nagle (1986) also introduced the terms $-a_{\text{head}}(\tilde{\ell}/b)^q$ and $\gamma\tilde{\ell}$ into the free energy to allow respectively for headgroup–headgroup and headgroup–water interactions. Quantities connected with the phase separation region, including the tricritical temperature and pressure, are tabulated for various parameter values. Other chain isomerism models and Nagle's earlier papers are reviewed by Wiegel and Kox (1980) and Bell et al. (1981). A more recent theory is due to Firpo et al. (1984).

Examples

8.1 Fig. 8.10 shows an arrow assignment due to Montroll (1964) for the triangular lattice. As in Figs. 8.2 and 8.3 the cell boundary is marked by a broken line and L and R sites are represented by • and ∘ respectively. Given that the fugacities $\mathfrak{z}_1$, $\mathfrak{z}_2$ and $\mathfrak{z}_3$ are associated with the horizontal, vertical and diagonal edges respectively, obtain $z(0,0)$, $z(\pm 1, 0)$, $z(0, \pm 1)$, $z(1,1)$ and $z(-1,-1)$. Hence derive the expression

$$\begin{aligned}\mathrm{Det}\{\zeta(\theta_1,\theta_2)\} = 4\{&\mathfrak{z}_1^2\cos^2\left(\tfrac{1}{2}\theta_1\right) + \mathfrak{z}_2^2\sin^2\left(\tfrac{1}{2}\theta_2\right)\\ &+ \mathfrak{z}_3^2\sin^2\left(\tfrac{1}{2}\theta_1 + \theta_2\right)\}\end{aligned}$$

and show that this then yields the dimensionless free energy density

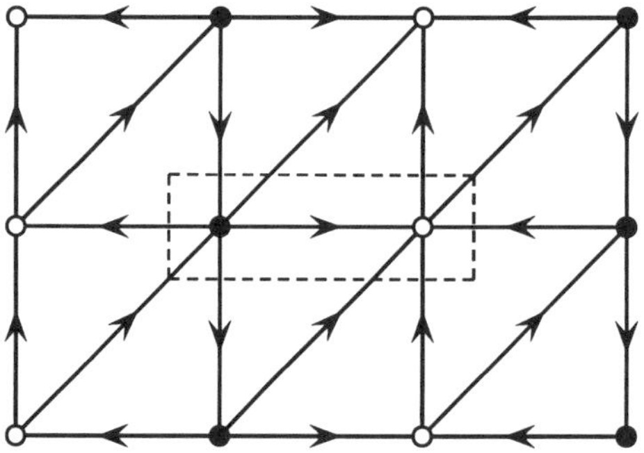

Fig. 8.10. An arrow assignment on the triangular lattice due to Montroll (1964).

$$\phi(\mathfrak{z}_1,\mathfrak{z}_2,\mathfrak{z}_3) = -\tfrac{1}{4}\ln(2) - \frac{1}{(4\pi)^2}\int_0^{2\pi} \mathrm{d}\theta_1 \int_0^{2\pi} \mathrm{d}\theta_2 \ln[\mathfrak{z}_1^2 + \mathfrak{z}_2^2 + \mathfrak{z}_3^2 + \mathfrak{z}_1^2\cos\theta_1 + \mathfrak{z}_2^2\cos\theta_2 + \mathfrak{z}_3^2\cos(\theta_2-\theta_1)]\,.$$

Treating the integrand in the same way as the integrand of (8.34) was treated in Sect. 8.2.2, show that in this case

$$Y^2(\mathfrak{z}_1,\mathfrak{z}_2,\mathfrak{z}_3;\cos\theta_1) - X^2(\mathfrak{z}_1,\mathfrak{z}_2,\mathfrak{z}_3;\cos\theta_1)$$

is positive for all values of $\mathfrak{z}_1$, $\mathfrak{z}_2$, $\mathfrak{z}_3$ and θ_1. (This indicates that no transition occurs.)

8.2 Consider a pure dimer assembly on the square lattice with the notation of Sect. 8.2. According to the maximum term method the equilibrium value $\overline{M}_1$ of M_1 at fixed N is given by determining the maximum term in (8.1) subject to the constraint $M_1 + M_2 = \frac{1}{2}N$. With $\Omega_N(M_1) = \Omega(N, M_1, \frac{1}{2}N - M_1)$ this is given by

$$\overline{\frac{\mathrm{d}\ln\Omega_N}{\mathrm{d}M_1}} + \ln\eta = 0\,,$$

where $\eta = \mathfrak{z}_1/\mathfrak{z}_2$. Now suppose that, with $\mathfrak{z}_1 = \mathfrak{z}_2 = 1$, there is a long range interaction energy $-w(M_1 - M_2)^2/N$ and that the equilibrium values of M_1 and M_2 are $\overline{M}_1$ and $\overline{M}_2$ respectively. By the maximum term method show that these equilibrium values correspond to a value η_e of η in the original model without interaction where

$$2w(\overline{M}_1 - \overline{M}_2) - NT\ln(\eta_\mathrm{e}) = 0\,.$$

Prove that the equilibrium equation for the system can be expressed in the form

$$2w[\arctan(\eta_\mathrm{e}) - \arctan(\eta_\mathrm{e}^{-1})] = \pi T\ln(\eta_\mathrm{e})\,.$$

Prove that symmetric pairs of solutions η_e, η_e^{-1}, with $\eta_\mathrm{e} \neq 1$ exist for $T < 2w/\pi$, but that for $T > 2w/\pi$ the only solution is $\eta_\mathrm{e} = 1$ given $\overline{M}_1 = \overline{M}_2$.

8.3 Show that the formula (8.98) for the dimensionless free energy density of the zero-field triangular Ising Model with unequal couplings K_1, K_2 and K_3, is unaffected by changing the signs of two of the couplings. For the model with $K_1 = -K_2 = -K_3 = K > 0$ show that, in the ground state, spins are oppositely aligned on two equal but non-interpenetrating sublattices. Establish equivalence with the isotropic ferromagnet, where $K_1 = K_2 = K_3 = K$, by considering the effect on the Hamiltonian of simultaneously reversing the spins on one sublattice and changing K_2 and K_3 from $-K$ to K.

A. Appendices

A.1 Fourier Transforms in d Dimensions

We present a summary of useful results related to Fourier transforms and series.[1]

A.1.1 Discrete Finite Lattices

Consider a d-dimensional hypercubic lattice $\mathcal{N}$ with lattice spacing a, N sites and periodic boundary conditions. The sites are given by

$$\boldsymbol{r} = a\Big(n^{(1)}, \ldots, n^{(d)}\Big)\,, \tag{A.1}$$

where $n^{(i)}$ ranges over all distinct integer values modulo $N^{(i)}$, for $i = 1, 2, \ldots, d$, and

$$N = \prod_{i=1}^{d} N^{(i)}\,. \tag{A.2}$$

Let $f(\boldsymbol{r})$ be a function defined at each point of $\mathcal{N}$. Because of the periodic boundary conditions $f(\boldsymbol{r})$ is invariant under the transformation

$$\boldsymbol{r} \to \boldsymbol{r} + a\Big(m^{(1)}N^{(1)}, \ldots, m^{(d)}N^{(d)}\Big) \tag{A.3}$$

for any integers $m^{(1)}, \ldots, m^{(d)}$ and the Fourier transform $f^*(\boldsymbol{k})$ of $f(\boldsymbol{r})$ is given by

$$f^*(\boldsymbol{k}) = \sum_{\{\boldsymbol{r}\}} f(\boldsymbol{r}) \exp(-\mathrm{i}\boldsymbol{k} \boldsymbol{.} \boldsymbol{r})\,, \tag{A.4}$$

$$f(\boldsymbol{r}) = \frac{1}{N} \sum_{\{\boldsymbol{k}\}} f^*(\boldsymbol{k}) \exp(\mathrm{i}\boldsymbol{k} \boldsymbol{.} \boldsymbol{r})\,, \tag{A.5}$$

where

[1] It is convenient to use the generic term *transform* to refer to both the discrete (series) and continuous (integral) cases.

$$\boldsymbol{k} = \frac{2\pi}{a}\left(\frac{k^{(1)}}{N^{(1)}}, \ldots, \frac{k^{(d)}}{N^{(d)}}\right), \tag{A.6}$$

and, in (A.5), $k^{(i)}$ ranges over all distinct integer values modulo $N^{(i)}$, for $i = 1, 2, \ldots, d$.

A.1.2 A Continuous Finite Volume

The case where the vector $\boldsymbol{r}$ can specify any point in a hyper-cubic volume $\mathcal{V}$ of edge-lengths $L^{(1)}, \ldots, L^{(d)}$, periodic boundary conditions and

$$|\mathcal{V}| = \widetilde{V} = \prod_{i=1}^{d} L^{(i)}, \tag{A.7}$$

can be obtained as the limit of the lattice case where $a \to 0$, $N_i \to \infty$ with $N^{(i)}a = L^{(i)}$, for $i = 1, 2, \ldots, d$. With $f^*(\boldsymbol{k}) = N f_1^*(\boldsymbol{k})$, equation (A.4) is replaced by an integral according to the prescription

$$\frac{1}{N}\sum_{\{\boldsymbol{r}\}} = \frac{a^d}{V}\sum_{\{\boldsymbol{r}\}} \to \frac{1}{V}\int_{\mathcal{V}} \mathrm{d}\mathcal{V},$$

giving, in place of (A.4) and (A.5),

$$f_1^*(\boldsymbol{k}) = \frac{1}{\widetilde{V}}\int_{\mathcal{V}} f(\boldsymbol{r})\exp(-\mathrm{i}\boldsymbol{k}\cdot\boldsymbol{r})\,\mathrm{d}\mathcal{V}, \tag{A.8}$$

$$f(\boldsymbol{r}) = \sum_{\{\boldsymbol{k}\}} f_1^*(\boldsymbol{k})\exp(\mathrm{i}\boldsymbol{k}\cdot\boldsymbol{r}), \tag{A.9}$$

where the definition of the values of $\boldsymbol{k}$ given by (A.6) is replaced by

$$\boldsymbol{k} = 2\pi\left(\frac{k^{(1)}}{L^{(1)}}, \ldots, \frac{k^{(d)}}{L^{(d)}}\right), \qquad k^{(i)} = 0, \pm 1, \pm 2, \ldots. \tag{A.10}$$

For the case $d = 1$ the vector $\boldsymbol{k}$ becomes a scalar of the form $k = 2n\pi/L^{(1)}$, $n = 0, \pm 1, \pm 2, \ldots$. The range of the scalar variable r can be taken as $[-L^{(1)}/2, L^{(1)}/2]$ and, with the change of variable $r = xL^{(1)}/(2\pi)$, (A.8) and (A.9) give the usual Fourier series formulae

$$f_1^*(n) = \frac{1}{2\pi}\int_{-\pi}^{\pi} f(x)\exp(-\mathrm{i}nx)\,\mathrm{d}x, \tag{A.11}$$

$$f(x) = \sum_{n=-\infty}^{\infty} f_1^*(n)\exp(\mathrm{i}nx). \tag{A.12}$$

The condition for the convergence of the series (A.9) and (A.12) is that $f(\boldsymbol{r})$ is piecewise smooth on every finite interval in all its component variables and equivalently $f(x)$ has this property in x (Courant and Hilbert 1962).

Suppose that $\varphi(x)$ is a continuous and continuously differentiable function of x so that the series

$$\sum_{n=-\infty}^{\infty} \varphi(2\pi n + y)\,, \qquad \sum_{n=-\infty}^{\infty} \frac{\partial \varphi(2\pi n + y)}{\partial y} \tag{A.13}$$

converge absolutely and uniformly for $-\pi \leq y < \pi$. Then the second series in (A.13) is the derivative of the first which can be transformed using (A.11) and (A.12) to give

$$\sum_{n=-\infty}^{\infty} \varphi(2\pi n + y) = \frac{1}{2\pi} \sum_{m=-\infty}^{\infty} \sum_{n=-\infty}^{\infty} \int_{-\pi}^{\pi} \varphi(2\pi n + x) \exp[\mathrm{i}m(y - x)]\mathrm{d}x\,. \tag{A.14}$$

Setting $y = 0$ in (A.14) and using the transformation

$$\begin{aligned}\sum_{n=-\infty}^{\infty} \int_{-\pi}^{\pi} \varphi(2\pi n + x) \exp(-\mathrm{i}mx)\mathrm{d}x &= \sum_{n=-\infty}^{\infty} \int_{2\pi n}^{2\pi(n+1)} \varphi(x) \exp(-\mathrm{i}mx)\mathrm{d}x \\ &= \int_{-\infty}^{\infty} \varphi(x) \exp(-\mathrm{i}mx)\mathrm{d}x,\end{aligned}$$

gives

$$\sum_{n=-\infty}^{\infty} \varphi(2\pi n) = \frac{1}{2\pi} \sum_{m=-\infty}^{\infty} \int_{-\infty}^{\infty} \varphi(x) \exp(-\mathrm{i}mx)\mathrm{d}x, \tag{A.15}$$

which is the *Poisson summation formula* (Courant and Hilbert 1962).

A.1.3 A Continuous Infinite Volume

Suppose now that $f_1^*(\boldsymbol{k}) = f_2^*(\boldsymbol{k})/\widetilde{V}$ and consider the limiting situation $L^{(i)} \to \infty$, for $i = 1, 2, \ldots, d$. The sum in (A.9) becomes an integral according to the prescription

$$\frac{1}{\widetilde{V}} \sum_{\{\boldsymbol{k}\}} = \frac{1}{(2\pi)^d} \prod_{i=1}^{d} \left(\frac{2\pi}{L_i}\right) \sum_{\{\boldsymbol{k}\}} \to \frac{1}{(2\pi)^d} \int_{\mathcal{K}} \mathrm{d}\mathcal{K}\,,$$

giving

$$f_2^*(\boldsymbol{k}) = \int_{\mathcal{R}} f(\boldsymbol{r}) \exp(-\mathrm{i}\boldsymbol{k} \boldsymbol{.} \boldsymbol{r})\, \mathrm{d}\mathcal{R}\,, \tag{A.16}$$

$$f(\boldsymbol{r}) = \frac{1}{(2\pi)^d} \int_{\mathcal{K}} f_2^*(\boldsymbol{k}) \exp(\mathrm{i}\boldsymbol{k} \boldsymbol{.} \boldsymbol{r})\, \mathrm{d}\mathcal{K}\,, \tag{A.17}$$

where $\mathcal{R}$ and $\mathcal{K}$ are the infinite d-dimensional spaces of the vectors $\boldsymbol{r}$ and $\boldsymbol{k}$ respectively.

The function f is dependent only on the magnitude r of the vector $\boldsymbol{r}$ if and only if f_2^* is a function only of the magnitude k of $\boldsymbol{k}$. In this case the integrals in (A.16) and (A.17) can be expressed in polar variables in the form

$$f_2^*(k) = \int_0^\infty \mathrm{d}r\, f(r) r \int_0^\pi \mathrm{d}\theta\, \mathcal{S}_{d-1}(r\mathrm{sin}\theta) \exp(-\mathrm{i}kr\mathrm{cos}\theta)\,, \tag{A.18}$$

$$f(r) = \frac{1}{(2\pi)^d} \int_0^\infty \mathrm{d}k\, f_2^*(k) k \int_0^\pi \mathrm{d}\theta\, \mathcal{S}_{d-1}(k\mathrm{sin}\theta) \exp(\mathrm{i}rk\mathrm{cos}\theta)\,, \tag{A.19}$$

where

$$\mathcal{S}_d(q) = 2\pi^{\frac{d}{2}} q^{d-1} / \Gamma\left(\tfrac{d}{2}\right), \tag{A.20}$$

is the surface area of a d-dimensional hypersphere of radius q, Γ being the gamma function. To evaluate the integrals (A.18) and (A.19) we need the formula

$$\int_0^\pi \sin^{2\nu}\theta \exp(\pm \mathrm{i}z\mathrm{cos}\theta)\mathrm{d}\theta = \pi^{\frac{1}{2}} \Gamma\left(\nu + \tfrac{1}{2}\right) (2/z)^\nu \mathrm{J}_\nu(z)\,, \qquad \Re(\nu) > -\tfrac{1}{2} \tag{A.21}$$

(Gradshteyn and Ryzhik 1980, p. 482), where $\mathrm{J}_\nu(z)$ is a *Bessel function of the first kind.* Substituting from (A.20) into (A.18) and (A.19) and using (A.21) we have, for all $d > 1$,

$$f_2^*(k) = k \int_0^\infty f(r) \left(\frac{2\pi r}{k}\right)^{\frac{d}{2}} \mathrm{J}_{\frac{d}{2}-1}(kr)\mathrm{d}r\,, \tag{A.22}$$

$$f(r) = r \int_0^\infty f_2^*(k) \left(\frac{k}{2\pi r}\right)^{\frac{d}{2}} \mathrm{J}_{\frac{d}{2}-1}(kr)\mathrm{d}k\,. \tag{A.23}$$

Since $\mathrm{J}_{-\frac{1}{2}}(z) = (2/z\pi)^{\frac{1}{2}} \cos(z)$ it follows directly from (A.16) and (A.17) that (A.22) and (A.23) apply also to $d = 1$.

A.1.4 A Special Case

A case of (A.22) and (A.23) needed in Sect. 3.3.2, is when

$$f_2^*(k) = \frac{G}{\kappa^2 + k^2}\,, \tag{A.24}$$

for some constants G and κ. Substituting from (A.24) into (A.23) gives

$$f(r) = \frac{F(\kappa r)}{r^{d-2}}\,, \tag{A.25}$$

where

$$F(q) = \frac{G}{(2\pi)^{\frac{d}{2}}} Q\left(\tfrac{d}{2}, \tfrac{d}{2} - 1; q\right) \tag{A.26}$$

and

$$Q(\mu,\nu;q) = \int_0^\infty \frac{k^\mu \mathrm{J}_\nu(k)}{k^2+q^2}\mathrm{d}k\,. \tag{A.27}$$

With real μ, ν and q the integral in (A.27) converges for $1-\nu<\mu<\frac{5}{2}$, when $q=0$, and for $-1-\nu<\mu<\frac{5}{2}$ when $q\neq 0$ (Gradshteyn and Ryzhik 1980, p. 684, 687). In particular

$$Q(\nu+1,\nu;q) = q^\nu \mathrm{K}_\nu(q)\,, \tag{A.28}$$

when $q\neq 0$ and $-1<\nu<\frac{3}{2}$. $\mathrm{K}_\nu(z)$ is the *Bessel function of complex argument* defined by

$$\mathrm{K}_\nu(z) = \tfrac{1}{2}\mathrm{i}\pi\exp(\mathrm{i}\pi\nu/2)[\mathrm{J}_\nu(\mathrm{i}z)+\mathrm{i}\mathrm{N}_\nu(\mathrm{i}z)]\,, \tag{A.29}$$

where $\mathrm{N}_\nu(z)$ is a *Bessel function of the second kind.* From (A.25)–(A.29)

$$f(r) = \frac{G}{(2\pi)^{\frac{d}{2}}}\left(\frac{\kappa}{r}\right)^{\frac{d}{2}-1}\mathrm{K}_{\frac{d}{2}-1}(\kappa r)\,, \tag{A.30}$$

when $\kappa\neq 0$ and $d<5$, with convergence to a finite limit as $\kappa\to 0$ when $2<d<5$. Since in the limit $|z|\to\infty$

$$\mathrm{K}_\nu(z) \simeq (\pi/2z)^{\frac{1}{2}}\exp(-z)\,, \tag{A.31}$$

(Gradshteyn and Ryzhik 1980, p. 963),

$$f(r) \simeq \frac{G\kappa^{\frac{d}{2}-\frac{3}{2}}}{2^{\frac{d}{2}+\frac{1}{2}}(\pi r)^{\frac{d}{2}-\frac{1}{2}}}\exp(-\kappa r)\,, \tag{A.32}$$

in the limit $r\to\infty$ with $\kappa\neq 0$. In fact (A.31) is exact for $\nu=\pm\frac{1}{2}$ (Gradshteyn and Ryzhik 1980, p. 967) and so (A.32) is exact for $d=1$ and $d=3$. In both these cases (A.32) gives the limiting behaviour as $\kappa\to 0$. Since $\mathrm{K}_0(z)\simeq -\ln(z/2)$ and $K_1(z)\simeq 1/z$ in the limit $|z|\to 0$ (Gradshteyn and Ryzhik 1980, p. 961)

$$f(r) \simeq -\frac{G}{2\pi}\ln\left(\frac{\kappa r}{2}\right), \qquad d=2\,, \tag{A.33}$$

$$f(r) = \frac{G}{(2\pi)^2}\frac{1}{r^2}, \qquad d=4, \tag{A.34}$$

in the limit $\kappa\to 0$. This completes the analysis for $d<5$. For $d\geq 5$ we must consider the singular behaviour of the integral (A.27) when $\mu\geq\frac{5}{2}$. We note that

$$Q(\mu,\nu;q) = Q(\mu,\nu;0) - q^2 Q(\mu-2,\nu;q)\,. \tag{A.35}$$

This equation can be iterated to produce a series in ascending powers of q, until the first argument in Q lies between zero and $\frac{3}{2}$ inclusively. The most singular contribution to this series is the first term and thus

$$Q(\mu,\nu;q) \simeq \lim_{k_0\to\infty} \int_0^{k_0} k^{\mu-2} J_\nu(k)\mathrm{d}k\,. \tag{A.36}$$

The solution of the integral in (A.36) for the required range of μ and ν is rather complicated and involves *Lommel functions* (Gradshteyn and Ryzhik 1980, p. 684). However, our interest is confined to the limit of large k_0, for which

$$\int_0^{k_0} k^{\mu-2}\mathrm{J}_\nu(k)\mathrm{d}k \simeq k_0^{\mu-4} + \left(\frac{2k_0}{\pi}\right)^{\frac{1}{2}} \left[(\mu+\nu-3)\cos\left(k_0 - \frac{\pi}{2}\nu - \frac{\pi}{4}\right) - k_0^{\mu-3}\cos\left(k_0 - \frac{\pi}{2}\nu + \frac{\pi}{4}\right)\right]. \tag{A.37}$$

Thus, when $d \geq 5$, from (A.25) and (A.26),

$$f(r) \simeq \frac{G}{r^{d-2}(2\pi)^{\frac{d}{2}}} \left\{k_0^{\frac{d}{2}-4} - \left(\frac{2}{\pi}\right)^{\frac{1}{2}} k_0^{\frac{d}{2}-\frac{5}{2}} \cos\left[k_0 - \frac{\pi}{4}(d+1)\right]\right\}, \tag{A.38}$$

in the limit $k_0 \to \infty$, where k_0 is understood as a high-frequency cutoff for the Fourier components of $f(r)$.

A.2 The Conformal Group

We confine our attention to the d-dimensional Euclidean space of the vectors $\boldsymbol{r} = \left(r^{(1)}, r^{(2)}, \ldots, r^{(d)}\right)$ measured in units of the lattice spacing. The line element $\mathrm{d}r$ is given by

$$[\mathrm{d}r]^2 = \mathbf{d}\boldsymbol{r}\,.\,\mathbf{d}\boldsymbol{r} = [\mathrm{d}r^{(1)}]^2 + [\mathrm{d}r^{(2)}]^2 + \cdots + [\mathrm{d}r^{(d)}]^2\,. \tag{A.39}$$

The *conformal group* is the set of transformations $\boldsymbol{r} \to \tilde{\boldsymbol{r}}$ which satisfy conditions of the form

$$[\mathrm{d}\tilde{r}]^2 = \Omega(\boldsymbol{r})[\mathrm{d}r]^2\,, \tag{A.40}$$

for some scalar fields $\Omega(\boldsymbol{r})$. Suppose that a transformation is given by

$$\tilde{r}^{(i)} = R^{(i)}(\boldsymbol{r}) \qquad i = 1, 2, \ldots, d\,. \tag{A.41}$$

Then, for the condition (A.40) to be satisfied, it must be the case that

$$\Omega(\boldsymbol{r})\delta^{\mathrm{Kr}}(j-k) = \sum_{i=1}^{d} \frac{\partial R^{(i)}}{\partial r^{(j)}} \frac{\partial R^{(i)}}{\partial r^{(k)}}\,, \tag{A.42}$$

where δ^{Kr} is the Kronecker delta function and $[\Omega(\boldsymbol{r})]^{\frac{d}{2}}$ is the Jacobian of the transformation (A.41). It is clear that the following transformations satisfy (A.42):

(i) *Translations* where

$$R^{(i)}(\boldsymbol{r}) = r^{(i)} + c^{(i)}, \qquad i = 1, 2, \ldots, d\,, \tag{A.43}$$

for some constants $c^{(i)}$ and $\Omega(\boldsymbol{r}) = 1$.

(ii) *Rotations* where

$$R^{i)}(\boldsymbol{r}) = \omega_i^{(1)} r^{(1)} + \cdots + \omega_i^{(d)} r^{(d)}, \qquad i = 1, \ldots, d, \tag{A.44}$$

the vectors $\boldsymbol{\omega}_i = (\omega_i^{(1)}, \ldots, \omega_i^{(d)})$ are orthonormal and $\Omega(\boldsymbol{r}) = 1$.

(iii) *Dilatations* where

$$R^{(i)}(\boldsymbol{r}) = \lambda^{-1} r^{(i)}, \qquad i = 1, 2, \ldots, d, \tag{A.45}$$

and $\Omega(\boldsymbol{r}) = \lambda^{-2}$.

(iv) *Inversions* where

$$R^{(i)}(\boldsymbol{r}) = r^{(i)}/r^2, \qquad i = 1, 2, \ldots, d, \tag{A.46}$$

$r = |\boldsymbol{r}|$ and $\Omega(\boldsymbol{r}) = 1/r^4$.

It can be shown (Ginsparg 1990) that, for $d > 2$, the conformal group is finite-dimensional and generated by the transformations (i)–(iv). Alternatively, inversion can be replaced as a generator of the group by the *special conformal transformation*

$$R^{(i)}(\boldsymbol{r}) = \frac{r^{(i)} + c^{(i)} r^2}{1 + 2\boldsymbol{r} \centerdot \boldsymbol{c} + c^2 r^2}, \tag{A.47}$$

for any constant vector $\boldsymbol{c} = (c^{(i)}, \ldots, c^{(d)})$, with $\Omega(\boldsymbol{r}) = (1 + 2\boldsymbol{r}.\boldsymbol{c} + c^2 r^2)^{-2}$. This can be seen to be composed of, a translation, an inversion and then a further translation.

For $d = 2$ we use the notation $\boldsymbol{r} = (x, y)$, with the transformation (A.41) given by

$$\tilde{x} = \mathsf{X}(x, y), \qquad \tilde{y} = \mathsf{Y}(x, y). \tag{A.48}$$

Equation (A.42) is satisfied if

$$\frac{\partial \mathsf{X}}{\partial x} = \frac{\partial \mathsf{Y}}{\partial y}, \qquad \frac{\partial \mathsf{X}}{\partial y} = -\frac{\partial \mathsf{Y}}{\partial x}. \tag{A.49}$$

These are the Cauchy–Riemann conditions for

$$\mathsf{Z}(z) = \mathsf{X}(x, y) + \mathrm{i}\mathsf{Y}(x, y) \tag{A.50}$$

to be an analytic function of the complex variable $z = x + \mathrm{i}y$. In this case, therefore, the conformal group is infinite-dimensional. From (A.42) the Jacobian of the transformation $z \to \mathsf{Z}(z)$ is

$$\Omega(z) = |\mathsf{Z}'(z)|^2. \tag{A.51}$$

A.3 Group Representation Theory

In this appendix we summarize the elements of group representation theory. For more detailed treatments of the theory of groups and their representations the reader is referred to McWeeny (1963), Falicov (1966), Leech and Newman (1969) and Serre (1977). Standard theorems which are contained in these texts are stated without proof.

A.3.1 Groups

A set of elements $\mathcal{Q} = \{\mathrm{Q}^{(0)}, \mathrm{Q}^{(1)}, \mathrm{Q}^{(2)}, \ldots\}$, with a binary operation, which we shall refer to as the *product*, forms a *group* if:

(i) The product $\mathrm{Q}^{(s)}\mathrm{Q}^{(s')}$ of two arbitrary elements belongs to $\mathcal{Q}$.

(ii) The identity I, which satisfies $\mathrm{I}\mathrm{Q}^{(s)} = \mathrm{Q}^{(s)}\mathrm{I} = \mathrm{Q}^{(s)}$, for all s, is a member of $\mathcal{Q}$, which is identified with $\mathrm{Q}^{(0)}$.

(iii) For every s there exists an s′ such that $\mathrm{Q}^{(s')}\mathrm{Q}^{(s)} = \mathrm{Q}^{(s)}\mathrm{Q}^{(s')} = \mathrm{I}$. The element $\mathrm{Q}^{(s')}$ is the *inverse* of $\mathrm{Q}^{(s)}$ and is denoted by $[\mathrm{Q}^{(s)}]^{-1}$.

(iv) For all s ,s′ and s″ the three elements $\mathrm{Q}^{(s)}$, $\mathrm{Q}^{(s')}$ and $\mathrm{Q}^{(s'')}$ satisfy the associative law $(\mathrm{Q}^{(s)}\mathrm{Q}^{(s')})\mathrm{Q}^{(s'')} = \mathrm{Q}^{(s)}(\mathrm{Q}^{(s')}\mathrm{Q}^{(s'')})$.

The number of elements in a group is its *order*. For the group $\mathcal{Q}$ this will be denoted by $n(\mathcal{Q})$, or, when there is no danger of ambiguity, simply by n. The order of a group is not necessarily finite, but in this appendix only finite groups are considered. Two elements $\mathrm{Q}^{(s)}$ and $\mathrm{Q}^{(s')}$ are said to be *conjugate* if there exists an element $\mathrm{Q}^{(s'')}$ in the group such that $\mathrm{Q}^{(s)} = [\mathrm{Q}^{(s'')}]^{-1}\mathrm{Q}^{(s')}\mathrm{Q}^{(s'')}$. It is clear that any element which commutes with all the other elements of the group is conjugate only to itself. The set of all elements conjugate to a particular element is called a *conjugacy class* of $\mathcal{Q}$. It is not difficult to show that any two elements are in the same conjugacy class if and only if they are conjugate and thus that the group decomposes into a set of mutually exclusive conjugacy classes. The number of conjugacy classes in $\mathcal{Q}$ is denoted by $\kappa(\mathcal{Q})$ or simply κ. If $\mathrm{Q}^{(s)}\mathrm{Q}^{(s')} = \mathrm{Q}^{(s')}\mathrm{Q}^{(s)}$ for all s and s′ then the group is *commutative* and $\kappa(\mathcal{Q}) = n(\mathcal{Q})$.

A.3.2 Representations

If, for all s, a square $M \times M$ dimensional matrix $\boldsymbol{R}(\mathrm{Q}^{(s)}) = \boldsymbol{Q}^{(s)}$ can be assigned to $\mathrm{Q}^{(s)}$ such that $\boldsymbol{Q}^{(s)}\boldsymbol{Q}^{(s')} = \boldsymbol{Q}^{(s'')}$ whenever $\mathrm{Q}^{(s)}\mathrm{Q}^{(s')} = \mathrm{Q}^{(s'')}$, then the set of matrices $\mathcal{R}(\mathcal{Q}) = \{\boldsymbol{I}, \boldsymbol{Q}^{(1)}, \ldots, \boldsymbol{Q}^{(n-1)}\}$ is an M-dimensional *representation* of $\mathcal{Q}$. It must, of course, be the case that $\boldsymbol{R}(\mathrm{I}) = \boldsymbol{I}$, the M-dimensional unit matrix, and $\boldsymbol{R}([\mathrm{Q}^{(s)}]^{-1}) = [\boldsymbol{Q}^{(s)}]^{-1}$. For ease of notation

$\boldsymbol{Q}^{(0)}$ is taken to denote $\boldsymbol{I}$, where necessary. It is clear that, for any group, one possible assignment is $\boldsymbol{R}(\mathsf{Q}^{(\mathsf{s})}) = (1)$, for ever s. This representation (usually denoted by A) is called the *one-dimensional symmetric representation.* If two representations, of the same dimension, $\mathcal{R}(\mathcal{Q})$ and $\mathcal{R}'(\mathcal{Q})$ are related by the transformation $\boldsymbol{R}'(\mathsf{Q}^{(\mathsf{s})}) = \boldsymbol{P}^{-1}\boldsymbol{R}(\mathsf{Q}^{(\mathsf{s})})\boldsymbol{P}$, for some non-singular matrix $\boldsymbol{P}$ and all s, then the representations are said to be *equivalent.* A representation $\mathcal{R}(\mathcal{Q})$ is said to be *reducible* if it is equivalent to another representation $\mathcal{R}'(\mathcal{Q})$ whose matrices have the block-diagonal form

$$\boldsymbol{R}'(\mathsf{Q}^{(\mathsf{s})}) = \boldsymbol{R}_1(\mathsf{Q}^{(\mathsf{s})}) \oplus \boldsymbol{R}_2(\mathsf{Q}^{(\mathsf{s})}) \oplus \boldsymbol{R}_3(\mathsf{Q}^{(\mathsf{s})}) \oplus \cdots , \tag{A.52}$$

where the blocks are of the same, non-zero, dimension for all $\mathsf{Q}^{(\mathsf{s})}$. It is clear that, for any i, the set of matrices $\boldsymbol{R}_i(\mathsf{Q}^{(\mathsf{s})})$, $\mathsf{s} = 0, 1, \ldots, \mathsf{n}-1$ is itself a representation. Any representation which is not reducible is call *irreducible* and it follows that any reducible representation is equivalent to a representation which has matrices consisting of blocks which are irreducible representations.

For any matrix $\boldsymbol{P}$, $\overline{\boldsymbol{P}}$ denotes the matrix obtained from $\boldsymbol{P}$ by taking the conjugate complex of every element and $\boldsymbol{P}^{(\mathrm{T})}$ denotes the transpose of $\boldsymbol{P}$. The matrix $\overline{\boldsymbol{P}}^{(\mathrm{T})}$ denoted by $\boldsymbol{P}^{\dagger}$ is call the *Hermitian transpose* of $\boldsymbol{P}$. A matrix $\boldsymbol{U}$ such that $\boldsymbol{U}^{-1} = \boldsymbol{U}^{\dagger}$ is called *unitary.* Unitary matrices play a particularly important role in the theory of group representations as can be seem from the following theorems:

Theorem A.3.1. *Any representation of a finite group in terms of non-singular matrices is equivalent to a representation in terms of unitary matrices, (called a* unitary representation*).*

Theorem A.3.2. *Any two equivalent unitary representations can be related by a unitary transformation.*

It follows from Theorem A.3.1 that all the irreducible representations can be expressed in terms of unitary matrices. For one-dimensional irreducible representations this form is, of course, the unique expression of the representation. A one-dimensional matrix (η) belonging to a one-dimensional representation of a group must satisfy the unitary condition $\eta^{-1} = \overline{\eta}$ from which it follows that η must be a complex root of unity.

The *character* $\chi(\mathsf{Q}^{(\mathsf{s})})$ of $\mathsf{Q}^{(\mathsf{s})}$ in the representation $\mathcal{R}(\mathcal{Q})$ is defined by the formula

$$\chi(\mathsf{Q}^{(\mathsf{s})}) = \mathrm{Trace}\{\boldsymbol{Q}^{(\mathsf{s})}\}\,. \tag{A.53}$$

The set of characters $\chi(\mathsf{Q}^{(\mathsf{s})})$, $\mathsf{s} = 0, 1, \ldots, \mathsf{n} - 1$ is called the *character system* of $\mathcal{Q}$ in $\mathcal{R}(\mathcal{Q})$. Since $\mathrm{Trace}\{\boldsymbol{P}^{-1}\boldsymbol{Q}^{(\mathsf{s})}\boldsymbol{P}\} = \mathrm{Trace}\{\boldsymbol{Q}^{(\mathsf{s})}\}$, it follows that all equivalent representations have the same character set. The same type of argument also shows that all operators of the group in the same conjugacy class have the same character. We can conveniently regard all equivalent representations as essentially the *same* representation *realized* in different forms. This representation has a character set, with a character given for each conjugacy class.

An important property of a group is the collection of character sets for the irreducible representations. An useful result in this respect is:

Theorem A.3.3. *The number of (inequivalent) irreducible representations of a group is equal to the number of conjugacy classes in the group.*

We denote the irreducible representations of $\mathcal{Q}$ by $\mathsf{E}(\mathsf{k})$, $\mathsf{k} = 1, 2, \ldots, \kappa$, where $\mathsf{E}(1) = \mathsf{A}$ is the one-dimensional symmetric representation, and $\mathsf{e}(\mathsf{k})$ is the dimension of $\mathsf{E}(\mathsf{k})$. If we denote the (unitary) matrix for $\mathsf{Q}^{(\mathsf{s})}$ in the irreducible representation $\mathsf{E}(\mathsf{k})$ by $\boldsymbol{R}^{(\mathsf{k})}(\mathsf{Q}^{(\mathsf{s})})$ then the character of $\mathsf{Q}^{(\mathsf{s})}$ in the representation $\mathsf{E}(\mathsf{k})$ is $\chi^{(\mathsf{k})}(\mathsf{Q}^{(\mathsf{s})}) = \mathrm{Trace}\{\boldsymbol{R}^{(\mathsf{k})}(\mathsf{Q}^{(\mathsf{s})})\}$. It is clear that

$$\chi^{(\mathsf{k})}(\mathsf{Q}^{(0)}) = \mathsf{e}(\mathsf{k}) \,, \qquad \mathsf{k} = 1, 2, \ldots, \kappa \tag{A.54}$$

and, since the modulus of the trace of a unitary matrix cannot exceed its dimension,

$$|\chi^{(\mathsf{k})}(\mathsf{Q}^{(\mathsf{s})})| \leq \mathsf{e}(\mathsf{k}) \,, \qquad \mathsf{s} = 1, 2, \ldots, \mathsf{n} - 1 \,, \qquad \mathsf{k} = 1, 2, \ldots, \kappa \tag{A.55}$$

with

$$|\chi^{(\mathsf{k})}(\mathsf{Q}^{(\mathsf{s})})| = 1 \,, \qquad \mathsf{s} = 0, 1, \ldots, \mathsf{n} - 1 \,, \qquad \text{if } \mathsf{e}(\mathsf{k}) = 1 \,. \tag{A.56}$$

It can be shown that:

Theorem A.3.4. *If the elements of* $\boldsymbol{R}^{(\mathsf{k})}(\mathsf{Q}^{(\mathsf{s})})$ *are denoted by* $R_{ij}^{(\mathsf{k})}(\mathsf{Q}^{(\mathsf{s})})$ *for* $i, j = 1, 2, \ldots, \mathsf{e}(\mathsf{k})$ *and* $\omega(\mathsf{s})$ *is the number of elements in the conjugacy class containing* $\mathsf{Q}^{(\mathsf{s})}$ *then*

$$\sum_{\mathsf{s}=0}^{n-1} \overline{R}_{ij}^{(\mathsf{k})}(\mathsf{Q}^{(\mathsf{s})}) R_{i'j'}^{(\mathsf{k}')}(\mathsf{Q}^{(\mathsf{s})}) = \frac{\mathsf{n}}{\mathsf{e}(\mathsf{k})} \delta^{\mathrm{Kr}}(\mathsf{k} - \mathsf{k}') \delta^{\mathrm{Kr}}(i - i') \delta^{\mathrm{Kr}}(j - j') \,, \tag{A.57}$$

$$\sum_{\mathsf{k}=1}^{\kappa} \overline{\chi}^{(\mathsf{k})}(\mathsf{Q}^{(\mathsf{s})}) \chi^{(\mathsf{k})}(\mathsf{Q}^{(\mathsf{s}')}) = \begin{cases} \mathsf{n}/\omega(\mathsf{s}) \,, & \mathsf{Q}^{(\mathsf{s})} \textit{ and } \mathsf{Q}^{(\mathsf{s}')} \textit{ conjugate} \,, \\ 0 \,, & \mathsf{Q}^{(\mathsf{s})} \textit{ and } \mathsf{Q}^{(\mathsf{s}')} \textit{ not conjugate} \,. \end{cases} \tag{A.58}$$

Equations (A.57) and (A.58) are the *Schur orthogonality relations.* From (A.57)

$$\sum_{s=0}^{n-1} \overline{\chi}^{(k)}(Q^{(s)}Q^{(s')})\chi^{(k')}(Q^{(s)}Q^{(s'')})$$

$$= \frac{n}{e(k)}\delta^{Kr}(k-k')\sum_{i=1}^{e(k)}\sum_{j=1}^{e(k)} \overline{R}_{ji}^{(k)}(Q^{(s')})R_{ji}^{(k)}(Q^{(s'')})$$

$$= \frac{n}{e(k)}\delta^{Kr}(k-k')\chi^{(k)}([Q^{(s')}]^{-1}Q^{(s'')}). \tag{A.59}$$

Since $Q^{(0)} = I$ is in a conjugacy class on its own, (A.58) with $s = s' = 0$ yields

$$\sum_{k=1}^{\kappa} e^2(k) = n. \tag{A.60}$$

Given that $\mathcal{R}(\mathcal{Q})$ is reducible, it follows from the above discussion that there exists a unitary matrix which will reduce all the matrices $\boldsymbol{Q}^{(s)}$ to block-diagonal form with blocks corresponding to the irreducible representations. The number of blocks $m(k)$ corresponding to $E(k)$ is called the *multiplicity* of $E(k)$. Since the blocks for any $E(k)$ can be collected together by a the operation of a permutation matrix, which is itself unitary, there exists a unitary matrix $\boldsymbol{U}$ such that

$$\boldsymbol{U}^{\dagger}\boldsymbol{Q}^{(s)}\boldsymbol{U} = \mathring{\boldsymbol{Q}}^{(s)} = \bigoplus_{k=1}^{\kappa} m(k)\boldsymbol{R}^{(k)}(Q^{(s)}), \qquad s = 0, 1, \ldots, n-1 \tag{A.61}$$

and

$$\mathring{\boldsymbol{Q}}^{(s)} = \mathring{\boldsymbol{Q}}^{(s)}(1) \oplus \mathring{\boldsymbol{Q}}^{(s)}(2) \oplus \cdots \oplus \mathring{\boldsymbol{Q}}^{(s)}(\kappa), \tag{A.62}$$

where $\mathring{\boldsymbol{Q}}^{(s)}(k)$ is the block corresponding to the representation $E(k)$. It is of dimension $w(k) = m(k)e(k)$ and itself consists of $m(k)$ identical blocks of dimension $e(k)$. The realization (A.62) of $\boldsymbol{Q}^{(s)}$ is called the *canonical form.* By taking the trace of equation (A.61) we have

$$\chi(Q^{(s)}) = \sum_{k=1}^{\kappa} m(k)\chi^{(k)}(Q^{(s)}). \tag{A.63}$$

It follows, from (A.54), (A.59) with $s' = s'' = 0$ and (A.63), that

$$m(k) = \frac{1}{n}\sum_{s=0}^{n-1} \overline{\chi}(Q^{(s)})\chi^{(k)}(Q^{(s)}). \tag{A.64}$$

The characters of the irreducible representations of a group are presented in the form of a character table (see the references given above or Conway et al. 1985). If, therefore, $\mathcal{Q}$ is a standard group (or the direct product of standard groups, see below) then (A.64) can be used to determine the multiplicities of the irreducible representations in the arbitrary representation $\mathcal{R}(\mathcal{Q})$.

Even if the character table of $\mathcal{Q}$ is not known we can still determine whether a particular representation is irreducible since, from (A.63) and (A.64),

$$\frac{1}{n}\sum_{s=0}^{n-1}\overline{\chi}(Q^{(s)})\chi(Q^{(s)}) = \sum_{k=1}^{\kappa} m^2(k)\,. \tag{A.65}$$

The right-hand side of this equation is unity if and only if only one irreducible representation appears with multiplicity one in $\mathcal{R}(\mathcal{Q})$.

The use of group representation theory in statistical mechanics arises because of the presence of certain symmetries in the Hamiltonian (see Sect. 4.9). These can be of different types. There may, for example, be a group of symmetry operations arising from rotations or translations of the lattice and another arising from permutations among the states of the microsystems. In such a case the total symmetry group $\mathcal{Q}$ will be the direct product $\mathcal{F}\times\mathcal{G}$ of two groups $\mathcal{F}=\{F^{(0)},F^{(1)},\ldots\}$ and $\mathcal{G}=\{G^{(0)},G^{(1)},\ldots\}$. The elements of $\mathcal{Q}$ will be all the pairs $F^{(i)}G^{(j)}$ of the elements of $\mathcal{F}$ and $\mathcal{G}$. Since these groups operating in different physical ways on the Hamiltonian $F^{(i)}G^{(j)}$ and $G^{(j)}F^{(i)}$ correspond to the same operator and $n(\mathcal{Q})=n(\mathcal{F})n(\mathcal{G})$ and $\kappa(\mathcal{Q})=\kappa(\mathcal{F})\kappa(\mathcal{G})$. It is, however, necessary to be rather careful when constructing the representation $\mathcal{R}(\mathcal{Q})$ of $\mathcal{Q}$ from the representations $\mathcal{R}(\mathcal{F})$ and $\mathcal{R}(\mathcal{G})$ of $\mathcal{F}$ and $\mathcal{G}$ respectively. This can be achieved either by setting

$$\boldsymbol{R}(F^{(i)}G^{(j)}) = \boldsymbol{R}(G^{(j)}F^{(i)}) = \boldsymbol{R}(F^{(i)})\otimes\boldsymbol{R}(G^{(j)}) \tag{A.66}$$

or

$$\boldsymbol{R}(F^{(i)}G^{(j)}) = \boldsymbol{R}(G^{(j)}F^{(i)}) = \boldsymbol{R}(G^{(j)})\otimes\boldsymbol{R}(F^{(i)}). \tag{A.67}$$

Since, in general,

$$\boldsymbol{R}(F^{(i)})\otimes\boldsymbol{R}(G^{(j)}) \neq \boldsymbol{R}(G^{(j)})\otimes\boldsymbol{R}(F^{(i)}) \tag{A.68}$$

(A.66) and (A.67) correspond to different representations of $\mathcal{R}(\mathcal{Q})$.[2] It is important to choose, in advance, a particular group ordering for the representation and we shall make the choice given by (A.66). The irreducible representations $E(k',k'')$ of $\mathcal{Q}$ are given by taking direct products $E_{\mathcal{F}}(k')\otimes E_{\mathcal{G}}(k'')$ of all the irreducible representations $E_{\mathcal{F}}(k')$ and $E_{\mathcal{G}}(k'')$ of $\mathcal{F}$ and $\mathcal{G}$. By taking the trace of (A.66) for irreducible representations it follows from the property of direct matrix products that

$$\chi^{(k',k'')}(F^{(i)}G^{(j)}) = \chi_{\mathcal{F}}^{(k')}(F^{(i)})\chi_{\mathcal{G}}^{(k'')}(G^{(j)})\,. \tag{A.69}$$

The character table of $\mathcal{Q}$ can, therefore, be constructed from those of $\mathcal{F}$ and $\mathcal{G}$.

The canonical form for $\boldsymbol{Q}^{(s)}$ given by (A.62) is unique. Suppose that $\boldsymbol{W}$ is a unitary matrix such that

---

[2] A useful summary of the properties of direct products of matrices is given by Lloyd (1953).

$$\boldsymbol{W}^{\dagger}\boldsymbol{Q}^{(s)}\boldsymbol{W} = \widetilde{\boldsymbol{Q}}^{(s)} = \bigoplus_{k=1}^{\kappa} \widetilde{\boldsymbol{Q}}^{(s)}(k)\,, \qquad s = 0, 1, \ldots, n-1\,, \tag{A.70}$$

where

$$\frac{1}{n}\sum_{s=0}^{n-1} \mathrm{Trace}\{\widetilde{\boldsymbol{Q}}^{(s)}(k)\}\overline{\chi}^{(k')}(Q^{(s)}) = \delta^{\mathrm{Kr}}(k-k')m(k)\,. \tag{A.71}$$

This represents a non-unique reduction of the matrices of $\mathcal{Q}$ to a set of κ reducible representations each of which involves only one irreducible representation. The final reduction to canonical form can be made if a set of unitary matrices $\boldsymbol{L}(k)$, $k = 1, 2, \ldots, \kappa$ can be found such that

$$\boldsymbol{L}^{\dagger}(k)\widetilde{\boldsymbol{Q}}^{(s)}(k)\boldsymbol{L}(k) = \mathring{\boldsymbol{Q}}^{(s)}(k)\,, \qquad s = 0, 1, \ldots, n-1\,. \tag{A.72}$$

Then

$$\boldsymbol{U} = \boldsymbol{W}\boldsymbol{L}\,, \tag{A.73}$$

where

$$\boldsymbol{L} = \boldsymbol{L}(1) \oplus \boldsymbol{L}(2) \oplus \cdots \oplus \boldsymbol{L}(\kappa)\,. \tag{A.74}$$

Procedures for obtaining the matrices $\boldsymbol{W}$ and $\boldsymbol{L}$ are discussed below. As we shall see, the matrix $\boldsymbol{W}$ can usually be obtained in a straightforward way. Computing the blocks of $\boldsymbol{L}$ is often more difficult. The following result is important for the discussion:

Theorem A.3.5. *For the elements of the matrices* $\widetilde{\boldsymbol{Q}}^{(s)}(k)$ *of (A.70)*

$$\sum_{s=0}^{n-1} \widetilde{Q}^{(s)}_{ij}(k)\overline{\widetilde{Q}^{(s)}_{i'j'}(k')} = 0 \tag{A.75}$$

for all i, j, i' *and* j', *whenever* $k \neq k'$.

Proof: From (A.72)

$$\widetilde{Q}^{(s)}_{ij}(k) = \sum_{p=1}^{w(k)}\sum_{q=1}^{w(k)} L_{ip}(k)\mathring{Q}^{(s)}_{pq}(k)\overline{L}_{jq}(k)\,. \tag{A.76}$$

Since, from (A.61), $\mathring{Q}^{(s)}_{pq}(k)$ is either zero or some element $R^{(k)}_{p'q'}(Q^{(s)})$, (A.75) follows from the orthogonality relation (A.57).

A.3.3 The Block Diagonalization of Transfer Matrices

In Sect. 4.9 group representation theory is applied to transfer matrix problems. The basic idea of this procedure can be described in the following way. The elements $V(\boldsymbol{\tau}_m, \boldsymbol{\tau}_{m'})$ of the transfer matrix $\boldsymbol{V}$ are parameterized by the states $\boldsymbol{\tau}_m$, $m = 1, 2, \ldots, M$ of two sets of sites on a lattice. The largest group

$\mathcal{Q} = \{\mathrm{I}, \mathrm{Q}^{(1)}, \mathrm{Q}^{(2)}, \dots, \mathrm{Q}^{(\mathrm{n}-1)}\}$, such that each $\mathrm{Q}^{(\mathrm{s})}$ maps a state $\boldsymbol{\tau}_m$ into another state $\mathrm{Q}^{(\mathrm{s})}\boldsymbol{\tau}_m$, with $\mathrm{Q}^{(\mathrm{s})}\boldsymbol{\tau}_m \neq \mathrm{Q}^{(\mathrm{s})}\boldsymbol{\tau}_{m'}$ if $\boldsymbol{\tau}_m \neq \boldsymbol{\tau}_{m'}$, and for which

$$V(\mathrm{Q}^{(\mathrm{s})}\boldsymbol{\tau}_m, \mathrm{Q}^{(\mathrm{s})}\boldsymbol{\tau}_{m'}) = V(\boldsymbol{\tau}_m, \boldsymbol{\tau}_{m'}) \tag{A.77}$$

is the *symmetry group of the transfer matrix*. The effect of the group $\mathcal{Q}$ on the states $\boldsymbol{\tau}_m$ is given, using the representation $\mathcal{R}(\mathcal{Q}) = \{\boldsymbol{I}, \boldsymbol{Q}^{(1)}, \dots, \boldsymbol{Q}^{(\mathrm{n}-1)}\}$ of M-dimensional matrices, by

$$(\mathrm{Q}^{(\mathrm{s})}\boldsymbol{\tau}_1, \mathrm{Q}^{(\mathrm{s})}\boldsymbol{\tau}_2, \dots, \mathrm{Q}^{(\mathrm{s})}\boldsymbol{\tau}_M) = (\boldsymbol{\tau}_1, \boldsymbol{\tau}_2, \dots, \boldsymbol{\tau}_M)\boldsymbol{Q}^{(\mathrm{s})}, \qquad \mathrm{s} = 0, 1, \dots, \mathrm{n}-1. \tag{A.78}$$

This representation of the group $\mathcal{Q}$ is called the *permutation representation* associated with the set $\boldsymbol{\tau}_m$, $m = 1, 2, \dots, M$ and the symmetry property of the transfer matrix can be represented in the form

$$[\boldsymbol{Q}^{(\mathrm{s})}]^{\dagger}\boldsymbol{V}\boldsymbol{Q}^{(\mathrm{s})} = \boldsymbol{V}, \qquad \mathrm{s} = 0, 1, \dots, \mathrm{n}-1. \tag{A.79}$$

If $M = \mathrm{n}$ and if the states can be labelled by the index $\mathrm{s} = 0, 1, \dots, \mathrm{n}-1$ so that $\mathrm{Q}^{(\mathrm{s})}\boldsymbol{\tau}_{\mathrm{s}'} = \boldsymbol{\tau}_{\mathrm{s}''}$ whenever $\mathrm{Q}^{(\mathrm{s})}\mathrm{Q}^{(\mathrm{s}')} = \mathrm{Q}^{(\mathrm{s}'')}$, then the representation is called *regular*. It is not difficult to show that:

Theorem A.3.6. *If the number of states $M = \mathrm{n}$ and if, for some arbitrarily chosen state denoted (without loss of generality) by $\boldsymbol{\tau}_1$, all the states $\mathrm{Q}^{(\mathrm{s})}\boldsymbol{\tau}_1$, $\mathrm{s} = 0, 1, \dots, \mathrm{n}-1$ are distinct then the permutation representation associated with $\boldsymbol{\tau}_{\mathrm{s}+1}$, $\mathrm{s} = 0, 1, \dots, \mathrm{n}-1$ defined by*

$$\boldsymbol{\tau}_{\mathrm{s}+1} = \mathrm{Q}^{(\mathrm{s})}\boldsymbol{\tau}_1 \tag{A.80}$$

is regular.

It is clear that, for the regular representation, $\mathrm{Q}^{(\mathrm{s}')}\boldsymbol{\tau}_{\mathrm{s}} \neq \boldsymbol{\tau}_{\mathrm{s}}$ unless $\mathrm{s}' = 0$ and thus

$$\begin{aligned} \chi(\mathrm{Q}^{(\mathrm{s})}) &= \mathrm{Trace}\{\boldsymbol{Q}^{(\mathrm{s})}\} = 0, \qquad \mathrm{s} = 1, 2, \dots, \mathrm{n}-1, \\ \chi(\mathrm{Q}^{(0)}) &= \mathrm{Trace}\{\boldsymbol{I}\} = \mathrm{n}. \end{aligned} \tag{A.81}$$

Then, from (A.64), $\mathrm{m}(\mathrm{k}) = \mathrm{e}(\mathrm{k})$; each irreducible representation occurs in the regular representation with a multiplicity equal to its dimension.

Equations (A.61) and (A.79) are important formulae for the application of group representation theory to a statistical mechanical system in transfer matrix form. It can be shown that, if the matrix $\boldsymbol{U}$ is known, it can be used, together with row and column permutations, to reduce $\boldsymbol{V}$ to a block-diagonal form, each block being identified with some representation $\mathrm{E}(\mathrm{k})$ and occurring $\mathrm{e}(\mathrm{k})$ times. This reduction is to the unique canonical form of $\boldsymbol{V}$. If this can be achieved it is ideal. However, as we have indicated above, the matrix $\boldsymbol{U}$ is often difficult to obtain. We shall therefore initially consider the more modest gains which can be achieved if the matrix $\boldsymbol{W}$, of (A.70) is all that is known. This can be characterized by the following result:

Theorem A.3.7. *If $\boldsymbol{W}$ has the effect on the matrices of the group $\mathfrak{Q}$ given by (A.70) and the transfer matrix satisfies the formula (A.79) then*

$$\boldsymbol{W}^{\dagger}\boldsymbol{V}\boldsymbol{W} = \widetilde{\boldsymbol{V}} = \widetilde{\boldsymbol{V}}(1) \oplus \widetilde{\boldsymbol{V}}(2) \oplus \cdots \oplus \widetilde{\boldsymbol{V}}(\kappa)\,. \tag{A.82}$$

Proof: From (A.70), (A.79) and the first equality of (A.82)

$$[\widetilde{\boldsymbol{Q}}^{(s)}]^{\dagger}\widetilde{\boldsymbol{V}}\widetilde{\boldsymbol{Q}}^{(s)} = \widetilde{\boldsymbol{V}}\,, \qquad s = 0, 1, \ldots, n-1\,. \tag{A.83}$$

Suppose that $\widetilde{\boldsymbol{V}}$ is expressed in blocks by

$$\widetilde{\boldsymbol{V}} = \begin{pmatrix} \widetilde{\boldsymbol{V}}(1,1) & \widetilde{\boldsymbol{V}}(1,2) & \cdots & \widetilde{\boldsymbol{V}}(1,\kappa) \\ \widetilde{\boldsymbol{V}}(2,1) & \widetilde{\boldsymbol{V}}(2,2) & \cdots & \widetilde{\boldsymbol{V}}(2,\kappa) \\ \vdots & \vdots & \ddots & \vdots \\ \widetilde{\boldsymbol{V}}(\kappa,1) & \widetilde{\boldsymbol{V}}(\kappa,2) & \cdots & \widetilde{\boldsymbol{V}}(\kappa,\kappa) \end{pmatrix}. \tag{A.84}$$

Then it follows from (A.83) that

$$[\widetilde{\boldsymbol{Q}}^{(s)}(k)]^{\dagger}\widetilde{\boldsymbol{V}}(k,k')\widetilde{\boldsymbol{Q}}^{(s)}(k') = \widetilde{\boldsymbol{V}}(k,k')\,, \qquad s = 0, 1, \ldots, n-1\,. \tag{A.85}$$

This gives

$$\widetilde{V}_{ij}(k,k') = \sum_{p=1}^{w(k)} \sum_{q=1}^{w(k')} \overline{\widetilde{Q}^{(s)}_{pi}(k)}\widetilde{V}_{pq}(k,k')\widetilde{Q}^{(s)}_{qj}(k')$$
$$s = 0, 1, \ldots, n-1\,. \tag{A.86}$$

Summing over all the operators $Q^{(s)}$ it follows from Theorem A.3.5 that the right-hand side of equation (A.86) is zero unless $k = k'$. Thus we have $\widetilde{\boldsymbol{V}}(k,k') = \boldsymbol{0}$ if $k \neq k'$ and we set $\widetilde{\boldsymbol{V}}(k,k) = \widetilde{\boldsymbol{V}}(k)$.

A.3.4 Equivalence Classes

For large M, the matrices $\widetilde{\boldsymbol{V}}(k)$ will need to be obtained using a computer. The obvious method for doing this, which involves storing $\boldsymbol{V}$, finding and storing $\boldsymbol{W}$, and using the formula (A.82), is not necessarily the most effective, particularly if the need is simply to obtain the block for one particular $E(k)$. A useful tool in this respect is the idea of *equivalence classes.* An equivalence class is a subset of the states $\boldsymbol{\tau}_m$, $m = 1, 2, \ldots, M$, which is closed under the operations of the group $\mathfrak{Q}$ and is not further divisible into closed subsets. Suppose that the states are ordered in equivalence classes $\mathfrak{E}(\alpha)$, $\alpha = 1, 2, \ldots, \gamma$, where $\mathfrak{E}(\alpha) = \{\boldsymbol{\tau}_1(\alpha), \boldsymbol{\tau}_2(\alpha), \ldots, \boldsymbol{\tau}_{\mathfrak{p}(\alpha)}(\alpha)\}$. The following result is useful in determining these equivalence classes:

Theorem A.3.8. *The equivalence class $\mathfrak{E}(\alpha)$ can be generated by applying all the elements of $\mathfrak{Q}$ to any one state of $\mathfrak{E}(\alpha)$. The number of states $\mathfrak{p}(\alpha)$ in $\mathfrak{E}(\alpha)$ is a divisor of n.*

Proof: To establish the first part of the theorem take an arbitrary $\boldsymbol{\tau}_m(\alpha) \in \mathfrak{E}(\alpha)$ and consider the set $\mathfrak{S} = \{\boldsymbol{\tau}_m(\alpha), \mathsf{Q}^{(1)}\boldsymbol{\tau}_m(\alpha), \ldots, \mathsf{Q}^{(n-1)}\boldsymbol{\tau}_m(\alpha)\}$. Clearly $\mathfrak{S}$ is closed under $\mathfrak{Q}$, since, for any i and j there is an i' such that $\mathsf{Q}^{(\mathsf{j})}\mathsf{Q}^{(\mathsf{i})}\boldsymbol{\tau}_m(\alpha) = \mathsf{Q}^{(\mathsf{i}')}\boldsymbol{\tau}_m(\alpha)$, and thus $\mathfrak{S} = \mathfrak{E}(\alpha)$.

If $\mathfrak{p}(\alpha) < n$, $\mathfrak{S}$ must contain some states which are identical. Suppose that the complete set of states in $\mathfrak{S}$ identical to $\boldsymbol{\tau}_m(\alpha)$ is $\mathsf{Q}^{(\mathsf{s}_q)}\boldsymbol{\tau}_m(\alpha)$, $q = 1, 2, \ldots, \eta$ and for some $\mathsf{Q}^{(\mathsf{j})}\boldsymbol{\tau}_m(\alpha) \neq \boldsymbol{\tau}_m(\alpha)$ the complete set of states in $\mathfrak{S}$ identical to $\mathsf{Q}^{(\mathsf{j})}\boldsymbol{\tau}_m(\alpha)$ is $\mathsf{Q}^{(\ell_r)}\boldsymbol{\tau}_m(\alpha)$, $r = 1, 2, \ldots, \eta'$. But the states $\mathsf{Q}^{(\mathsf{j})}\mathsf{Q}^{(\mathsf{s}_q)}\boldsymbol{\tau}_m(\alpha)$, $q = 1, 2, \ldots, \eta$ are identical to $\mathsf{Q}^{(\mathsf{j})}\boldsymbol{\tau}_m(\alpha)$, so $\eta' \geq \eta$. Similarly the states $[\mathsf{Q}^{(\mathsf{j})}]^{-1}\mathsf{Q}^{(\ell_r)}\boldsymbol{\tau}_m(\alpha)$, $r = 1, 2, \ldots, \eta'$ are identical to $\boldsymbol{\tau}_m(\alpha)$, so $\eta \geq \eta'$. Thus $\eta = \eta'$ and it follows that $\mathfrak{S}$ consists of $\mathfrak{p}(\alpha)$ sets each consisting of η identical states, $\mathfrak{p}(\alpha)$ thus being a divisor of n.

It follows from (A.78) that $\boldsymbol{Q}^{(\mathsf{s})}$ has the block-diagonal form

$$\boldsymbol{Q}^{(\mathsf{s})} = \boldsymbol{Q}^{(\mathsf{s})}(1) \oplus \boldsymbol{Q}^{(\mathsf{s})}(2) \oplus \cdots \oplus \boldsymbol{Q}^{(\mathsf{s})}(\gamma), \quad \mathsf{s} = 0, 1, \ldots, \mathsf{n}-1 \tag{A.87}$$

with $\mathcal{R}(\alpha; \mathfrak{Q}) = \{\boldsymbol{I}(\alpha), \boldsymbol{Q}^{(1)}(\alpha), \ldots, \boldsymbol{Q}^{(\mathsf{n}-1)}(\alpha)\}$, a $\mathfrak{p}(\alpha)$-dimensional representation of $\mathfrak{Q}$ operating on the equivalence class $\mathfrak{E}(\alpha)$, for $\alpha = 1, 2, \ldots, \gamma$. Formulae of the same form as (A.70) and (A.72) now apply to the matrices of each of these representations, namely

$$\boldsymbol{W}^{\dagger}(\alpha)\boldsymbol{Q}^{(\mathsf{s})}(\alpha)\boldsymbol{W}(\alpha) = \widetilde{\boldsymbol{Q}}^{(\mathsf{s})}(\alpha) = \bigoplus_{\mathsf{k}=1}^{\mathsf{K}} \widetilde{\boldsymbol{Q}}^{(\mathsf{s})}(\alpha; \mathsf{k}), \quad \mathsf{s} = 0, 1, \ldots, \mathsf{n}-1, \tag{A.88}$$

$$\boldsymbol{L}^{\dagger}(\alpha; \mathsf{k})\widetilde{\boldsymbol{Q}}^{(\mathsf{s})}(\alpha; \mathsf{k})\boldsymbol{L}(\alpha; \mathsf{k}) = \mathring{\boldsymbol{Q}}^{(\mathsf{s})}(\alpha; \mathsf{k}) = \bigoplus_{\mathsf{k}=1}^{\mathsf{K}} \mathsf{m}(\alpha; \mathsf{k})\boldsymbol{R}^{(\mathsf{k})}(\mathsf{Q}^{(\mathsf{s})}), \quad \mathsf{s} = 0, 1, \ldots, \mathsf{n}-1. \tag{A.89}$$

The matrix $\boldsymbol{W}(\alpha)$ have the effect of reducing $\boldsymbol{Q}^{(\mathsf{s})}(\alpha)$ to block-diagonal form with the blocks corresponding to the irreducible representations in order. But to achieve the reduction given in (A.70), we need a further ordering between equivalence classes. Thus

$$\boldsymbol{W} = [\boldsymbol{W}(1) \oplus \boldsymbol{W}(2) \oplus \cdots \oplus \boldsymbol{W}(\gamma)]\boldsymbol{\Pi}, \tag{A.90}$$

where $\boldsymbol{\Pi}$ is a permutation matrix, chosen to achieve the final ordering described above. The following theorem is useful in determining the multiplicities of the irreducible representations within equivalence classes:

Theorem A.3.9. *For any equivalence class* $\mathfrak{E}(\alpha)$ *the multiplicity* $\mathsf{m}(\alpha; \mathsf{k})$ *of the irreducible representation* $\mathsf{E}(\mathsf{k})$ *of dimension* $\mathsf{e}(\mathsf{k})$ *satisfies the condition*

$$\mathsf{m}(\alpha; \mathsf{k}) \leq \mathsf{e}(\mathsf{k}), \tag{A.91}$$

with $\mathsf{m}(\alpha; 1) = \mathsf{e}(1) = 1$.

Proof: From Theorem A.3.8 there are exactly $\mathfrak{n}/\mathfrak{p}(\alpha)$ values of s for which the $\mathsf{Q}^{(\mathsf{s})}\boldsymbol{\tau}_m(\alpha) = \boldsymbol{\tau}_{m'}(\alpha)$ is satisfied for any pair of states $\boldsymbol{\tau}_m(\alpha)$ and $\boldsymbol{\tau}_{m'}(\alpha)$ in $\mathfrak{E}(\alpha)$.

If $\mathfrak{p}(\alpha) = \mathfrak{n}$ it then follows from Theorems A.3.6 and A.3.8 that the representation associated with $\mathfrak{E}(\alpha)$ is regular. For such representations the characters are given by (A.81) and we have already seen that this means that the multiplicity of an irreducible representation is equal to its dimension.

If $\mathfrak{p}(\alpha) < \mathfrak{n}$ then among all the matrices of the representation $\mathcal{R}(\mathcal{Q})$ there will be $\mathfrak{p}(\alpha) \times \mathfrak{n}/\mathfrak{p}(\alpha) = \mathfrak{n}$ non-zero diagonal elements, all equal to unity. Thus the characters $\chi(\alpha; \mathsf{Q}^{(\mathsf{s})})$, $\mathsf{s} = 0, 1, \ldots, \mathfrak{n}-1$ of $\mathcal{R}(\alpha; \mathcal{Q})$ are non-negative integers satisfying

$$\sum_{\mathsf{s}=0}^{\mathfrak{n}-1} \chi(\alpha; \mathsf{Q}^{(\mathsf{s})}) = \mathfrak{n}. \tag{A.92}$$

The multiplicities are given by equation (A.64) applied to $\mathfrak{E}(\alpha)$ and, since $\chi^{(1)}(\mathsf{Q}^{(\mathsf{s})}) = 1$ for all $\mathsf{s} = 0, 1, \ldots, \mathfrak{n} - 1$, $\mathfrak{m}(\alpha; 1) = 1$. For $\mathsf{k} \neq 1$ it follows, from (A.55) and (A.64), that

$$\begin{aligned} \mathfrak{m}(\alpha; \mathsf{k}) &= \frac{1}{\mathfrak{n}} \sum_{\mathsf{s}=0}^{\mathfrak{n}-1} \overline{\chi}(\alpha; \mathsf{Q}^{(\mathsf{s})}) \chi^{(\mathsf{k})}(\mathsf{Q}^{(\mathsf{s})}) \\ &\leq \frac{1}{\mathfrak{n}} \sum_{\mathsf{s}=0}^{\mathfrak{n}-1} \overline{\chi}(\alpha; \mathsf{Q}^{(\mathsf{s})}) |\chi^{(\mathsf{k})}(\mathsf{Q}^{(\mathsf{s})})| \\ &\leq \frac{1}{\mathfrak{n}} \sum_{\mathsf{s}=0}^{\mathfrak{n}-1} \overline{\chi}(\alpha; \mathsf{Q}^{(\mathsf{s})}) \mathsf{e}(\mathsf{k}) = \mathsf{e}(\mathsf{k}). \end{aligned} \tag{A.93}$$

We now consider the problem of determining unitary matrices $\boldsymbol{W}(\alpha)$ which produce the block diagonalization given by (A.88). For this, and the subsequent task of obtaining the matrices $\boldsymbol{L}(\mathsf{k})$ of (A.72), the following results are useful:

Theorem A.3.10. *The operators*

$$\mathsf{P}_{j\ell}^{(\mathsf{k})} = \frac{\mathsf{e}(\mathsf{k})}{\mathfrak{n}} \sum_{\mathsf{s}=0}^{\mathfrak{n}-1} \overline{R}_{j\ell}^{(\mathsf{k})}(\mathsf{Q}^{(\mathsf{s})}) \mathsf{Q}^{(\mathsf{s})}$$
$$\ell = 1, 2, \ldots, \mathsf{e}(\mathsf{k}), \qquad \mathsf{k} = 1, 2, \ldots, \kappa \tag{A.94}$$

transform according to the formulae

$$\mathsf{Q}^{(\mathsf{s})} \mathsf{P}_{j\ell}^{(\mathsf{k})} = \sum_{i=1}^{\mathsf{e}(\mathsf{k})} R_{ij}^{(\mathsf{k})}(\mathsf{Q}^{(\mathsf{s})}) \mathsf{P}_{i\ell}^{(\mathsf{k})}, \tag{A.95}$$

$$\mathsf{P}_{j\ell}^{(\mathsf{k})} \mathsf{Q}^{(\mathsf{s})} = \sum_{i=1}^{\mathsf{e}(\mathsf{k})} R_{\ell i}^{(\mathsf{k})}(\mathsf{Q}^{(\mathsf{s})}) \mathsf{P}_{ji}^{(\mathsf{k})} \tag{A.96}$$

and satisfies the product rule

$$\mathsf{P}^{(\mathsf{k}')}_{j'\ell'}\mathsf{P}^{(\mathsf{k})}_{j\ell} = \delta^{\mathrm{Kr}}(\mathsf{k}-\mathsf{k}')\delta^{\mathrm{Kr}}(\ell'-j)\mathsf{P}^{(\mathsf{k})}_{j'\ell}\,. \tag{A.97}$$

Proof:

$$\begin{aligned}\mathsf{Q}^{(\mathsf{s})}\mathsf{P}^{(\mathsf{k})}_{j\ell} &= \frac{\mathsf{e}(\mathsf{k})}{\mathsf{n}}\sum_{\mathsf{s}'=0}^{\mathsf{n}-1}\overline{R}^{(\mathsf{k})}_{j\ell}(\mathsf{Q}^{(\mathsf{s}')})\mathsf{Q}^{(\mathsf{s})}\mathsf{Q}^{(\mathsf{s}')}\\ &= \frac{\mathsf{e}(\mathsf{k})}{\mathsf{n}}\sum_{\mathsf{s}'=0}^{\mathsf{n}-1}\overline{R}^{(\mathsf{k})}_{j\ell}([\mathsf{Q}^{(\mathsf{s})}]^{-1}\mathsf{Q}^{(\mathsf{s}')})\mathsf{Q}^{(\mathsf{s}')}\\ &= \frac{\mathsf{e}(\mathsf{k})}{\mathsf{n}}\sum_{i=1}^{\mathsf{e}(\mathsf{k})}\sum_{\mathsf{s}'=0}^{\mathsf{n}-1}R^{(\mathsf{k})}_{ij}(\mathsf{Q}^{(\mathsf{s})})\overline{R}^{(\mathsf{k})}_{i\ell}(\mathsf{Q}^{(\mathsf{s}')})\mathsf{Q}^{(\mathsf{s}')}\\ &= \sum_{i=1}^{\mathsf{e}(\mathsf{k})}R^{(\mathsf{k})}_{ij}(\mathsf{Q}^{(\mathsf{s})})\mathsf{P}^{(\mathsf{k})}_{i\ell}\,.\end{aligned} \tag{A.98}$$

The proof of (A.96) is similar to that of (A.95). To prove (A.97) express $\mathsf{P}^{(\mathsf{k}')}_{j'\ell'}$ in the form (A.94) and use (A.95) and the orthogonality formula (A.57).

Theorem A.3.11. *The operators*

$$\mathsf{T}^{(\mathsf{k})} = \sum_{j=1}^{\mathsf{e}(\mathsf{k})}\mathsf{P}^{(\mathsf{k})}_{jj} = \frac{\mathsf{e}(\mathsf{k})}{\mathsf{n}}\sum_{\mathsf{s}=0}^{\mathsf{n}-1}\overline{\chi}^{(\mathsf{k})}(\mathsf{Q}^{(\mathsf{s})})\mathsf{Q}^{(\mathsf{s})}\,, \qquad \mathsf{k}=1,2,\ldots,\kappa \tag{A.99}$$

satisfy the equations

$$\mathsf{Q}^{(\mathsf{s})}\mathsf{T}^{(\mathsf{k})} = \mathsf{T}^{(\mathsf{k})}\mathsf{Q}^{(\mathsf{s})}\,, \qquad \mathsf{s}=0,1,\ldots,\mathsf{n}-1\,, \qquad \mathsf{k}=1,2,\ldots,\kappa \tag{A.100}$$

and

$$\mathsf{T}^{(\mathsf{k}')}\mathsf{T}^{(\mathsf{k})} = \delta^{\mathrm{Kr}}(\mathsf{k}-\mathsf{k}')\mathsf{T}^{(\mathsf{k})}\,, \qquad \mathsf{k}=1,2,\ldots,\kappa\,. \tag{A.101}$$

Proof: Equation (A.100) follows by setting $\ell = j$ in (A.94) and (A.95) and summing over j. Equation (A.101) follows in a similar way from (A.96).

Theorem A.3.12. *With*

$$\boldsymbol{T}^{(\mathsf{k})}(\alpha) = \frac{\mathsf{e}(\mathsf{k})}{\mathsf{n}}\sum_{\mathsf{s}=0}^{\mathsf{n}-1}\overline{\chi}^{(\mathsf{k})}(\mathsf{Q}^{(\mathsf{s})})\boldsymbol{Q}^{(\mathsf{s})}(\alpha)\,, \qquad \mathsf{k}=1,2,\ldots,\kappa\,, \tag{A.102}$$

$$\boldsymbol{W}^{\dagger}(\alpha)\boldsymbol{T}^{(\mathsf{k})}(\alpha)\boldsymbol{W}(\alpha) = \bigoplus_{\mathsf{k}'=1}^{\kappa}\widetilde{\boldsymbol{T}}^{(\mathsf{k})}(\alpha;\mathsf{k}')\,, \tag{A.103}$$

where

$$\widetilde{\boldsymbol{T}}^{(\mathsf{k})}(\alpha;\mathsf{k}') = \delta^{\mathrm{Kr}}(\mathsf{k}-\mathsf{k}')\boldsymbol{I}(\alpha)\,. \tag{A.104}$$

Proof: From (A.88) and (A.102)

$$\widetilde{\boldsymbol{T}}^{(k)}(\alpha;k') = \frac{e(k)}{n}\sum_{s=0}^{n-1}\overline{\chi}^{(k)}(Q^{(s)})\widetilde{\boldsymbol{Q}}^{(s)}(\alpha;k') \tag{A.105}$$

and, from (A.89),

$$\boldsymbol{L}^{\dagger}(\alpha;k')\widetilde{\boldsymbol{T}}^{(k)}(\alpha;k')\boldsymbol{L}(\alpha;k') = \frac{e(k)}{n}\sum_{s=0}^{n-1}\overline{\chi}^{(k)}(Q^{(s)})\mathring{\boldsymbol{Q}}^{(s)}(\alpha;k')\,. \tag{A.106}$$

It follows from the orthogonality formula (A.57) that

$$\boldsymbol{L}^{\dagger}(\alpha;k')\widetilde{\boldsymbol{T}}^{(k)}(\alpha;k')\boldsymbol{L}(\alpha;k') = \delta^{\mathrm{Kr}}(k-k')\boldsymbol{I}(\alpha) \tag{A.107}$$

and (A.104) follows.

From (A.78), (A.87) and (A.88)

$$Q^{(s)}(\boldsymbol{\tau}_1(\alpha),\ldots,\boldsymbol{\tau}_{p(\alpha)}(\alpha))\boldsymbol{W}(\alpha) = (\boldsymbol{\tau}_1(\alpha),\ldots,\boldsymbol{\tau}_{p(\alpha)}(\alpha))\boldsymbol{W}(\alpha)\widetilde{\boldsymbol{Q}}^{(s)}(\alpha)\,. \tag{A.108}$$

The effect of $\boldsymbol{W}(\alpha)$ is to form linear combinations of the states of $\mathfrak{E}(\alpha)$, which transform according to representations each of which reduces to a multiple of a single irreducible representation. Thus a group of $w(\alpha;k) = m(\alpha;k)e(k)$ successive columns of $\boldsymbol{W}(\alpha)$ can be identified with the irreducible representation $E(k)$. It follows from Theorem A.3.12 that these columns are eigenvectors of $\boldsymbol{T}^{(k)}(\alpha)$ with eigenvalue one and all the other columns of $\boldsymbol{W}(\alpha)$ are eigenvectors of $\boldsymbol{T}^{(k)}(\alpha)$ with eigenvalue zero. However, from (A.101), we see that each of the columns of $\boldsymbol{T}^{(k')}(\alpha)$ is an eigenvector of $\boldsymbol{T}^{(k)}(\alpha)$ with eigenvalue one if $k' = k$ and zero if $k' \neq k$. The columns of the matrix $\boldsymbol{W}(\alpha)$ corresponding to the representation $E(k)$ are, therefore, obtained by finding a subset of $w(\alpha;k)$ of the $p(\alpha)$ columns of $\boldsymbol{T}^{(k)}(\alpha)$ which satisfied the required orthonormality. This is possible since the dimension of the space spanned by the columns is the number of non-zero eigenvalues which, from (A.103), is $w(\alpha;k)$. The simplest way to do this is to find all the functions

$$\psi^{(k)}(\alpha;m) = T^{(k)}\boldsymbol{\tau}_m(\alpha) = \frac{e(k)}{n}\sum_{s=0}^{n-1}\overline{\chi}^{(k)}(Q^{(s)})Q^{(s)}\boldsymbol{\tau}_m(\alpha)\,. \tag{A.109}$$

A selection is now made from the sets of coefficients of these linear sums of states to satisfy the orthonormality condition. Let the members of this set of columns be $(w_{1j}^{(k)}(\alpha), w_{2j}^{(k)}(\alpha),\ldots,w_{p(\alpha)j}^{(k)}(\alpha))$, $j = 1,2,\ldots,w(\alpha;k)$.

For the case of an irreducible representation for which $e(k) = 1$ it follows from Theorem A.3.9 that $w(\alpha;k) = 0$ or 1. In the latter case it follows, from (A.99) and (A.100), that

$$\begin{aligned}\mathsf{Q}^{(\mathsf{s})}\psi^{(\mathsf{k})}(\alpha;m) &= \frac{\mathsf{e}(\mathsf{k})}{\mathsf{n}}\sum_{\mathsf{s}'=0}^{\mathsf{n}-1}\overline{\chi}^{(\mathsf{k})}(\mathsf{Q}^{(\mathsf{s}')})\mathsf{Q}^{(\mathsf{s}')}\mathsf{Q}^{(\mathsf{s})}\boldsymbol{\tau}_m(\alpha)\\ &= \frac{\mathsf{e}(\mathsf{k})}{\mathsf{n}}\sum_{\mathsf{s}'=0}^{\mathsf{n}-1}\overline{\chi}^{(\mathsf{k})}(\mathsf{Q}^{(\mathsf{s}')}[\mathsf{Q}^{(\mathsf{s})}]^{-1})\mathsf{Q}^{(\mathsf{s}')}\boldsymbol{\tau}_m(\alpha)\\ &= \chi^{(\mathsf{k})}(\mathsf{Q}^{(\mathsf{s})})\psi^{(\mathsf{k})}(\alpha;m)\,. \end{aligned}\tag{A.110}$$

From Theorem A.3.8 it follows that, given any $\boldsymbol{\tau}_{m'}(\alpha)$, there exists an operator $\mathsf{Q}^{(\mathsf{s}')}$ such that $\boldsymbol{\tau}_{m'}(\alpha) = \mathsf{Q}^{(\mathsf{s}')}\boldsymbol{\tau}_m(\alpha)$. Then, from (A.100), (A.109) and (A.110),

$$\psi^{(\mathsf{k})}(\alpha;m') = \chi^{(\mathsf{k})}(\mathsf{Q}^{(\mathsf{s}')})\psi^{(\mathsf{k})}(\alpha;m)\,. \tag{A.111}$$

So, to within a multiplicative factor, all the functions $\psi^{(\mathsf{k})}(\alpha;m)$ are the same. In two cases it is straightforward to compute the column elements explicitly:

• *For the one-dimensional symmetric representation* $\mathsf{E}(1)$. Then $\chi(\mathsf{Q}^{(\mathsf{s})}) = 1$, $\mathsf{s} = 0, 1, \ldots, \mathsf{n}-1$ and

$$\psi^{(1)}(\alpha;m) = \frac{1}{\mathsf{n}}\sum_{\mathsf{s}=0}^{\mathsf{n}-1}\mathsf{Q}^{(\mathsf{s})}\boldsymbol{\tau}_m(\alpha) = \frac{1}{\mathfrak{p}(\alpha)}\sum_{j=1}^{\mathfrak{p}(\alpha)}\boldsymbol{\tau}_j(\alpha)\,. \tag{A.112}$$

Thus, on normalization,

$$w_{i1}^{(1)}(\alpha) = \frac{1}{\sqrt{\mathfrak{p}(\alpha)}}\,, \qquad i = 1, 2, \ldots, \mathfrak{p}(\alpha)\,. \tag{A.113}$$

• *When the states $\boldsymbol{\tau}_1(\alpha), \ldots, \boldsymbol{\tau}_{\mathfrak{p}(\alpha)}(\alpha)$ correspond to a regular representation of $\mathcal{Q}$.* With $\mathfrak{p}(\alpha) = \mathsf{n}$ and, as in Theorem A.3.6, $\boldsymbol{\tau}_{\mathsf{s}+1}(\alpha) = \mathsf{Q}^{(\mathsf{s})}\boldsymbol{\tau}_1(\alpha)$,

$$\begin{aligned}\psi^{(\mathsf{k})}(\alpha;m) &= \frac{\mathsf{e}(\mathsf{k})}{\mathsf{n}}\sum_{\mathsf{s}=0}^{\mathsf{n}-1}\overline{\chi}^{(\mathsf{k})}(\mathsf{Q}^{(\mathsf{s})})\mathsf{Q}^{(\mathsf{s})}\mathsf{Q}^{(m-1)}\boldsymbol{\tau}_1(\alpha)\\ &= \frac{\mathsf{e}(\mathsf{k})}{\mathsf{n}}\sum_{\mathsf{s}=0}^{\mathsf{n}-1}\overline{\chi}^{(\mathsf{k})}(\mathsf{Q}^{(\mathsf{s})}[\mathsf{Q}^{(m-1)}]^{-1})\boldsymbol{\tau}_1(\alpha),\\ &\qquad m = 1, 2, \ldots, \mathsf{n}\,. \end{aligned}\tag{A.114}$$

Regarding the states as a set of orthonormal vectors, it follows from (A.59) that

$$\psi^{(\mathsf{k})}(\alpha;m)\,.\,\overline{\psi^{(\mathsf{k}')}(\alpha;m')} = \delta^{\mathrm{Kr}}(\mathsf{k}-\mathsf{k}')\frac{\mathsf{e}(\mathsf{k})}{\mathsf{n}}\chi^{(\mathsf{k})}(\mathsf{Q}^{(m-1)}[\mathsf{Q}^{(m'-1)}]^{-1})\,. \tag{A.115}$$

In this case $w(\alpha;\mathsf{k}) = \mathsf{e}(\mathsf{k})^2$ for all irreducible representations. For a one-dimensional representation we take

$$w_{i1}^{(\mathsf{k})}(\alpha) = \frac{\overline{\chi}^{(\mathsf{k})}(\mathsf{Q}^{(i-1)})}{\sqrt{\mathsf{n}}}\,. \tag{A.116}$$

When $\mathfrak{e}(\mathsf{k}) > 1$, an orthonormal set of columns can be obtained by choosing a subset of the group operators such that $\chi^{(\mathsf{k})}(\mathrm{Q}^{(\mathsf{s})}[\mathrm{Q}^{(\mathsf{s}')}]^{-1}) = 0$ for all $\mathsf{s} \neq \mathsf{s}'$.

A.3.5 Using Equivalence Classes for Block Diagonalization

Suppose that $\boldsymbol{V}$ is expressed in the form

$$\boldsymbol{V} = \begin{pmatrix} \boldsymbol{V}(1,1) & \boldsymbol{V}(1,2) & \cdots & \boldsymbol{V}(1,\gamma) \\ \boldsymbol{V}(2,1) & \boldsymbol{V}(2,2) & \cdots & \boldsymbol{V}(2,\gamma) \\ \vdots & \vdots & \ddots & \vdots \\ \boldsymbol{V}(\gamma,1) & \boldsymbol{V}(\gamma,2) & \cdots & \boldsymbol{V}(\gamma,\gamma) \end{pmatrix}, \tag{A.117}$$

where $\boldsymbol{V}(\alpha,\alpha')$ is the submatrix with row and column numbers in the equivalence classes α and α' respectively. The matrix

$$\overset{*}{\boldsymbol{V}} = [\boldsymbol{W}(1) \oplus \boldsymbol{W}(2) \oplus \cdots \oplus \boldsymbol{W}(\gamma)]^{\dagger} \boldsymbol{V} [\boldsymbol{W}(1) \oplus \boldsymbol{W}(2) \oplus \cdots \oplus \boldsymbol{W}(\gamma)] \tag{A.118}$$

has submatrices $\overset{*}{\boldsymbol{V}}(\alpha,\alpha') = \boldsymbol{W}^{\dagger}(\alpha)\boldsymbol{V}(\alpha,\alpha')\boldsymbol{W}(\alpha')$ which, from (A.79), (A.87) and (A.88), satisfy the equations

$$[\widetilde{\boldsymbol{Q}}^{(\mathsf{s})}(\alpha)]^{\dagger} \overset{*}{\boldsymbol{V}}(\alpha,\alpha')\widetilde{\boldsymbol{Q}}^{(\mathsf{s})}(\alpha') = \overset{*}{\boldsymbol{V}}(\alpha,\alpha'), \qquad \mathsf{s} = 1,2,\ldots,\mathfrak{n}-1. \tag{A.119}$$

The row indices of $\overset{*}{\boldsymbol{V}}(\alpha,\alpha')$ can now be labelled by the pair (k, j) with $j = 1,2,\ldots,\mathfrak{w}(\alpha;\mathsf{k})$ and the column indices by the pair (k', j') with $j' = 1,2,\ldots$ $\ldots,\mathfrak{w}(\alpha';\mathsf{k}')$.[3] Then, using analysis similar to that for (A.85) and(A.86), it can be shown that

$$\langle \mathsf{k}, j|\overset{*}{\boldsymbol{V}}(\alpha,\alpha')|\mathsf{k}', j'\rangle = \delta^{\mathrm{Kr}}(\mathsf{k}-\mathsf{k}')\langle \mathsf{k}, j|\overset{*}{\boldsymbol{V}}(\alpha,\alpha')|\mathsf{k}, j'\rangle. \tag{A.120}$$

For every irreducible representation $\mathsf{E}(\mathsf{k})$ which occurs in both $\mathfrak{C}(\alpha)$ and $\mathfrak{C}(\alpha')$ there will be $\mathfrak{w}(\alpha;\mathsf{k}) \times \mathfrak{w}(\alpha';(\mathsf{k})$ elements of $\overset{*}{\boldsymbol{V}}(\alpha,\alpha')$ which can be non-zero. The permutation matrix $\boldsymbol{\Pi}$ of equation (A.90) collects them together to form the block $\widetilde{\boldsymbol{V}}(\mathsf{k})$ of $\widetilde{\boldsymbol{V}}$ given in equation (A.82). The (possibly) non-zero elements of $\overset{*}{\boldsymbol{V}}(\alpha,\alpha')$ are given by

$$\langle \mathsf{k}, j|\overset{*}{\boldsymbol{V}}(\alpha,\alpha')|\mathsf{k}, j'\rangle = \sum_{m=1}^{\mathfrak{p}(\alpha)} \sum_{m'=1}^{\mathfrak{p}(\alpha')} \overline{w}_{mj}^{(\mathsf{k})}(\alpha)\langle m|\boldsymbol{V}(\alpha,\alpha')|m'\rangle w_{m'j'}^{(\mathsf{k})}(\alpha'). \tag{A.121}$$

From (A.113), for the one-dimension symmetric representation, each submatrix contributes exactly one term given by

[3] As in Chap. 4 we shall, when convenient, use the notation $\langle r|\boldsymbol{X}|j\rangle$ to denote the r,j-th element of any matrix $\boldsymbol{X}$. With this notation we shall also use the pairs of suffices like (k, j) and (p, r).

$$\langle 1,1|\overset{*}{\boldsymbol{V}}(\alpha,\alpha')|1,1\rangle = \frac{1}{\sqrt{\mathfrak{p}(\alpha)\mathfrak{p}(\alpha')}} \sum_{m=1}^{\mathfrak{p}(\alpha)} \sum_{m'=1}^{\mathfrak{p}(\alpha')} \langle m|\boldsymbol{V}(\alpha,\alpha')|m'\rangle . \qquad (\text{A.122})$$

The reduction of the transfer matrix to the block-diagonal form given by (A.82) is often sufficient for our purpose. It is, however, possible, when $\mathfrak{e}(\mathsf{k}) > 1$, to effect a further block diagonalization of the submatrix $\widetilde{\boldsymbol{V}}(\mathsf{k})$, corresponding to the full reduction of the group elements to canonical form, given by (A.61). From (A.72) and (A.85)

$$[\mathring{\mathrm{Q}}^{(\mathsf{s})}(\mathsf{k})]^{\dagger} \mathring{\boldsymbol{V}}(\mathsf{k})\mathring{\mathrm{Q}}^{(\mathsf{s})}(\mathsf{k}) = \mathring{\boldsymbol{V}}(\mathsf{k}) , \qquad \mathsf{s} = 0,1,\ldots,\mathfrak{n}-1 , \qquad (\text{A.123})$$

where $\mathring{\mathrm{Q}}^{(\mathsf{s})}(\mathsf{k})$ consists of the direct sum of $\mathfrak{m}(\mathsf{k})$ copies of the matrix $\boldsymbol{R}^{(\mathsf{k})}(\mathrm{Q}^{(\mathsf{s})})$ and

$$\mathring{\boldsymbol{V}}(\mathsf{k}) = \boldsymbol{L}^{\dagger}(\mathsf{k})\widetilde{\boldsymbol{V}}(\mathsf{k})\boldsymbol{L}(\mathsf{k}) . \qquad (\text{A.124})$$

Since $\mathring{\mathrm{Q}}^{(\mathsf{s})}(\mathsf{k})$ is a multiple of the unit matrix when $\mathfrak{e}(\mathsf{k}) = 1$, (A.123) is a non-trivial symmetry property only when $\mathfrak{e}(\mathsf{k}) > 1$. Then the rows and columns of $\mathring{\boldsymbol{V}}(\mathsf{k})$ can be labelled by the pair (j,r) with $j = 1,2,\ldots,\mathfrak{e}(\mathsf{k})$ and $r = 1,2,\ldots,\mathfrak{m}(\mathsf{k})$. Equation (A.123) can now be expressed in the form

$$\langle j,r|\mathring{\boldsymbol{V}}(\mathsf{k})|j',r'\rangle = \sum_{p=1}^{\mathfrak{e}(\mathsf{k})} \sum_{p'=1}^{\mathfrak{e}(\mathsf{k})} \overline{R}^{(\mathsf{k})}_{pj}(\mathrm{Q}^{(\mathsf{s})})\langle p,r|\mathring{\boldsymbol{V}}(\mathsf{k})|p',r'\rangle R^{(\mathsf{k})}_{p'j'}(\mathrm{Q}^{(\mathsf{s})}) ,$$

$$\mathsf{s} = 1,2,\ldots,\mathfrak{n}-1 . \qquad (\text{A.125})$$

Summing over all the operators $\mathrm{Q}^{(\mathsf{s})}$ and using the formula (A.57) we obtain

$$\langle j,r|\mathring{\boldsymbol{V}}(\mathsf{k})|j',r'\rangle = \delta^{\mathrm{Kr}}(j-j')\langle j,r|\mathring{\boldsymbol{V}}(\mathsf{k})|j,r'\rangle . \qquad (\text{A.126})$$

The only distinct elements of $\mathring{\boldsymbol{V}}(\mathsf{k})$ correspond to different values of r and r'. When $\mathfrak{m}(\mathsf{k}) = 1$, $\mathring{\boldsymbol{V}}(\mathsf{k})$ is a multiple of the unit matrix. When $\mathfrak{e}(\mathsf{k}) > 1$ and $\mathfrak{m}(\mathsf{k}) > 1$ a further permutation of the rows and columns of $\mathring{\boldsymbol{V}}(\mathsf{k},\mathsf{k})$ by a unitary matrix $\boldsymbol{\Pi}(\mathsf{k})$ leads to the block diagonalization

$$[\boldsymbol{\Pi}(\mathsf{k})]^{\dagger}\mathring{\boldsymbol{V}}(\mathsf{k})\boldsymbol{\Pi}(\mathsf{k}) = \boldsymbol{V}'(\mathsf{k}) \oplus \boldsymbol{V}'(\mathsf{k}) \oplus \cdots \oplus \boldsymbol{V}'(\mathsf{k}) , \qquad (\text{A.127})$$

where the direct sum contains $\mathfrak{e}(\mathsf{k})$ identical terms each of which is the $\mathfrak{m}(\mathsf{k})$-dimensional matrix denoted by $\boldsymbol{V}'(\mathsf{k})$. It follows that the eigenvalue $\Lambda_r(\mathsf{k})$ of $\boldsymbol{V}$ has a degeneracy of (at least) $\mathfrak{e}(\mathsf{k})$ and the eigenvalue problem takes its most reduced form if the matrices $\boldsymbol{V}'(\mathsf{k})$ can be found. This involves a determination of the matrices $\boldsymbol{L}(\mathsf{k})$ which produce the canonical reduction given in (A.72). One way of doing this is first to compute the matrices $\widetilde{\mathcal{R}}^{(\mathsf{k})}(\mathcal{Q}) = \{\boldsymbol{I},\widetilde{\boldsymbol{Q}}^{(1)}(\mathsf{k}),\widetilde{\boldsymbol{Q}}^{(2)}(\mathsf{k}),\ldots,\widetilde{\boldsymbol{Q}}^{(\mathfrak{n}-1)}(\mathsf{k})\}$ of the $\mathfrak{w}(\mathsf{k})$-dimensional representation obtained from the initial reduction to block-diagonal form. This can achieved using the equivalence classes and the computed elements of

$\boldsymbol{W}(\alpha)$, as described above. The representation $\widetilde{\mathcal{R}}^{(\mathrm{k})}(\mathcal{Q})$ can be presented as the operations

$$\mathrm{Q}^{(\mathrm{s})}(\boldsymbol{\rho}_1^{(\mathrm{k})},\boldsymbol{\rho}_2^{(\mathrm{k})},\ldots,\boldsymbol{\rho}_{\mathrm{w(k)}}^{(\mathrm{k})}) = (\boldsymbol{\rho}_1^{(\mathrm{k})},\boldsymbol{\rho}_2^{(\mathrm{k})},\ldots,\boldsymbol{\rho}_{\mathrm{w(k)}}^{(\mathrm{k})})\widetilde{\boldsymbol{Q}}^{(\mathrm{s})}(\mathrm{k}) \qquad \text{(A.128)}$$

on the set of basis vectors $\boldsymbol{\rho}_\mu^{(\mathrm{k})}$, $\mu = 1,2,\ldots,\mathrm{w(k)}$. These vectors are linear combinations of the lattice states $\boldsymbol{\tau}_m$, $m = 1,2,\ldots,M$. In themselves they have no physical realization and $\widetilde{\mathcal{R}}^{(\mathrm{k})}(\mathcal{Q})$ is not in general a permutation representation. From (A.72)

$$\mathrm{Q}^{(\mathrm{s})}(\boldsymbol{\rho}_1^{(\mathrm{k})},\boldsymbol{\rho}_2^{(\mathrm{k})},\ldots,\boldsymbol{\rho}_{\mathrm{w(k)}}^{(\mathrm{k})})\boldsymbol{L}(\mathrm{k}) = (\boldsymbol{\rho}_1^{(\mathrm{k})},\boldsymbol{\rho}_2^{(\mathrm{k})},\ldots,\boldsymbol{\rho}_{\mathrm{w(k)}}^{(\mathrm{k})})\boldsymbol{L}(\mathrm{k})\mathring{\boldsymbol{Q}}^{(\mathrm{s})}(\mathrm{k})\,. \qquad \text{(A.129)}$$

As in the case of $\mathring{\boldsymbol{V}}(\mathrm{k})$ the columns of $\boldsymbol{L}(\mathrm{k})$ can be labelled by the pair (j,r) with $j = 1,2,\ldots,\mathrm{e(k)}$ and $r = 1,2,\ldots,\mathfrak{m}(\mathrm{k})$. Denoting these columns by $\boldsymbol{L}(\mathrm{k})|j,r\rangle$ and forming the linear combinations

$$\boldsymbol{\phi}_j^{(r)}(\mathrm{k}) = \sum_{\mu=1}^{\mathrm{w(k)}} \boldsymbol{\rho}_\mu^{(\mathrm{k})}\langle\mu|\boldsymbol{L}(\mathrm{k})|j,r\rangle)\,,$$
$$j = 1,\ldots,\mathrm{e(k)}\,, \qquad r = 1,2,\ldots,\mathfrak{m}(\mathrm{k})\,, \qquad \text{(A.130)}$$

it follows from (A.129) that

$$\mathrm{Q}^{(\mathrm{s})}\boldsymbol{\phi}_j^{(r)}(\mathrm{k}) = \sum_{i=1}^{\mathrm{e(k)}} R_{ij}^{(\mathrm{k})}(\mathrm{Q}^{(\mathrm{s})})\boldsymbol{\phi}_i^{(r)}(\mathrm{k})\,, \qquad \mathrm{s} = 0,1,\ldots,\mathfrak{n}-1\,. \qquad \text{(A.131)}$$

From (A.95) we see that the linear sum of the vectors $\boldsymbol{\rho}_\mu^{(\mathrm{k})}$ given by $\mathrm{P}_{j\ell}^{(\mathrm{k})}\boldsymbol{\rho}_\mu^{(\mathrm{k})}$, for any ℓ and μ, transforms like $\boldsymbol{\phi}_j^{(r)}(\mathrm{k})$ for any r. We can, therefore, determine the columns of $\boldsymbol{L}(\mathrm{k})$ by selecting linearly independent sets of coefficients generated by the operators $\mathrm{P}_{j\ell}^{(\mathrm{k})}$ given by (A.94) operating on the basis vectors $\boldsymbol{\rho}_\mu^{(\mathrm{k})}$.

A.3.6 The Transfer Matrix Eigen Problem

An important part of the use of transfer matrices in statistical mechanics, as described in Sect. 4.9, is the determination of the eigenvalues and eigenvectors of $\boldsymbol{V}$. The eigenvalues of $\widetilde{\boldsymbol{V}}$ given by (A.82), are the same as those of $\boldsymbol{V}$. Each eigenvalue can therefore be identified with some irreducible representation $\mathsf{E}(\mathrm{k})$ and these will, for convenience, be denoted in order of decreasing magnitude within representations by $\Lambda_\ell(\mathrm{k})$, $\ell = 1,2,\ldots,\mathrm{w(k)}$. It will not be assumed that $\boldsymbol{V}$ is necessarily symmetric and thus we have left and right eigenvalue equations,

$$\boldsymbol{x}_\ell(\mathrm{k})\boldsymbol{V} = \Lambda_\ell(\mathrm{k})\boldsymbol{x}_\ell(\mathrm{k})\,, \qquad \text{(A.132)}$$
$$\boldsymbol{V}[\boldsymbol{y}_\ell(\mathrm{k})]^{(\mathrm{T})} = [\boldsymbol{y}_\ell(\mathrm{k})]^{(\mathrm{T})}\Lambda_\ell(\mathrm{k})\,, \qquad \text{(A.133)}$$

respectively, where $\boldsymbol{x}_\ell(\mathsf{k})$ and $\boldsymbol{y}_\ell(\mathsf{k})$ denote the left and right eigenvectors, for the eigenvalue $\Lambda_\ell(\mathsf{k})$. It is simple to show that $\boldsymbol{x}_\ell(\mathsf{k})$ and $\boldsymbol{y}_{\ell'}(\mathsf{k}')$ are orthogonal when $\Lambda_\ell(\mathsf{k}) \neq \Lambda_{\ell'}(\mathsf{k}')$. We shall assume that, within subspaces of eigenvectors for degenerate eigenvalues, a choice of basis vectors can be made for which this condition also holds. Then

$$\boldsymbol{x}_\ell(\mathsf{k}).\boldsymbol{y}_{\ell'}(\mathsf{k}') = \delta^{\mathrm{Kr}}(\ell - \ell')\delta^{\mathrm{Kr}}(\mathsf{k} - \mathsf{k}') . \tag{A.134}$$

In general some of the eigenvalues of $\boldsymbol{V}$ will be complex. Since, however, $\boldsymbol{V}$ has only real elements, the eigenvectors corresponding to the conjugate $\overline{\Lambda_\ell(\mathsf{k})}$ of some complex eigenvalue $\Lambda_\ell(\mathsf{k})$ will be $\overline{\boldsymbol{x}_\ell(\mathsf{k})}$ and $\overline{\boldsymbol{y}_\ell(\mathsf{k})}$ and the elements of the eigenvectors corresponding to a real eigenvalue will be real. From (A.82) and (A.132), the eigenvalue equations for $\widetilde{\boldsymbol{V}}$ are

$$\tilde{\boldsymbol{x}}_\ell(\mathsf{k})\widetilde{\boldsymbol{V}} = \Lambda_\ell(\mathsf{k})\tilde{\boldsymbol{x}}_\ell(\mathsf{k}) , \tag{A.135}$$

$$\widetilde{\boldsymbol{V}}\tilde{\boldsymbol{y}}_\ell^{(\mathrm{T})}(\mathsf{k}) = \tilde{\boldsymbol{y}}_\ell^{(\mathrm{T})}(\mathsf{k})\Lambda_\ell(\mathsf{k}) , \tag{A.136}$$

where

$$\tilde{\boldsymbol{x}}_\ell(\mathsf{k}) = \boldsymbol{x}_\ell(\mathsf{k})\boldsymbol{W} , \tag{A.137}$$

$$\tilde{\boldsymbol{y}}_\ell(\mathsf{k}) = \boldsymbol{y}_\ell(\mathsf{k})\overline{\boldsymbol{W}} . \tag{A.138}$$

The eigenvectors $\tilde{\boldsymbol{x}}_\ell(\mathsf{k})$ and $\tilde{\boldsymbol{y}}_\ell(\mathsf{k})$ of $\widetilde{\boldsymbol{V}}$ have zero elements for all entries not belonging to the representation $\mathsf{E}(\mathsf{k})$. The non-zero elements can be obtained by solving the eigenvalue equation for $\widetilde{\boldsymbol{V}}(\mathsf{k})$. The following result is important for Sects. 4.9 and 4.10:

Theorem A.3.13. *Suppose that $B(\boldsymbol{\tau}_m, \boldsymbol{\tau}_{m'})$ is some function which, like $V(\boldsymbol{\tau}_m, \boldsymbol{\tau}_{m'})$, is invariant under the operations of the symmetry group $\mathcal{Q}$; that is*

$$B(\mathsf{Q}^{(\mathsf{s})}\boldsymbol{\tau}_m, \mathsf{Q}^{(\mathsf{s})}\boldsymbol{\tau}_{m'}) = B(\boldsymbol{\tau}_m, \boldsymbol{\tau}_{m'}) \qquad \mathsf{s} = 0, 1, \ldots, \mathfrak{n} - 1 \tag{A.139}$$

(see (A.77)). If $\boldsymbol{B}$ is the M-dimensional matrix with elements $B(\boldsymbol{\tau}_m, \boldsymbol{\tau}_{m'})$ then

$$\boldsymbol{x}_\ell(\mathsf{k})\boldsymbol{B}[\boldsymbol{y}_{\ell'}(\mathsf{k}')]^{(\mathrm{T})} = \delta^{\mathrm{Kr}}(\mathsf{k} - \mathsf{k}')f(\mathsf{k}, \ell, \ell') . \tag{A.140}$$

Proof: The matrix $\boldsymbol{B}$ will, like $\boldsymbol{V}$, be block diagonalized by $\boldsymbol{W}$;

$$\boldsymbol{W}^\dagger \boldsymbol{B}\boldsymbol{W} = \widetilde{\boldsymbol{B}} = \widetilde{\boldsymbol{B}}(1) \oplus \widetilde{\boldsymbol{B}}(2) \oplus \cdots \oplus \widetilde{\boldsymbol{B}}(\kappa) \tag{A.141}$$

(see (A.82)). Then

$$\boldsymbol{x}_\ell(\mathsf{k})\boldsymbol{B}[\boldsymbol{y}_{\ell'}(\mathsf{k}')]^{(\mathrm{T})} = \tilde{\boldsymbol{x}}_\ell(\mathsf{k})\widetilde{\boldsymbol{B}}[\tilde{\boldsymbol{y}}_{\ell'}(\mathsf{k}')]^{(\mathrm{T})} . \tag{A.142}$$

The vector $\tilde{\boldsymbol{x}}_\ell(\mathsf{k})\widetilde{\boldsymbol{B}}$ has non-zero elements only for entries belonging to the representation $\mathsf{E}(\mathsf{k})$ and the vector $\tilde{\boldsymbol{y}}_{\ell'}(\mathsf{k}')$ has non-zero elements only for entries belonging to the representation $\mathsf{E}(\mathsf{k}')$ and the theorem follows.

A case of particular interest is when $\boldsymbol{B}$ is the boundary matrix given by (4.178). Since the equivalence classes are invariant under the operation of $\mathcal{Q}$, it is clear that this form satisfies (A.139). The matrix $\boldsymbol{B}$ is of the same structure as that given for the transfer matrix in (A.117) with

$$\langle m|\boldsymbol{B}(\alpha,\alpha')|m'\rangle = \delta^{\mathrm{Kr}}(\alpha-\alpha')/\mathfrak{p}(\alpha)\,. \tag{A.143}$$

Since, for a particular value of k and α, the vectors $(w_{1j}^{(\mathsf{k})}(\alpha),\ldots,w_{\mathfrak{p}(\alpha)j}^{(\mathsf{k})}(\alpha))$, $j=1,2,\ldots,\mathsf{w}(\alpha;\mathsf{k})$ are orthogonal to $(1,1,\ldots,1)$ except when $\mathsf{k}=1$, $\alpha=1$ and $j=1$, it follows from (A.120)–(A.122) that

$$\langle \mathsf{k},j|\tilde{\boldsymbol{B}}(\alpha,\alpha')|\mathsf{k}',j'\rangle = \delta^{\mathrm{Kr}}(\mathsf{k}-\mathsf{k}')\delta^{\mathrm{Kr}}(j-j')\delta^{\mathrm{Kr}}(\alpha-\alpha')\,. \tag{A.144}$$

Collecting the elements together to form blocks corresponding to the representations, $\widetilde{\boldsymbol{B}}$ has the form (A.141) with $\tilde{\boldsymbol{B}}(\mathsf{k}) = \delta^{\mathrm{Kr}}(\mathsf{k}-1)\boldsymbol{I}$. Then, from (A.142),

$$\boldsymbol{x}_\ell(\mathsf{k})\boldsymbol{B}[\boldsymbol{y}_{\ell'}(\mathsf{k}')]^{(\mathrm{T})} = \delta^{\mathrm{Kr}}(\mathsf{k}-1). \tag{A.145}$$

The transfer matrix $\boldsymbol{V}$ has positive (non-zero) elements, so according to Perron's theorem (see Gantmacher 1979) it has a real positive eigenvalue $\Lambda_{\max}$ larger in magnitude than any other eigenvalue with left and right eigenvectors having real positive (non-zero) elements. It further follows from the orthogonality condition (A.134) that the largest eigenvalue is the only one with eigenvectors with real positive non-zero elements. From (A.90) and (A.137),

$$\tilde{\boldsymbol{x}}_{\max} = \boldsymbol{x}_{\max}[\boldsymbol{W}(1)\oplus\boldsymbol{W}(2)\oplus\cdots\oplus\boldsymbol{W}(\gamma)]\boldsymbol{\Pi}. \tag{A.146}$$

The effect of the matrices $\boldsymbol{W}(\alpha)$ is to form linear combinations of the elements of $\boldsymbol{x}_{\max}$ within each equivalence class, corresponding to each of the irreducible representations present in that class. These are then collected together by $\boldsymbol{\Pi}$. It follows from (A.113) that the elements of $\tilde{\boldsymbol{x}}_{\max}$ in the block corresponding to $\mathsf{E}(1)$ are all real positive non-zero. The largest eigenvalue belongs to the one-dimensional symmetric representation $\mathsf{E}(1)$ and in our notation is denoted by $\Lambda_1(1)$. This is particular useful information, since, if we wish to obtain only the largest eigenvalue, it can be obtained by solving the eigenvalue problem for $\widetilde{\boldsymbol{V}}(1)$ and this matrix can be obtained very simply from $\boldsymbol{V}$ using (A.122).

In Sect. 4.9 the systems are lattices with periodic boundary conditions and because of this the symmetry group will normally have a cyclic subgroup or itself be a cyclic group. In fact a cyclic subgroup can also sometimes arise from other symmetries as in, for example, the zero-field spin-$\frac{1}{2}$ Ising model which has the symmetry of the second-order cyclic group $\mathcal{C}_2$ $(=\mathcal{S}_2)$. Suppose that the cyclic group $\mathcal{C}_\eta = \{\mathrm{I},\mathrm{C},\mathrm{C}^2,\ldots,\mathrm{C}^{\eta-1}\}$ of order η is a subgroup of $\mathcal{Q}$. The M-dimensional representation $\mathcal{R}(\mathcal{C}_\eta) = \{\boldsymbol{I},\boldsymbol{C},\boldsymbol{C}^2,\ldots,\boldsymbol{C}^{\eta-1}\}$ of $\mathcal{C}_\eta$ is a subset of the matrices of $\mathcal{R}(\mathcal{Q})$. From (A.79), (A.132) and (A.133),

$$\boldsymbol{x}_\ell(\mathsf{k})\boldsymbol{C}^\dagger\boldsymbol{V} = \Lambda_\ell(\mathsf{k})\boldsymbol{C}^\dagger\boldsymbol{x}_\ell(\mathsf{k})\,, \tag{A.147}$$

$$\boldsymbol{V}\boldsymbol{C}[\boldsymbol{y}_\ell(\mathsf{k})]^{(\mathrm{T})} = \boldsymbol{C}[\boldsymbol{y}_\ell(\mathsf{k})]^{(\mathrm{T})}\Lambda_\ell(\mathsf{k})\,. \tag{A.148}$$

It follows from (A.147) that $\boldsymbol{x}_\ell(\mathsf{k})\boldsymbol{C}^\dagger$ is a left eigenvector of $\boldsymbol{V}$ with eigenvalue $\Lambda_\ell(\mathsf{k})$. Unless $\Lambda_\ell(\mathsf{k})$ is degenerate

$$\boldsymbol{x}_\ell(\mathsf{k})\boldsymbol{C}^\dagger = \omega_\ell(\mathsf{k})\boldsymbol{x}_\ell(\mathsf{k})\,, \tag{A.149}$$

where, by repeated application of $\boldsymbol{C}$, it follows that $\omega_\ell(\mathsf{k})$ is one of the complex η-roots of unity. We suppose that (A.149) also applies to cases of degenerate eigenvectors. This is simply equivalent to supposing that within the subspace of degenerate eigenvectors a basis can be chosen, whose elements satisfy (A.149). Assuming that this is compatible with the orthonormality condition (A.134) it follows, from (A.148), that

$$\boldsymbol{C}[\boldsymbol{y}_\ell(\mathsf{k})]^{(\mathrm{T})} = \overline{\omega_\ell(\mathsf{k})}[\boldsymbol{y}_\ell(\mathsf{k})]^{(\mathrm{T})}. \tag{A.150}$$

The effect of the matrices $\boldsymbol{C}$ and $\boldsymbol{C}^\dagger$ is simply to cyclically permute the elements of the eigenvectors within subsets of the equivalence classes. We have shown above that $\boldsymbol{x}_1(1)$ has real positive elements and it therefore follows that $\omega_1(1) = 1$.

A.4 Some Transformations in the Complex Plane

From the discussion of Ruelle's theorem in Sect. 4.6 we saw that a fundamental role was played by the polynomial $\lambda(\mathcal{Y}_\kappa, T, \{\mathfrak{z}_\kappa(\boldsymbol{r})\})$ given by equation (4.45), where $\mathcal{Y}_\kappa$ was a member of the covering $\mathfrak{Y}_{\mathcal{N}}$ of the lattice $\mathcal{N}$. When $\mathcal{Y}_\kappa$ consists of two lattice sites $\boldsymbol{r}$ and $\boldsymbol{r}'$

$$\lambda(\mathcal{Y}_\kappa, T, \{\mathfrak{z}_\kappa(\boldsymbol{r})\}) = f(\mathfrak{z}, \mathfrak{z}') = c_0 + c_1(\mathfrak{z} + \mathfrak{z}') + c_2\mathfrak{z}\mathfrak{z}'\,, \tag{A.151}$$

where the real positive numbers c_0, c_1 and c_2 are the Boltzmann weights when the pair of sites is occupied by 0, 1 and 2 particles respectively and $\mathfrak{z} = \mathfrak{z}_\kappa(\boldsymbol{r})$, $\mathfrak{z}' = \mathfrak{z}_\kappa(\boldsymbol{r}')$. We are interested in obtaining two closed sets $\mathbb{X}$ and $\mathbb{X}'$ in the complex plane $\mathbb{C}_{\mathfrak{z}}$, which do not contain 0, such that $f(\mathfrak{z}, \mathfrak{z}') \neq 0$ when $\mathfrak{z} \notin \mathbb{X}$ and $\mathfrak{z}' \notin \mathbb{X}'$. In the Ising model spin formulation $c_0 = c_2 \neq c_1$ and in the lattice fluid formulation $c_0 = c_1 = 1$, $c_2 \neq 1$. In the former case we may, without loss of generality, set $c_0 = c_2 = 1$ and the two cases are related by a multiplicative transformation of the variables $\mathfrak{z}$ and $\mathfrak{z}'$. We, therefore, begin by considering the function

$$f(\mathfrak{z}, \mathfrak{z}') = 1 + c(\mathfrak{z} + \mathfrak{z}') + \mathfrak{z}\mathfrak{z}', \tag{A.152}$$

where $f(\mathfrak{z}, \mathfrak{z}') = 0$ corresponds to the mapping

$$\mathfrak{z}' = -\frac{1 + c\mathfrak{z}}{c + \mathfrak{z}}\,. \tag{A.153}$$

With the change of variables

$$\zeta = \mathfrak{z} + c\,, \qquad \zeta' = \mathfrak{z}' + c\,, \tag{A.154}$$

this mapping takes the form

$$\zeta' = (c^2 - 1)/\zeta \,. \tag{A.155}$$

Excluding the case $c = 1$, which corresponds to no interaction for any site occupation, the fixed points

$$\zeta^{(\pm)} = \pm\sqrt{c^2 - 1} \tag{A.156}$$

of (A.155) lie on the real or imaginary axis of the complex plane $\mathbb{C}_\zeta$ of ζ according as $c > 1$ or $c < 1$, respectively. It is now not difficult to establish the following result:

Theorem A.4.1. *Let $\mathcal{V}$ be the family of circles in $\mathbb{C}_\zeta$ passing through $\zeta = \zeta^{(\pm)}$.*[4] *Any circle $\mathbb{V} \in \mathcal{V}$ is divided into two open arcs $\mathbb{V}^{(\pm)}$ by the fixed points. Every $\zeta \neq \zeta^{(\pm)}$ in $\mathbb{C}_\zeta$ belongs to exactly one circle $\mathbb{V} \in \mathcal{V}$ and, if $\zeta \in \mathbb{V}^{(\pm)}$, then $\zeta' \in \mathbb{V}^{(\mp)}$.*

It follows from this theorem that every circle $\mathbb{V} \in \mathcal{V}$ is invariant under the mapping (A.155). Since the arcs $\mathbb{V}_1^{(\pm)}$ of any other $\mathbb{V}_1 \in \mathcal{V}$ lie one in the interior and one in the exterior of $\mathbb{V}$, it follows that (A.155) is a mapping between the interior and exterior of $\mathbb{V}$.

Returning now to the complex plane $\mathbb{C}_{\mathfrak{z}}$ of the variable $\mathfrak{z}$, it follows that any circle through the points

$$\mathfrak{z}^{(\pm)} = -c \pm \sqrt{c^2 - 1} \tag{A.157}$$

is invariant under (A.153), with (A.157) as fixed points. The mapping is between interior and exterior points of the circle. We consider the two cases $c < 1$ and $c > 1$ separately:

(a) $c < 1$. In this case the fixed points are

$$\mathfrak{z}^{(\pm)} = -c \pm \mathrm{i}\sqrt{1 - c^2} \,. \tag{A.158}$$

The circle

$$\mathbb{V}(x) = \{\mathfrak{z} : |2c\mathfrak{z} + x| = \sqrt{4c^2(1 - x) + x^2}\} \tag{A.159}$$

passes through the points (A.158) for any real x. Apart from $\mathbb{V}(1)$, which passes through the origin, any circle $\mathbb{V}(x)$ can be used to define the regions $\mathbb{X}$ and $\mathbb{X}'$. Thus we see that $f(\mathfrak{z}, \mathfrak{z}') \neq 0$, when $\mathfrak{z} \notin \mathbb{X}$ and $\mathfrak{z}' \notin \mathbb{X}$, if

$$\mathbb{X} = \begin{cases} \{\mathfrak{z} : |2c\mathfrak{z} + x| \leq \sqrt{4c^2(1 - x) + x^2}\}\,, & \text{for any } x > 1 \,, \\ \{\mathfrak{z} : |2c\mathfrak{z} + x| \geq \sqrt{4c^2(1 - x) + x^2}\}\,, & \text{for any } x < 1 \,. \end{cases} \tag{A.160}$$

In particular the unit circle $|\mathfrak{z}| = 1$ corresponds to $x = 0$ in (A.159) and gives a region in the second category of (A.160). The regions in the first category correspond to discs with centres on $\mathbb{R}^{(-)}$ and lying entirely in the

[4] Including the limiting circle consisting of the axis through the fixed points.

region $\Re(\mathfrak{z}) < 0$. The angles made with $\mathbb{R}^{(-)}$ by tangents from the origin to the boundary of the disc are $\pm\arccos(2c\sqrt{x-1}/x)$. These angles take a minimum absolute value when $x = 2$, when the points of contact of the tangents and the boundary are the fixed points (A.158) lying on $|\mathfrak{z}| = 1$.

(b) $c > 1$. In this case the fixed points (A.157) lie on the negative real axis $\mathbb{R}^{(-)}$ of $\mathbb{C}_{\mathfrak{z}}$. The circle

$$\mathbb{V}(y) = \{\mathfrak{z} : |\mathfrak{z} + c + iy| = \sqrt{y^2 + c^2 - 1}\} \tag{A.161}$$

passes through the points (A.157) for any real y. The origin is exterior to all the circles of this family and thus the definition (A.160) can be replaced by

$$\mathbb{Y} = \{\mathfrak{z} : |\mathfrak{z} + c + iy| \le \sqrt{y^2 + c^2 - 1}\}, \qquad \text{for any } y . \tag{A.162}$$

The angles made with $\mathbb{R}^{(-)}$ by tangents from the origin to the boundary of the disc are $\arccos\left(c/\sqrt{y^2+c^2}\right) \pm \arccos\left(1/\sqrt{y^2+c^2}\right)$. The larger of these angles take its minimum value when $y = 0$.

We now consider the other case of interest for (A.151) where

$$f(\mathfrak{z},\mathfrak{z}') = 1 + \mathfrak{z} + \mathfrak{z}' + c\mathfrak{z}\mathfrak{z}' . \tag{A.163}$$

This equation can be expressed in the form

$$f(\mathfrak{z},\mathfrak{z}') = 1 + (1/\sqrt{c})(\sqrt{c}\mathfrak{z} + \sqrt{c}\mathfrak{z}') + (\sqrt{c}\mathfrak{z})(\sqrt{c}\mathfrak{z}') , \tag{A.164}$$

and the previous results apply with c replaced by $1/\sqrt{c}$ and $\mathfrak{z}$ replaced by $\sqrt{c}\,\mathfrak{z}$. The two cases are:

(a) $c < 1$. From (A.162) and (A.164), $f(\mathfrak{z},\mathfrak{z}') \neq 0$ when $\mathfrak{z} \notin \mathbb{Y}'$ and $\mathfrak{z}' \notin \mathbb{Y}'$ if

$$\mathbb{Y}' = \{\mathfrak{z} : |c\mathfrak{z} + 1 + iy| \le \sqrt{y^2 + 1 - c}\}, \qquad \text{for any } y . \tag{A.165}$$

The angles made with $\mathbb{R}^{(-)}$ by tangents from the origin to the boundary of the disc are $\arccos\left(1/\sqrt{y^2+1}\right) \pm \arccos\left(\sqrt{c}/\sqrt{y^2+1}\right)$. The larger of these angles take its minimum value when $y = 0$.

(b) $c > 1$. From (A.160), $f(\mathfrak{z},\mathfrak{z}') \neq 0$ when $\mathfrak{z} \notin \mathbb{X}'$ and $\mathfrak{z}' \notin \mathbb{X}'$ if

$$\mathbb{X}' = \begin{cases} \{\mathfrak{z} : |2\mathfrak{z} + x| \le \sqrt{4c^{-1}(1-x) + x^2}\}, & \text{for any } x > 1 , \\ \{\mathfrak{z} : |2\mathfrak{z} + x| \ge \sqrt{4c^{-1}(1-x) + x^2}\}, & \text{for any } x < 1 . \end{cases} \tag{A.166}$$

The angles made with $\mathbb{R}^{(-)}$ by tangents from the origin to the boundary of the discs in the first category are $\pm\arccos(2\sqrt{x-1}/x\sqrt{c})$. These angles take a minimum absolute value when $x = 2$.

A.5 Algebraic Functions

An algebraic function $\lambda(z)$ is the solution of an equation

$$f_n(\lambda, z) = 0\,, \tag{A.167}$$

where

$$f_n(\lambda, z) = g_n(z)\lambda^n + g_{n-1}(z)\lambda^{n-1} + \cdots + g_0(z)\,. \tag{A.168}$$

The coefficients $g_j(z)$, $j = 0, 1, \ldots, n$, are polynomials in z with complex coefficients, having no common factor apart from a constant and $f_n(\lambda, z)$ is not reducible[5] to a product of polynomials in λ, which themselves have coefficients which are polynomials in z. $f_n(\lambda, z)$ is called the *defining polynomial* of $\lambda(z)$ and the solutions $\lambda_1(z), \lambda_2(z), \ldots, \lambda_n(z)$ are called the *branches* of $\lambda(z)$. A point at which the values of two or more of the branches of $\lambda(z)$ are equal is a value of z for which (A.167) has repeated roots. The set of such points is given by the simultaneous solution of the equations

$$f_n(\lambda, z) = 0\,, \qquad \frac{\partial f_n(\lambda, z)}{\partial \lambda} \equiv f_n'(\lambda, z) = 0\,, \tag{A.169}$$

which is obtain by solving the equation

$$\mathrm{Res}\{f_n, f_n'; z\} = 0 \tag{A.170}$$

for the *resultant* $\mathrm{Res}\{f_n; f_n'; z\}$ of $f_n(\lambda, z)$ and $f_n'(\lambda, z)$. The *resultant* of two polynomials is given by their Sylvester determinant (Bliss 1966). Since in this case $g_n(z)$ is a factor in the determinant equation, (A.170) can be replaced by

$$\mathrm{Dis}\{f_n; z\} = 0\,, \tag{A.171}$$

where

$$\mathrm{Dis}\{f_n; z\} = (-1)^{\frac{1}{2}n(n-1)}[g_n(z)]^{-1}\mathrm{Res}\{f_n, f_n'; z\} \tag{A.172}$$

is called the *discriminant* (Hille 1962). Given that the coefficients $g_j(z)$ are polynomials of degree m at most, it is clear that $\mathrm{Dis}\{f_n; z\}$ is a polynomial in z of degree not exceeding $2m(n-1)$. The following theorem can be proved (Hille 1962):

Theorem A.5.1. *The branches $\lambda_1(z), \lambda_2(z), \ldots, \lambda_n(z)$ of the algebraic function $\lambda(z)$ defined by equations (A.167) and (A.168) are analytic functions in any simply-connected region of the complex plane which excludes (i) the zeros of $g_n(z)$, (ii) solutions of (A.171), (iii) the point at infinity.*

---

[5] The irreducibility of $f_n(\lambda, z)$ is not always taken as part of the definition of an algebraic function (Bliss 1966). We shall however, following Hille (1962), assume this to be included.

In general the zeros of $g_n(z)$ will be poles of one or more branches of $\lambda(z)$. However, in the application of the theory of this appendix in Sect. 4.10, the polynomials of interest have constant leading coefficients so the algebraic functions have no infinities. The solutions of (A.171) are points where two or more of the branches of $\lambda(z)$ have the same value. Suppose that z_0 is such a point and that the branches of $\lambda(z)$ are labelled in such a way that $\lambda_1(z), \lambda_2(z), \ldots, \lambda_p(z)$, $(1 < p \leq n)$ is the complete set of branches for which

$$\lambda_1(z_0) = \lambda_2(z_0) = \cdots = \lambda_p(z_0)\,. \tag{A.173}$$

There are now three possibilities which can occur:

(i) The branches which satisfy (A.173) can be ordered in such a way that, for small positive real r,

$$\lambda_1[z_0 + r\exp(2\ell \mathrm{i}\pi)] = \lambda_{\ell+1}(z_0 + r)\,, \tag{A.174}$$

where ℓ is an integer modulo p. It is clear that $\lambda_1(z)$ is an analytic function of the variable $(z - z_0)^{1/p}$ in a neighbourhood of z_0 and the power-series expansion

$$\sum_{q=0}^{\infty} c_q (z - z_0)^{q/p} \tag{A.175}$$

is applicable to all the branches $\lambda_1(z), \lambda_2(z), \ldots, \lambda_p(z)$ in a neighbourhood of z_0, when $(z - z_0)^{1/p}$ takes its p different values. The point z_0 is a *branch-point* of $\lambda(z)$ with *cycle number* p. To distinguish this case from case (iii) below, such a branch-point is called *monocyclic*.

(ii) The branches which satisfy (A.173) also satisfy the condition

$$\lambda_\ell(z_0 + r) = \lambda_\ell[z_0 + r\exp(2\mathrm{i}\ell\pi)]\,, \qquad \ell = 1, 2, \ldots, p\,, \tag{A.176}$$

for any integer r. This is a case where each of the branches is analytic at z_0, but they happen to have the same value there. Joyce (1988b) has referred to points of this type as *apparent singular points*.

(iii) The branches which satisfy (A.173) divide into distinct subsets containing $p_1, p_2, \ldots, p_k$ members where

$$\sum_{s=1}^{k} p_s = p\,, \tag{A.177}$$

and, within the s-th subset, the members have a cycle structure like that given by (A.174) and a power-series expansion like (A.175), with cycle number p_s. This can be thought of as a situation where k branch-points happen to coincide. We can also include within this case a combination of coincident branch-points and analytic components, where the latter correspond to instances with $p_s = 1$. A branch-point of this type is call *k-fold multicyclic*.

The reason why Theorem A.5.1 contains the restriction to simply-connected regions is now clear. In a multi-connected region around a branch-point some of the branches of $\lambda(z)$ will not be single-valued. The usual way to define a region within which the branches of $\lambda(z)$ are analytic is to define *cuts* in the complex plane, which prevent the construction of any curve along which a branch of $\lambda(z)$ passes continuously from one of its values to another. Within such a cut plane each branch of $\lambda(z)$ will be analytic and if we take n copies of the complex plane and join them in a suitable way along the edges of the cuts a *Riemann surface* of n *sheets* can be constructed on which $\lambda(z)$ is itself analytic.[6]

The set of cuts needed to construct the Riemann surface for any algebraic function is not unique and we shall describe one of the methods for doing this based on the idea of *connection curves* introduced by Wood (1987). These are defined by all values of z which are simultaneous solutions of the equations

$$f_n(\lambda, z) = 0\,, \qquad f_n(\varpi\lambda, z) \equiv f_n^{(\varpi)}(\lambda, z) = 0\,, \tag{A.178}$$

for all $\varpi = \exp(\mathrm{i}\vartheta)$, where ϑ is real and in the range $[0, 2\pi)$. The connection curves are then given by

$$\mathrm{Res}\{f_n, f_n^{(\varpi)}; z\} = 0\,. \tag{A.179}$$

It is clear that the resultant of $f_n(\lambda, z)$ and $f_n^{(\varpi)}(\lambda, z)$ is identically zero when $\varpi = 1$ with

$$\begin{aligned}\mathrm{Res}\{f_n, f_n^{(\varpi)}; z\} &= \mathrm{Res}\{f_n, f_n^{(\varpi)} - f_n; z\}\\ &= (\varpi - 1)^n \mathrm{Res}\left\{f_n, \frac{f_n^{(\varpi)} - f_n}{\varpi - 1}; z\right\}\end{aligned} \tag{A.180}$$

and

$$\frac{f_n^{(\varpi)}(\lambda, z) - f_n(\lambda, z)}{\varpi - 1} = \lambda h_n^{(\varpi)}(\lambda, z), \tag{A.181}$$

where

$$\begin{aligned}h_n^{(\varpi)}(\lambda, z) = g_n(z)\phi^{(n)}(\varpi)\lambda^{n-1} + g_{n-1}(z)\phi^{(n-1)}(\varpi)\lambda^{n-2}&\\ + \cdots + g_1(z)\,,&\end{aligned} \tag{A.182}$$

$$\phi^{(s)}(\varpi) = \frac{\varpi^s - 1}{\varpi - 1}\,. \tag{A.183}$$

The determinant $\mathrm{Res}\{f_n, \frac{f_n^{(\varpi)} - f_n}{\varpi - 1}; z\}$ has only one non-zero element in the last column, which is $g_0(z)$ in the n-th row. So

$$\mathrm{Res}\{f_n, f_n^{(\varpi)}; z\} = -g_0(z)(1 - \varpi)^n \mathrm{Res}\{f_n, h_n^{(\varpi)}; z\}\,. \tag{A.184}$$

The connection curves are thus given by

$$\mathrm{Res}\{f_n, h_n^{(\varpi)}; z\} = 0\,, \tag{A.185}$$

[6] For an account of the theory of Riemann surfaces see Beardon (1984).

and, since

$$\lim_{\varpi \to 1} \phi^s(\varpi) = s\,, \tag{A.186}$$

it follows, from (A.182), that

$$\lim_{\varpi \to 1} \mathrm{Res}\{f_n, h_n^{(\varpi)}; z\} = \mathrm{Res}\{f_n, f_n'; z\}\,. \tag{A.187}$$

The solutions of (A.171), which are the branch-points and the apparent singular points, correspond to points on the connection curves when $\varpi = 1$.

A.6 Elliptic Integrals, Functions and Nome Series

Some results for elliptic integrals and functions were necessary for the discussion of the exact solution of the two-dimensional Ising model (Volume 1, Chap. 8). These were given in Volume 1, Appendix A.2. Further results in this area are needed for the treatment of the eight-vertex model in Chap. 5 of the present volume. For easy of reference we repeat a number of the formulae given in Volume 1, Appendix A.2. An advanced treatment of elliptic functions is given in Baxter (1982a), Chap. 15. Comprehensive lists of formulae are given by Abramowitz and Segun (1965) and Gradshteyn and Ryzhik (1980).

A.6.1 Elliptic Integrals

Elliptic integrals of the first and second kind, respectively, are defined by

$$\begin{aligned} F(\arcsin(x_0)|m) &= \int_0^{x_0} \frac{\mathrm{d}x}{\sqrt{(1-x^2)(1-m^2x^2)}}\,, \\ E(\arcsin(x_0)|m) &= \int_0^{x_0} \sqrt{\frac{1-m^2x^2}{1-x^2}}\mathrm{d}x\,, \end{aligned} \tag{A.188}$$

where $0 \le m < 1$ is the *modulus. Complete elliptic integrals of the first and second kind*, respectively, are defined by

$$\mathcal{K}(m) = F\left(\tfrac{1}{2}\pi|m\right)\,, \qquad \mathcal{E}(m) = E\left(\tfrac{1}{2}\pi|m\right)\,. \tag{A.189}$$

From (A.188) and (A.189),

$$\mathcal{K}(0) = \mathcal{E}(0) = \tfrac{1}{2}\pi\,, \qquad \mathcal{E}(1) = 1\,. \tag{A.190}$$

We shall also need the result, (Gradshteyn and Ryzhik 1980, p. 905),

$$\mathcal{K}(m) \sim \ln\left(\frac{4}{m'}\right)\,, \qquad \text{as } m \to 1\,, \tag{A.191}$$

where $m' = \sqrt{1-m^2}$ is the *complementary modulus*. It can be shown (Volume 1, Appendix A.2) that

$$\mathcal{E}(m)\mathcal{K}(m') + \mathcal{E}(m')\mathcal{H}(m) - \mathcal{K}(m)\mathcal{K}(m') = \tfrac{1}{2}\pi\,, \tag{A.192}$$

$$\begin{aligned}\frac{d\mathcal{K}(m)}{dm} &= \frac{\mathcal{E}(m) - m'^2\mathcal{K}(m)}{mm'^2}\,,\\ \frac{d\mathcal{K}(m')}{dm} &= \frac{m^2\mathcal{K}(m') - \mathcal{E}(m')}{mm'^2}\,.\end{aligned} \tag{A.193}$$

A.6.2 Elliptic Functions

Writing

$$F(\arcsin(x_0)|m) = \mathrm{u}\,, \tag{A.194}$$

the inverse relation defines the *Jacobian elliptic function* sn by

$$x_0 = \mathrm{sn}(\mathrm{u}|m)\,, \tag{A.195}$$

and the *Jacobian elliptic functions* cn and dn are defined by

$$\begin{aligned}\mathrm{cn}(\mathrm{u}|m) &= \sqrt{1 - \mathrm{sn}^2(\mathrm{u}|m)}\,,\\ \mathrm{dn}(\mathrm{u}|m) &= \sqrt{1 - m^2\mathrm{sn}^2(\mathrm{u}|m)}\,.\end{aligned} \tag{A.196}$$

The elliptic functions are doubly-periodic (Gradshteyn and Ryzhik 1980, p. 909). For real u, $\mathrm{sn}(\mathrm{u}|m)$ and $\mathrm{cn}(\mathrm{u}|m)$ have period $4\mathcal{K}(m)$ and $\mathrm{dn}(\mathrm{u}|m)$ has period $2\mathcal{K}(m)$ with

$$\begin{aligned}&0 \le \mathrm{sn}(\mathrm{u}|m) \le 1, && 0 \le \mathrm{u} \le 2\mathcal{K}(m),\\ &-1 \le \mathrm{sn}(\mathrm{u}|m) \le 0, && 2\mathcal{K}(m) \le \mathrm{u} \le 4\mathcal{K}(m),\\ &m' \le \mathrm{dn}(\mathrm{u}|m) \le 1, && 0 \le \mathrm{u} \le 4\mathcal{K}(m),\\ &0 \le \mathrm{cn}(\mathrm{u}|m) \le \mathrm{dn}(\mathrm{u}|m), && \begin{cases} 0 \le \mathrm{u} \le \mathcal{K}(m),\\ 3\mathcal{K}(m) \le \mathrm{u} \le 4\mathcal{K}(m),\end{cases}\\ &-\mathrm{dn}(\mathrm{u}|m) \le \mathrm{cn}(\mathrm{u}|m) \le 0, && \mathcal{K}(m) \le \mathrm{u} \le 3\mathcal{K}(m).\end{aligned} \tag{A.197}$$

From (A.188), (A.189) and (A.194)–(A.196) it is not difficult to show that

$$\begin{aligned}&\mathrm{sn}(0|m) = 0\,, && \mathrm{sn}(\mathcal{K}(m)|m) = 1\,,\\ &\mathrm{sn}\left(\tfrac{1}{2}\mathrm{i}\mathcal{K}(m')|m\right) = \mathrm{i}/\sqrt{m}\,, &&\\ &\mathrm{cn}(0|m) = 1, && \mathrm{cn}(\mathcal{K}(m)|m) = 0\,,\\ &\mathrm{cn}\left(\tfrac{1}{2}\mathrm{i}\mathcal{K}(m')|m\right) = \sqrt{1 + m^{-1}}\,, &&\\ &\mathrm{dn}(0|m) = 1, && \mathrm{dn}(\mathcal{K}(m)|m) = m'\,,\\ &\mathrm{dn}\left(\tfrac{1}{2}\mathrm{i}\mathcal{K}(m')|m\right) = \sqrt{1 + m}, &&\end{aligned} \tag{A.198}$$

$$
\begin{aligned}
&\mathrm{sn}(\mathrm{u}|0) = \sin(\mathrm{u}), &&\mathrm{sn}(\mathrm{u}|1) = \tanh(\mathrm{u})\,, \\
&\mathrm{cn}(\mathrm{u}|0) = \cos(\mathrm{u}), &&\mathrm{cn}(\mathrm{u}|1) = \mathrm{sech}(\mathrm{u}) \\
&\mathrm{dn}(\mathrm{u}|0) = 1\,, &&\mathrm{dn}(\mathrm{u}|1) = \mathrm{sech}(\mathrm{u})\,,
\end{aligned}
\tag{A.199}
$$

and in Chap. 5 we also need the formulae

$$
\mathrm{cn}(\mathrm{u} \pm \mathrm{v}|m) = \frac{\mathrm{cn}(\mathrm{u}|m)\mathrm{cn}(\mathrm{v}|m) \mp \mathrm{sn}(\mathrm{u}|m)\mathrm{sn}(\mathrm{v}|m)\mathrm{dn}(\mathrm{u}|m)\mathrm{dn}(\mathrm{v}|m)}{1 - m^2\mathrm{sn}^2(\mathrm{u}|m)\mathrm{sn}(\mathrm{v}|m)}\,,
\tag{A.200}
$$

$$
\mathrm{dn}(\mathrm{u} \pm \mathrm{v}|m) = \frac{\mathrm{dn}(\mathrm{u}|m)\mathrm{dn}(\mathrm{v}|m) \mp \mathrm{sn}(\mathrm{u}|m)\mathrm{sn}(\mathrm{v}|m)\mathrm{cn}(\mathrm{u}|m)\mathrm{cn}(\mathrm{v}|m)}{1 - m^2\mathrm{sn}^2(\mathrm{u}|m)\mathrm{sn}(\mathrm{v}|m)}
\tag{A.201}
$$

(Gradshteyn and Ryzhik 1980, p. 916). For present purposes it is the elliptic function dn which is of particular importance. We shall need the following results

$$
\int_0^{2\mathcal{K}(m)} \mathrm{du}\,\mathrm{dn}^2(\mathrm{u}|m) = 2\mathcal{E}(m)\,,
\tag{A.202}
$$

which is straightforward to prove using (A.196) and the substitution (A.195), and

$$
\int_0^{2\mathcal{K}(m)} \mathrm{du} \int_0^{2\mathcal{K}(m)} \mathrm{dv}\,\mathrm{dn}(\mathrm{u}|m)\mathrm{dn}(\mathrm{u}-\mathrm{v}|m)\mathrm{dn}(\mathrm{v}|m)
= 2m'^2\mathcal{K}^2(m) + \tfrac{1}{2}\pi^2\,,
\tag{A.203}
$$

$$
\int_0^{2\mathcal{K}(m)} \mathrm{du}\,\mathrm{dn}(\mathrm{u}|m)\mathrm{dn}\left(\mathrm{u} + \tfrac{1}{2}\mathrm{i}\mathcal{K}(m')|m\right)
= \tfrac{1}{2}\sqrt{1+m}[\pi + 2(1-m)\mathcal{K}(m)]\,,
\tag{A.204}
$$

which are a little more difficult to prove and need the use of (A.201).

A.6.3 Nome Series

The *nome* x of m is defined by

$$
x = \exp[-\pi\mathcal{K}(m')/\mathcal{K}(m)]\,.
\tag{A.205}
$$

From (A.192) and (A.193),

$$
\frac{\mathrm{d}x}{\mathrm{d}m} = \frac{x\pi^2}{2\mathcal{K}^2(m)mm'^2}\,.
\tag{A.206}
$$

In terms of x,

$$\mathrm{dn}(\mathrm{u}|m) = \sqrt{m'} \prod_{n=1}^{\infty} \left\{ \frac{1 + 2x^{2n-1} \cos\left[\frac{\pi \mathrm{u}}{\mathcal{K}(m)}\right] + x^{4n-2}}{1 - 2x^{2n-1} \cos\left[\frac{\pi \mathrm{u}}{\mathcal{K}(m)}\right] + x^{4n-2}} \right\} \tag{A.207}$$

(Gradshteyn and Ryzhik 1980, p. 913) and

$$\mathrm{dn}(\mathrm{u}|m) = \frac{\pi}{2\mathcal{K}(m)} + \frac{2\pi}{\mathcal{K}(m)} \sum_{n=1}^{\infty} \frac{x^n}{1 + x^{2n}} \cos\left[\frac{n\pi \mathrm{u}}{\mathcal{K}(m)}\right] \tag{A.208}$$

(Gradshteyn and Ryzhik 1980, p. 911). Substituting $\mathrm{u} = \mathcal{K}(m)$ into (A.207) and using (A.198) gives

$$\prod_{n=1}^{\infty} \left(\frac{1 - x^{2n-1}}{1 + x^{2n-1}} \right)^4 = m' . \tag{A.209}$$

Substituting $\mathrm{u} = 0$ into (A.208) and again using (A.198) gives

$$\sum_{n=1}^{\infty} \frac{x^n}{1 + x^{2n}} = \frac{\mathcal{K}(m)}{2\pi} - \frac{1}{4} . \tag{A.210}$$

By substituting from (A.208) into (A.202) and noting that the cosines in (A.208) are orthogonal over the range of this integral, it follows that

$$\sum_{n=1}^{\infty} \frac{x^{2n}}{(1 + x^{2n})^2} = \frac{\mathcal{E}(m)\mathcal{K}(m)}{2\pi^2} - \frac{1}{8} . \tag{A.211}$$

In a similar way, by substituting from (A.208) into (A.203),

$$\sum_{n=1}^{\infty} \frac{x^{3n}}{(1 + x^{3n})^3} = \frac{m'^2 \mathcal{K}^3(m)}{4\pi^3} + \frac{\mathcal{K}(m)}{16\pi} - \frac{1}{16} . \tag{A.212}$$

We now use a modified version of the Landen transformation, which was defined in Volume 1, Appendix A.2. Let

$$m_1 = \frac{1 - m'}{1 + m'} , \qquad m_1' = \sqrt{1 - m_1^2} = \frac{2\sqrt{m'}}{1 + m'} \tag{A.213}$$

with

$$\mathcal{K}(m_1) = \frac{1}{2}(1 + m')\mathcal{K}(m) , \qquad \mathcal{E}(m_1) = \frac{\mathcal{E}(m) + m'\mathcal{K}(m)}{1 + m'} \tag{A.214}$$

(Gradshteyn and Ryzhik 1980, p. 908). It follows from (A.205) that x^2 is the nome of m_1. Replacing m by m_1 in (A.208), setting $\mathrm{u} = \mathrm{i}\mathcal{K}(m_1')/2$ and using (A.198), gives

$$\begin{aligned} \sum_{n=1}^{\infty} \frac{x^n(1 + x^{2n})}{1 + x^{4n}} &= \frac{\mathcal{K}(m_1)\sqrt{1 + m_1}}{\pi} - \frac{1}{2} \\ &= \sqrt{\frac{1 + m'}{2}} \frac{\mathcal{K}(m)}{\pi} - \frac{1}{2} , \end{aligned} \tag{A.215}$$

and replacing m by m_1 in (A.211) gives

$$\sum_{n=1}^{\infty}\frac{x^{4n}}{(1+x^{4n})^2}=\frac{\mathcal{E}(m_1)\mathcal{K}(m_1)}{2\pi^2}-\frac{1}{8}$$

$$=\frac{1}{4\pi^2}\mathcal{K}(m)[\mathcal{E}(m)+m'\mathcal{K}(m)]-\frac{1}{8}\,. \qquad (A.216)$$

Final, by replacing m by m_1 in (A.204), substituting from (A.208) and performing the integration, it can be shown that

$$\sum_{n=1}^{\infty}\frac{x^{3n}(1+x^{2n})}{(1+x^{4n})^2}=\frac{1}{4\pi^2}\mathcal{K}(m_1)\sqrt{1+m_1}[\pi+2(1-m_1)\mathcal{K}(m_1)]-\frac{1}{4}$$

$$=\frac{1}{4\pi^2}\mathcal{K}(m)\sqrt{\frac{1+m'}{2}}[\pi+2m'\mathcal{K}(m)]-\frac{1}{4}\,. \qquad (A.217)$$

A.7 Lattices and Graphs

Extensive treatments of the graph theory relevant to statistical mechanics are to be found in Sykes et al. (1966), Domb (1974a) and Temperley (1981). Here we summarize those features of the subject needed for the discussion of series expansions in Chap. 7.

A.7.1 Subgraphs

Consider a lattice $\mathcal{N}$ of N sites and assume it to be so large that boundary effects can be neglected. In graph theoretic language the sites of the lattice are *vertices* and an *edge* connects each nearest-neighbour pair of sites. The complete set of vertices and edges, which is effectively synonymous with $\mathcal{N}$ itself, is in this context called the *full lattice graph.* A *graph* $\mathfrak{g}$, which is a *subgraph* of $\mathcal{N}$, is a subset of the vertices of $\mathcal{N}$ with some, but not necessarily all of the nearest-neighbour pairs of sites in $\mathfrak{g}$ connected by *lines.* When any vertices of $\mathfrak{g}$, which are placed on a nearest-neighbour pair of lattice sites, are connected by a line of $\mathfrak{g}$ then $\mathfrak{g}$ is called a *section graph.*

The number of lines incident at a vertex of $\mathfrak{g}$ is called its *valency.* A vertex is called *odd* or *even* according to whether its valency is an odd or even number. Since each line connects two vertices, the sum of all the vertex valencies is equal to twice the number of lines. There is, therefore, either zero or an even number of odd vertices. When any two vertices of $\mathfrak{g}$ are linked by a path consisting of vertices and lines of $\mathfrak{g}$ it is *connected.* A graph which is not connected is *separated* and consists of several disjoint connected parts or *components*, where *disjoint* is defined as having no vertices in common. An *articulation point* is a vertex of a connected graph such that the removal of this vertex and the lines incident there divides the graph into disjoint parts. A connected graph without articulation points is called a *star*. A *multiply connected* graph is one for which any two vertices are linked by a least two paths consisting of its vertices and lines, which have no vertices in common except the end-points. It can be shown quite easily that a graph with three or more vertices is multiply connected if and only if it is a star. A connected graph, which is not a star, is called a *tree*. It can be regarded as formed from stars connected at articulation points. A tree with no closed circuits of lines is called a *Cayley tree* and, if all vertices have the same valency, it is a *Bethe lattice* (see Volume 1, Chap. 7). A Cayley tree with $n > 2$ and exactly two vertices of valency one is a *finite chain*. A Bethe lattice where each vertex is of valency two is an *infinite chain*.

The *topology* of a graph comprises the number of its vertices and the way in which they are linked by lines. Graphs with the same topology are called *isomorphic*. The topologies of graphs with n sites are denoted by $[n, k]$ where k is an index which distinguishes different topologies. An alternative designation of topologies is in terms of lines using (q, k), with the number of vertices denoted by $n(q, k)$. In the case of a section graph the topology is fixed once the *configuration*, that is the relative location of the vertices, is given. Then the lines are drawn between *every* nearest-neighbour pair. The number of lines in $[n, k]$ is denoted by $q[n, k]$. Since the topology of non-section graphs is not fixed by the configuration of sites we use the notation $[n, k]$ exclusively for section graphs using (q, k) for graphs in general. In fact, for the discussion of high-temperature series in Sect. 7.4, it is convenient to use the specification $(2s|q, k)$ for a topology, where $2s$ is the number of odd vertices in the graph and k, is used to distinguish between different topologies with the same number of odd vertices *and* lines. For reasons made clear in Sect. 7.4 the topologies $(0|q, k)$ are those of *zero-field graphs*,[7] with $(2|q, k)$ and $(4|q, k)$ being the topologies of *magnetic graphs* and *hypermagnetic graphs* respectively.

The number of ways that a section graph $[n, k]$ can be placed on the lattice is denoted by the *configuration number* $\Omega[n, k]$. Similar notations $\Omega(q, k)$ and $\Omega(2s|q, k)$ are used for the configuration numbers for the topologies (q, k) and $(2s|qk)$ respectively. The idea of *embedding* is sometimes used in treatments of lattice graphs (Sykes et al. 1966, Domb 1974a). When a subgraph of a graph $\mathfrak{G}$ is isomorphic to a graph $\mathfrak{g}$ it is said to be a *weak embedding* of $\mathfrak{g}$ in $\mathfrak{G}$. However, when the subgraph is a section graph it is called a *strong embedding* of $\mathfrak{g}$ in $\mathfrak{G}$. Thus $\Omega[n, k]$ is the number of possible strong embeddings in $\mathcal{N}$ of a graph $\mathfrak{g}$ with topology $[n, k]$, while $\Omega(2s|q, k)$ is the number of possible weak embeddings of a graph $\mathfrak{g}$ with topology $(2s|q, k)$ in $\mathcal{N}$. In this terminology, rather unfortunately, strong embeddings are subclasses of weak embeddings.

Subgraphs of the simple cubic lattice are shown in Fig. A.1. Here (a) and (b), with all vertices of valency two, and (c), with one vertex of valency four,

[7] An alternative name is *polygon graphs* (see Volume 1, Sect. 8.3).

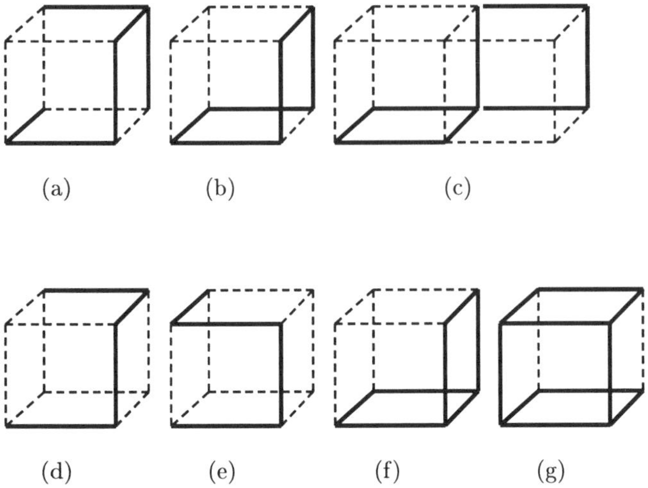

Fig. A.1. Subgraphs on the simple cubic lattice. The broken segments are lattice edges not included in the graphs.

are zero-field graphs. (d) and (e), with two vertices of valency one, and (f), with two vertices of valency three, are magnetic graphs. (e) with four vertices of valency three is a hypermagnetic graph. Chains like (d) and (e) make an important contribution to the zero-field susceptibility series, discussed in Sect. 7.4.2. Contributions from separated graphs must be included. A separated subgraph whose components are all zero-field graphs, say (a), (b) and (c), is itself a zero-field graph, since the total number of odd vertices is zero, while one with one magnetic component and the others zero-field, say (d), (a) and (b), is a magnetic graph since the total number of odd vertices is two. All the graphs of Fig. A.1 are weak embeddings, of which (a), (c), (d) and (f) are strong embeddings, while (b), (e) and (g) are not. It should be noted that many subgraphs $\mathfrak{g}$ can be isomorphic to other subgraphs of the same lattice graph, some of which are section graphs and some are not. Thus (a) and (b) in Fig. A.1 are both embeddings of a six-line polygon (hexagon), while (d) and (e) are both embeddings of a four-line open chain.

There is an important distinction between the graphs which contribute to low-temperature and high-temperature series. Since they represent actual configurations of reversed spins, low-temperature graphs must be section graphs, with any nearest-neighbour pair of vertices connected by a line; while high-temperature graphs, which represent only terms in a mathematical expansion, are free from this restriction. We now consider the problem of calculating configuration numbers, beginning with the case of section graphs.

A.7.2 Section Graphs

Because a section graph is completely determined by the location of its vertices the different topologies $[\mathfrak{n},\mathfrak{k}]$ for fixed $\mathfrak{n}$ satisfy

$$\sum_{\{\mathfrak{k}\}} \Omega[\mathfrak{n},\mathfrak{k}] = \frac{N(N-1)\cdots(N-\mathfrak{n}-1)}{\mathfrak{n}!} . \tag{A.218}$$

We denote the number of components of $[\mathfrak{n},\mathfrak{k}]$ by $\mathfrak{c}[\mathfrak{n},\mathfrak{k}]$ and define

$$\overline{\omega}[\mathfrak{n},\mathfrak{k}] = \Omega[\mathfrak{n},\mathfrak{k}]/N . \tag{A.219}$$

For a connected graph ($\mathfrak{c}[\mathfrak{n},\mathfrak{k}] = 1$), with one vertex placed on a given lattice site $\overline{\omega}[\mathfrak{n},\mathfrak{k}]$ is the number of ways of placing the remaining $\mathfrak{n}-1$ vertices, divided, when necessary, by a symmetry factor. This is independent of N and we set $\overline{\omega}[\mathfrak{n},\mathfrak{k}] = \omega[\mathfrak{n},\mathfrak{k}]$, called the *lattice constant.* For a separated graph, $\Omega[\mathfrak{n},\mathfrak{k}]$ is a polynomial in N of degree $\mathfrak{c}[\mathfrak{n},\mathfrak{k}]$ with no constant term. The quantity $\overline{\omega}[\mathfrak{n},\mathfrak{k}]$ is now a function of N. However, as we shall see, the only part of this expression which is needed is the constant term and it is this which is defined as the lattice constant $\omega[\mathfrak{n},\mathfrak{k}]$.

For the discussion of low-temperature series for the simple cubic Ising model in Sect. 7.3 we need $\overline{\omega}[\mathfrak{n},\mathfrak{k}]$ for all the graphs with $\mathfrak{n} \leq 3$ and for the connected graphs with $\mathfrak{n} = 4$. These are given in Table A.1 together with the number of lines $\mathfrak{q}[\mathfrak{n},\mathfrak{k}]$ and components $\mathfrak{c}[\mathfrak{n},\mathfrak{k}]$. For $\mathfrak{n} \leq 3$ the topologies are obvious from the table and the configuration numbers are easy to obtain, with the most difficult being $\overline{\omega}[3,3]$ for three disconnected points. The easiest way to find this is from (A.218) which gives

$$\overline{\omega}[3,3] = \frac{1}{6}(N-1)(N-2) - \overline{\omega}[3,1] + \overline{\omega}[3,2] . \tag{A.220}$$

Isomorphic subgraphs on the same lattice can have different shapes or geometries. Thus the two lines of $[3,1]$ can be either parallel or at right-angles. The graphs $[4,1]$, $[4,2]$ and $[4,3]$ are connected. Since it possesses no articulation

Table A.1. Section graphs on the simple cubic lattice.

$[\mathfrak{n},\mathfrak{k}]$	$\mathfrak{c}[\mathfrak{n},\mathfrak{k}]$	$\mathfrak{q}[\mathfrak{n},\mathfrak{k}]$	$\overline{\omega}[\mathfrak{n},\mathfrak{k}]$
$[1,1]$	1	0	1
$[2,1]$	1	1	3
$[2,2]$	2	0	$\frac{1}{2}(N-7)$
$[3,1]$	1	2	15
$[3,2]$	2	1	$3(N-12)$
$[3,3]$	3	0	$\frac{1}{6}(N^2-21N+128)$
$[4,1]$	1	4	3
$[4,2]$	1	4	20
$[4,3]$	1	4	63

Table A.2. Zero-field graphs on the square lattice.

$(0\|q,k)$	$c(0\|q,k)$	$n(0\|q,k)$	$\overline{\omega}(0\|q,k)$
$(0\|4,1)$	1	4	1
$(0\|6,1)$	1	6	7
$(0\|8,2)$	1	7	2
$(0\|8,3)$	2	8	$\frac{1}{2}(N-9)$
$(0\|10,1)1$	10	28	
$(0\|10,2)$	1	9	8
$(0\|10,3)$	2	10	$2(N-12)$

points the square $[4,1]$ is a multiply connected star. The graphs $[4,2]$ and $[4,3]$ are trees of three lines, $[4,2]$ having three vertices of valency one and one of valency three and $[4,3]$ being a chain. These cases show that graphs with the same values of n and $q[n,k]$ need not be isomorphic.

A.7.3 Zero-Field Graphs

As for section graphs we define $\overline{\omega}(2s|q,k) = \Omega(2s|q,k)/N$ with the lattice constant $\omega(2s|q,k)$ being the N-independent term in $\overline{\omega}(2s;q,k)$. Data for zero-field graphs on the square lattice are listed, for $q \le 10$, in Table A.2. In each case $k = 1$ denotes a polygon. The embedding numbers for these graphs of are not difficult to calculate, by direct combinatorial arguments. However, the task becomes more difficult for increasing values of q and s. To reduce the work involved and systematize procedures a number of ingenious relations have been deduced (Domb 1974a, 1974b). Although space does not permit a comprehensive treatment, we shall derive some important relations relevant to high-temperature Ising model series.

A.7.4 Magnetic Graphs and Coefficient Relations

For a lattice consisting of a ring of N sites the only magnetic graphs are (open) chains of q lines, where $q < N$. The number of ways the chain can be embedded in the lattice is N; so $\omega(2|q,1) = 1$. For a general lattice, the number of embeddings of a one-link chain is $\frac{1}{2}zN$; so $\overline{\omega}(2|1,1) = \frac{1}{2}z$. A second link can be attached in $z-1$ ways to either end of the chain giving a total of $z(z-1)N$ two-link chains. However, this double-counts, since the two links of each chain are equivalent and either could be placed first; thus $\overline{\omega}(2|2,1) = \frac{1}{2}z(z-1)$. Generalizing this argument gives

$$\overline{\omega}(2;q,1) = \tfrac{1}{2}z(z-1)^{q-1} = \omega(2;q,1)\,, \qquad q < q_f\,, \tag{A.221}$$

where q_f is the number of edges of a lattice face. Equation (A.221), therefore, gives the number of embeddings per site for a chain of any length q on a Bethe lattice.

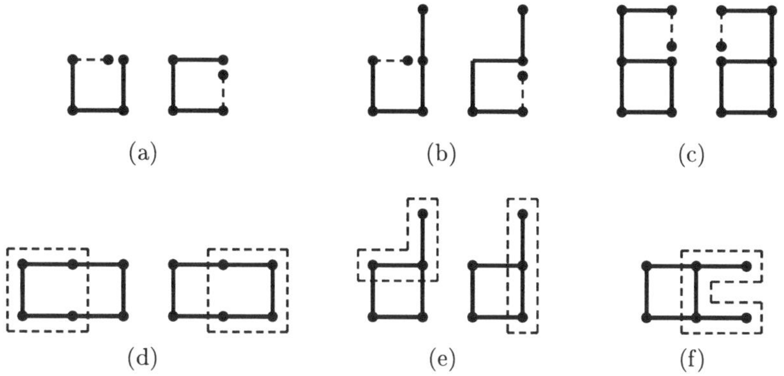

Fig. A.2. Diagrams **(a)**, **(b)** and **(c)** illustrate the closing of chains and 'tails' on the square lattice; diagrams **(d)**, **(e)** and **(f)** illustrate subgraph overlap.

For $\mathsf{q} < \mathsf{q}_{\mathsf{f}}$, it follows from (A.221) that $\varpi(2|\mathsf{q},1) = (z-1)\varpi(2|\mathsf{q}-1,1)$. However, when a further link is added to a $(\mathsf{q}_{\mathsf{f}}-1)$-link chain, the chain ends may meet to form a q-line polygon. The joining of the ends of a q-link chain to form a q-line polygon can take place in two ways at any of the q vertices of the polygon. This is illustrated in Fig. A.2(a), where the last added link is shown as a broken segment. For $\mathsf{q} > \mathsf{q}_{\mathsf{f}}$, the addition of another link to a $(\mathsf{q}-1)$-link chain may also produce graphs of the 'tadpole' type, where a 'tail' of chain links is attached to a polygon 'body'. The join can occur in two ways, as illustrated in Fig. A.2(b), but only at the vertex where the tail meets the body. A new link can be added to a $(\mathsf{q}-1)$-link open chain in $2(z-1)$ ways. If this is done for all $(\mathsf{q}-1)$-link chains then all q-link chains are generated twice and, from the considerations above, all q-line polygons 2q times and all q-line tadpoles twice. Thus

$$2(z-1)\Omega(2|\mathsf{q}-1,1) = 2\Omega(2|\mathsf{q},1) + 2\mathsf{q}\Omega(0|\mathsf{q},1) + 2\sum^{(\mathrm{tp})}\Omega(2|\mathsf{q},\mathsf{k})\,, \tag{A.222}$$

where the final sum is over tadpoles and $\Omega(0|\mathsf{q},1)$ is the number of q-line polygon embeddings. Dividing by $2N$, rearranging terms and noting that all graphs are connected gives the relationship

$$\omega(2|\mathsf{q},1) = (z-1)\omega(2|\mathsf{q}-1,1) - \mathsf{q}\omega(0|\mathsf{q},1) - \sum^{(\mathrm{tp})}\omega(2|\mathsf{q},\mathsf{k})\,, \tag{A.223}$$

for lattice constants.

We now consider separated graphs with connected components labelled $(2\mathsf{s}_1|\mathsf{q}_1,\mathsf{k}_1)$ and $(2\mathsf{s}_2|\mathsf{q}_2,\mathsf{k}_2)$ The number of simultaneous embeddings of $(2\mathsf{s}_1|\mathsf{q}_1,\mathsf{k}_1)$ and $(2\mathsf{s}_2|\mathsf{q}_2,\mathsf{k}_2)$ in the lattice graph is $\Omega(2\mathsf{s}_1|\mathsf{q}_1,\mathsf{k}_1)\Omega(2\mathsf{s}_2|\mathsf{q}_2,\mathsf{k}_2)$. These embeddings produce not only the separated graphs, which will be denoted by $(2\mathsf{s}_1+2\mathsf{s}_2|\mathsf{q}_1+\mathsf{q}_2,\mathsf{k}_\mathsf{s})$, but also overlap graphs, where some lattice

Table A.3. Magnetic graphs on the square lattice; $\mathrm{k} = 1$ denotes a chain.

$(2\|\mathrm{q},\mathrm{k})$	$\mathrm{c}(2\|\mathrm{q},\mathrm{k})$	$\mathrm{n}(2\|\mathrm{q},\mathrm{k})$	$\overline{\omega}(0\|\mathrm{q},\mathrm{k})$
(2|1, 1)	1	2	2
(2|2, 1)	1	3	6
(2|3, 1)	1	4	18
(2|4, 1)	1	5	50
(2|5, 1)	1	6	142
(2|5, 2)	1	5	1
(2|5, 3)	2	6	$2N - 12$
(2|6, 1)	1	7	390
(2|6, 2)	1	6	24
(2|6, 3)	1	6	4
(2|6, 4)	2	7	$6N - 48$
(2|7, 1)	1	8	1086
(2|7, 2)	1	6	2
(2|7, 3)	1	7	64
(2|7, 4)	1	7	24
(2|7, 5)	1	7	20
(2|7, 6)	2	8	$2N - 34$
(2|7, 7)	2	8	$18N - 176$
(2|9, 2)	1	9	12

sites are occupied by vertices of both $(2\mathrm{s}_1|\mathrm{q}_1,\mathrm{k}_1)$ and $(2\mathrm{s}_2|\mathrm{q}_2,\mathrm{k}_2)$. It follows that

$$\Omega(2\mathrm{s}_1|\mathrm{q}_1,\mathrm{k}_1)\Omega(2\mathrm{s}_2|\mathrm{q}_2,\mathrm{k}_2) = \Omega(2\mathrm{s}_1 + 2\mathrm{s}_2|\mathrm{q}_1 + \mathrm{q}_2,\mathrm{k}_\mathrm{s}) + \sum^{(\mathrm{olg})} \mathrm{m}'\Omega(2\mathrm{s}'|\mathrm{q}',\mathrm{k}')\,, \tag{A.224}$$

where the summation is over overlap graphs, $\mathrm{q}' \le \mathrm{q}_1 + \mathrm{q}_2$ and m' denotes the number of ways in which one particular $(2\mathrm{s}'|\mathrm{q}',\mathrm{k}')$ subgraph can be formed by an overlap of a $(2\mathrm{s}_1|\mathrm{q}_1,\mathrm{k}_1)$ and a $(2\mathrm{s}_2|\mathrm{q}_2,\mathrm{k}_2)$. By replacing $\Omega(2\mathrm{s}|\mathrm{q},\mathrm{k})$ by $N\omega(2\mathrm{s}|\mathrm{q},\mathrm{k})$ for the connected subgraphs in (A.224) and dividing by N we obtain

$$\overline{\omega}(2\mathrm{s}_1 + 2\mathrm{s}_2|\mathrm{q}_1 + \mathrm{q}_2,\mathrm{k}_\mathrm{s}) = N\omega(2\mathrm{s}_1\mathrm{q}_1,\mathrm{k}_1)\omega(2\mathrm{s}_2|\mathrm{q}_2,\mathrm{k}_2) - \sum^{(\mathrm{olg})} \mathrm{m}'\omega(2\mathrm{s}'|\mathrm{q}',\mathrm{k}')\,. \tag{A.225}$$

As an example consider the separated zero-field graph $(0|10,3)$ on the square lattice whose components are the zero-field graphs $(0|4,1)$, which is the 1×1 square, and $(0|6,1)$, which is the 2×1 rectangle. In this case the overlap graphs are created by overlapping at a single vertex to give the zero-field graph $(0|10,2)$, by superimposing $(0|4,1)$ on one of the squares of $(0|6,1)$ to give $(2|7,2)$ and by overlapping by a single line to give $(2|9,2)$. For $(0|10,2)$ and $(2|9,2)$ $\mathrm{m}' = 1$, and for $(2|7,2)$ $\mathrm{m}' = 2$. Hence, by substituting from

Tables A.2 and A.3

$$\begin{aligned}\overline{\omega}(0;10,3) &= N\omega(0;6,1)\omega(0;4,1) - \omega(0;10,2)\\ &\quad - 2\omega(2;7,2) - \omega(2;9,2)\\ &= 2(N-12)\,,\end{aligned} \tag{A.226}$$

which is the result given in Table A.2. When $(2s_2|q_2,k_2)$ is the same as $(2s_1|q_1,k_1)$ the two graphs can be completely superimposed and a symmetry factor $\frac{1}{2}$ must be introduced; (A.225) is replaced by

$$\begin{aligned}\overline{\omega}(4s_1|2q_1,k_s) &= \tfrac{1}{2}N\omega^2(2s_1|q_1,k_1) - \tfrac{1}{2}\omega(2s_1|q_1,k_1)\\ &\quad - \sum^{(\mathrm{olg})} m'\omega(3s'|q',k')\,.\end{aligned} \tag{A.227}$$

The contribution of a separated subgraph $(2s|q,k_s)$ to the appropriate term in a high-temperature series is the lattice constant $\omega(2s|(q,k_s)$, which is N-independent term $\overline{\omega}(q,k_s)$. Thus from (A.225) and (A.227)

$$\omega(2s_1+2s_2|q_1+q_2,k_s) = \begin{cases} -\sum^{(\mathrm{olg})} m'\omega(2s'|q',k')\,, \\ \quad \text{for } (2s_1|q_1,k_1) \neq (2s_2|q_2,k_2)\;, \\ -\frac{1}{2}\omega(s_1|q_1,k_1) - \sum^{(\mathrm{olg})} m'\omega(2s'|q',k')\,, \\ \quad \text{for } (2s_1|q_1,k_1) = (2s_2|q_2,k_2)\;. \end{cases} \tag{A.228}$$

The lattice constants of separated graphs are thus linearly dependent on the lattice constants of connected graphs. A similar development is possible, at least in principle, for separated subgraphs with any number of components. It follows that all high-temperature series coefficients depend linearly on connected lattice constants. Relations also exist which express the lattice constants of subgraphs with articulation points in terms of the lattice constants of stars (Domb 1974a).

Table A.3 lists the data for all magnetic graphs, for $q \leq 7$, on the square lattice, as well as $(2|9,2)$, whose lattice constant was used in one of the examples treated above. The lattice constants for the connected graphs which are not chains are not difficult to obtain by straightforward combinatorial arguments. For the chains $\omega(2|q,1)$ is given by (A.221), for $q = 1,2,3$, and by (A.223) for $q \geq 4$. Thus, since $z = 4$,

Table A.4. Zero-field and magnetic graphs on the honeycomb lattice.

$(2\|q,k)$	$c(2\|q,k)$	$n(2\|q,k)$	$\overline{\omega}(0\|q,k)$
$(0\|6,1)$	1	6	$\frac{1}{2}$
$(0\|10,1)$	1	10	$\frac{3}{2}$
$(0\|12,1)$	1	12	1
$(0\|12,2)$	2	12	$\frac{1}{8}(N-14)$
$(0\|14,1)$	1	14	$\frac{9}{2}$
$(0\|14,2)$	1	14	$\frac{3}{2}$
$(2\|11,1)$	1	10	$\frac{3}{2}$
$(2\|13,1)$	1	12	$\frac{3}{2}$
$(2\|14,1)$	1	13	3
$(2\|14,2)$	1	13	3

$$\begin{aligned}
\omega(2|4,1) &= 3\omega(2|3,1) - 4\omega(0|4,1)\,,\\
\omega(2|5,1) &= 3\omega(2|4,1) - \omega(2|5,2)\,,\\
\omega(2|6,1) &= 3\omega(2|5,1) - 6\omega(0|6,1) - \omega(2|6,2)\,,\\
\omega(2|7,1) &= 3\omega(2|6,1) - \omega(2|7,3) - \omega(2|7,5)\,.
\end{aligned} \tag{A.229}$$

Substituting in (A.229) from the numbers shown in Tables A.2 and A.3 gives the required lattice constants.

The numbers $\overline{\omega}(2s|q,k)$ for the separated magnetic subgraphs are obtained using (A.225). As an example consider $(2|7,6)$ which has connected parts consisting of a single link chain $(2|1,1)$ and the 1×2 square $(0|6,1)$. The overlap graphs are:

(i) $(0|6,1)$ with $(2|1,1)$ placed on any one of its $m'=6$ links,

(ii) $(2|7,2)$, which is $(0|6,1)$ with the midpoints of the longer sides joined by the link $(2,1,1)$ in $m'=1$ way,

(iii) $(2|7,5)$, which is the tadpole formed by attaching $(2|1,1)$ to $(0|6,1)$ in the $m'=1$ topologically distinct way.

Thus

$$\begin{aligned}
\overline{\omega}(2|7,6) &= N\omega(2|1,1)\omega(0|6,1) - 6\omega(0|6,1)\\
&\quad - \omega(2|7,2) - \omega(2|7,5)\\
&= 2N^2 - 34N\,.
\end{aligned} \tag{A.230}$$

Data for the honeycomb lattice can be obtained in a similar way and are listed in Table A.4. In this case lattice constants can take half-integer values. Consider, for example, $(0|6,1)$ which is a single hexagon. One vertex can be fixed on the lattice in N ways and then the hexagon placed in three ways. But this counts each placement six times so $\Omega = N \times 3/6 = N/2$.

A.7.5 Multi-Bonded Graphs

A graph on the lattice $\mathcal{N}$ is called *multi-bonded* when some nearest-neighbour pairs of vertices are joined by more than one line. If there are α lines for a particular nearest-neighbour pair (bond) then it is said to have *multiplicity* α. The topology of a multi-bonded graph of $\mathring{q}$ lines is denoted by $(\mathring{q}, \mathring{k})$, where as usual the second index is used to distinguish different topologies with the same $\mathring{q}$. If all bonds are reduced to single lines the resulting graph of q lines is called the *silhouette* of the original graph. Its topology is denoted by (q, k), where $q \leq \mathring{q}$.

The number of ways in which the $(\mathring{q}, \mathring{k})$ multi-bonded graph can be placed on the lattice is denoted by $\mathring{\Omega}(\mathring{q}, \mathring{k})$. The placement can be regarded as occurring in two stages. First, the silhouette graph (q, k) with q lines and topology k is weakly embedded in the lattice. This can be done in $\Omega(q, k)$ ways. The $\mathring{q} - q$ extra lines are placed between the vertices to give the topology $\mathring{k}$ and the number of ways of doing this is denoted by $S(q, k; \mathring{q}, \mathring{k})$ so that

$$\mathring{\Omega}(\mathring{q}, \mathring{k}) = \Omega(q, k) S(q, k; \mathring{q}, \mathring{k}) . \tag{A.231}$$

The last factor is independent of the lattice type. Examples of multi-bonded graphs on a single square are given in Table A.5. The first three rows correspond to ways of putting two lines on a square, with the first row being a ‘genuine’ multi-bonded configurations and the second and third being the topologically distinct ways of arranging single lines. The fourth and fifth rows give the topologically distinct ways of arranging six lines on square silhouette.

We denote the bond multiplicities for topology $\mathring{k}$ by $\alpha_1, \alpha_2, \ldots, \alpha_q$. The α's, regarded as an unordered set, do not completely determine the topology. There are, for example, in rows four and five of Table A.5 two bonds of multiplicity two and two single lines. It is, therefore, necessary to specify in addition the vertex valencies denoted by $\gamma_1, \gamma_2, \ldots, \gamma_n$, where

$$\gamma_1 + \gamma_2 + \cdots + \gamma_n = 2(\alpha_1 + \alpha_2 + \cdots + \alpha_q) = 2\mathring{q} . \tag{A.232}$$

Table A.5. Multi-bonded graphs on a square.

$(\mathring{q}, \mathring{k})$	(q, k)	$\mathring{\Omega}(q, k)$	$S(q, k; \mathring{q}, \mathring{k})$
(2, 1)	(1, 1)	4	1
(2, 2)	(2, 1)	2	1
(2, 3)	(2, 2)	4	1
(6, 1)	(4, 1)	2	2
(6, 2)	(4, 1)	4	4

A.7.6 Polygons and Triangulation

Consider a two dimensional graph $\mathfrak{G}$ of N vertices labelled $i = 1, 2, \ldots, N$. A *polygon* $\mathcal{C}$ of length ℓ on $\mathfrak{G}$ is an ordered sequence of distinct vertices $(\alpha_1, \alpha_2, \ldots, \alpha_\ell)$ and pairs (α_i, α_{i+1}), where (α_i, α_{i+1}) for $i = 1, 2, \ldots, N-1$ and (α_N, α_1) are all the lines of $\mathfrak{G}$. A polygon with $\ell = 2$ is a single line. The open areas of the plane bounded but not cut by polygons are the *faces* of $\mathfrak{G}$ and the boundaries of the faces are *elementary polygons.*

A polygon $\mathcal{C}$ of length $\ell > 2$ defines a graph $\mathfrak{g}(\mathcal{C})$ consisting of the polygon itself together with the lines and vertices contained within its *interior*, which is the open area bounded by $\mathfrak{g}(\mathcal{C})$. Let e and v be the number of lines and vertices of $\mathfrak{g}(\mathcal{C})$. Then the number f of faces bounded by lines of $\mathfrak{g}(\mathcal{C})$ is given by *Euler's formula*, (Temperley 1981),

$$\mathsf{f} = 1 - \mathsf{v} + \mathsf{e} . \tag{A.233}$$

Since $\mathsf{e}_\mathsf{i} = \mathsf{e} - \ell$ and $\mathsf{v}_\mathsf{i} = \mathsf{v} - \ell$ are the number of edges and vertices in the interior of $\mathcal{C}$ respectively

$$\mathsf{f} + \mathsf{v}_\mathsf{i} - \mathsf{e}_\mathsf{i} = 1 . \tag{A.234}$$

Suppose now that $\mathfrak{G}$ is triangular; that is every elementary polygon is of length three. Then every edge of $\mathfrak{g}(\mathcal{C})$ which is part of $\mathcal{C}$ is a side of exactly one triangular face and every interior edge is a side of exactly two faces. So $3\mathsf{f} = \ell + 2\mathsf{e}_\mathsf{i}$. Substituting into (A.234) gives the number of faces

$$\mathsf{f} = \ell + 2\mathsf{v}_\mathsf{i} - 2 \tag{A.235}$$

of $\mathfrak{g}(\mathcal{C})$ in terms of the length of $\mathcal{C}$ and the number of interior vertices.

A general planar graph $\mathfrak{G}$ is turned into a triangular graph $\mathfrak{G}'$ (*triangulated*) by adding extra edges, but no extra vertices so that each face of $\mathfrak{G}'$ is a triangle. It is clear that this process can be carried out for any plane graph and that it can, in general, be carried out in a number of different ways. It follows, however, from (A.235), that if a face of $\mathfrak{G}$ is bounded by an elementary polygon of length ℓ then it will be divided into $\ell - 2$ faces by any process of triangulation. Suppose the faces of $\mathfrak{g}(\mathcal{C})$ are bounded by the elementary polygons $\mathcal{C}_k$, $k = 1, 2, \ldots \mathsf{f}$ of lengths ℓ_k, $k = 1, 2, \ldots, \mathsf{f}$. The *triangulation number*

$$\Upsilon(\mathcal{C}) = \sum_{k=1}^{\mathsf{f}} (\ell_j - 2) \tag{A.236}$$

of $\mathcal{C}$ is the number of triangular faces interior to $\mathcal{C}$ in $\mathfrak{G}'$.

A.7.7 Oriented Graphs

A graph $\mathfrak{G}$ is *oriented* if each line of the graph is given a direction by marking it with an arrow. Every polygon on $\mathfrak{G}$ has a clockwise and an anticlockwise *orientation* and an arrow on a line of $\mathcal{C}$ is said to be clockwise or anticlockwise with respect to $\mathcal{C}$ according to whether it is in the direction of the clockwise or anticlockwise orientation of $\mathcal{C}$. A polygon is *clockwise odd or even* according as it has an odd or an even number of clockwise arrows.

With the faces of $\mathcal{C}$ labelled $k = 1, 2, \ldots, \mathrm{f}$ and bounded by the elementary polygons $\mathcal{C}_k$, let ω_k be the number of clockwise arrows on $\mathcal{C}_k$ and ω the number of clockwise arrows on $\mathcal{C}$. Any arrow which is clockwise with respect to $\mathcal{C}$ is also clockwise with respect to one of the f elementary polygons. Any interior edge separates two of the f faces and the arrow on this edge is clockwise with respect to one of the corresponding elementary polygons but anticlockwise with respect to the other. Thus we have, using (A.234),

$$\sum_{k=1}^{\mathrm{f}} \omega_k = \omega + \mathrm{e_i} = \omega + \mathrm{f} + \mathrm{v_i} - 1\,, \tag{A.237}$$

and it follows that

$$\omega = \sum_{k=1}^{\mathrm{f}} (\omega_k - 1) - (\mathrm{v_i} - 1)\,. \tag{A.238}$$

A.8 The Weak-Graph Transformation

We first concentrate on the square lattice and then generalize to any regular lattice. Consider a lattice partition function whose terms are characterized by the set $\{\mathfrak{g}\}$ of subgraphs, composed of arbitrary sets of lines, which are weak embeddings in the lattice $\mathcal{N}$. On the square lattice this gives rise to sixteen vertex types as shown in Fig. A.3. Although a site without incident lines is not a subgraph vertex in the sense of Chap. 7, it will be labelled as a 'vertex of type 1' here in conformity with the nomenclature of the six- and eight-vertex models (see Fig. 5.1). Suppose that each vertex of type p contributes an energy term e_{p} and define

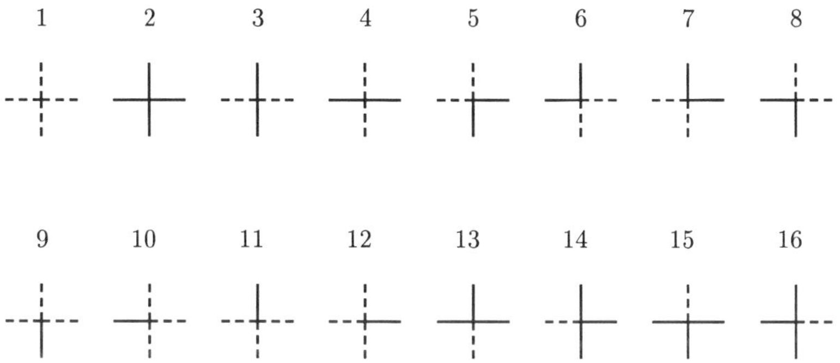

Fig. A.3. The sixteen vertex types on the square lattice.

$$\mathfrak{Z}(\mathrm{p}) = \exp(-e_{\mathrm{p}}/T)\,, \qquad \mathrm{p} = 1, \ldots, 16\,. \tag{A.239}$$

The partition function is now given by

$$Z(\mathfrak{Z}(1), \ldots, \mathfrak{Z}(16)) = \sum_{\{\mathfrak{g}\}} \prod_{\mathrm{i}=1}^{N} \mathfrak{Z}(\mathrm{p_i})\,, \tag{A.240}$$

where the product is over all the sites of the lattice, labelled with the index i, and the sum is over the set of graphs specified above. Examples of systems of this kind are the six- and eight-vertex models, where $\mathfrak{Z}(\mathrm{p}) = 0$ for $\mathrm{p} = 7, \ldots, 16$ and $\mathrm{p} = 9, \ldots, 16$, respectively (see Volume 1, Chap. 10 and Chap. 5 of this volume). The high-temperature series expansion of Sect. 7.4 can be obtained by putting

$$\mathfrak{Z}(\mathrm{p}) = \begin{cases} v^{\frac{1}{2}\gamma_{\mathrm{p}}}\,, & \gamma_{\mathrm{p}} \text{ even}, \\ \tau v^{\frac{1}{2}\gamma_{\mathrm{p}}}\,, & \gamma_{\mathrm{p}} \text{ odd}\,, \end{cases} \tag{A.241}$$

where γ_{p} is the valency at a type-p vertex and v and τ are defined by (7.49).

A different expression for the partition function will now be developed. For a pair (i, j) of lattice sites connected by an edge the vertex types $\mathrm{p_i}$ and $\mathrm{p_j}$ are not arbitrary; they are compatible only if they both imply a line, or if neither implies a line, on the edge (i, j). For instance, if j lies to the right of i then, from Fig. A.3, $\mathrm{p_j} = 1, 3, 5, 7, 10, 11, 12, 13$ are compatible with $\mathrm{p_i} = 1$, while $\mathrm{p_j} = 2, 4, 6, 8, 9, 14, 15, 16$ are incompatible. We now define a function $\mathfrak{C}_\mathrm{j}(\mathrm{p_i})$ such that $\mathfrak{C}_\mathrm{j}(\mathrm{p_i}) = -1$ when the value of $\mathrm{p_i}$ implies a line on the edge (i, j) and $\mathfrak{C}_\mathrm{j}(\mathrm{p_i}) = +1$ otherwise. Hence $1 + \mathfrak{C}_\mathrm{i}(\mathrm{p_j})\mathfrak{C}_\mathrm{j}(\mathrm{p_i}) = 2$ or 0, respectively, according to whether $\mathrm{p_i}$ and $\mathrm{p_j}$ are compatible or incompatible. The partition function can now be expressed as a sum over *all* values $\mathrm{p_i} = 1, \ldots, 16$, for $\mathrm{i} = 1, \ldots, N$,

$$Z(\mathfrak{Z}(1), \ldots, \mathfrak{Z}(16)) = \sum_{\{\mathrm{p_i}\}} \mathfrak{P} \prod_{\mathrm{i}=1}^{N} \left[\tfrac{1}{4}\mathfrak{Z}(\mathrm{p_i})\right]\,, \tag{A.242}$$

$$\mathfrak{P} = \prod_{\{\mathrm{i,j}\}} [1 + \mathfrak{C}_\mathrm{i}(\mathrm{p_j})\mathfrak{C}_\mathrm{j}(\mathrm{p_i})] \tag{A.243}$$

with each edge (i, j) appearing only once in the product in (A.243). From the considerations above, $\mathfrak{P} = 0$ unless $\mathrm{p_i}$ and $\mathrm{p_j}$ are compatible for all edges (i, j) of the lattice. It follows that the impermissible distributions of $\mathrm{p_i}$, $\mathrm{i} = 1, \ldots, N$ are given zero weight. For permissible distributions $\mathfrak{P} = 2^{2N} = 4^N$ and cancels with the 4^{-N} from the product of N factors $\frac{1}{4}\mathfrak{Z}(\mathrm{p_i})$.

The essential step in the weak-graph transformation is an expansion of $\mathfrak{P}$ in (A.243) into its 2^{2N} terms. This gives

$$Z(\mathfrak{Z}(1), \ldots, \mathfrak{Z}(16)) = \sum_{\{\mathfrak{g}'\}} W(\mathfrak{g}')\,, \tag{A.244}$$

$$W(\mathfrak{g}') = \sum_{\{p_i\}} \prod{}' \left[\mathfrak{C}_i(p_j)\mathfrak{C}_j(p_i)\right] \prod_{i=1}^{N} \left[\tfrac{1}{4}\mathfrak{Z}(p_i)\right] . \tag{A.245}$$

The product $\prod' [\mathfrak{C}_i(p_j)\mathfrak{C}_j(p_i)]$ is a term in the expansion of $\mathfrak{P}$ and the factors $\mathfrak{C}_i(p_j)\mathfrak{C}_j(p_i)$ in the product correspond to an arbitrary set of edges (i, j). The subgraph $\mathfrak{g}'$ is obtained by placing a line on each edge (i, j) of that set. Hence the summation in (A.244), like that in (A.240), is over all subgraphs weakly embedded in $\mathcal{N}$ and composed of arbitrary sets of lines. The formula (A.245) is now rearranged by first summing over $p_i = 1, \ldots, 16$ for each site i and then taking the product over sites. Like the general term in (A.240), $W(\mathfrak{g}')$ becomes a product of N factors, each depending on the vertex type p_i' implied by $\mathfrak{g}'$ at a given site i. These factors are

$$\begin{aligned} \mathfrak{N}(1) &= \sum_{p_i=1}^{16} \left[\tfrac{1}{4}\mathfrak{Z}(p_i)\right] , \\ \mathfrak{N}(p_i') &= \sum_{p_i=1}^{16} \left[\prod_{\{j\}}^{(p_i')} \mathfrak{C}_j(p_i)\right] \left[\tfrac{1}{4}\mathfrak{Z}(p_i)\right] , \qquad p_i' \neq 1 , \end{aligned} \tag{A.246}$$

where the product is over the edges (i, j) occupied by lines at a vertex of type p_i'. The $\mathfrak{N}(p')$ are, therefore, linearly dependent on the $\mathfrak{Z}(p)$. Let

$$\mathfrak{N}(p') = \sum_{p=1}^{16} \mathfrak{A}(p', p)\mathfrak{Z}(p) , \qquad p' = 1, \ldots, 16 . \tag{A.247}$$

From the definition of $\mathfrak{C}_i(p_j)$ and from (A.246), the matrix elements $\mathfrak{A}(p', p)$ can be obtained by superposing the vertex configurations p' and p and putting $\mathfrak{A}(p', p) = \frac{1}{4}(-1)^{m}$, where m is the number of edges occupied by lines for *both* p and p'. From Fig. A.3, writing $+$ and $-$ for $+1$ and -1

$$\mathfrak{A}(p', p) = \frac{1}{4} \left(\begin{array}{cccccccccccccccc}
+ & + & + & + & + & + & + & + & + & + & + & + & + & + & + & + \\
+ & + & + & + & + & + & + & + & - & - & - & - & - & - & - & - \\
+ & + & + & + & - & - & - & - & + & - & + & - & + & - & + & - \\
+ & + & + & + & - & - & - & - & - & + & - & + & - & + & - & + \\
+ & + & - & - & + & + & - & - & + & + & - & - & + & + & - & - \\
+ & + & - & - & + & + & - & - & - & - & + & + & - & - & + & + \\
+ & + & - & - & - & - & + & + & + & - & - & + & + & - & - & + \\
+ & + & - & - & - & - & + & + & - & + & + & - & - & + & + & - \\
+ & - & + & - & + & - & + & - & - & + & + & + & + & - & - & - \\
+ & - & - & + & + & - & - & + & + & - & + & + & - & + & - & - \\
+ & - & + & - & - & + & - & + & + & + & - & + & - & - & + & - \\
+ & - & - & + & - & + & + & - & + & + & + & - & - & - & - & + \\
+ & - & + & - & + & - & + & - & + & - & - & - & - & + & + & + \\
+ & - & - & + & + & - & - & + & - & + & - & - & + & - & + & + \\
+ & - & + & - & - & + & - & + & - & - & + & - & + & + & - & + \\
+ & - & - & + & - & + & + & - & - & - & - & + & + & + & + & -
\end{array}\right)
\begin{array}{c} 1 \\ 2 \\ 3 \\ 4 \\ 5 \\ 6 \\ 7 \\ 8 \\ 9 \\ 10 \\ 11 \\ 12 \\ 13 \\ 14 \\ 15 \\ 16 \end{array}
\tag{A.248}$$

$$\begin{array}{cccccccccccccccc} 1 & 2 & 3 & 4 & 5 & 6 & 7 & 8 & 9 & 10 & 11 & 12 & 13 & 14 & 15 & 16 \end{array}$$

where, for clarity, the rows and columns are labelled $1,\ldots,16$. This example of the weak-graph transformation is completed by writing

$$Z(\mathfrak{N}(1),\ldots,\mathfrak{N}(16)) = Z(\mathfrak{Z}(1),\ldots,\mathfrak{Z}(16)), \tag{A.249}$$

where the left-hand side has the same form as the right-hand side, as given by (A.240).[8]

The method described above is now generalized to any regular lattice, so that the number of vertex types becomes 2^{z}, z being the coordination number. Also, to give more flexibility, a free variable y is introduced into the formula (A.242) for the partition function (Nagle and Temperley 1968). We replace $\frac{1}{4}\mathfrak{Z}(\mathrm{p_i})$ by $2^{-\frac{1}{2}\mathrm{z}}\mathfrak{Z}(\mathrm{p_i})y^{\gamma_\mathrm{i}}$, where γ_i is the valency of the vertex at site i, and put $\mathfrak{C}_\mathrm{j}(\mathrm{p_i}) = -y^{-1}$, when the value of $\mathrm{p_i}$ implies a line on the edge (i,j) and $\mathfrak{C}_\mathrm{j}(\mathrm{p_i}) = y$ otherwise. Thus we retain the property that $1+\mathfrak{C}_\mathrm{j}(\mathrm{p_i})\mathfrak{C}_\mathrm{i}(\mathrm{p_j}) = 0$ when $\mathrm{p_i}$ and $\mathrm{p_j}$ are incompatible and the impermissible distributions (i.e. those which do not correspond to a subgraph $\mathfrak{g}'$) are still eliminated from (A.242). Noting that

$$\sum_{\mathrm{i}=1}^{N} \gamma_\mathrm{i} = 2\mathrm{q}\,, \tag{A.250}$$

where q is the number of lines in subgraph $\mathfrak{g}'$, (A.242) becomes

$$\begin{aligned} Z(\mathfrak{Z}(1),\ldots,\mathfrak{Z}(2^{\mathrm{z}});y) &= \sum_{\{\mathfrak{g}\}} \mathfrak{P} \prod_{\mathrm{i}=1}^{N} \left[2^{-\frac{1}{2}\mathrm{z}}\mathfrak{Z}(\mathrm{p_i})y^{\gamma_\mathrm{i}}\right] \\ &= \sum_{\{\mathfrak{g}\}} \mathfrak{P}y^{2\mathrm{q}} \prod_{\mathrm{i}=1}^{N} \left[2^{-\frac{1}{2}\mathrm{z}}\mathfrak{Z}(\mathrm{p_i})\right], \end{aligned} \tag{A.251}$$

where, for permissible distributions,

$$\mathfrak{P} = (1+y^{-2})^{\mathrm{q}}(1+y^2)^{\frac{1}{2}\mathrm{z}N-\mathrm{q}} = y^{-2\mathrm{q}}(1+y^2)^{\frac{1}{2}\mathrm{z}N}, \tag{A.252}$$

$\frac{1}{2}\mathrm{z}N-\mathrm{q}$ being the number of lattice edges not occupied by lines. Substituting (A.252) into (A.251) and comparing with (A.240) gives

$$Z(\mathfrak{Z}(1),\ldots,\mathfrak{Z}(2^{\mathrm{z}});y) = \left[\tfrac{1}{2}(1+y^2)\right]^{\frac{1}{2}\mathrm{z}N} Z(\mathfrak{Z}(1),\ldots,\mathfrak{Z}(2^{\mathrm{z}}))\,. \tag{A.253}$$

After replacing $\mathfrak{Z}(\mathrm{p_i})$ by $\mathfrak{Z}(\mathrm{p_i})y^{\gamma_\mathrm{i}}$, the factor $\mathfrak{P}$, given by (A.243), can be expanded to a sum of terms each corresponding to a subgraph $\mathfrak{g}'$. Applying the same reasoning as before a new set of vertex weights $\mathfrak{N}(\mathrm{p}')$ are obtained with

$$\mathfrak{N}(\mathrm{p}') = \sum_{\mathrm{p}=1}^{2^{\mathrm{z}}} \mathfrak{A}(\mathrm{p}',\mathrm{p})\mathfrak{Z}(\mathrm{p})y^{\gamma_\mathrm{p}}\,, \qquad \mathrm{p}' = 1,\ldots,2^{\mathrm{z}}\,. \tag{A.254}$$

[8] For further discussion of the weak-graph transformation, see Nagle (1968, 1974) and Temperley (1981).

With the new definition of $\mathfrak{C}_j(\mathrm{p}_i)$,

$$\mathfrak{A}(\mathrm{p}',\mathrm{p}) = (-y^{-1})^{\mathrm{m}} y^{\gamma_{\mathrm{p}'}-\mathrm{m}} = (-1)^{\mathrm{m}} y^{\gamma_{\mathrm{p}'}-2\mathrm{m}}, \tag{A.255}$$

m being the number of lines of p' which coincide with a line of p and $\gamma_{\mathrm{p}'} - \mathrm{m}$ the number which do not. Using (A.253), we have

$$Z(\mathfrak{N}(1),\ldots,\mathfrak{N}(2^{\mathrm{z}})) = \left[\tfrac{1}{2}(1+y^2)\right]^{\frac{1}{2}\mathrm{z}N} Z(\mathfrak{Z}(1),\ldots,\mathfrak{Z}(2^{\mathrm{z}}))\,. \tag{A.256}$$

We now specialize to the case where the vertex weights depends only on the valencies (as, for example, in (A.241)) and denote these weights by $\mathfrak{Z}_\gamma$ and $\mathfrak{N}_\gamma$. For the square lattice (Fig. A.3) this implies that

$$\begin{aligned}
&\mathfrak{Z}_0 = \mathfrak{Z}(1)\,, \qquad \mathfrak{Z}_1 = \mathfrak{Z}(9) = \mathfrak{Z}(10) = \mathfrak{Z}(11) = \mathfrak{Z}(12)\,,\\
&\mathfrak{Z}_2 = \mathfrak{Z}(3) = \mathfrak{Z}(4) = \mathfrak{Z}(5) = \mathfrak{Z}(6) = \mathfrak{Z}(7) = \mathfrak{Z}(8)\,,\\
&\mathfrak{Z}_3 = \mathfrak{Z}(13) = \mathfrak{Z}(14) = \mathfrak{Z}(15) = \mathfrak{Z}(16)\,, \qquad \mathfrak{Z}_4 = \mathfrak{Z}(2)\,.
\end{aligned} \tag{A.257}$$

It is easily seen that this property is preserved by the transformation given by (A.248). This remains true for all weak-graph transformations. The total number of vertices of valency γ is $\binom{\mathrm{z}}{\gamma}$ and the number of these having m lines coincident with those of a given vertex of valency γ' is $\binom{\mathrm{z}-\gamma'}{\gamma-\mathrm{m}}\binom{\gamma'}{\mathrm{m}}$. Hence, using (A.254) and (A.255),

$$\mathfrak{N}_{\gamma'} = \sum_{\gamma=0}^{\mathrm{z}} \mathfrak{F}_\gamma(\mathrm{z},\gamma';y)\mathfrak{Z}_\gamma\,, \tag{A.258}$$

where

$$\mathfrak{F}_\gamma(\mathrm{z},\gamma';y) = \sum_{\mathrm{m}=0}^{\gamma} (-1)^{\mathrm{m}} \binom{\mathrm{z}-\gamma'}{\gamma-\mathrm{m}}\binom{\gamma'}{\mathrm{m}} y^{\gamma'+\gamma-2\mathrm{m}}, \tag{A.259}$$

in which the usual convention, that $\binom{a}{b} = 0$, if $b < 0$ or $a < b$, is adopted.

For the discussion of Sect. 7.4.3 it is useful to obtain explicit expressions for $\mathfrak{N}_{\gamma'}$ in the case where the vertex weights $\mathfrak{Z}_\gamma$ are given by (A.241). Equation (A.258) can, using (A.241), be expressed in the form

$$\mathfrak{N}_{\gamma'} = \sum_{\{\gamma\,\text{even}\}} \mathfrak{F}_\gamma(\mathrm{z},\gamma';y)v^{\frac{1}{2}\gamma} + \tau \sum_{\{\gamma\,\text{odd}\}} \mathfrak{F}_\gamma(\mathrm{z},\gamma';y)v^{\frac{1}{2}\gamma}\,, \tag{A.260}$$

where the sums now range over all positive even or odd values. Substituting from (A.259) into (A.260) and interchanging the order of summation, the series can be summed by elementary analytic procedures to give

$$\begin{aligned}
\mathfrak{N}_{\gamma'} = {}& \tfrac{1}{2}(1+\tau)(1+yv^{\frac{1}{2}})^{\mathrm{z}-\gamma'}(y-v^{\frac{1}{2}})^{\gamma'}\\
&+ \tfrac{1}{2}(1-\tau)(1-yv^{\frac{1}{2}})^{\mathrm{z}-\gamma'}(y+v^{\frac{1}{2}})^{\gamma'}\,.
\end{aligned} \tag{A.261}$$

A.9 The Generalized Moment-Cumulant Relations

The eigenvalues of any quantum operator $\hat{\mathcal{O}}$ are the diagonal elements in a representation in which $\hat{\mathcal{O}}$ is given by a diagonal matrix. Since the sum of the diagonal elements of the matrix representation of an operator is invariant under a unitary transformation between representations, $\text{Trace}\{\hat{\mathcal{O}}\}$ unambiguously denotes the trace of any matrix representation of $\hat{\mathcal{O}}$. For later reference we note that for any two operators $\hat{\mathcal{O}}_1$ and $\hat{\mathcal{O}}_2$

$$\text{Trace}\{\hat{\mathcal{O}}_1 + \hat{\mathcal{O}}_2\} = \text{Trace}\{\hat{\mathcal{O}}_1\} + \text{Trace}\{\hat{\mathcal{O}}_2\}\,, \tag{A.262}$$

$$\text{Trace}\{\hat{\mathcal{O}}_1\hat{\mathcal{O}}_2\} = \text{Trace}\{\hat{\mathcal{O}}_2\hat{\mathcal{O}}_1\}\,, \tag{A.263}$$

with

$$\text{Trace}\{\hat{\mathcal{O}}_1\hat{\mathcal{O}}_2\} = \text{Trace}\{\hat{\mathcal{O}}_1\}\text{Trace}\{\hat{\mathcal{O}}_2\} \tag{A.264}$$

if and only if $\hat{\mathcal{O}}_1$ and $\hat{\mathcal{O}}_2$ commute. It follows from (A.263) that, for any product $\hat{\mathcal{O}}_1\hat{\mathcal{O}}_2\hat{\mathcal{O}}_3\cdots\hat{\mathcal{O}}_m$ of a finite number of operators, $\text{Trace}\{\hat{\mathcal{O}}_1\hat{\mathcal{O}}_2\hat{\mathcal{O}}_3\cdots\hat{\mathcal{O}}_m\}$ is invariant under cyclic permutation.

Consider now a quantum system with Hamiltonian $\widehat{H}$ and corresponding dimensionless Hamiltonian $\widehat{\mathsf{H}} = \widehat{H}/T$. Given that $\widehat{\mathsf{H}}$ satisfies certain analytic conditions (see, for example, Ruelle 1969), the operator $\exp(-\widehat{\mathsf{H}})$ can be formally defined in terms of the power series of the exponential. The probability function defined in (1.12) for the distribution of microstates of a classical system is now replaced by the *density operator*

$$\hat{\mathsf{p}} = \frac{\exp(-\widehat{\mathsf{H}})}{Z}\,, \tag{A.265}$$

where Z is the partition function given by

$$Z = \text{Trace}\{\exp(-\widehat{\mathsf{H}})\}\,, \tag{A.266}$$

and

$$\text{Trace}\{\hat{\mathsf{p}}\} = 1\,. \tag{A.267}$$

In Sect. 1.3 we introduced the use of the angle brackets $\langle\cdots\rangle$ to denote the statistical mechanical expectation value (or ensemble average) of a statistical mechanical quantity. Using the density operator defined in (A.265) the corresponding formula for the expectation value of a quantum operator $\hat{\mathcal{O}}$ is

$$\langle\hat{\mathcal{O}}\rangle = \text{Trace}\{\hat{\mathcal{O}}\hat{\mathsf{p}}\}\,. \tag{A.268}$$

Let the dimensionless Hamiltonian for a particular system be given in terms of a set of couplings $\kappa_1, \kappa_2, \ldots, \kappa_m$ and a set of quantum operators $\hat{\phi}_1, \hat{\phi}_2, \ldots, \hat{\phi}_m$ by

$$\widehat{\mathsf{H}}(\kappa_i; \hat{\phi}_i) = \widehat{\mathsf{H}}_0 + \widehat{\mathsf{H}}_1(\kappa_i; \hat{\phi}_i)\,, \tag{A.269}$$

where $\widehat{\mathsf{H}}_0$ is independent of the couplings κ_i and

$$\widehat{H}_1(K_i;\hat{\phi}_i) = -\sum_{i=1}^{m} K_i\hat{\phi}_i\,. \tag{A.270}$$

Then, from (A.266),

$$\ln Z(K_i) = \ln Z^+(K_i) + \ln Z_0\,, \tag{A.271}$$

where

$$Z^+(K_i) = \text{Trace}\{\exp\left[-\widehat{H}_1(K_i;\hat{\phi}_i)\right]\hat{\rho}_0\}\,, \tag{A.272}$$

$$Z_0 = \text{Trace}\{\exp(-\widehat{H}_0)\} \tag{A.273}$$

and

$$\hat{\rho}_0 = \frac{\exp(-\widehat{H}_0)}{Z_0} \tag{A.274}$$

is the density operator for the case when all K_i are set to zero. From (A.265), (A.266), (A.269)–(A.274), equation (A.268) can be expressed in the form

$$\langle\hat{\mathcal{O}}\rangle = \frac{\text{Trace}\{\hat{\mathcal{O}}\exp\left[-\widehat{H}_1(K_i;\hat{\phi}_i)\right]\hat{\rho}_0\}}{Z^+(K_i)}\,. \tag{A.275}$$

We use the notation $\langle\cdots\rangle_0$ to denote the expectation value when all K_i are set equal to zero. Equation (A.275) then gives

$$\langle\hat{\mathcal{O}}\rangle_0 = \text{Trace}\{\hat{\mathcal{O}}\hat{\rho}_0\}\,. \tag{A.276}$$

Using $\mathcal{D}_j$ to denote partial differentiation with respect to K_j we adopt the notation

$$\langle\!\langle\hat{\phi}_p\hat{\phi}_{p-1}\cdots\hat{\phi}_1\rangle\!\rangle = \frac{\text{Trace}\{\mathcal{D}_p\mathcal{D}_{p-1}\cdots\mathcal{D}_1\exp\left[-\widehat{H}_1(K_i;\hat{\phi}_i)\right]\hat{\rho}_0\}}{Z^+(K_i)}\,, \tag{A.277}$$

$$\begin{aligned}\lceil\hat{\phi}_p\hat{\phi}_{p-1}\cdots\hat{\phi}_1\rfloor &= \mathcal{D}_p\mathcal{D}_{p-1}\cdots\mathcal{D}_1\ln Z(K_i)\\ &= \mathcal{D}_p\mathcal{D}_{p-1}\cdots\mathcal{D}_1\ln Z^+(K_i)\,.\end{aligned} \tag{A.278}$$

Because the labelling of the operators and couplings on the right-hand side of (A.270) is arbitrary the set of operators $\hat{\phi}_1,\hat{\phi}_2,\ldots,\hat{\phi}_p$ and the corresponding $\mathcal{D}_1,\mathcal{D}_2,\ldots,\mathcal{D}_p$ will be understood as any ordered subset of not necessarily distinct members of the complete set. From (A.277) and (A.278),

$$\lceil\hat{\phi}_1\rfloor = \frac{\text{Trace}\{\mathcal{D}_1\exp\left[-\widehat{H}_1(K_i;\hat{\phi}_i)\right]\hat{\rho}_0\}}{Z^+(K_i)} = \langle\!\langle\hat{\phi}_1\rangle\!\rangle\,, \tag{A.279}$$

$$\mathcal{D}_{p+1}\lceil\hat{\phi}_p\hat{\phi}_{p-1}\cdots\hat{\phi}_1\rfloor = \lceil\hat{\phi}_{p+1}\hat{\phi}_p\cdots\hat{\phi}_1\rfloor\,, \tag{A.280}$$

$$\mathcal{D}_{p+1}\langle\!\langle\hat{\phi}_p\hat{\phi}_{p-1}\cdots\hat{\phi}_1\rangle\!\rangle = \langle\!\langle\hat{\phi}_{p+1}\hat{\phi}_p\cdots\hat{\phi}_1\rangle\!\rangle - \langle\!\langle\hat{\phi}_{p+1}\rangle\!\rangle\langle\!\langle\hat{\phi}_p\hat{\phi}_{p-1}\cdots\hat{\phi}_1\rangle\!\rangle\,. \tag{A.281}$$

From (A.279)–(A.281)

$$\lceil\hat{\phi}_2\hat{\phi}_1\rfloor = \mathcal{D}_2\langle\!\langle\hat{\phi}_1\rangle\!\rangle, = \langle\!\langle\hat{\phi}_2\hat{\phi}_1\rangle\!\rangle - \langle\!\langle\hat{\phi}_2\rangle\!\rangle\langle\!\langle\hat{\phi}_1\rangle\!\rangle\,. \tag{A.282}$$

Equations (A.279) and (A.282) are the $p = 1$ and $p = 2$ cases of the general formula

$$\lceil\hat{\phi}_p\cdots\hat{\phi}_1\rfloor = \sum_{\ell=1}^{p}(-1)^{\ell-1}(\ell-1)!\sum_{\{\pi_p(\ell)\}}\langle\!\langle\hat{\phi}^{(p)}\cdots\hat{\phi}^{(j)}\rangle\!\rangle\cdots$$
$$\cdots\langle\!\langle\hat{\phi}^{(t)}\cdots\hat{\phi}^{(r)}\rangle\!\rangle\cdots\langle\!\langle\cdots\hat{\phi}^{(1)}\rangle\!\rangle\,, \tag{A.283}$$

where $\pi_p(\ell)$ denotes a partition of $\hat{\phi}_p,\ldots,\hat{\phi}_1$ into ℓ subsets with the operators $\hat{\phi}_i$ ordered within each $\langle\!\langle\cdots\rangle\!\rangle$ in descending order with respect to their indices from left to right. The corresponding result for $p+1$ can be derived from (A.283) by applying the differential operator $\mathcal{D}_{p+1}$ and using (A.282). Equation (A.283) therefore follows by mathematical induction. The inverse formula

$$\langle\!\langle\hat{\phi}_p\cdots\hat{\phi}_1\rangle\!\rangle = \sum_{\ell=1}^{p}\sum_{\{\pi_p(\ell)\}}\lceil\hat{\phi}^{(p)}\cdots\hat{\phi}^{(j)}\rfloor\cdots\lceil\hat{\phi}^{(t)}\cdots\hat{\phi}^{(r)}\rfloor\cdots\lceil\cdots\hat{\phi}^{(1)}\rfloor \tag{A.284}$$

can be established in a similar way. We now derive the form for $\langle\!\langle\hat{\phi}_p\cdots\hat{\phi}_1\rangle\!\rangle$ in the case where all K_i are set equal to zero. Consider first

$$\mathcal{D}_p\cdots\mathcal{D}_1\exp\left[-\widehat{H}_1(K_i;\hat{\phi}_i)\right] = \mathcal{D}_p\cdots\mathcal{D}_1\sum_{j=0}^{\infty}\frac{\left[-\widehat{H}_1(K_i;\hat{\phi}_i)\right]^j}{j!}\,. \tag{A.285}$$

Since $\widehat{H}_1(K_i;\hat{\phi}_i)$ is a linear function of the couplings the application of the differential operators will eliminate all the terms in the summation on the right-hand side of (A.285) with $j < p$. In addition, since $\widehat{H}_1(0;\hat{\phi}_i) = 0$, if we set all K_i to zero, all terms with $j > p$ will also be eliminated. Using as before the subscript '0' to indicate this condition it follows that

$$\left[\mathcal{D}_p\cdots\mathcal{D}_1\exp\left(-\widehat{H}_1(K_i;\hat{\phi}_i)\right)\right]_0 = \frac{1}{p!}\sum_{\{\mu_p\}}\hat{\phi}^{(p)}\cdots\hat{\phi}^{(1)}, \tag{A.286}$$

where the summation is over all permutations μ_p of products of the p operators $\hat{\phi}_1,\ldots,\hat{\phi}_p$ including counting terms with the appropriate multiplicity when they corresponding to the interchange of operators which happen to be the same. Alternatively equation (A.286) can be expressed as

$$\left[\mathcal{D}_p\cdots\mathcal{D}_1\exp\left(-\widehat{H}_1(K_i;\hat{\phi}_i)\right)\right]_0 = \frac{1}{|\nu_p|}\sum_{\{\nu_p\}}\hat{\phi}^{(p)}\cdots\hat{\phi}^{(1)}, \tag{A.287}$$

where the summation is over all the $|\nu_p|$ distinguishable permutations ν_p of the (not necessarily distinct) operators. From (A.276) and (A.277),

$$\langle\!\langle \hat{\phi}_p \hat{\phi}_{p-1} \cdots \hat{\phi}_1 \rangle\!\rangle_0 = \frac{1}{|\nu_p|} \sum_{\{\nu_p\}} \langle \hat{\phi}^{(p)} \cdots \hat{\phi}^{(1)} \rangle_0 \tag{A.288}$$

and, using this formula,

$$\lceil \hat{\phi}_p \cdots \hat{\phi}_1 \rfloor_0 = \sum_{\ell=1}^{p} (-1)^{\ell-1} (\ell-1)! \sum_{\{\pi_p(\ell)\}} \langle\!\langle \hat{\phi}^{(p)} \cdots \hat{\phi}^{(j)} \rangle\!\rangle_0 \cdots$$

$$\cdots \langle\!\langle \hat{\phi}^{(t)} \cdots \hat{\phi}^{(r)} \rangle\!\rangle_0 \cdots \langle\!\langle \cdots \hat{\phi}^{(1)} \rangle\!\rangle_0 , \tag{A.289}$$

$$\langle\!\langle \hat{\phi}_p \cdots \hat{\phi}_1 \rangle\!\rangle_0 = \sum_{\ell=1}^{p} \sum_{\{\pi_p(\ell)\}} \lceil \hat{\phi}^{(p)} \cdots \hat{\phi}^{(j)} \rfloor_0 \cdots \lceil \hat{\phi}^{(t)} \cdots$$

$$\cdots \hat{\phi}^{(r)} \rfloor_0 \cdots \lceil \cdots \hat{\phi}^{(1)} \rfloor_0 \tag{A.290}$$

are the *generalized moment-cumulant relations.* If the operators $\hat{\phi}_1, \ldots, \hat{\phi}_m$ all commute then, for any arbitrary subset $\hat{\phi}_1, \ldots, \hat{\phi}_p$,

$$\langle\!\langle \hat{\phi}_p \cdots \hat{\phi}_1 \rangle\!\rangle_0 = \langle \hat{\phi}_p \cdots \hat{\phi}_1 \rangle_0 . \tag{A.291}$$

Even when they do not commute, if $\widehat{H}_0 = 0$, the number of terms in the summation in (A.288) can be further reduced by taking into account the invariance of the trace under cyclic permutation.

A.10 Kastelyn's Theorem

In this appendix we provide results needed to evaluate the dimer partition functions of the two dimensional lattices considered in Chap. 8. We follow the development of Lieb and Loss (1993), referring to that work for proofs of some of the results.

A.10.1 The Canonical Flux Distribution

Let $\mathcal{N}$ be a lattice of N sites labelled $j = 1, 2, \ldots, N$ and let $\boldsymbol{T}$ be the Hermitian matrix with elements $\langle i|\boldsymbol{T}|j\rangle = t(i,j)$ given by[9]

$$t(i,j) = \begin{cases} \overline{t(j,i)} \neq 0 , & i \text{ and } j \text{ a nearest-neighbour pair,} \\ 0 , & i \text{ and } j \text{ not a nearest-neighbour pair.} \end{cases} \tag{A.292}$$

The polygon $\mathcal{C}$, defined as in Sect. A.7.6, is of length ℓ and has the vertices $(\alpha_1, \alpha_2, \ldots, \alpha_\ell)$. The *flux* $\Phi_{\mathcal{C}}(\boldsymbol{T})$ of $\boldsymbol{T}$ through $\mathcal{C}$, with the orientation specified by the order of the vertices, is given by

[9] For physical reasons associated with the problem they are concerned with Lieb and Loss (1993) call $\boldsymbol{T}$ a *hopping matrix* and for convenience we shall also use this terminology.

$$\Phi_{\mathcal{C}}(\boldsymbol{T}) = \operatorname{Arg}\{t(\alpha_1,\alpha_2)\cdots t(\alpha_{N-1},\alpha_N)t(\alpha_N,\alpha_1)\}\,. \tag{A.293}$$

If the orientation of $\mathcal{C}$ is reversed by specifying its vertices in the reverse order $(\alpha_\ell,\alpha_{\ell-1},\ldots,\alpha_1)$ then the flux changes sign. The polygon $\mathcal{C}$ will circumscribe a certain number of faces $f_1, f_2, \ldots, f_{\mathsf{f}}$. Choosing the orientation of the elementary circuit $\mathcal{C}_k$ around f_k to be the same as that of $\mathcal{C}$, for all $k = 1, 2, \ldots, \mathsf{f}$, it is obvious that

$$\Phi_{\mathcal{C}}(\boldsymbol{T}) = \sum_{k=1}^{\mathsf{f}} \Phi_{\mathcal{C}_k}(\boldsymbol{T})\,. \tag{A.294}$$

It can also be shown (Lieb and Loss 1993) that:

Theorem A.10.1. *If a number $\Phi^{(k)}$ is assigned to the face f_k of the lattice for $k = 1, 2, \ldots$ then there is an assignment of real numbers $\theta(i,j)$ to the nearest-neighbour pairs of the lattice with $\theta(j,i) = -\theta(i,j)$ and $0 \le |\theta(i,j)| < 2\pi$ such that the hopping matrix $\boldsymbol{T}$ with elements*

$$t(i,j) = |t(i,j)|\exp[\mathrm{i}\theta(i,j)] \tag{A.295}$$

has the property

$$\Phi_{\mathcal{C}_k}(\boldsymbol{T}) = \Phi^{(k)} \tag{A.296}$$

for every elementary polygon of the lattice.

In Appendix A.7.6 we defined the triangulation number $\Upsilon(\mathcal{C})$ for any polygon $\mathcal{C}$ on a plane graph. Now according to Theorem A.10.1 there exist Hermitian matrices $\boldsymbol{T}$ with elements of the form (A.295) such that

$$\Phi_{\mathcal{C}}(\boldsymbol{T}) = \frac{\pi}{2}\sum_{k=1}^{\mathsf{f}} \Upsilon(\mathcal{C}_k) \tag{A.297}$$

for every polygon on the lattice. Hopping matrices of this form are said to satisfy the *canonical flux distribution.*

A.10.2 The Dimer Partition Function

A *spanning subgraph* $\mathfrak{g}$ of the lattice $\mathcal{N}$ is one for which every lattice site is a vertex of $\mathfrak{g}$. A *dimer arrangement* on $\mathcal{N}$ is a subgraph consisting of (non-overlapping) components each of which is a dimer, that is a subgraph with exactly one line and two vertices of valency one. A dimer covering of $\mathcal{N}$ is a dimer arrangement which is a spanning subgraph. The dimer partition function is defined by (8.7) for a symmetric matrix $\boldsymbol{Q}$ of non-negative elements. We now generalize the definition to

$$Z(N;\boldsymbol{Z}) = \sum_{\{\Pi\}} |v(\pi_1,\pi_2)||v(\pi_3,\pi_4)|\cdots|v(\pi_{N-1},\pi_N)| \tag{A.298}$$

for any matrix $\boldsymbol{Z}$ with elements $\langle i|\boldsymbol{Z}|j\rangle = v(i,j)$ satisfying

$$|v(i,j)| = \begin{cases} |v(j,i)| \neq 0\,, & i \text{ and } j \text{ a nearest-neighbour pair,} \\ 0\,, & i \text{ and } j \text{ not a nearest-neighbour pair.} \end{cases} \tag{A.299}$$

The following results can now be established (Lieb and Loss 1993):

Theorem A.10.2. *If $\boldsymbol{T}$ is given by (A.292) and satisfies the canonical flux distribution condition (A.297) then*

(i) $\mathrm{Det}\{\boldsymbol{T}\} = (-1)^{N/2}[Z(N;\boldsymbol{T})]^2\,,$ (A.300)

(ii) there exists a gauge transformation[10] $\boldsymbol{U}$ *such that*

$$\boldsymbol{U}^{\dagger}\boldsymbol{T}\boldsymbol{U} = \mathrm{i}\boldsymbol{Z}\,, \tag{A.301}$$

where $\boldsymbol{Z}$ is a real antisymmetric matrix.

It follows from Theorem A.10.1, that a hopping matrix $\boldsymbol{T}$, satisfying the canonical flux distribution, can be constructed from any symmetric matrix $\boldsymbol{Q}$, which satisfies (8.5), by taking $t(i,j) = q(i,j)\exp[\mathrm{i}\theta(i,j)]$ with a suitable choice of the antisymmetric arguments $\theta(i,j)$. The matrix $\boldsymbol{Z}$ is then related to $\boldsymbol{Q}$ by (8.8). Since, from (A.301),

$$\mathrm{Det}\{\boldsymbol{T}\} = (-1)^{N/2}\mathrm{Det}\{\boldsymbol{Z}\}\,, \tag{A.302}$$

and $Z(N;\boldsymbol{T}) = Z(N;\boldsymbol{Q})$, (8.9) follows from (A.300).

A.10.3 Superposition Polynomials and Pfaffians

We now consider the problem of determining the signs in (8.8) for the antisymmetric matrix $\boldsymbol{Z}$. Following Kasteleyn (1961) we use the *Pfaffian* of $\boldsymbol{Z}$ defined by

$$\mathrm{Pfaf}\{\boldsymbol{Z}\} = \sum_{\{\Pi\}} \epsilon(\Pi) z(\pi_1,\pi_2) z(\pi_3,\pi_4)\cdots z(\pi_{N-1},\pi_N)\,, \tag{A.303}$$

where the sum is over all partitions $\Pi = (\pi_1,\pi_2,\ldots,\pi_N)$ satisfying (8.3) and (8.4), with $\epsilon(\Pi)$ being the *signature*[11] of Π. *Pfaff's theorem* states that:

Theorem A.10.3. *The determinant and Pfaffian of the $N\times N$ antisymmetric matrix $\boldsymbol{Z}$ satisfy the relationship*

$$\mathrm{Det}\{\boldsymbol{Z}\} = [\mathrm{Pfaf}\{\boldsymbol{Z}\}]^2\,. \tag{A.304}$$

---

[10] The matrix $\boldsymbol{U}$ of a gauge transformation is of the form $\langle j|\boldsymbol{U}|k\rangle = \exp[\mathrm{i}\varphi(j)]\delta^{\mathrm{Kr}}(j-k)$, where $\varphi(j)$ is real.

[11] $\epsilon(\Pi) = 1$ or -1 according to whether Π is of even or odd parity, that is whether Π is obtained from $\Pi_0 = (1,2,\ldots,N)$ by an even or odd number of transpositions. A transposition is an exchange of the position of two indices, with the order of the remaining indices unchanged. Any transposition can be effected by an odd number of transpositions of adjacent indices.

A result which can be proved either by induction on the order N (Green and Hurst 1964) or directly (Kasteleyn 1967). It follows from (8.8), (8.9) and (A.304) that[12]

$$\text{Pfaf}\{\boldsymbol{Z}\} = Z(N;\boldsymbol{Z}). \tag{A.305}$$

This result is known as *Kastelyn's theorem* (Kasteleyn 1961). Now let

$$\mathcal{P}(\boldsymbol{Z};\Pi) = \epsilon(\Pi)z(\pi_1,\pi_2)z(\pi_3,\pi_4)\cdots z(\pi_{N-1}.\pi_N). \tag{A.306}$$

This is an arbitrary term of the Pfaffian (A.303). The correct choice of signs for the elements of $\boldsymbol{Z}$ to satisfy (8.9) is, from (A.305), the one for which $\mathcal{P}(\boldsymbol{Z};\Pi) > 0$ for all permutations Π. We have introduced a standard dimer covering of $\mathcal{N}$, labelled so that the dimers are the lines $(1,2),(3,4),\ldots,(N-1,N)$ with $\Pi_0 = (1,2,\ldots,N)$ being the identity permutation. Whatever the eventual choice of signs for the elements of $\boldsymbol{Z}$ we can choose $\epsilon(\Pi_0)$, so that $\mathcal{P}(\boldsymbol{Z};\Pi_0) > 0$ and thus we need to ensure that

$$\text{Sign}\{\mathcal{P}(\boldsymbol{Z};\Pi)\} = \text{Sign}\{\mathcal{P}(\boldsymbol{Z};\Pi_0)\}, \qquad \forall\ \mathcal{P}(\boldsymbol{Z};\Pi) \neq 0. \tag{A.307}$$

Now consider the product

$$\epsilon(\mathrm{T})z(\tau_1,\tau_2)z(\tau_3,\tau_4)\cdots z(\tau_{N-1},\tau_N), \tag{A.308}$$

where $(\tau_1,\tau_2),(\tau_3,\tau_4),\ldots,(\tau_{N-1},\tau_N)$ is a dimer covering but the indices in the permutation $\mathrm{T} = (\tau_1,\tau_2,\tau_3,\tau_4,\ldots)$ are otherwise unordered. Suppose that for one of the pairs $\tau_{2k} < \tau_{2k-1}$. The transposition of these indices multiplies $\epsilon(\mathrm{T})$ by -1; but, from (8.8), $z(\tau_{2k},\tau_{2k-1}) = -z(\tau_{2k-1},\tau_{2k})$, so the product (A.308) remains unchanged. Again $z(\tau_{2k},\tau_{2k-1})$ can be exchanged with $z(\tau_{2j},\tau_{2j-1})$ for any k and j without affecting the sign of the product (A.308) since it involves two transpositions. It follows that the indices can be rearranged to satisfy (8.3) and (8.4) without altering the value of (A.308), which is thus equal to the Pfaffian term $\mathcal{P}(\boldsymbol{Z};\Pi)$ corresponding to the same dimer configuration on the lattice. This degree of flexibility in the order of the indices is helpful when the assignment of signs to edges is discussed.

The relationship between an arbitrary dimer configuration and the standard configuration can be represented by a *superposition diagram.* In such a diagram full and broken lines respectively are used to denote the dimers of the arbitrary and standard configurations. There will thus be exactly one full and one broken line incident at each site. Fig. A.4 shows the superposition diagram corresponding to the dimer configuration of Fig. 8.1. Some nearest-neighbour site pairs are connected by both a full and a broken line and will be called *doubly connected.* Since the lattice is assumed to have a finite number of sites it is clear that every site of the lattice which is not doubly connected is a vertex of exactly one (closed) polygon of an even number of lines, which are alternately full and broken. These are called *superposition polygons.* Examples are the polygons through sites $(9,10,16,15,9)$ and

[12] In order to achieve the positive sign it may be necessary to change the definition of $\boldsymbol{Z}$ by multiplying the first row and column by -1.

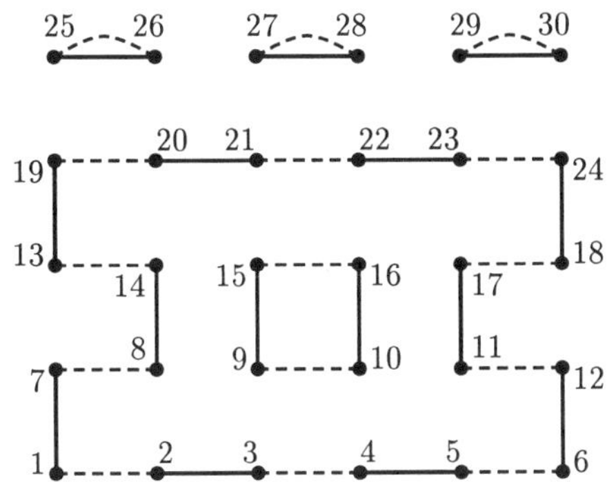

Fig. A.4. The superposition diagram corresponding to Fig. 8.1.

$(1, 2, 3, 4, \ldots, 14, 8, 7, 1)$ in Fig. A.4. A superposition diagram consists of doubly connected nearest-neighbour pairs and superposition polygons. Any sites enclosed inside a given superposition polygon are arranged in doubly connected pairs and/or smaller polygons so that the number of such sites is even. The larger polygon in Fig. A.4 encloses four sites, themselves forming a polygon with two full and two broken edges.

Now set up an alternative permutation T_0 for the standard configuration, starting with the indices of the doubly connected pairs of sites and continuing with a sequence $(\mu_1, \mu_2, \mu_3, \ldots, \mu_{2n})$, representing the pairs connected by broken lines in anticlockwise order, for each superposition polygon. For the superposition diagram Fig. A.4

$$\mathsf{T}_0 = (25, 26|27, 28|29, 30|9, 10, 16, 15|1, 2, 3, 4, \ldots, 13, 14, 8, 7)\,. \qquad \text{(A.309)}$$

Next perform cyclic permutation $(\mu_1, \ldots, \mu_{2n-1}, \mu_{2n}) \to (\mu_{2n}, \mu_1, \ldots, \mu_{2n-1})$ on each polygon, so that it now represents the indices of the site pairs connected by full lines in anticlockwise order. We now have a permutation T representing the arbitrary dimer configuration. For Fig. A.4

$$\mathsf{T} = (25, 26|27, 28|29, 30|15, 9, 10, 16|7, 1, 2, 3, \ldots, 19, 13, 14, 8)\,. \qquad \text{(A.310)}$$

A cycle of $2n$ indices can be achieved by $2n-1$ adjacent transpositions and thus multiplies the parity factor $\epsilon(\mathsf{T}_0)$ by $(-1)^{2n-1} = -1$. For instance, for the indices of the sites of the small polygon in Fig. A.4, the transpositions from T_0 to T are

$$(9, 10, 16, 15) \to (9, 10, 15, 16) \to (9, 15, 10, 16) \to (15, 9, 10, 16)\,,$$

yielding a factor $(-1)^3 = -1$. Thus

$$\epsilon(\mathsf{T}) = (-1)^{\mathsf{s}}\epsilon(\mathsf{T}_0)\,, \qquad \text{(A.311)}$$

where s is the number of superposition polygons in the diagram, which can be even or odd. Equation (A.307) applies if the product (A.308) and the similar product for T_0 have the same sign. For this, a sufficient condition is that, for each superposition polygon,

$$\begin{aligned}&\mathrm{Sign}\{z(\mu_1,\mu_2)z(\mu_3,\mu_4)\cdots z(\mu_{2n-1},\mu_{2n})\}\\&=-\mathrm{Sign}\{z(\mu_{2n},\mu_1)z(\mu_2,\mu_3)\cdots z(\mu_{2n-2},\mu_{2n-1})\}\,,\end{aligned}\qquad\text{(A.312)}$$

since each factor (-1) in (A.311) is then neutralized by another factor (-1). Condition (A.312) can be re-expressed in the form

$$\mathrm{Sign}\{z(\mu_1,\mu_2)z(\mu_2,\mu_3)z(\mu_3,\mu_4)\cdots z(\mu_{2n-1},\mu_{2n})z(\mu_{2n},\mu_1)\}=-1\,,\qquad\text{(A.313)}$$

where $(\mu_1,\mu_2),(\mu_2,\mu_3),\ldots,(\mu_{2n},\mu_1)$ specify the edges, taken in anticlockwise order, of any lattice polygon which can be a superposition polygon.[13] Condition (A.313) is sufficient to ensure that all the terms $\mathcal{P}(\boldsymbol{Z};\Pi)$ in (A.303) are positive and (8.9) is then valid.

We can now develop a topological criterion for ascribing appropriate signs to the non-zero $z(i,j)=\pm q(i,j)$. Put an arrow on the edge connecting sites i and j which indicates the order of the indices for a positive factor. Equation (A.313) is equivalent to the condition that any superposition polygon is clockwise odd (Appendix A.7.7). Suppose that a polygon $\mathcal{C}$ encloses a subgraph $\mathfrak{g}(\mathcal{C})$ with f faces, bounded by elementary polygons $\mathcal{C}_k$, and v_i interior sites. The number ω_k of clockwise arrows on $\mathcal{C}_k$ and the number ω of clockwise arrows on $\mathcal{C}$ are related by (A.238). For a superposition polygon v_i is an even number since it encloses only other superposition polynomials and doubly connected lines. It follows that ω is odd if all ω_k are odd. Thus, when the lattice $\mathcal{N}$ has arrows indicating the signs of the elements of $\boldsymbol{Z}$ oriented in such a way that all elementary polygons are clockwise odd, then any superposition polygon is clockwise odd and the consequential terms of the Pfaffian are all positive. This establishes Theorem 8.2.2.

A.11 Determinants of Cyclic Matrices

Let $\boldsymbol{Z}$ be an $N\times N$ dimensional matrix with elements $z(\boldsymbol{j},\boldsymbol{\ell})$, indexed by two-dimensional vectors of the form

$$\boldsymbol{j}=(j_1,j_2)\,,\qquad j_1=1,2,\ldots,N_1\,,\qquad j_2=1,2,\ldots,N_2\qquad\text{(A.314)}$$

with $N=N_1N_2$. The matrix $\boldsymbol{Z}$ is said to be *cyclic* if

$$\begin{aligned}&z(\boldsymbol{j},\boldsymbol{\ell})=z(\boldsymbol{\ell}-\boldsymbol{j})\,,\\&z(\boldsymbol{j}+\boldsymbol{n})=z(\boldsymbol{j})\,,\qquad \boldsymbol{n}=(n_1N_1,n_2N_2)\end{aligned}\qquad\text{(A.315)}$$

for any integers n_1 and n_2. Let $\boldsymbol{\Omega}$ be the $N\times N$ dimensional matrix with elements

$$\omega(\boldsymbol{j},\boldsymbol{\ell})=\frac{1}{\sqrt{N}}\exp\left[2\pi\mathrm{i}\left(\frac{j_1\ell_1}{N_1}+\frac{j_2\ell_2}{N_2}\right)\right]\,,\qquad\text{(A.316)}$$

[13] Provided consistency is maintained, either anticlockwise or clockwise order can be used.

where vectors $\boldsymbol{j}$ and $\boldsymbol{\ell}$ are of the form (A.314). It is not difficult to show that $\boldsymbol{\Omega}$ is a unitary matrix and

Theorem A.11.1. *The cyclic matrix* $\boldsymbol{Z}$ *is diagonalized by the transformation* $\boldsymbol{\Omega}^{-1}\boldsymbol{Z}\boldsymbol{\Omega}$ *and*

$$\mathrm{Det}\{\boldsymbol{Z}\} = \prod_{\{\boldsymbol{k}\}} \zeta(\boldsymbol{k}) , \tag{A.317}$$

where

$$\zeta(\boldsymbol{k}) = \sum_{\ell_1=1}^{N_1} \sum_{\ell_2=1}^{N_2} z(\ell_1, \ell_2) \exp\left[2\pi \mathrm{i} \left(\frac{\ell_1 k_1}{N_1} + \frac{\ell_2 k_2}{N_2}\right)\right] . \tag{A.318}$$

Proof: The matrix $\boldsymbol{Z}' = \boldsymbol{\Omega}^{-1}\boldsymbol{Z}\boldsymbol{\Omega}$ has the elements

$$\begin{aligned} z'(\boldsymbol{j},\boldsymbol{k}) &= \sum_{\{\boldsymbol{\ell}\}} \sum_{\{\boldsymbol{\ell}'\}} \overline{\omega(\boldsymbol{\ell},\boldsymbol{j})} z(\boldsymbol{\ell}' - \boldsymbol{\ell}) \omega(\boldsymbol{\ell}',\boldsymbol{k}) \\ &= \sqrt{N} \sum_{\{\boldsymbol{\ell}\}} \overline{\omega(\boldsymbol{\ell},\boldsymbol{j})} \omega(\boldsymbol{\ell},\boldsymbol{k}) \sum_{\{\boldsymbol{\ell}'\}} z(\boldsymbol{\ell}' - \boldsymbol{\ell}) \omega(\boldsymbol{\ell}' - \boldsymbol{\ell},\boldsymbol{k}) . \end{aligned} \tag{A.319}$$

Since $z(\boldsymbol{j})\omega(\boldsymbol{j},\boldsymbol{k}) = z(\boldsymbol{j}+\boldsymbol{n})\omega(\boldsymbol{j}+\boldsymbol{n},\boldsymbol{j})$, for any of the vectors $\boldsymbol{n}$ defined in (A.315), $\boldsymbol{\ell}' - \boldsymbol{\ell}$ can be replaced by $\boldsymbol{\ell}'$ in the second summation, giving, from (A.317) and (A.318),

$$\begin{aligned} z'(\boldsymbol{j},\boldsymbol{k}) &= \sqrt{N} \sum_{\{\boldsymbol{\ell}\}} \overline{\omega(\boldsymbol{\ell},\boldsymbol{j})} \omega(\boldsymbol{\ell},\boldsymbol{k}) \sum_{\{\boldsymbol{\ell}'\}} z(\boldsymbol{\ell}') \omega(\boldsymbol{\ell}',\boldsymbol{k}) \\ &= \delta^{\mathrm{Kr}}(\boldsymbol{j} - \boldsymbol{k}) \zeta(\boldsymbol{k}) . \end{aligned} \tag{A.320}$$

The result (A.317) then follows from $\mathrm{Det}\{\boldsymbol{Z}\} = \mathrm{Det}\{\boldsymbol{Z}'\}$. These results can be generalized by replacing each element $z(\boldsymbol{j},\boldsymbol{\ell})$ of the $N \times N$ dimensional matrix $\boldsymbol{Z}$ by a $\eta \times \eta$ dimensional submatrix, or block, $\boldsymbol{z}(\boldsymbol{j},\boldsymbol{\ell})$ with

$$\boldsymbol{z}(\boldsymbol{j},\boldsymbol{\ell}) = \boldsymbol{z}(\boldsymbol{\ell} - \boldsymbol{j}) , \qquad \boldsymbol{z}(\boldsymbol{j} + \boldsymbol{n}) = \boldsymbol{z}(\boldsymbol{j}) . \tag{A.321}$$

Such a matrix $\boldsymbol{Z}$ is called a *cyclic block matrix.* With $\omega(\boldsymbol{j},\boldsymbol{\ell})$ replaced by $\boldsymbol{\omega}(\boldsymbol{j},\boldsymbol{\ell}) = \omega(\boldsymbol{j},\boldsymbol{\ell})\boldsymbol{I}_\eta$, where $\boldsymbol{I}_\eta$ is the unit matrix of dimension η, the blocks $\boldsymbol{z}'(\boldsymbol{j},\boldsymbol{\ell})$ of $\boldsymbol{Z}' = \boldsymbol{\Omega}^{-1}\boldsymbol{Z}\boldsymbol{\Omega}$ are given by

$$\boldsymbol{z}'(\boldsymbol{j},\boldsymbol{k}) = \delta^{\mathrm{Kr}}(\boldsymbol{j} - \boldsymbol{k}) \boldsymbol{\zeta}(\boldsymbol{k}) , \tag{A.322}$$

where

$$\boldsymbol{\zeta}(\boldsymbol{k}) = \sum_{\{\boldsymbol{\ell}\}} \boldsymbol{z}(\boldsymbol{\ell}) \exp\left[2\pi \mathrm{i} \left(\frac{\ell_1 k_1}{N_1} + \frac{\ell_2 k_2}{N_2}\right)\right] . \tag{A.323}$$

Equation (A.317) is then replaced by

$$\mathrm{Det}\{\boldsymbol{Z}\} = \prod_{\{\boldsymbol{k}\}} \mathrm{Det}\{\boldsymbol{\zeta}(\boldsymbol{k})\} . \tag{A.324}$$

Now consider the limit $N_s \to \infty$, for $s = 1, 2$. In (A.323) the vector $\boldsymbol{k}$ is replaced by the vector $\boldsymbol{\theta} = (\theta_1, \theta_2)$, where $\theta_s = 2\pi i k_s/N_s$, $s = 1, 2$. In the limit $N_s \to \infty$, θ_s becomes a continuous variable in the range $[0, 2\pi]$. From (A.324) we obtain

$$\lim_{N_1\to\infty} \lim_{N_2\to\infty} \frac{\ln[\mathrm{Det}\{\boldsymbol{Z}\}]}{N_1 N_2} = \frac{1}{(2\pi)^2} \int_0^{2\pi} \mathrm{d}\theta_1 \int_0^{2\pi} \mathrm{d}\theta_2 \ln[\mathrm{Det}\{\boldsymbol{\zeta}(\theta_1, \theta_2)\}]\,. \tag{A.325}$$

A.12 The T Matrix

For a lattice $\mathcal{N}$ of N sites we generate a sequence $\mathcal{S}_n$ of section subgraphs in the following way:

(i) Choose a *largest* type of subgraph $\mathfrak{g}_{\mathrm{max}}$.

(ii) Find all topologically distinct subgraphs obtained as overlap graphs of copies of $\mathfrak{g}_{\mathrm{max}}$, then as overlap graphs of these overlaps and so on until the process is exhausted. All the graphs of this sequence will be section graphs,

(iii) Label the members of $\mathcal{S}_n$ in a dictionary ordering according to their topologies $[\mathfrak{n}, \mathfrak{k}]$. Thus $\mathfrak{g}_1$ will be a single site with topology $[1, 1]$ and, with n denoting the number of members of $\mathcal{S}_n$, $\mathfrak{g}_n = \mathfrak{g}_{\mathrm{max}}$. It is clear that $\mathfrak{g}_i$ can be an overlap graph of two $\mathfrak{g}_j$ only if $j \geq i$.

In this new labelling we denote the lattice constant $\omega[\mathfrak{n}, \mathfrak{k}]$ of the graph $\mathfrak{g}_j$ with topology $[\mathfrak{n}, \mathfrak{k}]$ by ω_j. Let $\Omega_i^{(j)}$ be the number of ways of arranging $\mathfrak{g}_i$ on $\mathfrak{g}_j$. A *configuration* of $\mathcal{N}$ is an arrangement of section graphs on $\mathcal{N}$. A configuration on $\mathfrak{g}_j$ is an arrangement of section graphs on $\mathfrak{g}_j$. We label the topologically distinct configurations on $\mathfrak{g}_j$ with the index ℓ and denote the graph $\mathfrak{g}_j$ in configuration ℓ by $\mathfrak{g}_j(\ell)$. Let $\lambda^{(j)}(\ell)$ be the number of configurations on $\mathfrak{g}_j$ which are topologically the same as ℓ. For some configuration $\mathcal{C}$ of the lattice, let $N\omega_j p^{(j)}(\ell)$ be the number of graphs of type $\mathfrak{g}_j$ which are in a particular one of the topologically equivalent configurations ℓ. The set of *distribution number* $\{p^{(n)}(\ell)\}$ for $\mathfrak{g}_n$ define the configuration $\mathcal{C}$ with the distribution numbers $\{p^{(j)}(\ell)\}$ for all $j = 1, 2, \ldots, n-1$ being given by consistency conditions (Hijmans and de Boer 1955). Assuming that each of these configurations occur the same number of times, the distribution numbers satisfy the normalization conditions

$$\sum_{\ell} \lambda^{(j)}(\ell) p^{(j)}(\ell) = 1\,, \qquad j = 1, 2, \ldots, n\,. \tag{A.326}$$

Let $\mathcal{E}\{\mathfrak{g}_j(\ell)\}$ be a property of $\mathfrak{g}_j(\ell)$ which is additive on non-overlapping copies of $\mathfrak{g}_j$ and is such that if $\mathfrak{g}_i(k) = \mathfrak{g}_j(\ell) \cap \mathfrak{g}_{j'}(\ell')$

$$\mathcal{E}\{\mathfrak{g}_j(\ell) \cup \mathfrak{g}_{j'}(\ell')\} = \mathcal{E}\{\mathfrak{g}_j(\ell)\} + \mathcal{E}\{\{\mathfrak{g}_{j'}(\ell')\} - \mathcal{E}\{\mathfrak{g}_j(\ell) \cap \{\mathfrak{g}_j(\ell)\}\,. \tag{A.327}$$

Now suppose that the only contributions to $\mathcal{E}$ for any graph $\mathfrak{G}$ are given by members of the sequence $\mathcal{S}_n$. We want to obtain an expression for $\mathcal{E}\{\mathcal{C}\}$, the value for the property $\mathcal{E}$ for the whole lattice when it is in configuration $\mathcal{C}$ determined by $p^{(n)}(\ell)$, $\ell = 1, 2, \ldots$. If we were simply to count contributions from all the graphs $\mathfrak{g}_n$ in their various configurations and ignore any overlaps

$$\mathcal{E}(\mathcal{C}) = N\omega_n \sum_{\ell} \lambda^{(n)}(\ell) p^{(n)}(\ell) \mathcal{E}\{\mathfrak{g}_n(\ell)\}\,. \tag{A.328}$$

This is clearly incorrect and once we begin to taking into account overlaps the process carries on through the sequence $\mathcal{S}_n$ down to $\mathfrak{g}_1$. The correct expression must be of the form

$$\mathcal{E}(\mathcal{C}) = N \sum_{j=1}^{n} \chi_j^{(n)} \sum_{\ell} \lambda^{(j)}(\ell) p^{(j)}(\ell) \mathcal{E}\{\mathfrak{g}_j(\ell)\}\,, \tag{A.329}$$

for some set of numbers $\chi_j(n)$, $j = 1, 2, \ldots, n$. To determine these numbers we observe that they must be independent of the property $\mathcal{E}$. In particular we can use $\mathcal{E}\{\mathfrak{g}_j(\ell)\} = \Omega_i^{(j)}$, the number of ways of arranging $\mathfrak{g}_i$ on $\mathfrak{g}_j$. Then $\mathcal{E}\{\mathcal{C}\} = N\omega_i$ and substituting into (A.329) and using (A.326)

$$\omega_i = \sum_{j=1}^{n} \chi_j^{(n)} \Omega_i^{(j)}, \qquad i = 1, 2, \ldots, n\,. \tag{A.330}$$

The result (A.329) can be applied directly to the energy of the system and, with an approximation of the same kind as that applied in standard mean-field cluster approximations (see Volume 1, Sect. 7.2), also to the entropy. Then by summing over all configurations $\mathcal{C}$

$$\Phi_n(\mathcal{N}) = \sum_{j=1}^{n} \chi_j^{(n)} \Phi(\mathfrak{g}_j)\,, \tag{A.331}$$

where $\Phi(\mathfrak{g}_j)$ is the dimensionless free energy of a system on a graph $\mathfrak{g}_j$ and $\Phi_n(\mathcal{N})$ is the dimensionless free energy of the same type of system on $\mathcal{N}$ but approximated by truncating the sequence of graphs at a maximum graph $\mathfrak{g}_n$. If the sequence were continued up to a maximum graph which was the lattice itself then $\chi_j^{(n)} = \delta^{\mathrm{Kr}}(n-j)$ and (A.331) becomes an identity. Equation (A.330) can be expressed in matrix form[14]

$$\omega = \Omega \chi\,, \tag{A.332}$$

where the triangular array

[14] Contrary to the convention of A.3 we now use vector notation in column form.

$$\Omega = \begin{pmatrix} \Omega_1^{(1)} & \Omega_1^{(2)} & \cdots & \Omega_1^{(n-1)} & \Omega_1^{(n)} \\ 0 & \Omega_2^{(2)} & \cdots & \Omega_2^{(n-1)} & \Omega_2^{(n)} \\ 0 & 0 & \cdots & \Omega_3^{(n-1)} & \Omega_3^{(n)} \\ 0 & 0 & \vdots & \vdots & \vdots \\ 0 & 0 & \vdots & 0 & \Omega_n^{(n)} \end{pmatrix} \tag{A.333}$$

is the *T matrix* of Domb (1974a) and

$$\omega = \begin{pmatrix} \omega_1 \\ \omega_2 \\ \vdots \\ \omega_n \end{pmatrix}, \qquad \chi = \begin{pmatrix} \chi_1^{(n)} \\ \chi_2^{(n)} \\ \vdots \\ \chi_n^{(n)} \end{pmatrix}. \tag{A.334}$$

As a simple example on the square lattice let $\mathfrak{g}_{\max}$ be a square block of four nearest-neighbour squares. This generates the sequence:

$\mathfrak{g}_1$ a single site,

$\mathfrak{g}_2$ a single line,

$\mathfrak{g}_3$ a single square,

$\mathfrak{g}_4$ a single connected co-linear pair of lines,

$\mathfrak{g}_5$ a block of two squares,

$\mathfrak{g}_6$ a square block of four squares.

Then

$$\Omega = \begin{pmatrix} 1 & 2 & 4 & 3 & 6 & 9 \\ 0 & 1 & 4 & 2 & 7 & 12 \\ 0 & 0 & 1 & 0 & 2 & 4 \\ 0 & 0 & 0 & 1 & 2 & 6 \\ 0 & 0 & 0 & 0 & 1 & 4 \\ 0 & 0 & 0 & 0 & 0 & 1 \end{pmatrix}, \qquad \omega = \begin{pmatrix} 1 \\ 2 \\ 1 \\ 2 \\ 2 \\ 1 \end{pmatrix}. \tag{A.335}$$

Solving (A.332)

$$\begin{aligned} &\chi_1^{(6)} = 0\,, \qquad \chi_2^{(6)} = 0\,, \qquad \chi_3^{(6)} = 1\,, \\ &\chi_4^{(6)} = 0\,, \qquad \chi_5^{(6)} = -2\,, \qquad \chi_6^{(6)} = 1\,. \end{aligned} \tag{A.336}$$

A general procedure for obtaining results of this kind on the square lattice was given by Enting (1978). The first step is to count the two graphs arising from the vertical and horizontal orientations of oblong block separately. (In the case given above $\mathfrak{g}_j$ for $j = 2, 4, 5$ are split to increase n to 9.) Now $\omega_j = 1$ for all $j = 1, 2, \ldots, 9$. Suppose a rectangular p sites by q sites block on the square lattice is denoted by $[p, q]$. Then it is not difficult to see that the elements of the T matrix Ω are given by

$$\langle [a,b]|\Omega|[p,q]\rangle = \begin{cases} (p-a+1)(q-b+1)\,, \text{if } a \le p \text{ and } b \le q \;, \\ 0\,, \qquad\qquad\qquad\qquad \text{otherwise} \;. \end{cases} \tag{A.337}$$

It can then be shown (Enting 1978) that

$$\langle [a,b]|\Omega^{-1}|[p,q]\rangle = \eta(a,p)\eta(b,q)\,, \tag{A.338}$$

where

$$\eta(a,p) = \begin{cases} 1\,, & \text{if } p = a \text{ or if } a+2 = p > 2 \;, \\ -2\,, & \text{if } a+1 = p > 1 \;, \\ 0\,, & \text{otherwise} \;. \end{cases} \tag{A.339}$$

Then, from (A.332),

$$\langle [a,b]|\chi = \sum_{[p,q]} \eta(a,p)\eta(b,q)\,, \tag{A.340}$$

where the sum is over all the blocks $[p, q]$ in the sequence. In the case where the sequence was derived by choosing a largest member $[k_{\mathrm{h}}, k_{\mathrm{v}}]$ with a fixed number of horizontal sites k_{h} and vertical sites k_{v} the summations in (A.340) can be taken independently over the ranges $p = 1, 2, \ldots, k_{\mathrm{h}}$ and $q = 1, 2, \ldots, k_{\mathrm{v}}$. A case of this kind is the example given above where $k_{\mathrm{h}} = k_{\mathrm{v}} = 3$. We can then, for example, calculate

$$\begin{aligned} \chi_5^{(6)} &= \langle [2,3]|\chi + \langle [3,2]|\chi \\ &= 2\Big[\sum_{p=1}^{3} \eta(2,p)\Big]\Big[\sum_{q=1}^{3} \eta(3,q)\Big] \\ &= 2[\eta(2,2) + \eta(2,3)]\eta(3,3) \\ &= 2(1-2) = -2\,. \end{aligned} \tag{A.341}$$

A more complicated form of constraint on the blocks is to impose an upper bound on the circumference, $p + q \le r$. A procedure for calculating the parameters $\chi_j^{(n)}$ in this case has been given by Enting (1978).

References and Author Index

The page numbers where the references are cited in the text are given in square brackets.

Abraham D.B. and Heilmann O.J. (1980): Interacting dimers on the simple cubic lattice as a model for liquid crystals. J. Phys. A: Math. Gen. **13**, 1051–1062, [114].

Abramowitz M. and Segun I.A. (1965): Handbook of Mathematical Functions. Dover, New York, U.S.A. [185,281,310,366].

Adler J., Brandt A., Janke W. and Shmulyian S. (1995): Three-state Potts model on the triangular lattice. J. Phys. A: Math. Gen. **28**, 5117–5129, [108].

Adler J. and Enting I.G. (1984): The two-dimensional spin-1 Ising system and related models. J. Phys. A: Math. Gen. **17**, 2233–2345, [295,300].

Adler J., Moshe M. and Privman V. (1981): Confluent singularities in directed bond percolation. J. Phys. A: Math. Gen. **14**, L363–L367, [257].

Affleck I. (1986): Universal term in the free energy at a critical point and the conformal anomaly. Phys. Rev. Lett. **56**, 746–748, [85].

Aharony A. (1974): Two-scale-factor universality and the ε expansion. Phys. Rev. B **9**, 2107–2109, [74].

Alexander S. (1975): Lattice gas transitions of He on grafoil. A continuous transition with cubic terms. Phys. Lett. A **54**, 353–354, [238,247].

Amit D.J., (1978): Field Theory, the Renormalization Group and Critical Phenomena. McGraw-Hill, New York, U.S.A. [93,203].

Arisue H. and Tabata K. (1997): Low-temperature series for the square lattice Potts model by the improved finite-lattice method. J. Phys. A: Math. Gen. **30**, 3313–3327, [290,295].

Arrott A. and Noakes J.E. (1967): Approximate equation of state for nickel near its critical temperature. Phys. Rev. Lett. **19**, 786–789, [71].

Asano T. (1968): Generalization of the Lee-Yang theorem. Prog. Theor. Phys. **40**, 1328–1336, [115].

Asano T. (1970): Theorems on the partition functions of the Heisenberg ferromagnets. J. Phys. Soc. Japan **29**, 350–359, [115,128].

Ashcroft N.W. and Mermin N.D. (1976): Solid State Physics. Holt-Saunders, New York, U.S.A. [253].

Au-Yang H. and Fisher M.E. (1975): Bounded and inhomogeneous Ising models. II Specific heat scaling function for a strip. Phys. Rev. B **11**, 3469–3487, [79].

Baker G.A. Jr. (1961): Application of the Padé approximant method to the investigation of some magnetic properties of the Ising model. Phys. Rev. **124**, 768–774, [256].

Baker G.A. Jr. (1965): The theory and application of the Padé approximant method. In Advances in Theoretical Physics: Vol.1, (Ed. K.A. Brueckner) 1–58, Academic Press, New York, U.S.A. [256].

Baker G.A. Jr. and Hunter D.L. (1973): Methods of series analysis. II Generalized and extended methods with application to the Ising model. Phys. Rev. B **7**, 3377–3392, [256,257].

Barber M.N. (1983): Finite-size scaling. In Phase Transitions and Critical Phenomena. (Eds. C. Domb and J. L. Lebowitz), Vol. 8, 145–266. Academic Press, London, U.K., [75,79,81,242].

Barber M.N. and Fisher M.E.(1973a): Critical phenomena in systems of finite thickness. I The spherical model. Ann. Phys. **77**, 1–78, [79].

Barber M.N. and Fisher M.E.(1973b): Critical phenomena in systems of finite thickness. III Specific heat of an ideal boson film. Phys. Rev. A **8**, 1124–1135, [79].

Barouch E., McCoy B.M. and Wu T.T. (1973): Zero-field susceptibility of the two-dimensional Ising model near T_c. Phys. Rev. Lett. **31**, 1409–1417, [299].

Baxter R.J. (1972): Partition function of the eight-vertex lattice model. Ann. Phys. **70**, 193–228, [182,198].

Baxter R.J. (1973): Potts model at the critical temperature. J. Phys. C: Solid State Phys. **6**, L445–L448, [108,299].

Baxter R.J. (1974): Ising model on a triangular lattice with three-spin interactions. II Free energy and correlation length. Aust. J. Phys. **27**, 369–381, [162].

Baxter R.J. (1980): Hard hexagons: exact solution. J. Phys. A: Math. Gen. **13**, L61–L70, [135].

Baxter R.J. (1981): Rogers-Ramanujan identities in the hard hexagon model. J. Stat. Phys. **26**, 427–452, [135].

Baxter R.J. (1982a): Exactly Solved Models in Statistical Mechanics, Academic Press, London, U.K. [72,113,135,181,182,184,186,187,190,199,234,266,366].

Baxter R.J. (1982b): Critical antiferromagnetic square lattice Potts model. Proc. Roy. Soc. A **383**, 43–54, [108].

Baxter R.J. and Enting I.G. (1979): Series expansions for corner transfer matrices: the square lattice Ising model. J. Stat. Phys. **21**, 103–123, [300].

Baxter R.J., Enting I.G. and Tsang S.K. (1980): Hard-square lattice gas. J. Stat. Phys. **22**, 465–489, [135].

Baxter R.J., Temperley H.N.V. and Ashley S.E. (1978): Triangular Potts model at its transition temperature, and related problems. Proc. Roy. Soc. A **358**, 535–559, [108,299].

Baxter R.J. and Wu, F.Y. (1973): Exact solution of an Ising model with three-spin interactions on a triangular lattice. Phys. Rev. Lett. **31**, 1294–1303, [162].

Baxter R.J. and Wu, F.Y. (1974): Ising model on a triangular lattice with three-spin interactions. I The eigenvalue problem. Aust. J. Phys. **27**, 357–367, [162].

Beardon A.F. (1984): A Primer on Riemann Surfaces. LMS Lecture Note Series:78. C.U.P., Cambridge, U.K. [365].

Bell G.M. (1972): Statistical mechanics of water: lattice model with directed bonding. J. Phys. C: Solid State Phys. **5**, 889–905, [114,303].

Bell G.M. (1974): Ising ferrimagnetic models: I. J. Phys. C: Solid State Phys. **7**, 1174–1188, [18].

Bell G.M., Combs L.L. and Dunne L.J. (1981): Theory of cooperative phenomena in lipid systems. Chem. Rev. **81**, 15–48, [332].

Bell G.M. and Lavis D.A. (1970): Two-dimensional bonded lattice fluids. II Orientable molecule model. J. Phys. A: Math. Gen. **3**, 568–581, [114,303].

Bell G.M. and Lavis D.A. (1989): Statistical Mechanics of Lattice Models. Vol.1: Closed Form and Exact Theories of Cooperative Phenomena. Ellis Horwood, Chichester, U.K. [V].

Bell G.M., Mingins J. and Taylor R.A.G. (1978): Second-order phase changes in phospholipid monolayer at the oil-water interface and related phenomena. J. Chem. Soc. Faraday Trans. II **74**, 223–234, [327].

Berlin T.H. and Kac M. (1952): The spherical model of a ferromagnet. Phys. Rev. **86**, 821–835, [279].

Bethe H.A. (1931): Theorie der Metalle. Erster Teil. Eigenwerte und Eigenfunktionen der lineären atomischen Kette. Z. Physik, **71**, 205–226. English translation in Mattis (1993), 689–716, [253].

Binder K., (1983): Critical behaviour at surfaces. In Phase Transitions and Critical Phenomena. (Eds. C. Domb and J. L. Lebowitz), Vol. 8, 1–144. Academic Press, London, U.K., [75].

Binder K.(Ed.): (1986): Monte Carlo Methods in Statistical Physics. Topics in Current Physics,**7**. Springer, Berlin, Heidelberg, Germany, [248].

Binder K. and Landau D.P. (1984): Finite-size scaling at first-order phase transitions. Phys. Rev. B **30**, 1477–1485, [75].

Binney J.J., Dowrick N.J., Fisher A.J. and Newman M.E.J. (1993): The Theory of Critical Phenomena. O.U.P., Oxford, U.K. [11,203].

Black J.L. and Emery V.J. (1981): Critical properties of two-dimensional models. Phys. Rev. B **23**, 429–432, [296].

Bliss G.A. (1966): Algebraic Functions. Dover, New York, U.S.A. [363].

Blöte H.W.J, Nightingale M.P. and Derrida B. (1981): Critical exponents of two-dimensional Potts and bond percolation models. J. Phys. A: Math. Gen. **14**, L45–L49, [245,247].

Blöte H.W.J. and Nightingale M.P. (1982): Critical behaviour of the two-dimensional Potts model with a continuous number of states; a finite-size scaling analysis. Physica A **112**, 405–465, [245].

Blöte H.W.J., Cardy J.L. and Nightingale M.P. (1986): Conformal invariance, the central charge and universal finite-size amplitudes at criticality. Phys. Rev. Lett. **56**, 742–745, [85].

Blöte H.W.J. and Swendsen R.H. (1979): First-order phase tansitions and the three-state Potts model. Phys. Rev. Lett. **43**, 799–802, [299].

Blume M., Emery V.J. and Griffiths R.B. (1971): Ising model for the λ transition and phase separation on ^{3}He-^{4}He mixtures. Phys. Rev. A **4**, 1071–1077, [103].

Brascamp H.J. and Kunz H. (1974): Zeros of the partition function for the Ising model in the complex temperature plane. J. Math. Phys. **15**, 65–66, [115].

Bretz M., (1977): Ordered helium films on highly uniform graphite – finite-size effects, critical parameters and the three-state Potts model. Phys. Rev. Lett. **38**, 501–505, [238].

Brézin E. (1982): An investigation of finite-size scaling. J. de Phys. **43**, 15–22, [241].

Brézin E. (1983): Finite-size scaling. Ann. N. Y. Sci. **410**, 339–349, [79].

Briggs K.M., Enting I.G. and Guttmann A.J. (1994): Series studies of the Potts model. II: Bulk series for the square lattice. J. Phys. A: Math. Gen. **27**, 1503–1523, [290,295,300].

Buckingham M.J. (1972): Thermodynamics. In Phase Transitions and Critical Phenomena. (Eds. C. Domb and M. S. Green), Vol. 2, 1–38. Academic Press, London, U.K., [16,53].

Burgoyne P.N. (1963): Remarks on the combinatorial approach to the Ising problem. J. Math. Phys. **4**, 1320–1326, [264].

Burkhardt T.W. (1976a): Kadanoff's lower-bound renormalization transformation. Phys. Rev. B **13**, 3187–3191, [241].

Burkhardt T.W. (1976b): Application of Kadanoff's lower-bound renormalization transformation to the Blume–Capel model. Phys. Rev. B **14**, 1196–1201, [241].

Burkhardt T.W. and Eisenriegler E. (1978): Renormalization group approach to surface critical behaviour in the Ising model. Phys. Rev. B **17**, 318–323, [241].

Burkhardt T.W. and Knops H.J.F. (1977): Renormalization group results for the Blume–Capel model in two and three dimensions. Phys. Rev. B **15**, 1602–1605, [241].

Burkhardt T.W., Knops H.J.F. and den Nijs M. (1976): Renormalization group results for the three-state Potts model. J. Phys. A: Math. Gen. **9**, L179–L181, [241].

Burkhardt T.W. and Swendsen R.H. (1976): Critical temperatures of the spin-s Ising model. Phys. Rev. B **13**, 3071–3073, [245].

Burkhardt T.W. and van Leeuwen J.M.J.(Eds.): (1982): Real-Space Renormalization. Topics in Current Physics, **30**. Springer, Berlin, Heidelberg, Germany, [203,247].

Cardy J.L. (1984): Conformal invariance and universality in finite-size scaling. J. Phys. A: Math. Gen. **17**, L385–L387, [81,85].

Cardy J.L. (1987): Conformal invariance. In Phase Transitions and Critical Phenomena. (Eds. C. Domb and J. L. Lebowitz), Vol. 11, 55–126. Academic Press, London, U.K., [16,81,83,85].

Cardy J.L. (1988): Finite-Size Scaling. Current Physics – Sources and Comments, **2**. North-Holland, Amsterdam, New York, [75].

Cardy J.L. (1996): Scaling and Renormalization in Statistical Physics. C.U.P. Cambridge, U.K. [203].

Chandler D. (1987): Introduction to Modern Statistical Mechanics. O.U.P. Oxford, U.K. [1].

Chen J.-H., Fisher M.E. and Nickel B.G. (1982): Unbiased estimation of corrections to scaling by partial differential approximants. Phys. Rev. Lett. **48**, 630–634, [298].

Christe P. and Henkel M. (1993): Introduction to Conformal Invariance and Its Applications to Critical Phenomena. Lecture Notes in Physics, **16**. Springer, Berlin, Heidelberg, Germany, [16,81].

Conway J.H., Curtis R.T., Norton S.P., Parker R.A. and Wilson R.A. (1985): Atlas of Finite Groups O.U.P. Oxford, U.K. [345].

Courant R. and Hilbert D. (1962): Methods of Mathematical Physics. Interscience, New York, U.S.A. [336,337].

De Boer J. (1974): Van der Waals in his time and the present revival. Physica **73**, 1–27, [21].

De Gennes P.G. and Prost J. (1993): The Physics of Liquid Crystals. O.U.P. Oxford, U.K. [15].

De Neef T. (1975): Ph.D. Thesis. Eindhoven University of Technology, [290,294].

De Neef T. and Enting I.G. (1977): Series expansions from the finite-lattice method. J. Phys. A: Math. Gen. **10**, 801–805, [290,294,299].

Den Nijs M.P.M. (1979): A relationship between the temperature exponents of the eight-vertex and q-state Potts model. J. Phys. A: Math. Gen. **12**, 1857–1868, [295].

Devaney R.L. (1989): Chaotic Dynamic Systems. Addison-Wesley, California, U.S.A. [207].

Dobrushin R.L. (1965): Existence of a phase transition in two-dimensional and three-dimensional Ising models. Theor. Prob. Appl. **10**, 193–213, [114].

Dobrushin R.L. (1968a): Gibbsian random fields for lattice systems with pairwise interactions. Funct. Anal. Appl. **2**, 292–301, [114,119].

Dobrushin R.L. (1968b): The problem of uniqueness of a Gibbsian random field and the problem of phase transitions. Funct. Anal. Appl. **2**, 302–312, [114,119].

Domb C. (1960): On the theory of cooperative phenomena in crystals. Phil. Mag. Supp. **9**, 149–361, [154,290].

Domb C. (1974a): Graph theory and embeddings. In Phase Transitions and Critical Phenomena. (Eds. C. Domb and M. S. Green), Vol. 3, 1–95. Academic Press, London, U.K., [267,275,290,370,371,374,377,398].

Domb C. (1974b): Ising model. In Phase Transitions and Critical Phenomena. (Eds. C. Domb and M. S. Green), Vol. 3, 357–484. Academic Press, London, U.K., [261,374].

Domb C. (1974c): Star lattice constant expansions for magnetic models. J. Phys. A: Math. Gen. **7**, L45–L47, [275].

Domb C. (1974d): Configurational studies of the Potts models. J. Phys. A: Math. Gen. **7**, 1335–1348, [253].

Domb C. (1976): Finite-cluster patition functions for the D-vector model. J. Phys. A: Math. Gen. **9**, 983–998, [293,300].

Domb C., (1985): Critical phenomena: a brief historical survey. Contemporary Physics **26**, 49–72, [V].

Domb C. and Green M.S. (Eds.) Phase Transitions and Critical Phenomena, Vol. 6, Academic Press, London, U.K. [203].

Domb C. and Hunter D.L. (1965): On the critical behaviour of ferromagnets. Proc. Phys. Soc. **86**, 1147–1151, [16].

Domb C. and Sykes M.F. (1956): On metastable approximations in cooperative assemblies. Proc. Roy. Soc. A **235**, 247–259, [254].

Domb C. and Sykes M.F. (1957a): On the susceptibility of a ferromagnetic above the Curie point. Proc. Roy. Soc. A **240**, 214–228, [254].

Domb C. and Sykes M.F. (1957b): The calculation of lattice constants in crystal statistics. Phil. Mag. **2**, 733–749, [254].

Edmonds A.R. (1957): Angular Momentum in Quantum Mechanics. Princeton U.P., Princeton, U.S.A. [285].

Ehrenfest P. (1933): Phase changes in the ordinary and extended sense classified according to the corresponding singularity of the thermodynamic potential. Proc. Acad. Sci. Amsterdam **36**, 153–157, [15].

Eisenberg D. and Kauzmann W. (1969): The Structure and Properties of Water. Clarendon Press, Oxford, U.K. [15].

English P.S., Hunter D.L. and Domb C. (1979): Extension of the high-temperature free energy series for the classical vector model of ferromagnetism in general spin dimensionality. J. Phys. A: Math. Gen. **12**, 2111–2130, [293,300].

Enting I.G. (1978): Generalized Möbius functions for rectangles on the square lattice. J. Phys. A: Math. Gen. **11**, 563–568, [294,398,399].

Enting I.G. and Baxter R.J. (1977): A special series expansion technique for the square lattice. J. Phys. A: Math. Gen. **10**, L117–L119, [294].

Enting I.G. and Baxter R.J. (1980): An investigation of the high-temperature series expansion for the square lattice Ising model. J. Phys. A: Math. Gen. **13**, 3723–3734, [252].

Enting I.G., Guttmann A.J. and Jensen I. (1994): Low-temperature series expansions for the spin-1 Ising model. J. Phys. A: Math. Gen. **27**, 6987–7006, [290,295,300].

Enting I.G. and Wu F.Y. (1982): Triangular lattice Potts models. J. Stat. Phys. **28**, 351–373, [108].

Essam J.W. and Fisher M.E. (1963): Padé approximant studies of the lattice gas and Ising ferromagnet below the critical point. J. Chem. Phys. **38**, 802–812, [54].

Falicov L.M. (1966): Group Theory and Its Applications. U. Chicago Press, Chicago, U.S.A. [342].

Fan C. and Wu F.Y. (1970): General lattice statistical model of phase transitions. Phys. Rev. B **2**, 723–733, [174,199].

Ferdinand A.E. and Fisher M.E. (1969): Bounded and inhomogeneous Ising models. I Specific heat anomaly on a finite lattice. Phys. Rev. **185**, 832–846, [79].

Ferer M., Moore M.A. and Wortis M. (1973): Critical indices and amplitudes of classical planar models in finite field for temperatures greater that T_c. Phys. Rev. **8**, 5205–5212, [74].

Firpo J.L., Legre J.P., Bois A.G. and Baret J.F. (1984): Equilibrium and non-equilibrium critical behaviour of amphiphilic monolayers at the LE-LC transition – a model with broken symmetry and melted chain packing. J. Chim. Phys. et Phys. Chim. Biol. **81**, 113–120, [332].

Fischer P., Lebech B., Meier G., Rainford B.D. and Vogt O. (1978): Magnetic phase transitions of CeSb: I Zero applied magnetic field. J. Phys. C: Solid State Phys. **11**, 345–364, [15].

Fisher M.E. (1959): Transformations of Ising models. Phys. Rev. **113**, 969–981, [270].

Fisher M.E. (1961): Statstical mechanics of dimers on a plane lattice. Phys. Rev. **124**, 1664–1672, [304,309,310].

Fisher M.E. (1965): The nature of critical points. In Lectures in Theoretical Physics, Vol.7c, (Ed. W.E. Brittin), 1–157. U. Colorado Press, Boulder, U.S.A. [115,156].

Fisher M.E. (1966): On the dimer solution of planar Ising models. J. Math. Phys. **7**, 1776–1781, [321].

Fisher M.E. (1967): Theory of equilibrium critical phenomena. Rep. Prog. Phys. **30**, 615–730, [15].

Fisher M.E. (1968): Renormalization of critical exponents by hidden variables. Phys. Rev. **176**, 257–272, [23].

Fisher M.E. (1969): Rigorous inequalities for critical-point correlation exponents. Phys. Rev. **180**, 594–600, [58].

Fisher M.E. (1971): in Critical Phenomena, Proceedings of the 51st Enrico Fermi Summer School, Varenna, Italy, (Ed. M.S. Green), Academic Press, London, U.K. [75,76,78].

Fisher M.E. (1983): Scaling, universality and renormalization group theory. In Critical Phenomena, (Ed. F.J.W.Hahne) 1–139, Springer, Berlin, Heidelberg, Germany, [70,72,102].

Fisher M.E. and Barber M.N. (1972): Scaling theory for finite-size effects in the critical region. Phys. Rev. Lett. **28**, 1516–1519, [75,163].

Fisher M.E. and Berker A.N. (1982): Scaling for first-order transitions in thermodynamic and finite systems. Phys. Rev. B **26**, 2507–2513, [32,75].

Fisher M.E. and Chen J.-H. (1985): The validity of hyperscaling in 3-dimensions for scalar spin systems. J. de Phys. **46**, 1645–1654, [298].

Fisher M.E. and Ferdinand A.E. (1967): Interfacial, boundary and size effects at critical points. Phys. Rev. Lett. **19**, 169–172, [76,78,79,163].

Fisher M.E. and Privman V. (1985): First-order transitions breaking $O(n)$ symmetry: finite-size scaling. Phys. Rev. B **32**, 447–464, [76].

Fox P.F. and Guttmann A.J. (1973): Low-temperature critical behaviour of the Ising model with $S > \frac{1}{2}$. J. Phys. C: Solid State Phys. **6**, 913–930, [261,300].

Gantmacher F.R. (1979): Theory of Matrices. Chelsea, New York, U.S.A. [359].

Gartenhaus S. and Scott McCollough W. (1988): Higher-order corrections to the quadratic Ising lattice susceptibility at criticality. Phys. Lett. A **127**, 315–318, [299].

Gaunt D.S. and Fisher M.E. (1965): Hard-sphere lattice gases. I Plane-square lattice. J. Chem. Phys. **43**, 2840–2863, [135].

Gaunt D.S. and Guttmann A.J. (1965): Asymptotic analysis of coefficients. In Phase Transitions and Critical Phenomena. (Eds. C. Domb and M. S. Green), Vol. 3, 181–243. Academic Press, London, U.K., [254,299].

Gaunt D.S. and Sykes M.F. (1973): Estimation of critical indices for the three-dimensional Ising model. J. Phys. A: Math. Gen. **6**, 1517–1526, [299].

George M.J. and Rehr J.J. (1984): Two-series approach to partial differential approximants: three dimensional Ising models. Phys. Rev. Lett. **53**, 2063–2066, [298].

Ginsparg P. (1990): Applied conformal field theory. In Fields, Strings and Critical Phenomena., Proceedings of 49th Summer School in Theoretical Physics, Les Houches, 1988, (Ed. E. Brezin and J. Zinn-Justin), 6–168. North-Holland, Amsterdam, New York, [341].

Ginzburg V.L. (1960): Some remarks on phase transitions of the second kind and the microscopic theory of ferroelectric materials. Soviet Phys. JETP. **2**, 1824–1834, [103].

Ginzburg V.L. and Landau L.D. (1950): 1950 On the theory of superconductivity. Reprinted in (1965): Collected Papers of L.D. Landau, (Ed. D. ter Haar), 546–568, Pergamon, Oxford, U.K. [93].

Gradshteyn I.S. and Ryzhik I.M. (1980): Table of Integrals, Series, and Products, (4th Ed.). Academic Press, London, U.K. [98,185,338–340,366–369].

Graves-Morris P.T. (Ed.): (1973): Padé Approximants and their Applications. Academic Press, London, U.K. [258].

Graves-Morris P.T. (1988): The critical index of the magnetic susceptibility of 3D Ising models. J. Phys. A: Math. Gen. **21**, 1867–1881, [298].

Green H.S. and Hurst C.A. (1964): Order-Disorder Phenomena, Interscience. London, U.K. [392].

Green M.S., Vicentini-Missoni M. and Levelt Sengers J.M.H. (1967): Scaling-law equation of state for gases in the critical region. Phys. Rev. Lett. **18**, 1113–1117, [71,72].

Grest G.S. (1981): Monte Carlo study of the antiferromagnetic Potts model on frustrated lattices. J. Phys. A: Math. Gen. **14**, L217–L221, [108].

Griffiths R.B. (1964): Peierls proof of spontaneous magnetization in a two-dimensional Ising model. Phys. Rev. A **136**, 437–439, [114].

Griffiths R.B. (1965): Ferromagnets and simple fluids near the critical point: some thermodynamic inequalities. J. Chem. Phys. **43**, 1958–1968, [46].

Griffiths R.B. (1970): Thermodynamics near the two-fluid critical mixing point in He^3-He^4. Phys. Rev. Lett. **24**, 715–717, [20].

Griffiths R.B. (1972): Rigorous results and theorems. In Phase Transitions and Critical Phenomena. (Eds. C. Domb and M. S. Green), Vol. 1, 7–109. Academic Press, London, U.K., [5,113,116,260].

Griffiths R.B. (1973): Proposal for notation at tricritical points. Phys. Rev. B **7**, 545–551, [62].

Griffiths R.B. (1981): Mathematical properties of renormalization group transformations. Physica A **106**, 59–69, [209,210].

Griffiths R.B. and Pearce P.A. (1978): Position-space renormalization group transformations: some proofs and some problems. Phys. Rev. Lett. **41**, 917–920, [209].

Griffiths R.B. and Pearce P.A. (1979): Mathematical properties of position-space renormalization group transformations. J. Stat. Phys. **20**, 499–545, [209,210].

Griffiths R.B. and Wheeler J.C. (1970): Critical points in multicomponent systems. Phys. Rev. A **2**, 1047–1064, [58].

Guttmann A.J. (1989): Asymptotic analysis of power-series expansions. In Phase Transitions and Critical Phenomena. (Eds. C. Domb and J. L. Lebowitz), Vol. 13, 1–234. Academic Press, London, U.K., [251,254,255,258,296,298].

Guttmann A.J. and Enting I.G. (1993): Series studies of the Potts model: I The simple cubic Ising model. J. Phys. A: Math. Gen. **26**, 807–821, [290,294,295,297,300].

Guttmann A.J. and Enting I.G. (1994): Series studies of the Potts model: III The 3-state model on the simple cubic lattice. J. Phys. A: Math. Gen. **27**, 5801–5812, [290,294,295].

Guttmann A.J. and Joyce G.S. (1972): On a new method of series analysis in lattice statistics. J. Phys. A: Math. Gen. **5**, L81–L84, [258].

Hankey A. and Stanley H.E. (1972): Systematic application of generalized homogeneous functions to static scaling, dynamic scaling and universality. Phys. Rev. B **6**, 3515–3542, [16,23,27].

Hankey A., Stanley H.E. and Chang T.S. (1972): Geometric predictions of scaling at tricritical points. Phys. Rev. Lett. **29**, 278–281, [16,63].

Heilmann O.J. (1972): Existence of phase transitions in certain lattice gases with repulsive potential. Lett. Nuovo Cim. **3**, 95–98, [114].

Heilmann O.J. (1980): Existence of an ordered phase for the repulsive lattice gas on the FCC lattice. J. Phys. A: Math. Gen. **13**, 1803–1810, [114].

Heilmann O.J. and Huckaby D.A. (1979): Phase transitions in lattice gas models for water. J. Stat. Phys. **20**, 371–383, [114,118,119].

Heilmann O.J. and Lieb E.H. (1972): Theory of monomer-dimer systems. Comm. Math. Phys. **25**, 190–232, [115].

Hermann H.J. (1979): Monte Carlo simulation of the three-dimensional Potts model. Z. Physik B **35**, 171–175, [299].

Hijmans J. and de Boer J. (1955): An approximation method for order-disorder problems I–III. Physica **21**, 471–484,485–498,499–516, [293,294,396].

Hilhorst H.J., Schick M. and van Leeuwen J.M.J. (1978): Differential form of real-space renormalization: exact results for two-dimensional Ising models. Phys. Rev. Lett. **40**, 1605–1608, [225].

Hilhorst H.J., Schick M. and van Leeuwen J.M.J. (1979): Exact renormalization group equations for the two-dimensional Ising model. Phys. Rev. B **19**, 2749–2763, [225].

Hille E. (1959): Analytic Function Theory, Vol. 1. Chelsea, New York, U.S.A. [254].

Hille E. (1962): Analytic Function Theory, Vol. 2. Chelsea, New York, U.S.A. [363].

Ho J.T. and Litster J.D. (1969): Magnetic equation of state of $CrBr_3$ near the critical point. Phys. Rev. Lett. **22**, 603–606, [71,72].

Hsu Shih-chieh and Gunton J.D. (1977): Study of the cumulant expansion for renormalization group transformations on two- and three-dimensional Ising models. Phys. Rev. B **15**, 2688–2693, [232].

Huang K. (1963): Statistical Mechanics. Academic Press, London, U.K. [1].

Ising E. (1925): Beitrag zur Theorie des Ferromagnetismus. Z. Physik, **31**, 253–258, [9].

Itzykson C., Pearson R.B. and Zuber J.B. (1983): Distribution of zeros in Ising and gauge models. Nucl. Phys. B **220**, 415–433, [162,163].

Jensen I., Guttmann A.J. and Enting I.G. (1996): Low-temperature series expansions for the square lattice Ising model with $S > 1$. J. Phys. A: Math. Gen. **29**, 3805–3815, [290,295,300].

Jensen I. and Guttmann A.J. (1996): Extrapolation procedure for the low-temperature series for the square lattice spin-1 Ising model. J. Phys. A: Math. Gen. **29**, 3817–3836, [290].

Jones G.L. (1966): Complex temperatures and phase transitions. J. Math. Phys. **7**, 2000–2005, [115].

Josephson B.D. (1967): Inequality for the specific heat. Proc. Phys. Soc. **92**, 269–275, 276–284, [58].

Joyce G.S. (1967): Classical Heisenberg model. Phys. Rev. **155**, 478–495, [293].

Joyce G.S. (1972): Critical properties of the spherical model. In Phase Transitions and Critical Phenomena. (Eds. C. Domb and M. S. Green), Vol. 2, 375–442. Academic Press, London, U.K., [279].

Joyce G.S. (1975a): Analytic properties of the Ising model with triplet interactions on the triangular lattice. Proc. Roy. Soc. A **343**, 45–62, [162].

Joyce G.S. (1975b): On the magnetization of the triangular lattice Ising model with triplet interactions. Proc. Roy. Soc. A **345**, 277–293, [162].

Joyce G.S. (1988a): Exact results for the activity and isothermal compressibility of the hard-hexagon model. J. Phys. A: Math. Gen. **21**, L983–L988, [135].

Joyce G.S. (1988b): On the hard-hexagon model and the theory of modular functions. Phil. Trans. Roy. Soc. **325A**, 643–702, [135,364].

Joyce G.S. and Bowers R.G. (1966): Cluster series for the infinite spin Heisenberg model. Proc. Phys. Soc. **88**, 1053–1055, [293].

Jüngling K. (1975): Exact solution of a nonplanar two-dimensional Ising model with short range two-spin interaction. J. Phys. C: Solid State Phys. **8**, L169–L171, [171,172].

Jüngling K.and Obermair G. (1974): Note on universality and the eight-vertex model. J. Phys. C: Solid State Phys. **7**, L363–L365, [171,172,200].

Kac M. and Thompson C.S. (1971): Spherical model and the infinite spin dimensionality limit. Physica Norvegica **5**, 163–168, [279].

Kac M. and Ward J.C. (1952): A combinatorial solution of the two-dimer Ising model. Phys. Rev. **88**, 1332–1337, [264].

Kadanoff L.P. (1966): Scaling laws for Ising models near T_c. Physics **2**, 263–272, [16,24,25,27,92,203].

Kadanoff L.P. (1969a): Operator algebra and the determination of critical indices. Phys. Rev. Lett. **23**, 1430–1433, [40].

Kadanoff L.P. (1969b): Correlations along a line in the two-dimensional Ising model. Phys. Rev. **188**, 859–863, [40].

Kadanoff L.P. (1975): Variational principles and approximate renormalization group calculations. Phys. Rev. Lett. **34**, 1005–1008, [232,239,241].

Kadanoff L.P. (1976a): Scaling, universality and operator algebras. In Phase Transitions and Critical Phenomena. (Eds. C. Domb and M. S. Green), Vol. 5a, 1–34. Academic Press, London, U.K., [22,23].

Kadanoff L.P. (1976b): Notes on Migdal's recursion formulas. Ann. Phys. **100**, 359–394, [232–234].

Kadanoff L.P. and Houghton A. (1975): Numerical evaluations of the critical properties of the two-dimensional Ising model. Phys. Rev. B **11**, 377–386, [208,210,239].

Kadanoff L.P., Houghton A. and Yalabik M.C. (1976): Variational approximations for renormalization group transformations. J. Stat. Phys. **14**, 171–203, [232,239,241].

Kadanoff L.P. and Wegner F.J. (1971): Some critical properties of the eight-vertex model. Phys. Rev. B **4**, 3989–3993, [40,170,191,199].

Kasteleyn P.W. (1961): The statistics of dimers on a lattice. I The number of dimer arrangements on a quadratic lattice. Physica **27**, 1209–1225, [304,310,391,392].

Kasteleyn P.W. (1963): Dimer statistics and phase transitions. J. Math. Phys. **4**, 287–293, [315,316].

Kasteleyn P.W. (1967): Graph theory and crystal statistics. In Graph Theory and Theoretical Physics, (Ed. F. Harary), 43–110. Academic Press, London, U.K. [392].

Kaufman B. (1949): Crystal statistics. II Partition function evaluated by spinor analysis. Phys. Rev. **76**, 1232–1243, [136,154,156].

Kaufman B. and Onsager L. (1949): Crystal statistics. III Short-range order in a binary Ising lattice. Phys. Rev. **76**, 1244–1252, [136].

Kaufman M. and Griffiths R.B. (1981): Exactly soluable Ising models on hierarchical lattices. Phys. Rev. B **24**, 496–498, [225].

Kim D. and Joseph R.I. (1975): High-temperature series study of the q-component Potts model in two and three dimensions. J. Phys. A: Math. Gen. **8**, 891–904, [299].

Kincaid J.M. and Cohen E.G.D. (1975): Phase diagrams of liquid helium mixtures and metamagnets: experiment and mean-field theory. Phys. Rep. **22**, 57–143, [18,20,59,60].

Kinzel W. and Schick M. (1981): Phenomenological scaling approach to the triangular Ising antiferromagnet. Phys. Rev. B **23**, 3435–3441, [243,246,247].

Knak Jensen S.J. and Mouritsen O.G. (1979): Is the phase transition of the three-state Potts model continuous in three dimensions?. Phys. Rev. Lett. **43**, 1736–1739, [299].

Kong X.P., Au-Yang H. and Perk J.H.H. (1986): New results for the susceptibility of the two-dimensional Ising model at criticality. Phys. Lett. A **116**, 54–56, [299].

Kouvel J.S. and Comly J.B. (1968): Magnetic equation of state for nickel near its Curie point. Phys. Rev. Lett. **20**, 1237–1239, [71].

Kouvel J.S. and Rodbell D.S. (1967): Magnetic equation of state for CrO_2 and nickel near their Curie points. Phys. Rev. Lett. **18**, 215–218, [71].

Landau L.D. (1937): On the theory of phase transitions. Reprinted in (1965): Collected Papers of L.D. Landau, (Ed. D. ter Haar), 193–216. Pergamon, Oxford, U.K. [21].

Landau L.D. and Lifschitz E.M. (1980): Statistical Physics, Part 1, (3rd Ed.). Pergamon, Oxford, U.K. [91].

Landsberg P.T. (1978): Thermodynamics and Statistical Mechanics. O.U.P., Oxford, [1].

Lavis D.A. (1976): An exact matrix calculation for a two-dimensional model of the steam-water-ice system. Bulk and boundary properties. J. Phys. A: Math. Gen. **9**, 2077–2095, [76,137].

Lavis D.A. (1996): The derivation of the free energy of the Ising model from that of the eight-vertex model. Rep. Math. Phys. **37**, 147–155, [193].

Lavis D.A. and Bell G.M. (1999): Statistical Mechanics of Lattice Systems, Vol.1: Closed-Form and Exact Solutions. Springer, Berlin, Heidelberg, Germany, [V].

Lavis D.A. and Quinn A.G. (1983): An Ising ferrimagnetic model on a triangular lattice. J. Phys. C: Solid State Phys. **16**, 3547–3562, [18,239,245].

Lavis D.A. and Quinn A.G., (1987): Transfer matrix and phenomenological renormalization methods applied to a triangular Ising ferrimagnetic model. J. Phys. C: Solid State Phys. **20**, 2129–2138, [18,245,247].

Lavis D.A., Southern B.W. and Bell G.M. (1982): Phase transitions in monolayers at air/water and oil/water interfaces. J. Phys. C: Solid State Phys. **15**, 1077–1088, [239].

Lavis D.A. and Southern B.W. (1984): Renormalization group study of a three-dimensional lattice model with directional bonding. J. Stat. Phys. **35**, 489–506, [114,239].

Lawrie I.D. and Sarbach S. (1984): Theory of tricritical points. In Phase Transitions and Critical Phenomena. (Eds. C. Domb and J. L. Lebowitz), Vol. 9, 2–161. Academic Press, London, U.K., [19,21,108].

Leech J.W. and Newman D.J. (1969): How to Use Groups. Methuen, London, U.K. [154,342].

Lee T.D. and Yang C.N. (1952): Statistical theory of equations of state and phase transitions. II Lattice gas and Ising model. Phys. Rev. **87**, 410–419, [113,114,127,149].

Le Guillou J.C. and Zinn-Justin J. (1980): Critical exponents from field theory. Phys. Rev. B **21**, 3976–3998, [296].

Levelt Sengers J.M.H. (1974): From van der Waals' equation to the scaling laws. Physica **73**, 73–106, [21,75].

Lieb E.H. (1967a): Residual entropy of square ice. Phys. Rev. **162**, 162–171, [167].

Lieb E.H. (1967b): Exact solution of the F model of an antiferroelectric. Phys. Rev. Lett. **18**, 1046–1048, [167].

Lieb E.H. (1967c): Exact solution of the two-dimensional Slater KDP model of a ferroelectric. Phys. Rev. Lett. **19**, 108–110, [167].

Lieb E.H. and Loss M. (1993): Fluxes, Laplacians and Kasteleyn's theorem. Duke Math. J. **71**, 337–363, [305,389–391].

Lieb E.H. and Ruelle D. (1972): A property of zeros of the partition function for Ising spin systems. J. Math. Phys. **13**, 781–784, [115].

Lieb E.H. and Sokal A.D. (1981): A general Lee-Yang theorem for one-component and multicomponent ferromagnets. Comm. Math. Phys. **80**, 153–179, [115].

Lieb E.H. and Wu F.Y. (1972): Two-dimensional ferroelectric models. In Phase Transitions and Critical Phenomena. (Eds. C. Domb and M. S. Green), Vol. 1, 331–487. Academic Press, London, U.K., [19,171,198,324,326].

Lloyd E.H. (1953): The direct product of matrices. Math. Gaz. **37**, 29–33, [284,346].

Luban M. (1976): Generalized Landau theories. In Phase Transitions and Critical Phenomena. (Eds. C. Domb and M. S. Green), Vol. 5a, 35–86. Academic Press, London, U.K., [93].

Ma S.-K. (1976a): Modern Theory of Critical Phenomena. Benjamin, Mass., U.S.A. [101,203,248].

Ma S.-K., (1976b): Renormalization group by Monte Carlo methods. Phys. Rev. Lett. **37**, 461–464, [248].

Ma S.-K. (1985): Statistical Mechanics. World Scientific, Philadelphia, U.S.A. [1].

Martin P.P. (1991): Potts Models and Related Problems in Statistical Mechanics. World Scientific, Philadelphia, U.S.A. [105].

Martin P.P. and Maillard J.M. (1986): Zeros of the partition function for the triangular lattice three-state Potts model. J. Phys. A: Math. Gen. **19**, L547–L551, [115].

Mattis D.C. (Ed.) (1993): The Many-Body Problem: An Encyclopedia of Exactly Solved Models in One Dimension. World Scientific, Singapore, [403]

McWeeny R. (1963): Symmetry. Pergamon, Oxford, U.K. [154,342].

Mermin N.D. and Wagner H. (1966): Absence of ferromagnetism or antiferromagnetism in one- and two-dimensional isotropic Heisenberg models. Phys. Rev. Lett. **17**, 1133–1136, [253].

Metcalf B.D. (1973): Phase diagram of a nearest-neighbour triangular antiferromagnet in an external field. Phys. Lett. A **45**, 1–3, [237,238].

Migdal A.A. (1975): Phase transitions in gauge and spin-lattice theory. Soviet Phys. JETP. **42**, 743–746, [233].

Mittag L. and Stephen M.J. (1974): Mean-field theory of the many component Potts model. J. Phys. A: Math. Gen. **7**, L109–L112, [108].

Mojumder M.A. (1991): (2,0) superstring theory and the critical exponents of the 3D Ising model. Mod. Phys. Lett. **6**, 2687–2691, [296].

Montroll E.W., (1964): Lattice statistics. In Applied Combinatorial Mathematics, (Ed. E.F. Beckenbach), 96–143. Wiley, New York U.S.A. [306,308,310,332].

Munkres J.R. (1984): Elements of Algebraic Topology. Addison-Wesley, London, U.K. [119].

Nagle J.F. (1968): Weak-graph method for obtaining formal series expansions for lattice statistical problems. J. Math. Phys. **9**, 1007–1026, [384].

Nagle J.F. (1974): Ferroelectric models. In Phase Transitions and Critical Phenomena. (Eds. C. Domb and M. S. Green), Vol. 3, 653–666. Academic Press, London, U.K., [384].

Nagle J.F. (1975a): Chain model theory of lipid monolayer transitions. J. Chem. Phys. **63**, 1255–1261, [327,330,331].

Nagle J.F. (1975b): Critical points for dimer models with $\frac{3}{2}$-order transitions. Phys. Rev. Lett. **34**, 1150–1153, [327,329].

Nagle J.F. (1986): Theory of lipid monolayer and bilayer chain-melting phase transitions. Faraday Disc. Chem. Soc. **81**, 151–162, [327,332].

Nagle J.F. and Temperley H.N.V. (1968): Combinatorial theorem for graphs on a lattice. J. Math. Phys. **9**, 1020–1026, [267,384].

Nagle J.F., Yokoi C.S.O. and Bhattacharjee S.M. (1989): Dimer modules on anisotropic lattices. In Phase Transitions and Critical Phenomena. (Eds. C. Domb and J. L. Lebowitz), Vol. 13, 235–297. Academic Press, London, U.K., [23,317,321,325,327].

Nauenberg M. (1975): Renormalization group solution of the one-dimensional Ising model. J. Math. Phys. **16**, 703–705, [220].

Nauenberg M. and Nienhuis B. (1974a): Critical surface for square Ising spin lattice. Phys. Rev. Lett. **33**, 944–946, [208,235].

Nauenberg M. and Nienhuis B. (1974b): Renormalization group approach to the solution of general Ising models. Phys. Rev. Lett. **33**, 1598–1601, [208,235].

Nelson D.R. and Fisher M.E. (1975): Soluble renormalization groups and scaling fields for low-dimensional Ising systems. Ann. Phys. **91**, 226–274, [209,216,221].

Newman C.M. (1975): Inequalities for Ising models and field theories which obey the Lee-Yang theorem. Comm. Math. Phys. **41**, 1–9, [115].

Nickel B.G. (1982): The problem of confluent singularities. In Phase Transitions: Cargese 1980, (Ed. M. Levy, J.C. Le Guillou and J. Zinn-Justin), 291–324, Plenum, New York, U.S.A [297].

Nickel B.G. (1991): Cofluent singularities in 3D continuum ϕ^4 theory: resolving critical point discrepancies. Physica A **171**, 189–196, [296].

Nickel B.G. and Rehr J.J. (1990): Hight-temperature series for scalar-field lattice nodels: generation and analysis. J. Stat. Phys. **61**, 1–50, [298].

Niemeijer, Th. and van Leeuwen, J.M.J. (1973): Wilson theory for spin systems on a triangular lattice. Phys. Rev. Lett. **31**, 1411–1414, [208].

Niemeijer, Th. and van Leeuwen, J.M.J. (1974): Wilson theory for two-dimensional Ising spin systems. Physica **71**, 17–40, [208,226].

Niemeijer, Th. and van Leeuwen, J.M.J. (1976): Renormalization theory for Ising-like spin systems. In Phase Transitions and Critical Phenomena. (Eds. C. Domb and M. S. Green), Vol. 6, 425–505. Academic Press, London, U.K., [209,226,235].

Nienhuis B. and Nauenberg M. (1975): First-order phase transitions in renormalization group theory. Phys. Rev. Lett. **35**, 477–479, [32].

Nienhuis B., Riedel E.K. and Schick M. (1980): Magnetic exponents of the two-dimensional q-state Potts model. J. Phys. A: Math. Gen. **13**, L189–L192, [295].

Nightingale M.P. (1976): Scaling theory and finite systems. Physica A **83**, 561–572, [242–244].

Nightingale M.P. (1977): Non-universality for Ising-like spin systems. Phys. Lett. A **59**, 486–488, [243].

Nightingale M.P. and Blöte H.W.J. (1980): Finite-size scaling and critical point exponents of the Potts model. Physica A **104**, 352–357, [245].

Nightingale M.P. and 'T Hooft A.H. (1974): Scaling theory and logarithmic singularities. Physica **77**, 390–402, [16,28,33].

Nishimori H. and Griffiths R.B. (1983): Structure and motion of the Lee-Yang zeros. J. Math. Phys. **24**, 2637–2647, [115].

Onsager L. (1944): Crystal statistics. I Two-dimensional model with order-disorder transition. Phys. Rev. **65**, 117–149, [22,136,154].

O'Rourke M.J., Baxter R.J. and Bazhanov V.V. (1995): Numerical results for the three-state critical Potts model on finite rectangular lattices. J. Stat. Phys. **78**, 665–680, [115].

Park H. (1994): Three-state Potts model on a triangular lattice. Phys. Rev. B **49**, 881–887, [108].

Patashinskii A.Z. and Pokrovskii V.L. (1966): Behaviour of ordered systems near the transition point. Soviet Phys. JETP. **23**, 292–297, [16].

Pawley G.S., Swendsen R.H., Wallace D.J. and Wilson K.G. (1984): Monte Carlo renormalization group calculations of critical behaviour in the simple cubic Ising model. Phys. Rev. B **29**, 4030–4040, [248].

Pearson R. (1980): Conjecture for the extended Potts model magnetic eigenvalue. Phys. Rev. B **22**, 2579–2580, [295].

Peierls R. (1936): On Ising's model of ferromagnetism. Proc. Camb. Phil. Soc. **32**, 477–481, [114].

Pfeuty P. and Toulouse G. (1977): Introduction to the Renormalization Group and to Critical Phenomena. Wiley, London, U.K. [31].

Pearce P.A. and Thompson C.J. (1977): The spherical limit for n-vector correlations. J. Stat. Phys. **17**, 189–196, [279].

Pippard A.B. (1957): Elements of Classical Thermodynamics. C.U.P. Cambridge, U.K. [1,15].

Polyakov, A.M. (1970): Conformal symmetry of critical fluctuations. Soviet Phys. JETP Lett. **12**, 381–383, [16,81,83,84].

Potts R.B. (1952): Some generalized order-disorder transformations. Proc. Camb. Phil. Soc. **48**, 106–169, [105,253].

Privman V. and Fisher M.E. (1983): Finite-size effects at first-order transitions. J. Stat. Phys. **33**, 385–417, [75].

Privman V. and Fisher M.E. (1984): Universal critical amplitudes in finite-size scaling. Phys. Rev. B **30**, 322–327, [74,77,81].

Pynn R. and Skjeltorp A. (Eds.) (1984): Multicritical Phenomena. Plenum, New York, U.S.A. [21].

Ree F.H. and Chesnut D.A. (1966): Phase transition of a hard-core lattice gas. The square lattice with nearest-neighbour exclusion. J. Chem. Phys. **45**, 3983–4003, [76,145].

Riedel E.K. (1972): Scaling approach to tricritical phase transitions. Phys. Rev. Lett. **28**, 675–678, [63].

Riedel E.K. and Wegner F.J. (1974): Effective critical and tricritical exponents. Phys. Rev. B **9**, 294–315, [32].

Rigby M. (1970): The van der Waals fluid: a renaissance. Quart. Rev. Chem. Soc. **24**, 416–432, [21].

Rohrer, H. (1975): Properties of $GdAlO_3$ near the spin flop bicritical point. Phys. Rev. Lett. **34**, 1638–1641, [19].

Roomany H.H. and Wyld H.W. (1981): Finite-lattice Hamiltonian results for the phase structure of the $Z(q)$ models and the q-state Potts models. Phys. Rev. B **23**, 1357–1361, [247].

Rossat-Mignod J., Burlet P., Bartholin H., Vogt O. and Lagnier R. (1980): Specific heat analysis of the magnetic phase diagram of CeSb. J. Phys. C: Solid State Phys. **13**, 6381–6389, [15].

Rowlinson J.S. (1973): Legacy of van der Waals. Nature **244**, 414–417, [21].

Rowlinson J.S. (1979): English translation of J.D. van der Waals' "The thermodynamic theory of capilliarity under the hypothesis of a continuous variation of density". J. Stat. Phys. **20**, 197–244, [419].

Rowlinson J.S. (1988): English translation (with introductory essay) of J.D. van der Waals: On the continuity of the gaseous and liquid states of matter, Studies in Statistical Mechanics, Vol. 14. North-Holland, Amsterdam, New York, [419].

Ruelle D. (1969): Statistical Mechanics. Benjamin, New York, U.S.A. [5,113,116,117,386].

Ruelle D. (1971): Extension of the Lee-Yang circle theorem. Phys. Rev. Lett. **26**, 303–304, [115,127,129,149].

Runnels L.K. and Combs L.L. (1971): Exact finite method of lattice statistics. I square and triangular lattice gases of hard molecules. J. Chem. Phys. **45**, 2482–2492, [76,135].

Runnels L.K. and Hubbard J.B. (1972): Applications of the Yang-Lee-Ruelle theory to hard-core lattice gases. J. Stat. Phys. **6**, 1–20, [134,135].

Rushbrooke G.S. (1963): On the thermodynamics of the critical region for the Ising model. J. Chem. Phys. **39**, 842–843, [46].

Rushbrooke G.S. (1964): On the theory of randomly dilute Ising and Heisenberg ferromagnetics. J. Math. Phys. **5**, 1106–1116, [291].

Rushbrooke G.S., Baker G.A. and Wood P.J. (1974): Heisenberg model. In Phase Transitions and Critical Phenomena. (Eds. C. Domb and M. S. Green), Vol. 3, 245–356. Academic Press, London, U.K., [254,276,277,286,289,291].

Rushbrooke G.S. and Wood P.J. (1955): On the high-temperture susceptibility for the Heisenberg model of a ferromagnet. Proc. Phys. Soc. **68**, 1162–1169, [300,302].

Rushbrooke G.S. and Wood P.J. (1958): On the curie points and high-temperature susceptibilities of Heisenberg model ferromagnetics. Mol. Phys. **1**, 257–283, [286,289].

Saito Y. (1981): Monte Carlo study of the three-state Potts model with two- and three-body interactions. J. Phys. A: Math. Gen. **15**, 1885–1892, [108].

Sarbach S. and Fisher M.E. (1978): Tricriticality and the failure of scaling in the many-component limit. Phys. Rev. B **18**, 2350–2363, [111].

Schäfer L. (1976): Conformal covariance in the framework of Wilson's renormalization group approach. J. Phys. A: Math. Gen. **9**, 377–395, [83].

Schick M. and Griffths R.B. (1977): Antiferromagnetic ordering in the three-state Potts model. J. Phys. A: Math. Gen. **10**, 2123–2131, [108,238,239].

Schick M., Walker J.S. and Wortis M. (1976): Antiferromagnetic triangular Ising model. Phys. Lett. A **58**, 479–480, [235,237,238].

Schick M., Walker J.S. and Wortis M. (1977): Phase diagram of the triangular Ising model: renormalization group calculation with application to adsorbed monolayers. Phys. Rev. B **16**, 2205–2219, [235,236,238].

Serre J-P. (1977): Linear Representations of Finite Groups. Springer, Berlin, Heidelberg, Germany, [342].

Sherman S. (1960): Combinatorial aspects of the Ising model for ferromagnetism. I A conjecture of Feynman on paths and graphs. J. Math. Phys. **1**, 202–217, [264].

Sherman S. (1962): Combinatorial aspects of the Ising model for ferromagnetism. II An analogue to the Witt identity. Bull. Am. Math. Soc. **68**, 225–229, [264].

Sinai Ya.G. (1982): Theory of Phase Transitions: Rigorous Results. Pergamon, Oxford, U.K. [113].

Sneddon L. (1978): Critical properties of two-dimensional spin s Ising systems with lattice anisotropy. J. Phys. C: Solid State Phys. **11**, 2823–2828, [243,245,246].

Sneddon L. and Barber M.N. (1977): Decimation transformations for two-dimensional Ising spin systems. J. Phys. C: Solid State Phys. **10**, 2653–2664, [210,232].

Southern B.W. (1978): Kadanoff's variational renormalization group method: the Ising model on the square and triangular lattices. J. Phys. A: Math. Gen. **11**, L1–L4, [241].

Southern B.W. and Lavis D.A. (1979): A model for adsorbed monolayers of orientable molecules. J. Phys. C: Solid State Phys. **12**, 5333–5343, [239].

Southern B.W. and Lavis D.A. (1980): Renormalization group study of a two-dimensional lattice model with directional bonding. J. Phys. A: Math. Gen. **13**, 251–262, [215,239].

Stanley H.E. (1967): High-temperature expansions for the classical Heisenberg model. II Zero-field susceptibility. Phys. Rev. **158**, 546–551, [300,302].

Stanley H.E. (1968a): Dependence of critical properties on dimensionality of spins. Phys. Rev. Lett. **20**, 589–592, [279].

Stanley H.E. (1968b): Spherical model as the limit of infinite spin dimensionality. Phys. Rev. **176**, 718–722, [279].

Stanley H.E. (1987): Introduction to Phase Transitions and Critical Phenomena. O.U.P. Oxford, U.K. [53,296].

Stanley H.E. (1974): *D*-vector model. In Phase Transitions and Critical Phenomena. (Eds. C. Domb and M. S. Green), Vol. 3, 485–567. Academic Press, London, U.K., [279,282,289].

Stauffer D., Ferer M. and Wortis M. (1972): Universality of second-order phase transitions: the scale factor for the correlation length. Phys. Rev. Lett. **29**, 345–349, [73,74].

Sutherland B. (1970): Two-dimensional hydrogen bonded crystals without the ice rule. J. Math. Phys. **11**, 3183–3186, [179].

Suzuki M. (1974): New universality of critical exponents. Prog. Theor. Phys. **51**, 1992–1993, [199].

Suzuki M. (1977): Static and dynamic finite-size scaling theory based on the renormalization group approach. Prog. Theor. Phys. **58**, 1142–??, [241].

Suzuki M. and Fisher M.E. (1971): Zeros of the partition function for the Heisenberg, ferroelectric and general Ising models. J. Math. Phys. **12**, 235–246, [115,167].

Swendsen R.H. (1979a): Monte Carlo renormalization group. Phys. Rev. Lett. **42**, 859–861, [248].

Swendsen R.H. (1979b): Monte Carlo renormalization group studies of the $d = 2$ Ising model. Phys. Rev. B **20**, 2080–2087, [248].

Sykes M.F. (1961): Some counting theorems in the theory of the Ising model and the excluded volume problem. J. Math. Phys. **2**, 52–62, [267].

Sykes M.F., Essam J.W. and Gaunt D.S. (1965): Derivation of low-temperature expansions for the Ising model of a ferromagnet and an antiferromagnet. J. Math. Phys. **6**, 283–298, [261].

Sykes M.F., Essam J.W., Heap B.R. and Hiley B.J. (1966): Lattice constant systems and graph theory. J. Math. Phys. **7**, 1557–1572, [370,371].

Sykes M.F., Gaunt D.S., Essam J.W. and Hunter D.L. (1973a): Derivation of low-temperature expansions for the Ising model. II General theory. J. Math. Phys. **14**, 1060–1065, [261].

Sykes M.F., Gaunt D.S., Mattingly S.R., Essam J.W. and Elliott C.J. (1973b): Derivation of low-temperature expansions for the Ising model. III Two-dimensional lattices – field grouping. J. Math. Phys. **14**, 1066–1070, [261].

Sykes M.F., Gaunt D.S., Martin J.L., Mattingly S.R. and Essam J.W. (1973c): Derivation of low-temperature expansions for the Ising model. IV Two-dimensional lattices – temperature grouping. J. Math. Phys. **14**, 1071–1073, [261].

Sykes M.F., Gaunt D.S., Essam J.W., Heap B.R., Elliott C.J. and Mattingly S.R. (1973d): Derivation of low-temperature expansions for the Ising model. V Three-dimensional lattices – field grouping. J. Phys. A: Math. Gen. **6**, 1498–1506, [261].

Sykes M.F., Gaunt D.S., Essam J.W. and Elliott C.J. (1973e): Derivation of low-temperature expansions for the Ising model. VI Three-dimensional lattices – temperature grouping. J. Phys. A: Math. Gen. **6**, 1507–1516, [261,299].

Sykes M.F., Watts M.G. and Gaunt D.S. (1975a): Derivation of low-temperature expansions for the Ising model. VII The honeycomb-triangular code system. J. Phys. A: Math. Gen. **8**, 1441–1447, [261].

Sykes M.F., Watts M.G. and Gaunt D.S. (1975b): Derivation of low-temperature expansions for the Ising model. VIII Ferromagnetic and antiferromagnetic polynomials for the honeycomb-triangular system. J. Phys. A: Math. Gen. **8**, 1448–1460, [261].

Sykes M.F., McKenzie S., Watts M.G. and Gaunt D.S. (1975c): Derivation of low-temperature expansions for the Ising model. IX High-field polynomials for the honeycomb-triangular system. J. Phys. A: Math. Gen. **8**, 1461–1468, [261].

Sykes M.F. and Watts M.G. (1975d): Derivation of low-temperature series expansions for the Ising model with triplet interactions on the plane triangular lattice. J. Phys. A: Math. Gen. **8**, 1469–1479, [261].

Syozi I. (1972): Transformation of Ising models. In Phase Transitions and Critical Phenomena. (Eds. C. Domb and M. S. Green), Vol. 1, 270–329. Academic Press, London, U.K., [270].

Szegö G. (1975): Orthogonal Polynomials. (4th Ed.)A.M.S. Colloquium Pubs. **23**, [254].

Temperley H.N.V. (1981): Graph Theory and Applications. Ellis Horwood, Chichester, U.K. [370,380,384].

Temperley H.N.V. and Fisher M.E. (1961): Dimer problem in statistical mechanics – an exact result. Phil. Mag. **6**, 1061–1063, [304].

ter Haar D. and Wergeland H.N.S. (1966): Elements of Thermodynamics. Addison-Wesley, Reading, Mass. U.S.A. [1].

Thompson C.S. (1988): Classical Equilibrium Statistical Mechanics. O.U.P. Oxford, U.K. [1,118].

Titchmarch E.C. (1939): The Theory of Functions,(2nd Ed.). O.U.P. Oxford, U.K. [125].

Tjon J.A. (1974): Numerical study of the renormalization group equations in the four-cell approximation. Phys. Lett. A **49**, 289–290, [235].

Tracy C.A. and McCoy B.M. (1973): Neutron scattering and the correlation functions of the Ising model near T_c. Phys. Rev. Lett. **31**, 1500–1504, [299].

Van der Waals J.D. (1873) Over de Continuiteit van der Gas- en Vloeistoftoestand, Doctoral Thesis, Leiden; (for an English translation see Rowlinson (1988)), [21].

Van der Waals J.D. (1893): English translation by Rowlinson (1979), [21].

Van Leeuwen J.M.J. (1975): Renormalization theory for spin systems. In Fundamental Problems in Statistical Mechanics, Vol.3, (Ed. E.G.D. Cohen), 81–101. North-Holland, Amsterdam, New York, [210].

Vohwinkel C. (1993): Yet another way to obtain low-temperature expansions for discrete spin systems. Phys. Lett. B **301**, 208–212, [261,300].

Wang Jian-Sheng, Swendsen R.H. and Kotecky R. (1989): Antiferromagnetic Potts models. Phys. Rev. Lett. **63**, 109–112, [108].

Widom B. (1964): Degree of the critical isotherm. J. Chem. Phys. **41**, 1633–1634, [53].

Widom B. (1965): Equation of state in the neighbourhood of the critical point. J. Chem. Phys. **43**, 3898–3905, [16].

Wiegel F.W. and Kox A.J. (1980): Theory of lipid monolayers. Adv. Chem. Phys. **41**, 195–228, [332].

Williams N.O. and Lavis D.A. (1996): Finite-size scaling of branch-points in lattice models. Phys. Lett. A **217**, 275–279, [163].

Wilson K.G. (1975): The renormalization group: critical phenomena and the Kondo problem. Rev. Mod. Phys. **47**, 773–840, [203,210,222,223].

Wilson K.G. and Kogut J. (1974): The renormalization group and the ε expansion. Phys. Rep. C **12**, 75–200, [248].

Witten L. (1954): A generalization of Yang and Lee's theory of condensation. Phys. Rev. **93**, 1131–1135, [116].

Wood D.W. (1985): The exact location of partition function zeros, a new method for statistical mechanics. J. Phys. A: Math. Gen. **18**, L917–L921, [116,156].

Wood D.W. (1987): The algebraic construction of partition function zeros: universality and algebraic cycles. J. Phys. A: Math. Gen. **20**, 3471–3493, [148,153,162,365].

Wood D.W. and Ball J.K. (1990): On a point of order. J. Stat. Phys. **58**, 599–615, [148,159].

Wood D.W. and Goldfinch M. (1980): Vertex models for the hard-square and hard-hexagon gases, and critical parameters for the scaling transformation. J. Phys. A: Math. Gen. **13**, 2718–2794, [135].

Wood D.W. and Osbaldstin A.H. (1982): Phase equilibria and the scaling transformation. J. Phys. A: Math. Gen. **15**, 3579–3591, [247].

Wood D.W. and Osbaldstin A.H. (1983): Multiple phase coexistence and the scaling transformation. J. Phys. A: Math. Gen. **16**, 1019–1033, [247].

Wood D.W. and Turnbull R.W. (1988): $z^2 - 11z - 1$ as an algebraic invariant for the hard-hexagon model. J. Phys. A: Math. Gen. **21**, L989–L994, [148].

Wood D.W. Turnbull R.W. and Ball J.K. (1987): Algebraic approximations to the locus of partition function zeros. J. Phys. A: Math. Gen. **20**, 3495–3521, [148,159,162].

Wood D.W. Turnbull R.W. and Ball J.K. (1989): An observation on the partition function zeros of the hard hexagon model. J. Phys. A: Math. Gen. **22**, L105–L109, [148].

Wu F.Y. (1968): Remarks on the modified potassium dihydrogen phosphate model of a ferroelectric. Phys. Rev. **168**, 539–543, [312,317].

Wu F.Y. (1969): Exact solution of a model of an antiferroelectric transition. Phys. Rev. **183**, 604–607, [176,202].

Wu F.Y. (1971): Ising model with four-spin interactions. Phys. Rev. B **4**, 2312–2314, [170].

Wu F.Y. (1982): The Potts model. Rev. Mod. Phys. **54**, 235–268, [105,116].

Wu T.T., McCoy B.B., Tracy C.A. and Barouch E. (1976): Spin-spin correlation functions for the two-dimensional Ising model: exact results in the scaling region. Phys. Rev. B **13**, 316–374, [299].

Yang C.N. and Lee T.D. (1952): Statistical theory of equations of state and phase transitions. I Theory of condensation. Phys. Rev. **87**, 404–409, [114,116,125,149].

Yeomans J.M. (1992): Statistical Mechanics of Phase Transitions. O.U.P. Oxford, U.K. [53].

Young S. (1892): On the generalizations of van der Waals regarding "corresponding" temperatures, pressures and volumes. Phil. Mag. **33**, 153–185, [21].

Young A.P. and Lavis D.A. (1979): Critical behaviour of a two-dimensional bonded lattice fluid. J. Phys. A: Math. Gen. **12**, 229–243, [239].

Ziman J.M. (1979): Models of Disorder. C.U.P. Cambridge, U.K. [15,24].

Subject Index

Springer and the environment

At Springer we firmly believe that an international science publisher has a special obligation to the environment, and our corporate policies consistently reflect this conviction.

We also expect our business partners – paper mills, printers, packaging manufacturers, etc. – to commit themselves to using materials and production processes that do not harm the environment. The paper in this book is made from low- or no-chlorine pulp and is acid free, in conformance with international standards for paper permanency.

Springer

Printing: Saladruck, Berlin
Binding: Buchbinderei Lüderitz & Bauer, Berlin